Telecommunications Engineering

Telecommunications Engineering

Third
Edition

J. Dunlop

and

D. G. Smith

*Department of Electronic
and Electrical Engineering,
University of Strathclyde,
Glasgow*

Stanley Thornes (Publishers) Ltd

First language edition published by Chapman & Hall in 1984
Second edition published by Chapman & Hall in 1989
Third edition published by Chapman & Hall in 1994

Reprinted in 1998 by:
Stanley Thornes (Publishers) Ltd
Ellenborough House
Wellington Street
CHELTENHAM
GL50 1YW
United Kingdom

98 99 00 01 02 / 10 9 8 7 6 5 4 3 2 1

A catalogue record for this book is available from the British Library

ISBN 0–7487–4044–9   1001816602

Typeset by Thomson Press (India) Ltd, New Delhi
Printed and bound in Croatia by Zrinski d.d., Cakovec

Contents

Preface to the third edition

During the lifetime of the second edition telecommunications systems have continued to develop rapidly, with major advances occurring in mobile communications and broadband digital networks and services. Digital systems have become even more prevalent and sophisticated signal processing techniques are now common at increasingly higher bit rates. These advances have prompted the inclusion of a new chapter on mobile communications and a considerable revision of many others. Chapter 10 has been extended to include ISDN and broadband digital communications.

Several readers have commented that a discussion of the transient performance of the phase locked loop would be useful and this has now been included in Chapter 2.

Chapter 3 has been extended to give a deeper insight into non-linear coding of voice waveforms for PCM and a section has been included on NICAM, which has been adopted for the digital transmission of television sound.

Chapter 5 now includes an introduction to coding techniques for burst errors which dominate in modern mobile communications systems.

Chapter 11 has been completely revised to address the major developments which have occurred in television since the second edition and a new section on satellite television has been included.

Chapter 15 is a completely new chapter on Mobile Communication Systems covering first generation analogue and second generation digital systems.

Preface to the second edition

Since the first edition telecommunications have changed considerably. In particular, digital systems have become more common. To reflect such changes two new chapters have been added to the text and several have been modified, some considerably.

Chapter 3 has been extended to give a detailed description of the A law compression characteristic used in PCM systems and substantial extra material is included on PCM transmission techniques. A section is also included on other digital transmission formats such as differential PCM and delta modulation.

Chapter 10 introduces several elements of switching, both analogue and digital. In the latter case the fundamentals of both time and space switching are discussed. The methods used to analyse the switching performance are covered although the discussion is limited to simple cases in order to highlight the underlying concepts. Some of the material on older signalling systems has been omitted from this edition.

Chapter 13 deals with the topic of packet transmission which is finding increasing use in both wide area and local networks. This chapter provides the theoretical background for scheduled and random access transmission and draws attention to the limitations of these theoretical descriptions and to the need for using reliable computer models for estimating the performance of practical systems. This chapter also introduces the concept of the 7-layer Open Systems Interconnection reference model and illustrates how some of the OSI layers are incorporated in packet switched systems.

An introduction to satellite communications is given in Chapter 14. So as not to over-complicate the concepts involved, discussion is limited to geo-stationary systems. At the time of writing, the development of direct broadcasting of television programmes by satellite is at a very early stage, but there is little doubt that it will become an increasingly important application.

The authors are most grateful to the many readers who have made constructive suggestions for improvement to the text and who have identified several errors that existed in the first edition. It is hoped that the errors have been corrected and that a number of areas of difficulty have been removed by additional explanation or discussion.

Preface to
the first edition

The influence of telecommunications has increased steadily since the introduction of telegraphy, radio and telephony. Now, most people are directly dependent on one or more of its many facets for the efficient execution of their work, at home and in leisure.

Consequently, as a subject for study it has become more and more important, finding its way into a large range of higher education courses, given at a variety of levels. For many students, telecommunications will be presented as an area of which they should be aware. The course they follow will include the essential features and principles of communicating by electromagnetic energy, without developing them to any great depth. For others, however, the subject is of more specialized interest; they will start with an overview course and proceed to specialize in some aspects at a later time. This book has been written with both types of student in mind. It brings together a broader range of material than is usually found in one text, and combines an analytical approach to important concepts with a descriptive account of system design. In several places the approximate nature of analysis has been stressed, and also the need to exercise engineering judgement in its application. The intention has been to avoid too much detail, so that the text will stand on its own as a general undergraduate-level introduction, and it will also provide a strong foundation for those who will eventually develop more specialized interests.

It has been assumed that the reader is familiar with basic concepts in electronic engineering, electromagnetic theory, probability theory and differential calculus.

Chapter 1 begins with the theoretical description of signals and the channels through which they are transmitted. Emphasis is placed on numerical methods of analysis such as the discrete Fourier transform, and the relationship between the time and frequency domain representations is covered in detail. This chapter also deals with the description and transmission of information-bearing signals.

Chapter 2 is concerned with analogue modulation theory. In this chapter there is a strong link between the theoretical concepts of modulation theory and the practical significance of this theory. The chapter assumes that the reader has a realistic knowledge of electronic circuit techniques.

Chapter 3 is devoted to discrete signals and in particular the coding and transmission of analogue signals in digital format. This chapter also emphasizes the relationship between the theoretical concepts and their practical significance.

Chapters 4 and 5 are concerned with the performance of telecommunications systems in noise. Chapter 4 covers the performance of analogue systems and concentrates on the spectral properties of noise. Chapter 5 covers the performance of digital systems and is based on the statistical properties of noise. This chapter also deals in detail with the practical implication of error correcting codes, a topic which is often ignored by more specialized texts in digital communications.

In Chapter 6 the elements of high-frequency transmission-line theory are discussed, with particular emphasis on lossless lines. The purpose is to introduce the concepts of impedance, reflection and standing waves, and to show how the designer can influence the behaviour of the line.

Basic antenna analysis, and examples of some commonly used arrays and microwave antennas, are introduced in Chapter 7, while Chapters 8 and 9 describe the essential features of waveguide-based microwave components. A fairly full treatment of the propagation of signals along waveguide is considered from both the descriptive and field-theory analysis points of view.

Telephone system equipment represents the largest part of a country's investment in telecommunications, yet teletraffic theory and basic system design do not always form part of a telecommunications class syllabus. Chapter 10 is a comprehensive chapter on traditional switching systems and the techniques used in their analysis. Care has been taken to limit the theoretical discussion to simple cases, to enable the underlying concepts to be emphasized.

Chapter 11 is devoted to television systems. In a text of this nature such a coverage must be selective. We have endeavoured to cover the main topics in modern colour television systems from the measurement of light to the transmission of teletext information. The three main television systems, NTSC, PAL and SECAM, are covered but the major part of this chapter is devoted to the PAL system.

One of the outstanding major developments in recent years has been the production of optical fibres of extremely low loss, making optical communication systems very attractive, both technically and commercially. Chapter 12 discusses the main features of these systems, without introducing any of the analytical techniques used by specialists. The chapter is intended to give an impression of the exciting future for this new technology.

It cannot be claimed that this is a universal text; some omissions will not turn out to be justified, and topics which appear to be of only specialized interest now may suddenly assume a much more general importance. However, it is hoped that a coverage has been provided in one volume which will find acceptance by many students who are taking an interest in this stimulating and expanding field of engineering.

List of symbols and abbreviations

a	normalized propagation delay
A	telephone traffic, in erlangs
A_e	effective aperture of an antenna
AAL	ATM adaption layer
ADC	analogue to digital conversion
ADPCM	adaptive differential pulse code modulation
AGCH	GSM Access Grant Channel
AM	amplitude modulation
AMI	alternate mark inversion
AMPS	advanced mobile phone system
ARQ	automatic repeat request
ASK	amplitude shift keying
ATDM	asynchronous time division multiplexing
ATM	asynchronous transfer mode
α	attenuation coefficient of a transmission line
α	traffic offered per free source
B	bandwidth of a signal or channel
	speech (64 kb/sec) channel in ISDN
B	call congestion
B_c	coherence bandwidth
BCCH	GSM Broadcast Control Channel
B-ISDN	broadband ISDN
BSC	base station controller
BST	base station transceiver
β	phase constant of a transmission line
β	modulation index
c	velocity of light in free space
C	capacitance per unit length of transmission line
CCCH	GSM Common Control Channel
CDMA	code division multiple access
CFP	cordless fixed part
CIR	carrier to interference ratio
$C(n)$	discrete spectrum
C_n	nth harmonic in a Fourier series
CPP	cordless portable part

CRC	cyclic redundancy check
CSMA/CD	carrier sense multiple access with collision detection
CVSDM	continuously variable slope delta modulation
D	re-use distance
	diameter, or largest dimension, of an antenna
	signalling channel in ISDN
DECT	digital European cordless telecommunication standard
DFT	discrete Fourier transform
δ_s	skin depth
$\partial(t - t_0)$	impulse function at t_0
Δ	delay spread
Δf	elemental bandwidth
Δf_c	carrier deviation
Δf_n	noise bandwidth
DPSK	differential phase shift keying
DQDB	distributed queue dual bus
DSB-AM	double sideband amplitude modulation
DSB-SC-AM	double sideband supressed carrier modulation
DUP	data user part
E	time congestion
$E(f)$	energy density spectrum
ERP	effective radiated power
EIRP	effective isotropic radiated power
$E_n(A)$	Erlang's loss function
$E(N, s, \alpha)$	Engset's loss function
ε	permittivity
ε_0	permittivity of free space
ε_r	relative permittivity of dielectric
$\varepsilon(t)$	phase error
$\varepsilon(s)$	Laplace transform of $\varepsilon(t)$
η_0	free space characteristic
η	single-sided power spectral density of white noise
F	noise figure of a network
f_0	fundamental frequency of a periodic wave
f_c	cut-off frequency, carrier frequency
FACCH	GSM Fast Associated Control Channel
FCCH	GSM Frequency Correction Channel
FDM	frequency division multiplex
FDMA	frequency division multiple access
FET	field effect transistor
FISU	fill in signalling unit
FFT	fast Fourier transform
FM	frequency modulation
FOCC	TACS Forward Control Channel
FSK	frequency shift keying
FVC	TACS Forward Voice Channel
$F_v(v)$	cumulative distribution function

G	conductance per unit length of transmission line
	normalized offered traffic
	generator polynomial
$G(f)$	power spectral density
$G(i)$	probability of any i devices being busy
GMSK	Gaussian minimum shift keying
GSM	global system for mobile communications
γ	propagation coefficient of a transmission line
n	thickness of dielectric in microstrip line
	height of an antenna
H	magnetic field
H_{av}	entropy of a message (bits/symbol)
$H(f)$	Fourier transform of $h(t)$
$h(k)$	discrete signal
$H(s)$	Laplace transform of $h(t)$
$h(t)$	general function of time
HLR	home location register
$[i]$	probability that a network is in state i
I_k	interference power
IDFT	inverse discrete Fourier transform
IF	intermediate frequency
ISDN	integrated services digital network
ISUP	ISDN user part
$I_0(x)$	modified Bessel function
$J_n(\beta)$	Bessel functions of the first kind
K	cluster size
k_c	$2\pi/\lambda_c$
L	inductance per unit length of transmission line
LAN	local area network
LED	light emitting diode
LAP	link access procedure
LAP-D	local access protocol for D Channel
λ	likelihood ratio
	mean packet arrival rate
	wavelength
λ_c	cut-off wavelength
λ_{cmn}	cut-off wavelength of the TE_{mn} or TM_{mn} mode
λ_g	guide wavelength
λ_i	call arrival rate in state i
λ_0	free space wavelength
$L(f)$	frequency domain output of a network
$l(t)$	time domain output of a network
m	depth of modulation
MAC	mixed analogue components TV standard
	medium access control
	TACS mobile attenuation code
MAN	metropolitan area network

MAP	manufacturing automation protocol
MSC	mobile switching centre
MSU	message signalling unit
MTP	message transfer part
μ	permeability
μ_i	call departure rate in state i
μ_0	permeability of free space
n	refractive index of glass fibre
N	number of devices
	electron density in ionosphere (electrons/m^3)
n_0	refractive index of freespace
N_i	normalized noise power at the input of a network
N_0	normalized noise power at the output of a network
$N(A_0)$	level crossing rate
NICAM	nearly instantaneously companded audio multiplex
NMT	nordic mobile telephone system
NNI	network–network interface
NTP	network service part
NTSC	National Television Systems Committee
$n(t)$	elemental noise voltage
OSI	open systems interconnection reference model
ω_n	natural (radian) frequency
P	power in a signal
P_a	power/unit area
P_c	error probability
P_r	received power
P_t	transmitted power
p_{nm}	root of $J'_n(k_c, a)$
PAD	packet assembler/disassembler
PAL	phase alternation line by line
PCM	pulse code modulation
PCH	GSM paging channel
PDH	plesiochronous digital hierarchy
PDU	ATM protocol data unit
$P(f)$	transfer function of a network
PM	phase modulation
POTS	plain old telephone system
PSK	phase shift keying
$p(t)$	impulse response of a network
$pv^{(v)}$	probability density function
ψ	angle of reflection coefficient
q	co-channel interference reduction factor
Q	resonator quality factor
QPSK	quaternary phase shift keying
QPSX	queued packet switched exchange
R	resistance per unit length of transmission line
RACH	GSM access channel
RECC	TACS Reverse Control Channel

$R_h(\tau)$	autocorrelation function of $h(t)$
RVC	TACS Reverse Voice Channel
ρ	reflection coefficient at transmission line load
s	mean call holding time
S	normalized throughput
	number of traffic sources
S	voltage standing wave ratio (VSWR)
S_c	normalized carrier power
S_i	normalized signal power at the input of a network
S_0	normalized signal power at the output of a network
SACCH	GSM Slow Associated Control Channel
SAPI	service access point identifier
SAT	TACS supervisory audio tone
SCART	Syndicat des Constructeurs d'Appareils Radio récepteurs et Teléviseurs
SCCP	signalling connection control part
SCH	GSM Synchronization Channel
SDCCH	GSM Slow Dedicated Control Channel
SDH	synchronous digital hierarchy
SECAM	séquential couleur à mémoire
SQNR	single to quantization noise ratio (power)
SNR	signal to noise ratio (power)
SONET	synchronous optical network
SP	signalling point
SS7	CCITT signalling system number 7
SSB-AM	signal sideband amplitude modulation
ST	TACS signalling tone
STM-n	synchronous transport mode-level n
STP	signalling transfer point
σ	rms voltage of a random signal
T	period of a periodic wave
θ_i	angle of incidence of radio wave to ionosphere
t_a	mean access delay
t_d	mean packet delivery time
t_p	end to end propagation delay
t_r	token rotation time
t_t	token transmission time
t_s	scan time
t_{sl}	walk time
T_c	effective noise temperature of a network or antenna
T_s	standard noise temperature (290K)
TACS	total access communication system
TCH	GSM Traffic Channel
TDD	time division duplex
TDM	time division multiplexing
TDMA	time division multiple access
TE_{mn}	transverse electric waveguide mode
TEI	terminal end point identifier

TM_{mn}	transverse magnetic waveguide mode
TU	tributary unit
TUP	telephone user part
τ	dummy time variable
$\tau(A_0)$	average fade duration below the level A_0
$u(t)$	unit step function
UNI	user-network interface
v	transmission line wave velocity
V	peak voltage of a waveform
V_1	incident (forward) voltage on a transmission line
V_2	reflected (backward) voltage on a transmission line
v_g	group velocity
v_{ph}	phase velocity
VAD	voice activity detector
$v(t)$	general function of time
$v_m(t)$	modulating waveform
$v_c(t)$	carrier waveform
$V_n(t)$	bandlimited noise voltage
VC	virtual container
VCI	virtual circuit identifier
VPI	virtual path identifier
VSB-AM	vestigial sideband amplitude modulation
W	highest frequency component in a signal
	width of strip in microstrip line
WAN	wide area network
X^n	power of 2 in a generator polynomial
$x(t)$	amplitude of in-phase noise component
y	mean call arrival rate
$y(t)$	amplitude of quadrature noise component
Z_L	transmission line load impedance
Z_O	characteristic impedance of transmission line
Z_O	characteristic impedance of microstrip line

Signals and channels $\boxed{1}$

1.1 INTRODUCTION

Telecommunication engineering is concerned with the transmission of information between two distant points. Intuitively we may say that a signal contains information if it tells us something we did not already know. This definition is too imprecise for telecommunications studies, and we shall devote a section of this chapter to a formal description of information. For the present it is sufficient to say that a signal that contains information varies in an unpredictable or random manner. We have thus specified a primary characteristic of the signals in telecommunications systems; they are random in nature.

These random signals can be broadly subdivided into discrete signals that have a fixed number of possible values, and continuous signals that have any value between given limits. Whichever type of signal we deal with, the telecommunication system that it uses can be represented by the generalized model of Fig. 1.1. The central feature of this model is the transmission medium or channel. Some examples of channels are coaxial cables, radio links, optical fibres and ultrasonic transmission through solids and liquids. It is clear from these examples that the characteristics of channels can vary widely. The common feature of all channels, however, is that they modify or distort the waveform of the transmitted signal. In some cases the distortion can be so severe that the signal becomes totally unrecognizable.

In many instances it is possible to minimize distortion by careful choice of the transmitted signal waveform. To do this the telecommunications engineer must be able to define and analyse the properties of both the signals and the channels over which they are transmitted. In this chapter we shall concentrate on the techniques used in signal and linear systems analysis, although we should point out that many telecommunications systems do have non-linear characteristics.

1.2 THE FREQUENCY AND TIME DOMAINS

The analysis of linear systems is relatively straightforward if the applied signals are sinusoidal. We have already indicated that the signals encountered in telecommunications systems are random in nature and, as such, are non-deterministic. It is often possible to approximate such signals by periodic

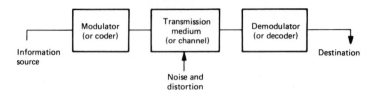

Fig. 1.1 Basic elements of a telecommunications system.

functions that themselves can be decomposed into a sum of sinusoidal components. The signal waveforms are functions of time and the variation of signal amplitude with time is known as the 'time domain representation' of the signal. Alternatively, if a signal is decomposed into a sum of sinusoidal components, the amplitude and phase of these components can be expressed as a function of frequency. This leads us to the 'frequency domain representation' of the signal.

The relationship between frequnecy domain and time domain is an extremely important one and is specified by Fourier's theorem. The response of a linear system to a signal can be determined in the time domain by using the principle of convolution, and in the frequency domain by applying the principle of superposition to the responses produced by the individual sinusoidal components. We will consider the frequency domain first, as this makes use of the theorems of linear network analysis which will be familiar to readers with an electronics background. Time domain analysis is considered in detail in Section 1.11. Frequency domain analysis will be introduced using traditional Fourier methods and we will then develop the discrete Fourier transform (DFT) which is now an essential tool in computer aided analysis of modern telecommunications systems.

1.3   CONTINUOUS FOURIER ANALYSIS

Fourier's theorem states that any single-valued periodic function, which has a repetition interval T, can be represented by an infinite series of sine and cosine terms which are harmonics of $f_0 = 1/T$. The theorem is given by Eqn (1.1).

$$h(t) = \frac{a_0}{T} + \frac{2}{T} \sum_{n=1}^{\infty} (a_n \cos 2\pi n f_0 t + b_n \sin 2\pi n f_0 t) \tag{1.1}$$

where $f_0 = 1/T$ is the fundamental frequency. The response of a linear system to a waveform $h(t)$ that is not a simple harmonic function is found by summing the responses produced by the individual sinusoidal components of which $h(t)$ is composed. The term a_0/T is known as the dc component and is the mean value of $h(t)$.

$$\frac{a_0}{T} = \frac{1}{T} \int_{-T/2}^{T/2} h(t)\,dt$$

i.e.

$$a_0 = \int_{-T/2}^{T/2} h(t)\,dt \tag{1.2}$$

The amplitudes of the sine and cosine terms are given by

$$a_n = \int_{-T/2}^{T/2} h(t) \cos (2\pi n f_0 t) \, dt$$

$$b_n = \int_{-T/2}^{T/2} h(t) \sin (2\pi n f_0 t) \, dt \qquad (1.3)$$

The Fourier series thus contains an infinite number of sine and cosine terms. This can be reduced to a more compact form as follows; let

$$x(t) = a_n \cos (2\pi n f_0 t) + b_n \sin (2\pi n f_0 t)$$

$$\cos \phi_n = \frac{a_n}{\sqrt{(a_n^2 + b_n^2)}}$$

$$\sin \phi_n = \frac{-b_n}{\sqrt{(a_n^2 + b_n^2)}}$$

Hence $\phi_n = \tan^{-1} [-b_n/a_n]$ and

$$x(t) = (a_n^2 + b_n^2)^{1/2} [\cos (2\pi n f_0 t) \cos \phi_n - \sin (2\pi n f_0 t) \sin \phi_n]$$

i.e.

$$x(t) = (a_n^2 + b_n^2)^{1/2} \cos (2\pi n f_0 t + \phi_n)$$

Hence the Fourier series can be modified to

$$h(t) = \frac{a_0}{T} + \frac{2}{T} \sum_{n=1}^{\infty} C_n \cos (2\pi n f_0 t + \phi_n)$$

where

$$C_n = (a_n^2 + b_n^2)^{1/2} \quad \text{and} \quad \phi_n = \tan^{-1} \left[\frac{-b_n}{a_n} \right] \qquad (1.4)$$

A graph of C_n against frequency is known as the **amplitude spectrum** of $h(t)$ and a graph of ϕ_n against frequency is known as the **phase spectrum** of $h(t)$.
Note that if the voltage developed across a $1\,\Omega$ resistance is

$$v(t) = \frac{2C_n}{T} \cos (2\pi n f_0 t + \phi_n)$$

the average power dissipated in the resistance is

$$P = \frac{2}{T^2} C_n^2 = \frac{2}{T^2} (a_n^2 + b_n^2) \qquad (1.5)$$

A graph of C_n^2 against frequency is known as the **power spectrum** of $h(t)$. The total power developed in a $1\,\Omega$ resistance by $h(t)$ is thus given by

$$P_T = \frac{a_2^0}{T^2} + \frac{2}{T^2} \sum_{n=1}^{\infty} C_n^2 \qquad (1.6)$$

The mean square value of $h(t)$ is given by

$$\sigma^2 = \frac{1}{T}\int_{-T/2}^{T/2} |h(t)|^2 \, dt$$

This is effectively the power dissipated when a voltage equal to $h(t)$ is developed across a resistance of 1Ω. The frequency and time domain representations of $h(t)$ are thus related by Eqn (1.7). This equation is formally known as Parseval's theorem.

$$\frac{1}{T}\int_{-T/2}^{T/2} |h(t)|^2 \, dt = \frac{a_0^2}{T^2} + \frac{2}{T^2}\cdot\sum_{n=1}^{\infty} C_n^2 \qquad (1.7)$$

EXAMPLE: Find the amplitude and power spectrum of the periodic rectangular pulse train of Fig. 1.2.

The zero frequency (mean) value is a_0/T where

$$a_0 \int_{-T/2}^{T/2} h(t)\,dt = \int_{-t_{1/2}}^{t_{1/2}} A\,dt = At_1$$

$$a_n \int_{-T/2}^{T/2} h(t)\cos(2\pi nf_0 t)\,dt = \int_{-t_{1/2}}^{t_{1/2}} A\cos(2\pi nf_0 t)\,dt$$

i.e.

$$a_n = \frac{A}{2\pi nf_0}[\sin(\pi nf_0 t_1) - \sin(-\pi nf_0 t_1)]$$

hence

$$a_n = \frac{A}{\pi nf}\sin(\pi nf_0 t_1)$$

similarly

$$b_n = \frac{A}{2\pi nf_0}[\cos(\pi nf_0 t_1) - \cos(-\pi nf_0 t_1)] = 0$$

Hence in this example $C_n = a_n$.

The amplitude spectrum is $C_n = At_1(\sin \pi nf_0 t_1)/\pi nf_0 t_1$, which is often written $C_n = At_1 \operatorname{sinc}(\pi nf_0 t_1)$.

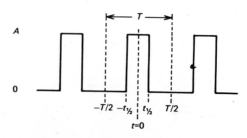

Fig. 1.2 Rectangular periodic pulse train.

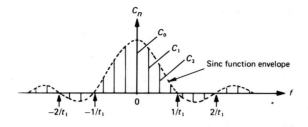

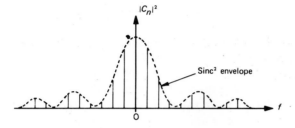

Fig. 1.3 Amplitude and power spectrum of a periodic pulse train.

The amplitude spectrum is plotted in Fig. 1.3 and it should be noted that the envelope of this spectrum is a sinc function that has unity value when $\pi f t_1 = 0$ and zero value when $\pi f t_1 = m\pi$, i.e. when $f = m/t_1$. In this particular example $\phi_n = 0$ indicating all harmonics are in phase. The power spectrum of $h(t)$ is simply the square of the amplitude spectrum.

A convenient alternative form of Eqn (1.1) can be developed by writing the sine and cosine terms in exponential notation, i.e.

$$\cos(2\pi n f_0 t) = [\exp(j2\pi n f_0 t) - \exp(-j2\pi n f_0 t)]/2$$

$$\sin(2\pi n f_0 t) = [\exp(j2\pi n f_0 t) - \exp(-j2\pi n f_0 t)]/2j$$

Substitution in Eqn (1.1) gives

$$h(t) = \frac{a_0}{T} + \frac{1}{T}\sum_{n=1}^{\infty}(a_n - jb_n)\exp(j2\pi n f_0 t) + (a_n + jb_n)\exp(-j2\pi n f_0 t)$$

$$(a_n - jb_n) = \int_{-T/2}^{T/2} h(t)[\cos(2\pi n f_0 t) - j\sin(2\pi n f_0 t)]\,dt$$

i.e. if $C_n = (a_n - jb_n)$, then

$$C_n = \int_{-T/2}^{T/2} h(t)\exp(-j2\pi n f_0 t)\,dt \qquad (1.8)$$

The complex conjugate of C_n is $C_n^* = (a_n + jb_n)$, and

$$C_n^* = \int_{-T/2}^{T/2} h(t)\exp(j2\pi n f_0 t)\,dt$$

i.e. $C_n^* = C_{-n}$.

Hence

$$h(t) = \frac{a_0}{T} + \frac{1}{T} \sum_{n=1}^{\infty} [C_n \exp(j2\pi n f_0 t) + C_{-n} \exp(-j2\pi n f_0 t)]$$

and

$$C_0 = \int_{-T/2}^{T/2} h(t) \exp(j^0) \, dt = a_0$$

so that $h(t)$ can be written

$$h(t) = \frac{1}{T} \sum_{n=-\infty}^{\infty} C_n \exp(j2\pi n f_0 t) \tag{1.9}$$

This is the exponential form of the Fourier series and the limits of the summation are now $n = \pm\infty$. The spectrum that contains both positive and negative components is known as a double-sided spectrum.

The negative frequencies are a direct result of expressing sine and cosine in complex exponential form. In Eqn (1.8) C_n is a complex quantity and can be separated into a magnitude and phase characteristic, i.e. $C_n = |C_n| \exp(j\phi_n)$; i.e.

$$h(t) = \frac{1}{T} \sum_{n=-\infty}^{\infty} |C_n| \exp[j(2\pi n f_0 t + \phi_n)]$$

But since $|C_n| = |C_{-n}|$ then

$$h(t) = \frac{C_0}{T} + \frac{1}{T} \sum_{n=1}^{\infty} |C_n| \exp[j(2\pi n f_0 t + \phi_n)] + |C_n| \exp[-j(2\pi n f_0 t + \phi_n)] \cdot$$

i.e.

$$h(t) = \frac{C_0}{T} + \frac{2}{T} \sum_{n=1}^{\infty} |C_n| \cos(2\pi n f_0 t + \phi_n)$$

Equations (1.9) and (1.4) are therefore equivalent, but some care is required in interpreting Eqn (1.9). The harmonic amplitude C_n/T is exactly half the value given by Eqn (1.4), but it is defined for both negative and positive values of n. The correct amplitude is obtained by summing the equal coefficients which are obtained for negative and positive values of n. This is quite reasonable because only one frequency component actually exists.

The power of any frequency is derived from Eqn (1.9) in a similar way. Since C_n is a complex quantity the power at any value of n is $(C_n/T) \cdot (C_n^*/T)$, i.e.

$$(C_n/T) \cdot (C_n^*/T) = (a_n - jb_n)(a_n + jb_n)/T^2 = (a_n^2 + b_n^2)/T$$

Both negative and positive values of n will contribute an equal amount of power; the total power at any frequency is thus $2(a_n^2 + b_n^2)/T^2$. This of course agrees with Eqn (1.5), since physically only a single component exists at any one frequency. We can write Parseval's theorem for the exponential series as

$$\frac{1}{T} \int_{-T/2}^{T/2} |h(t)|^2 \, dt = \frac{1}{T^2} \sum_{n=-\infty}^{\infty} |C_n|^2 \tag{1.10}$$

Note that only a single value of n appears at $n = 0$.

EXAMPLE: Evaluate the amplitude spectrum of the waveform in Fig. 1.2 using the exponential series

$$h(t) = \frac{1}{T} \sum_{n=-\infty}^{\infty} C_n \exp j(2\pi n f_0 t) \quad \text{where} \quad C_n = \int_{-t_{1/2}}^{t_{1/2}} A \exp(-j2\pi n f_0 t)\, dt$$

i.e.

$$C_n = \left[\frac{-A}{j2\pi n f_0} \exp(-j2\pi n f_0 t) \right]_{-t_{1/2}}^{t_{1/2}}$$

$$= \frac{A}{\pi n f_0} \left[\frac{\exp(-j\pi n f_0 t_1) - \exp(-j\pi n f_0 t_1)}{2j} \right]$$

i.e.

$$C_n = \frac{A}{\pi n f_0} \sin(\pi n f_0 t_1)$$

or

$$C_n = A t_1 \operatorname{sinc}(\pi n f_0 t_1)$$

This is identical to the equation obtained from the cosine series.

1.4 ODD AND EVEN FUNCTIONS

The waveform $h(t)$ is defined as an even function if $h(t) = h(-t)$; it has the property of being symmetrical about the $t = 0$ axis. If $h(t) = -h(-t)$ the waveform is an odd function and has skew symmetry about the $t = 0$ axis. A function that has no symmetry about the $t = 0$ axis is neither odd nor even; the sawtooth waveform of Fig. 1.4(c) is an example of such a waveform.

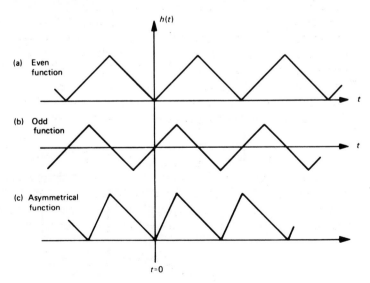

Fig. 1.4 Examples of periodic functions.

Odd and even functions have properties that may be used to simplify Fourier analysis, e.g.

if

$$h(t) = \frac{a_0}{T} + \frac{2}{T} \sum_{n=1}^{\infty} [a_n \cos(2\pi n f_0 t) + b_n (\sin 2\pi n f_0 t)]$$

then

$$h(-t) = \frac{a_0}{T} + \frac{2}{T} \sum_{n=1}^{\infty} [a_n \cos(2\pi n f_0 t) - b_n \sin(2\pi n f_0 t)]$$

If $h(t) = h(-t)$ this can only be true if $b_n = 0$, i.e. the Fourier series of an even function has cosine terms only. Alternatively, all phase angles ϕ_n in Eqn (1.4) are 0 or $\pm \pi$ and all values of C_n in Eqn (1.8) are real. If $h(t) = -h(-t)$ then $a_n = 0$, i.e. the Fourier series of an odd function contains only sine terms. Alternatively, all phase angles ϕ_n in Eqn (1.4) are $\pm \pi/2$ and all values of C_n in Eqn (1.8) are imaginary. If $h(t)$ has no symmetry about $t = 0$ the Fourier series contains both sines and cosines, the phase angles ϕ_n of Eqn (1.8) are given by $\tan^{-1}(-b_n/a_n)$, and all values of C_n in Eqn (1.8) are complex.

Many waveforms that are not symmetrical about $t = 0$ can be made either odd or even by shifting the waveform relative to the $t = 0$ axis. The shifting process is illustrated in Fig. 1.5. It is of interest to examine the effect of such a shift on the Fourier series. We shall consider the Fourier series of the shifted waveform $h(t - t_s)$. Let $(t - t_s) = t_x$; the amplitude spectrum is thus given by

$$C_n = \int_{-t_{1/2}}^{t_{1/2}} A \exp(-j2\pi n f_0 t_x) \, dt_x$$

This evaluates to $C_n = At_1 \operatorname{sinc}(\pi n f_0 t_1)$; hence shifting the time axis does not affect the amplitude spectrum. It does, however, affect the phase spectrum:

$$h(t - t_s) = \frac{1}{T} \sum_{n=-\infty}^{\infty} C_n \exp(j2\pi n f_0 t) \exp(-j2\pi n f_0 t_s)$$

This effectively adds a phase shift of $\phi_s = -2\pi n f_0 t_s$ to each component in the series. Interpreted in another way, a time delay of t_s is equivalent to a phase shift of $2\pi f_0 t_s$ in the fundamental, $4\pi f_0 t_s$ in the second harmonic, etc.

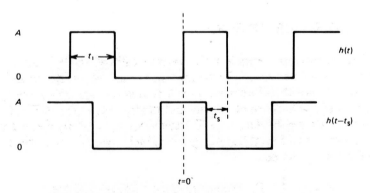

Fig. 1.5 Time shifting.

1.5 WAVEFORM SYNTHESIS

This can be regarded as the inverse of Fourier analysis. In effect the Fourier series indicates that any periodic waveform can be synthesized by adding an infinite number of cosine waves with specific amplitudes and phases. In most practical cases a very good approximation to a given periodic waveform can be obtained by truncating the series to only a few terms. As the number of terms in the series is increased the mean square error between the synthesized waveform and the desired waveform decreases. A difficulty does arise in the vicinity of a discontinuity, however. As the number of terms in the series tends to infinity the mean value of the synthesized waveform approaches the mean value of the desired waveform at the discontinuity. The amplitude of the synthesized waveform on either side of the discontinuity is subject to error, which is not reduced when the number of terms in the series is increased. This error is known as Gibb's phenomenon, and is illustrated for a synthesized rectangular wave in Fig. 1.6.

This is in fact a convergence property of the Fourier series. The Fourier series converges to the mean value of $h(t)$ at discontinuities in the waveform of $h(t)$. The conditions required for the convergence of the series are

(i) $h(t)$ must have a finite number of maxima and minima in the interval T;
(ii) $h(t)$ must have a finite number of discontinuities in the interval T;
(iii) $h(t)$ must satisfy the inequality $\int_0^T |h(t)| \, dt < \infty$.

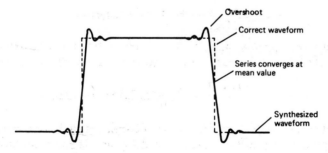

Fig. 1.6 Gibb's phenomenon in waveform synthesis.

1.6 THE FOURIER INTEGRAL

The Fourier series representation of $h(t)$ is only valid when $h(t)$ is periodic. We have already indicated that information-bearing signals change in a random fashion and do not therefore belong to this category. The amplitude spectra of non-periodic signals are obtained from the Fourier integral. The Fourier integral may be developed from the Fourier series by allowing the period T to approach infinity. In Fig. 1.2 allowing $T \rightarrow \infty$ means that $h(t)$ becomes a single pulse of width t_1 seconds.

$$h(t) = \frac{1}{T} \sum_{n=-\infty}^{\infty} C_n \exp(j2\pi n f_0 t) \quad \text{where } f_0 = 1/T$$

and

$$C_n = \int_{-T/2}^{T/2} h(t) \exp(-j2\pi n f_0 t)\, dt$$

If we let Δf be the spacing between harmonics in the Fourier series then $\Delta f = (n+1)f_0 - nf_0 = 1/T$.

The Fourier series may thus be written

$$h(t) = \sum_{n=-\infty}^{\infty} C_n \exp(j2\pi n f_0 t)\, \Delta f$$

As $T \to \infty$ then $\Delta f \to 0$ and the discrete harmonics in the series merge, and an amplitude spectrum that is a continuous function of frequency results, i.e.

$$\lim_{T \to \infty} C_n = H(f)$$

The harmonic number n now has all possible values and the summation of the series can thus be replaced by an integral, i.e. nf_0 is replaced by a continuous function f and

$$h(t) = \int_{-\infty}^{\infty} H(f) \exp(j2\pi ft)\, df$$

$$H(f) = \int_{-\infty}^{\infty} h(t) \exp(-j2\pi ft)\, dt$$

These two integrals are known as the **Fourier transform pair**. To illustrate the use of the Fourier transform, assume $h(t)$ is a single pulse of amplitude A and duration t_1 seconds.

$$H(f) = \int_{-\infty}^{\infty} h(t) \exp(-j2\pi ft)\, dt$$

$$= \int_{-t_1/2}^{t_1/2} A \exp(j2\pi ft)\, dt \qquad (1.11)$$

$$= \frac{A}{\pi f} \sin \pi f t_1$$

i.e.

$$H(f) = A t_1 \operatorname{sinc}(\pi f t_1)$$

A rectangular pulse in the time domain thus has a Fourier transform that is a sinc function in the frequency domain. The converse is also true, i.e. a sinc pulse in the time domain has a Fourier transform that is a rectangular function in the frequency domain. Consider the sinc pulse of Fig. 1.7(b).

$$h(t) = V \frac{\sin(2\pi f_1 t)}{2\pi f_1 t} \qquad \text{where } f_1 = 1/t_1$$

The Fourier transform is

$$H(f) = V \int_{-\infty}^{\infty} \frac{\sin(2\pi f_1 t)}{2\pi f_1 t} \exp(-j2\pi ft)\, dt$$

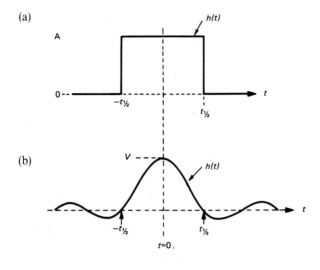

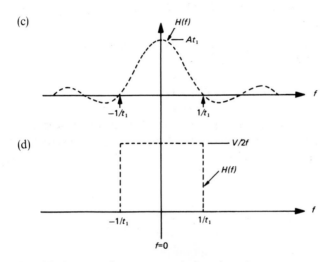

Fig. 1.7 Relationship between frequency and time domain.

Recalling that $\exp(-j2\pi ft) = \cos(2\pi ft) - j\sin(2\pi ft)$, we may write

$$H(f) = \frac{V}{2\pi f_1}\int_{-\infty}^{\infty}\frac{\sin(2\pi f_1 t)\cos(2\pi ft)}{t}\,dt$$

$$-j\frac{V}{2\pi f_1}\int_{-\infty}^{\infty}\frac{\sin(2\pi f_1 t)\sin(2\pi ft)}{t}\,dt \qquad (1.12)$$

The integral of an odd function betwen $\pm\infty$ is zero; thus the second integral of Eqn (1.12) vanishes. Using the trigonometric relationship $\cos\phi\sin\theta = \frac{1}{2}[\sin(\phi+\theta) - \sin(\phi-\theta)]$ we can say

$$H(f) = \frac{V}{2\pi f_1}\int_0^{\infty}\frac{\sin[2\pi(f+f_1)t]}{t}\,dt - \frac{V}{2\pi f_1}\int_0^{\infty}\frac{\sin[2\pi(f-f_1)t]}{t}\,dt \qquad (1.13)$$

At this point we make use of the standard integral

$$\int_0^\infty \frac{\sin ax}{x}\, dx = \pi/2 \quad \text{for } a > 0$$
$$0 \quad \text{for } a = 0$$
$$-\pi/2 \quad \text{for } a < 0$$

Re-writing Eqn (1.13) as

$$H(f)\frac{V}{2\pi f_1}(I_1 - I_2)$$

there are three frequency ranges of interest:

$$-\infty < f < f_1 \text{ gives } I_1 = -\pi/2, \quad I_2 = -\pi/2$$
$$-f_1 < f < f_1 \text{ gives } I_1 = \pi/2, \quad I_2 = -\pi/2$$
$$f_1 < f < \infty \text{ gives } I_1 = \pi/2, \quad I_2 = \pi/2$$

Hence $H(f) = V/2f_1$ for $-f_1 < f < f_1 = 0$ else. The resulting $H(f)$ is shown in Fig. 1.7 (d).

Comparing Fig. 1.7(a) with Fig. 1.3 shows another important relationship. The envelope of the amplitude spectrum of a single pulse is identical to the envelope of the amplitude spectrum of a periodic pulse train of the same pulse width. This relationship is not restricted to rectangular pulses and is useful in determining the spectral envelope of signals composed of randomly occurring pulses, such as are encountered in digital communications systems.

The fact that all frequencies are present in the amplitude spectrum of a non-periodic signal requires careful interpretation when considering the power dissipated by such signals.

1.7 POWER AND ENERGY DENSITY SPECTRUM

The power density spectrum of a non-periodic signal is developed in a similar way to the amplitude spectrum. If we assume that $h(t)$ is a periodic function we may write Eqn (1.10) as

$$\frac{1}{T}\int_{-T/2}^{T/2} |h(t)|^2\, dt = \frac{1}{T}\sum_{n=-\infty}^{\infty} |C_n|^2 \Delta f$$

As $T \to \infty$ for non-periodic signals, this equation can be written in the limit as

$$\frac{1}{T}\int_{-\infty}^{\infty} |h(t)|^2\, dt = \frac{1}{T}\int_{-\infty}^{\infty} |H(f)|^2\, df \tag{1.14}$$

The power spectrum of a non-periodic signal is then defined as

$$G(f) = \frac{|H(f)|^2}{T} \tag{1.15}$$

The power spectral density is a measure of the distribution of power as a function of frequency. It is a useful concept for random signals, such as noise, that have a finite power and are eternal; T is then the period of measurement. When T is large the power is independent of the value of T. If signals exist for a finite time only, the power spectrum approaches zero as $T \to \infty$. When dealing with such signals the concept of energy density spectrum is more meaningful. Equation (1.14) can also be written

$$\frac{1}{T} \int_{-\infty}^{\infty} |h(t)|^2 \, dt = \int_{-\infty}^{\infty} |H(f)|^2 \, df \qquad (1.16)$$

This is Parseval's theorem for non-periodic signals. The LHS of the equation represents the total energy dissipated in a $1\,\Omega$ resistance by a voltage equal in amplitude to $h(t)$. It is clear, therefore, that $|H(f)|^2$ is an energy density, that is, a measure of the distribution of the energy of $h(t)$ with frequency. The double-sided (defined for $\pm f$) energy density spectrum of $h(t)$ is

$$E(f) = |H(f)|^2 \qquad (1.17)$$

The total energy within the frequency range f_1 to f_2 is

$$E \int_{-f_2}^{-f_1} E(f) \, df + \int_{f_1}^{f_2} E(f) \, df \quad \text{joules} \qquad (1.18)$$

i.e. half the energy is contributed by the negative components. In particular, as $f_2 \to f_1$, the total energy $\to 0$. Thus although the energy density spectrum of a non-periodic signal is continuous, the energy at a specific frequency is zero.

1.8   SIGNAL TRANSMISSION THROUGH LINEAR SYSTEMS

We noted in Section 1.1 that all communications channels have the common feature of modifying or distorting the waveforms of signals transmitted through them. The amount of distortion produced by a channel with a given transfer function (attenuation and phase shift as a function of frequency) is readily calculated using Fourier transform techniques.

If we assume that $P(f)$ is the channel transfer function (often a voltage ratio in electrical networks) we can obtain the amplitude spectrum of the signal at the channel output by multiplying the amplitude spectrum of the input signal by the network transfer function, i.e.

$$L(f) = H(f) \cdot P(f) \qquad (1.19)$$

We can then obtain the output signal $l(t)$ by taking the Fourier transform of Eqn (1.19), i.e.

$$l(t) = \int_{-\infty}^{\infty} H(f) \cdot P(f) \exp(j2\pi ft) \, dt \qquad (1.20)$$

Note that $P(f) = |P(f)| \exp(-j2\pi ft)$ where $|P(f)|$ represents attenuation as a function of frequency (i.e. the frequency response of the channel) and $\phi(f)$ represents the phase shift produced. Both $|P(f)|$ and $\phi(f)$ produce signal distortion. Phase distortion is normally neglected when speech and music signals are transmitted over a channel, but it assumes special significance for digital transmission. This topic is covered further in Chapter 3.

When considering signal transmission through networks, we are often concerned with the loss of signal power or energy that occurs during transmission. If $G(f)$ is the power spectral density of a signal and $P(f)$ is the transfer function of a channel the power spectral density at the output is

$$G_0(f) = G_i(f) \cdot |P(f)|^2 \qquad (1.21)$$

The total power in a given frequency range f_1 to f_2 at the channel output is

$$W = \int_{-f_2}^{-f_1} G_0(f)\,df + \int_{f_1}^{f_2} G_0(f)\,df \quad \text{watts}$$

i.e.

$$W = \int_{-f_2}^{-f_1} G_i(f)|P(f)|^2\,df + \int_{f_1}^{f_2} G_i(f)|P(f)|^2\,df \qquad (1.22)$$

In most practical cases Eqn (1.22) can only be solved by numerical integration. In such cases the use of the discrete Fourier transform, which is discussed in Section 1.10, is particularly useful. To illustrate the application of Eqn (1.22) we will consider a specific example in which the integrals can be evaluated in closed form.

EXAMPLE: A sinc pulse of amplitude V and zero crossings at intervals of $\pm nt_1/2$ is passed through a low-pass RC network of the type shown in Fig. 1.8. If the value of t_1 for the pulse is 2 ms, find the cut-off frequency of the filter in order that 60% of the pulse energy is transmitted.

We calculate the incident energy of the pulse using Parseval's theorem, i.e.

$$E = \int_{-\infty}^{\infty} |h(t)|^2\,dt = \int_{-\infty}^{\infty} |H(f)|^2\,df$$

We have shown that for the sinc pulse $H(f) = V/2f_1$ for $-f_1 < f < f_1$ and zero otherwise.

In this example $f_1 = 1/t_1$ and the energy is

$$E_i = \int_{-f_1}^{f_1} \left(\frac{V}{2f_1}\right)^2\,df = \frac{V^2}{2f_1} \quad \text{joules}$$

The network transfer function is $P(f) = 1/[1 + j(f/f_c)]$ where $f_c = 1/2\pi RC$ is the network cut-off frequency. We note from Fig. 1.8 that the network response extends to negative frequencies because $H(f)$ is a double-sided spectrum. For this network $|P(f)|^2 = f_c^2/(f_c^2 + f^2)$.

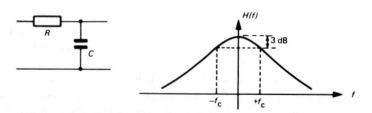

Fig. 1.8 Low-pass RC network.

The energy in the transmitted signal is then

$$E_0 = \int_{-f_1}^{f_1} \left(\frac{V}{2f_1}\right)^2 \frac{f_c^2}{f_c^2 + f^2}\, \mathrm{d}f$$

$$= 2f_c^2 \left(\frac{V}{2f_1}\right)^2 \frac{1}{f_c} \tan^{-1}\left(\frac{f_1}{f_c}\right)$$

But $E_0 = 0.6\, E_i$. Hence

$$\frac{V^2}{2f_1^2} f_c \tan^{-1}\left(\frac{f_1}{f_c}\right) = \frac{0.6\, V^2}{2f_1}$$

or

$$\tan^{-1}\left(\frac{f_1}{f_c}\right) = 0.6\frac{f_1}{f_c}$$

The solution to this equation is $f_1/f_c \simeq 1.755$, i.e.

$$f_c = \frac{1}{1.755 t_1} = 285\,\mathrm{Hz}$$

1.9   THE IMPULSE FUNCTION

If we consider the rectangular pulse of Fig. 1.7(a) and let $A = 1/t_1$, the pulse area becomes unity, i.e. area $= (1/t_1)\cdot t_1 = 1$. If t_1 is allowed to approach zero, then in order to preserve unit area the pulse amplitude A is allowed to approach infinity. Such a pulse cannot be produced practically, but it is extremely useful for analytical purposes and is known as the unit impulse. The unit impulse function is formally defined by Eqn (1.23):

$$\int_a^b \delta(t - t_0)\, \mathrm{d}t = 1 \quad \text{for } a < t_0 < b, \quad = 0 \text{ else} \qquad (1.23)$$

The impulse exists only at time t_0 and has zero value for all other values of t. The amplitude of the impulse at $t = t_0$ is undefined; instead the impulse is defined in terms of its area (or weight) at time $t = t_0$. If any continuous function $h(t)$ is multiplied by an impulse with unit weight at time $t = t_0$ the resulting function is given by

$$\int_a^b h(t)\, \delta(t - t_0)\, \mathrm{d}t = h(t_0) \quad \text{for } a < t_0 < b, \quad = 0 \text{ else} \qquad (1.24)$$

Hence multiplying $h(t)$ by an impulse function at $t = t_0$ and performing the integration of Eqn (1.24) is equivalent to taking an instantaneous sample of $h(t)$ at $t = t_0$. The impulse function is defined only in integral form and expressions such as $h(t_0) = h(t)\cdot \delta(t - t_0)$ are strictly meaningless. However, it is common practice to express the integral equation (1.24) in this form, the process of integration being implicit. The Fourier transform of an impulse function is particularly important. The Fourier transform of the unit impulse defined by Eqn (1.23) is

$$\Delta(f) = \int_{-\infty}^{\infty} \delta(t-t_0) \exp(-j2\pi ft) \, dt = \exp(-j2\pi ft_0) \qquad (1.25)$$

This means that $|\Delta(f)|$ has unity value for all values of f. The function $\exp(-j2\pi ft)$ represents the phase of each component in $\Delta(f)$, i.e. $\phi = 2\pi ft_0$.

If, instead of a single impulse, we consider a periodic train of impulses separated by a period T, the amplitude spectrum is obtained from the Fourier series. The amplitude of the nth harmonic is then

$$C_n = \int_{-\infty}^{\infty} h(t) \exp(-j2\pi f_0 t) \, dt \quad \text{where } h(t) = \delta(t - nT)$$

i.e.

$$C_n = \int_{-\infty}^{\infty} \delta(t - nT) \exp(-j2\pi f_0 t) \, dt = \exp(-j2\pi) \qquad (1.26)$$

Hence each component in the Fourier series has unity value and a phase of 2π radians. This periodic train of impulses is used to obtain regularly spaced samples of a continuous waveform $h(t)$ and is of fundamental importance in the digital transmission of analogue signals.

Now that we have defined the impulse (or delta) function we can show that the Fourier integral can also be used to define the amplitude spectrum of a periodic signal and is therefore a general transform. If

$$h(t) = \frac{1}{T} \sum_{n=-\infty}^{\infty} C_n \exp(j2\pi n f_0 t)$$

the Fourier transform of $h(t)$ is

$$H(f) = \frac{1}{T} \int_{-\infty}^{\infty} \sum_{n=-\infty}^{\infty} C_n \exp(j2\pi(f - nf_0)t) \, dt$$

$$= \frac{1}{T} \sum_{n=-\infty}^{\infty} C_n \int_{-\infty}^{\infty} \exp[j2\pi(f - nf_0)t] \, dt$$

$$= \frac{1}{T} \sum_{n=-\infty}^{\infty} C_n \delta(f - nf_0) \qquad (1.27)$$

The Fourier transform of a periodic signal is thus a set of impulses located at harmonics of the fundamental frequency $f_0 = 1/T$.

1.10   THE DISCRETE FOURIER TRANSFORM (DFT)

We pointed out in Section 1.8 the extensive use made of computer-aided analysis in the study of modern telecommunication systems. Computers cannot handle continuous signals but can process signals that are defined at discrete intervals of time. The DFT is an extension of the continuous Fourier transform designed specifically to operate on signals that have been sampled at regular intervals of time. The sampling process may be regarded as multiplying the continuous signal by a periodic series of impulses. We have shown [Eqn (1.27)] that the spectrum of such a periodic signal is a series of harmonics all of equal amplitude. When such a spectrum is multiplied by the spectrum of a continuous signal, each component in the continuous signal will form sum and difference frequencies with the harmonics of the periodic impulse train. If we assume that the impulses are separated by an interval T_s and that the maximum frequency component of the continuous signal is W Hz, the amplitude spectrum of the sampled signal will take the form of Fig. 1.9. It will be noted from Fig. 1.9 that, provided the sampling frequency $f_s (= 1/T_s)$ is at least $2W$, there will be no overlap (aliasing) between the signal spectrum and the first lower sideband of the sampled signal spectrum. The original signal is defined by its amplitude spectrum which is preserved in the sampled version provided that $f_s \geqslant 2W$. This is in fact a statement of the 'sampling theorem' that we consider in more detail in Chaper 3.

The DFT is developed for a periodic signal $h(t)$ with no components at or above a frequency $f_x = x/T$, x being an integer and T being the period of the signal waveform. An example of such a signal and its spectrum is given in Fig. 1.10, which also contains the sampled version of $h(t)$, denoted $h(k\Delta t)$, and its amplitude spectrum $C(n\Delta f)$. The sampling frequency $(f_s = 1/\Delta t)$ is chosen to equal $2f_x$ which avoids aliasing.

The Fourier series for $h(k\Delta t)$ is

$$h(k\Delta t) = \frac{1}{T} \sum_{n=-(x-1)}^{x-1} C_n \exp(j2\pi nk\Delta t/T) \qquad (1.28)$$

but over the range $-(x-1) \leqslant n \leqslant (x-1)$ the coefficients C_n are identical to $C(n\Delta f)$; hence

$$h(k\Delta t) = \frac{1}{T} \sum_{n=-(x-1)}^{(x-1)} C(n\Delta f) \exp(j2\pi nk\Delta t/T) \qquad (1.29)$$

If there is a total of N samples in the interval T, then $T = N\Delta t$ and the range of k is $0, \pm 1, \pm 2 \ldots \pm [(N/2) - 1]$. Since we can also write $\Delta t = 1/2f_x = T/2x$ then $N = 2x$ and Eqn (1.29) becomes

$$h(k\Delta t) = \frac{1}{N\Delta t} \sum_{n=-N/2+1}^{N/2-1} C(n\Delta f) \exp(j2\pi nk/N) \qquad (1.30)$$

We observe from Fig. 1.10 that $C(n\Delta f)$ is periodic and thus we can change the

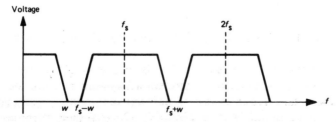

Fig. 1.9 Spectrum of a sampled signal.

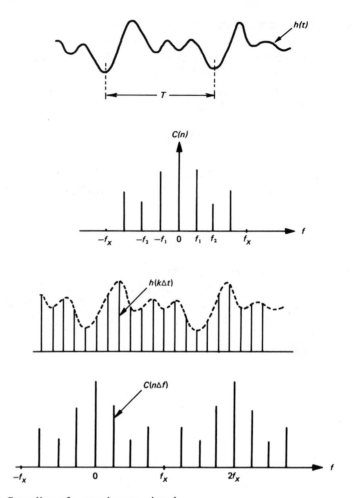

Fig. 1.10 Sampling of a continuous signal.

range of n in Eqn (1.30) to make $C(n\Delta f)$ symmetrical about the frequency f_x; thus

$$h(k\Delta t) = \frac{1}{N\Delta t} \sum_{n=0}^{N-1} C(n\Delta f) \exp(j2\pi nk/N)$$

This equation is usually written using the notation of Eqn (1.31), i.e.

$$h(k) = \frac{1}{N} \sum_{n=0}^{N-1} C(n) \exp(j2\pi nk/N) \tag{1.31}$$

The multiplying factor $1/\Delta t$ is often omitted, as in Eqn (1.31). This does not affect the relative values of $h(k)$, but it should be included for an accurate representation of $h(k)$. The amplitude spectrum $C(n\Delta f)$ is obtained using Eqn (1.8) and noting that, as $h(t)$ exists only for discrete values of t, the integral can be replaced by a summation:

$$C(n\Delta f) = \sum_{k=-N/2-1}^{N/2-1} h(k\Delta t) \exp(-j2\pi nk\Delta t/T)\Delta t \qquad (1.32)$$

Here we note that $\Delta t = T/N$ and $h(k\Delta t)$ is a periodic function so that the limits of the summation may be changed to give

$$C(n\Delta f) = \Delta t \sum_{k=0}^{N-1} h(k\Delta t) \exp(-j2\pi nk/N) \qquad (1.33)$$

This equation is usually written in the notation of Eqn (1.34) and it should be noted that once again it is customary to omit the multiplying factor, which in this case is Δt:

$$C(n) = \sum_{k=0}^{N-1} h(k) \exp(-j2\pi nk/N) \qquad (1.34)$$

Equation (1.34) is known as the discrete Fourier transform (DFT) of $h(t)$, and Eqn (1.31) is known as the inverse discrete Fourier transform (IDFT) of $C(n\Delta f)$. It should be noted that in both Eqns (1.31) and (1.34) there is no explicit frequency or time scale as the coefficients k, n and N simply have numerical values.

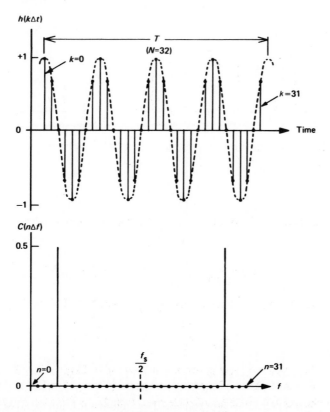

Fig. 1.11 DFT of a periodic band-limited waveform.

Some care is required in the use of the DFT because, as we have shown, it is valid only for the special case of a band-limited periodic signal. A waveform of this type and its DFT is shown in Fig. 1.11. In this figure $h(t)$ is a single tone with four complete cycles in the interval T. The DFT has a component at a value of $n = 4$ (i.e. the fourth harmonic of the fundamental frequency $f_0 = 1/T$) and a second component at a value of $n = 28$. This second component is the equivalent of $n = -4$ resulting from the change of range in Eqn (1.30). We note, therefore, that in the special case of a band-limited periodic function the DFT produces the correct spectrum of $h(t)$. In all other cases the DFT will produce only an approximation to the amplitude spectrum of $h(t)$.

Consider next the DFT of the waveform of Fig. 1.12. In this case the interval T contains 3.5 cycles of $h(t)$. The DFT requires the signal to be periodic with period T, and this means that discontinuities must now exist at the extremities of the interval T.

In other words, the periodic signal is no longer band-limited, and a form of distortion known as leakage is introduced into the spectrum. This form of distortion is considered in more detail in Section 1.11 after the concept of convolution has been introduced. It suffices here to note that sampling a non-band-limited signal produces a discrete spectrum of the form shown in Fig. 1.12. This is clearly an approximation to the original spectrum; the approximation can be made more accurate by increasing the

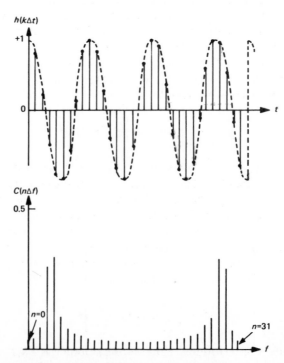

Fig. 1.12 DFT of a periodic signal truncated by an interval T not equal to a multiple of the signal period.

interval of observation T (for non-periodic signals) or by making T equal to a multiple of the period (for periodic signals). In addition, increasing the sampling frequency always reduces aliasing which is produced by sampling non-band-limited signals.

The DFT can be calculated directly from Eqn (1.34) but it will be noticed that each coefficient $C(n)$ requires N complex multiplications and additions. There are N spectral coefficients so that a total of N^2 complex multiplications and additions will be required in the complete DFT calculation. Multiplication is a relatively slow process in a general purpose digital processor, and for this reason the DFT is usually calculated by using the 'fast Fourier transform' algorithm. This is an algorithm designed to reduce the number of multiplications required to evaluate the DFT. The algorithm achieves a reduction from N^2 to $N \log_2 N$ multiplications by dividing Eqn (1.34) into the sum of several smaller sequences.[1] This reduction can be very significant when N is a very large number. (It should be noted that special purpose processors are now available which can perform multiplication in one machine cycle.)

1.11 TIME DOMAIN ANALYSIS

The time domain and frequency domain are uniquely linked by the Fourier transform and consequently the frequency domain analysis of the previous sections can also be undertaken in the time domain. To illustrate this point, consider Eqn (1.19) which relates the spectrum at a network output to the product of the input spectrum and the network transfer function. If the input to the network $h(t)$ is a unit impulse we have shown in Eqn (1.25) that the spectrum $\Delta(f)$ has unity value for all f. Hence the spectrum at the network output is simply $L(f) = P(f)$ where $P(f)$ is the network transfer function. The response in the time domain is the Fourier transform of $P(f)$, and is known as the impulse response, i.e.

$$p(t) = \int_{-\infty}^{\infty} P(f) \exp(j\, 2\pi ft)\, \mathrm{d}f \qquad (1.35)$$

Having defined impulse response we now make use of Eqn (1.24), which states that the value of a signal $h(t)$ at any time t_0 is obtained by multiplying $h(t)$ by a unit impulse centred at t_0. The signal $h(t)$ can thus be regarded as an infinite number of impulses, the weight of each impulse being equal to the instantaneous value of $h(t)$. Each of these impulses will produce an impulse response and the network output is then obtained by the superposition of the individual impulse responses. The response of a linear network $l(t)$ to an input signal $h(t)$ is given in terms of the network impulse response by Eqn (1.36):

$$l(t) = \int_{-\infty}^{\infty} h(\tau) p(t - \tau)\, \mathrm{d}\tau \qquad (1.36)$$

In this equation τ is a dummy time variable and both $h(\tau)$ and $p(\tau)$ are continuous functions. Equation (1.36) therefore states that the output of a linear network at time t is given by the sum of all values of the input $h(\tau)$

weighted by the appropriate value of $p(\tau)$ at time t. The integral in Eqn (1.36) is known as the convolution integral and the equation is often written as

$$l(\tau) = h(t) * p(t) \tag{1.37}$$

where the symbol $*$ denotes convolution. Comparing Eqn (1.37) with Eqn (1.19) we note the important relationship that multiplication in the frequency domain is equivalent to convolution in the time domain. The converse is also true; that is, multiplication in the time domain is equivalent to convolution in the frequency domain. We will now consider some examples of convolution.

The first example concerns frequency domain convolution. If we consider the waveform of Fig. 1.12 we note that in selecting a time window of T s we are in effect multiplying the continuous signal $h(t)$ by a rectangular pulse of unity amplitude and duration T. This is equivalent to convolving the amplitude spectrum of $h(t)$ with the spectrum of the rectangular window function which, as we have already seen, is a sinc function. The spectrum of $h(t)$ is actually a delta function at $\pm f_0$ since only a single frequency is present. The convolution integral is thus

$$\int_{-\infty}^{\infty} H(f)\delta(f-f_0)\,df + \int_{-\infty}^{\infty} H(f)\delta(f+f_0)\,df = H(-f_0) + H(f_0) \tag{1.38}$$

The original spectrum centred at $f = 0$ is thus transferred to frequencies $\pm f_0$. The procedure is illustrated in Fig. 1.13 and it is interesting to compare this spectrum with the DFT of Fig. 1.12. The effect of truncating the signal $h(t)$ in the time domain causes a spreading of the spectrum (leakage) in the frequency domain.

We next consider the transmission of a rectangular pulse through a low-pass RC network of the form shown in Fig. 1.8. The impulse response of this network is $p(t) = (1/RC)\exp(-t/RC)$. This may be proved as follows:

$$P(f) = \int_{-\infty}^{\infty} p(t)\exp(-j2\pi ft)\,dt$$

Since $p(t)$ is the impulse response of a real network it must have a value of zero for $t < 0$; hence

$$P(f) = \frac{1}{RC}\int_{0}^{\infty} \exp(-t/RC)\exp(-j2\pi ft)\,dt$$

$$= \frac{-1}{RC(j2\pi f + 1/RC)}[\exp\{-(j2\pi f + 1/RC)t\}]_0^\infty$$

i.e.

$$P(f) = \frac{1}{1 + j2\pi fRC} = \frac{1}{1 + j(f/f_c)}$$

which agrees with the expression obtained by network analysis.

Before proceeding further we shall consider the physical interpretation of Eqn (1.36). In this equation t represents the present instant in time and we

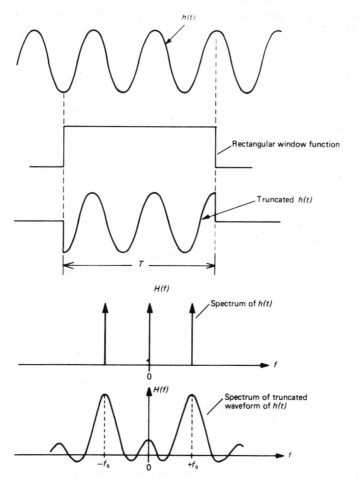

Fig. 1.13 Convolution in the frequency domain.

note that for practical filters $p(t - \tau)$ must be zero for $t < \tau$; in other words, the impulse response must be zero for all time before an impulse occurs. The impulse response of an ideal filter exists for all values of τ and is therefore unrealizable (see Section 3.3). If we confine our interest to practical networks, then it is clear that the impulse response $p(t - \tau)$ scans the signal $h(\tau)$ and produces a weighted sum of past inputs. The values of $h(\tau)$ closest to the present (i.e. $\tau \simeq t$) will have a greater effect on the output than values occurring a long time in the past ($\tau \ll t$). For practical networks Eqn (1.36) becomes

$$l(t) = \int_{-\infty}^{t} h(t)\, p(t - \tau)\, d\tau \tag{1.39}$$

We can now consider the response of the RC network to a rectangular pulse of width t_1. This is split into a positive step of unit amplitude at $\tau = 0$ followed by a negative step of unit amplitude at $\tau = t_1$, as shown in Fig. 1.14. Since the

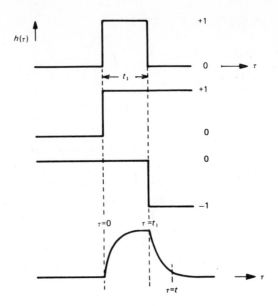

Fig. 1.14 Response of an RC network to a rectangular pulse.

system is linear the output is obtained by superposition. Considering the positive step first, since $h(\tau) = 0$ for $\tau < 0$, then $l(t) = \int_0^t p(t - \tau) d\tau$, i.e.

$$l(t) = \frac{1}{RC} \int_0^t \exp[-(t - \tau)/RC] dt$$

$$= \exp \frac{(-t/RC)}{RC} d\tau \int_0^t \exp(\tau/RC) d\tau$$

Therefore

$$l(t) = -\exp(-t/RC)[\exp(t/RC) - 1] = 1 - \exp(-t/RC)$$

This is the step response of the filter, i.e. the output for $0 < t < t_1$. For the negative step

$$l(t) = -\int_{-t_1}^t \exp[-(t - \tau)/RC] d\tau$$

i.e.

$$l(t) = \exp[(t_1 - t)/RC] - 1$$

The output of the filter at times in excess of $t = t_1$ is therefore

$$l(t) = \exp[(t_1 - t)/RC] - \exp(-t/RC)$$

As we have already pointed out, time domain analysis is equivalent to frequency domain analysis and it is not possible to give a general rule as to which technique is more appropriate to particular situations. Time domain analysis is frequently used in digital systems, especially in specifying network characteristics to minimize signal distortion.

1.12   CORRELATION FUNCTIONS

These functions have particular application in the time domain specification of signals that vary in an unpredictable manner (i.e. information-bearing signals, noise, etc.). The autocorrelation function of a waveform $h(t)$ is defined as

$$R_h(\tau) = \lim_{T \to \infty} \frac{1}{T} \int_{-T/2}^{T/2} h(t) \, h(t + \tau) \, dt \qquad (1.40)$$

The autocorrelation function is the average of the product $h(t) \, h(t + \tau)$ and will clearly depend on the value of τ. The function $h(t + \tau)$ is a replica of $h(t)$ delayed by an interval τ. The numerical value of $R_h(\tau)$ is a measure of the similarity (or correlation) of $h(t)$ and $h(t + \tau)$. If there is no similarity between $h(t)$ and $h(t + \tau)$ and each has zero mean value then $R_h(\tau)$ is zero. Completely random signals, such as white noise, have an autocorrelation function equal to zero. The maximum value of $R_h(\tau)$ for any signal will occur when $\tau = 0$. In these circumstances

$$R_h(0) = \lim_{T \to \infty} \frac{1}{T} \int_{-T/2}^{T/2} h^2(t) \, dt$$

which is the mean square value of $h(t)$.

The autocorrelation function is widely used in signal analysis for recognizing signals in the presence of noise and also for estimating the power spectral density of random signals. We will derive, as an example, the autocorrelation function of a periodic pulse train and a random binary data signal. Consider the periodic waveform of Fig. 1.15: it is only necessary to average the signal over one period and thus

$$R_h(\tau) = \frac{1}{T} \int_{-T/2}^{T/2} h(t) \, h(t + \tau) \, dt$$

The waveform of $h(t) \, h(t + \tau)$ is shown in Fig. 1.15. It is a periodic pulse train of amplitude A^2, pulse duration $t_1 - |\tau|$ and period T. The autocorrelation function is thus

$$R_h(\tau) = \frac{A^2}{T}(t_1 - |\tau|) \quad \text{for} -t_1 < \tau < t_1$$

but because $h(t + T) = h(t)$ the autocorrelation function is periodic. The value of $R_h(\tau)$ is plotted as a function of τ in Fig. 1.15.

We note that if $h(t)$ is a single pulse the autocorrelation function is modified to

$$R_h(\tau) = \frac{1}{t_1} \int_{-t_1/2}^{t_1/2} h(t) \, h(t + \tau) \, dt$$

which evaluates to

$$R_h(\tau) = A^2 \left(1 - \frac{|\tau|}{t_1}\right) \quad \text{for} -t_1 < \tau < t_1$$

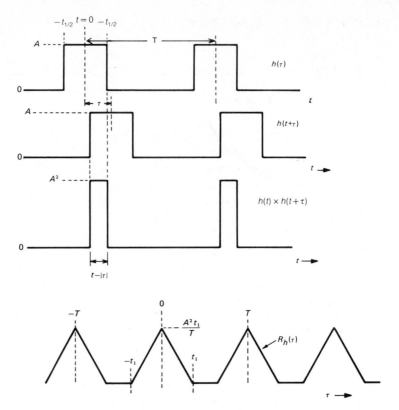

Fig. 1.15 Autocorrelation function of a periodic pulse train.

In this case $R_h(\tau)$ is a single triangular pulse of peak amplitude A^2. The autocorrelation function of a random binary pulse train is obtained in a similar way. Assume such a signal is composed of pulses of amplitude A volts and 0 volts and each of the duration t_1, both of equal probability. During any time interval T there will be an equal number of pulses of amplitude A volts and 0 volts. The average value of $h(t)\,h(t+\tau)$ must therefore be

$$R_h(\tau) = \frac{A^2}{2}\left(1 - \frac{|\tau|}{t_1}\right) \quad \text{for} -t_1 < \tau < t_1$$

and this is the autocorrelation function of a random binary pulse train.

Equation (1.40) is very similar to the convolution integral of Eqn (1.36). Remembering that convolution in the time domain is equivalent to multiplication in the frequency domain, the Fourier transform of $R_h(\tau)$ is equal to the Fourier transform of $h(t)$ multiplied by the Fourier transform of $h(t+\tau)$. The amplitude spectrum of $h(t)$ is identical to the amplitude spectrum of $h(t+\tau)$, i.e.

$$\int_{-\infty}^{\infty} R_h(\tau)\exp(j2\pi f\tau)\,d\tau = |H(f)|^2 \qquad (1.41)$$

Thus the power spectral density of any signal is the Fourier transform of its autocorrelation function. The power spectral density of the random data signal is thus

$$G(f) = \int_{-\infty}^{\infty} \frac{A^2}{2}\left(1 - \frac{|\tau|}{t_1}\right)\exp(j2\pi f\tau)\,d\tau$$

i.e.

$$G(f) = \int_{-\infty}^{\infty} \frac{A^2}{t_1}\left(1 - \frac{|\tau|}{t_1}\right)[\cos(2\pi f\tau) + j\sin(2\pi f\tau)]\,d\tau \qquad (1.42)$$

Since $R_h(\tau)$ is an even function of τ the imaginary terms in Eqn (1.42) vanish and

$$G(f) = \int_{-\infty}^{\infty} \frac{A^2}{2}\left(1 - \frac{|\tau|}{t_1}\right)\cos(2\pi f\tau)\,d\tau$$

which evaluates to $G(f) = (At_1)^2 \operatorname{sinc}^2(\pi f t_1)$.

The autocorrelation function is a measure of the degree of similarity between $h(t)$ and a delayed version of the same waveform. The cross-correlation function is a measure of the degree of similarity between two different waveforms $h(t)$ and $g(t)$. The cross-correlation function is defined as

$$R_{hg}(\tau) = \lim_{T \to \infty} \frac{1}{T}\int_{-T/2}^{T/2} h(t)\,g(t + \tau)\,dt \qquad (1.43)$$

This function finds specific application in the detection of signals at low signal-to-noise ratios. Correlation detection is considered in detail in Chapter 5.

It should be clear from the previous two sections that both time domain and frequency domain techniques are important tools in the analysis of telecommunications systems. They should be regarded as complementary, as it is not possible to give a general rule as to which technique is more appropriate to a particular situation.

1.13   INFORMATION CONTENT OF SIGNALS

In previous sections we have considered signals in terms of waveforms and spectra. In this section we consider the information content of signals and show how it is related to the information capacity of communication channels.

Information is conveyed by a signal that changes in an unpredictable fashion. It is important to have some method of evaluating the information content of a signal because this will determine whether or not the signal can be transmitted over a particular channel. The information content of a signal is measured in bits which, as we shall show later, is not necessarily related to the number of binary digits required to transmit it. The information capacity of a communication channel is limited by bandwidth, which determines the maximum signalling speed, and by noise, which determines the number of distinguishable signal levels.

We shall consider the specific example of a teleprinter which is restricted to transmitting the four signals $ABCD$. If one of these symbols is transmitted there are four (4^1) possible messages which are A or B or C or D. If two symbols are sent there are 16 (4^2) possible messages, viz.

$$AA \quad \text{or} \quad AB \quad \text{or} \quad AC \quad \text{or} \quad AD$$
$$\text{or} \quad BA \quad \text{or} \quad BB \quad \text{or} \quad BC \quad \text{or} \quad BD$$
$$\text{or} \quad CA \quad \text{or} \quad CB \quad \text{or} \quad CC \quad \text{or} \quad CD$$
$$\text{or} \quad DA \quad \text{or} \quad DB \quad \text{or} \quad DC \quad \text{or} \quad DD$$

If P symbols are sent the number of possible messages is 4^P. If the teleprinter can transmit n different symbols, the number of different messages that could be transmitted when P symbols are sent is n^P. Obviously the greater the number of possible messages the less predictable is any particular message. Intuitively we would argue that the more unpredictable a particular message the more information it contains. It is reasonable to assume that the information content is a function of the **unpredictability** of a message. In algebraic form the information content H is

$$H \propto f(n^P) \tag{1.44}$$

Equation (1.44) can be made a function of time by assuming that one symbol is transmitted every t_1 seconds. The total number of symbols transmitted in T seconds is thus T/t_1 and the information content of such a message of T seconds duration would be

$$H \propto f(n^{T/t_1}) \tag{1.45}$$

It is reasonable to assume that a similar message of duration $2T$ seconds would contain twice as much information as a message of duration T seconds; in other words $f(n^{T/t_1})$ should be linearly related to T. This defines the function f as a logarithm, i.e.

$$H \propto \log_x(n^{T/t_1})$$

or

$$H = K \frac{T}{t_1} \log_x n \tag{1.46}$$

We are still required to define the numerical values of K and x. The constant of proportionality is taken as unity and the base of the logarithm is specified by defining the unit of information. To illustrate this idea consider the simplest possible system, i.e. a source that can send only two possible symbols, A or B. The simplest possible message will occur when only one of the two possible symbols is sent. The information content of such a message is defined as 1 bit.

In such a system, T/t_1 symbols are sent in T seconds and the information transmitted is T/t_1 bits, i.e.

$$H = \frac{T}{t_1} \log_x n \quad \text{where } n = 2$$

or

$$\frac{T}{t_1}\log_x 2 = \frac{T}{t_1}$$

Hence

$$x = 2$$

The information transmitted by a source that can send n different symbols is

$$H = \frac{T}{t_1}\log_2 n \quad \text{bits} \tag{1.47}$$

The information rate is $(1/t_1)\log_2 n$ bits/s (also b/s) and the information per symbol is $\log_2 n$ bits. In arriving at this result we have made the implicit assumption that each of the n different symbols has equal probability of being sent. When the probability is not equal our definition of information informs us that symbols that occur least frequently contain a greater amount of information than symbols that occur very frequently. Before considering probability in detail it is important to note that the symbol example chosen is not restricted to alphabetic characters.

Consider the example of a voltage pulse, and assume that each pulse can have any one of eight different voltages. A typical signal is shown in Fig. 1.16. If we assume that each of the eight levels is equi-probable the information per pulse is

$$H = \log_2 8 = 3 \text{ bits/pulse}$$

It is not always convenient to use base 2 logarithms so, making use of the relationship

$$\log_2 n = \frac{\log_{10} n}{\log_{10} 2}$$

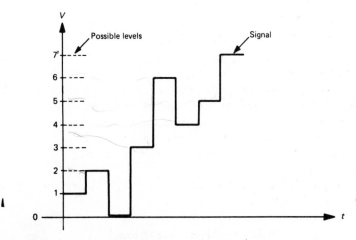

Fig. 1.16 Information-bearing signal.

then

$$H = 3.32 \log_{10} n \quad \text{bits/pulse} \tag{1.48}$$

The information capacity of a channel must be greater than the information rate of the transmitted signal in order for reliable communication to occur. We show in Section 3.7 that an ideal low-pass channel of bandwidth B can transmit pulses at a maximum rate of $2B$ per second. If we use the pulse analogy and assume that m different pulse levels can be distinguished at the channel output, the maximum rate at which information can be transmitted over the channel becomes

$$C = 2B \log_2 m \quad \text{bits/s} \tag{1.49}$$

This relationship is known as Hartley's law. If the capacity of a channel is known it is possible to determine the rate at which information can be transmitted and, consequently, the time required to transmit a given amount of information. Hartley's law does not give any indication of how the value of m is determined. This depends on the signal-to-noise ratio at the channel output, and to pursue this further it is necessary to introduce the significance of probability in information theory.

This may be illustrated by assuming that an information source can send n equi-likely symbols, each of which belongs to one of two groups. It is further assumed that the receiver is not interested in the value of a particular received symbol, rather it is concerned with knowing only to which group the received symbol belongs. If group 1 contains n_1 symbols and group 2 contains n_2 symbols there are two messages which are of interest to the receiver and these have a probability of occurrence of $P_1 = n_1/n$ and $P_2 = n_2/n$, respectively. The information per symbol for n equi-likely symbols is $\log_2 n$ bits, hence the total information in n symbols is $n \log_2 n$. It follows that the total information in group 1 (which is not of interest to the receiver) is $n_1 \log_2 n_1$ and that the total information in group 2 (which is also not of interest to the receiver) is $n_2 \log_2 n_2$. Thus the useful information H may be defined as the total information less the information which is not of interest, i.e.

$$H = n \log_2 n - n_1 \log_2 n_1 - n_2 \log_2 n_2 \tag{1.50}$$

The average information is thus $H_{av} = H/n$ or

$$H_{av} = \left(\frac{n_1 + n_2}{n} \right) \log_2 n - \frac{n_1}{n} \log_2 n_1 - \frac{n_2}{n} \log_2 n_2$$

or

$$H_{av} = \frac{n_1}{n} (\log_2 n - \log_2 n_1) + \frac{n_2}{n} (\log_2 n - \text{long}_2 n_2)$$

i.e.

$$H_{av} = \frac{-n_1}{n} \log_2 \frac{n_1}{n} - \frac{n_2}{n} \log_2 \frac{n_2}{n}$$

or

$$H_{av} = -P_1 \log_2 P_1 - P_2 \log_2 P_2 \qquad (1.51)$$

The more unpredictable an event the more information it contains; for instance let $P_1 = 0.8$, which means that $P_2 = 0.2$ since $P_1 + P_2 = 1$.

The information associated with the first event is

$$H_1 = -\log_2 0.8 = -3.32 \log_{10} 0.8 = 0.32 \text{ bits}$$

The information associated with the second event is

$$H_2 = -\log_2 0.2 = -3.32 \log_{10} 0.2 = 2.32 \text{ bits}$$

This agrees with our concept of information. The average information in this case would be

$$H_{av} = (0.8 \times 0.32) + (0.2 \times 2.32) = 0.72 \text{ bits/symbol}$$

which is considerably less than the information transmitted by the symbol with lower probability. It is of interest to determine the maximum value of H_{av}, and to do this we eliminate P_2 from Eqn. (1.51), i.e.

$$H_{av} = -P_1 \log_2 P_1 + (P_1 - 1) \log_2 (1 - P_1) \qquad (1.52)$$

To find the maximum value of H_{av} we differentiate with respect to P_1 and set the result equal to zero:

$$\frac{dH_{av}}{dP_1} = -P_1 \frac{1}{P_1} - \log_2 P_1 + (P_1 - 1) \frac{-1}{1 - P_1} + \log_2 (1 - P_1)$$

i.e.

$$\frac{dH_{av}}{dP_1} = \log_2 (1 - P_1) - \log_2 (P_1)$$

which is zero when $(1 - P_1) = P_2$ or $P_1 = P_2 = 0.5$. The average information is a maximum when the symbols are equi-probable.

In developing Eqn (1.50), the original n symbols were divided into two separate groups. This idea can be extended for any number of groups up to a maximum of n. When the number of groups equals the number of symbols, we are in effect saying that each individual symbol has its own probability of occurrence and the average information is

$$H_{av} = -\sum_{i=1}^{n} P_i \log_2 P_i \qquad (1.53)$$

Equation (1.53) is similar to an equation in statistical mechanics that defines a quantity known as 'entropy'. For this reason H_{av} is usually known as the entropy of a message. In particular, if all symbols are equi-probable, $P_i = 1/n$ and Eqn (1.53) becomes

$$H_{av} = -\sum_{i=1}^{n} P_i \log_2 n = \log_2 n \quad \text{since} \sum_{i=1}^{n} P_i = 1$$

Extension of the analysis for maximum entropy produces the same result as for the two-symbol case; that is, the entropy of a message is a maximum when all symbols are equi-probable. In any other situation the entropy will be less than the maximum and the message is said to contain 'redundancy'.

When all symbols are equi-probable, the average information is a maximum and it is not possible to make other than a pure guess at what the next symbol will be after a number have been received. In certain circumstances this can be a serious problem because if an error occurs during transmission the receiver will not be aware of it. When all symbols are not equi-probable, it becomes feasible to predict what the next symbol in a received sequence should be. The redundancy in such a message is defined as

$$R = \frac{H_{av(max)} - H_{av}}{H_{av(max)}} \times 100\% \qquad (1.54)$$

The significance of redundancy in a message will be illustrated by reference to the English language. If we assumed that all letters in the English alphabet were equi-probable, the average information per letter would be $\log_2 26 = 4.7$ bits. If the relative frequencies of occurrence of individual letters are taken into account (E has a probability of 0.1073, Z has a probability of 0.006) the figure works out as $H_{av} = 4.15$ bits/letter. This gives a redundancy of 11.7%. The redundancy is actually much higher than this because of the interdependence between letters, words and groups of words within English text. For example, if the letter Q occurs in a message it is almost certain that the next letter will be U. The U contains no information because it can be guessed with almost 100% certainty. If the letters IN have been received, the probability that the next letter will be G is much higher than the probability that it will be Z. There are many examples of this interdependence, which can be extended to words and sentences. When all these issues are considered the redundancy of English is estimated at 47%. The overall effect of redundancy is twofold; it reduces the rate of transmission of information but at the same time it allows the receiver to detect, and sometimes correct, errors.

Consider the received message

Thi ship wilp arrive on September 28

It is clear that we can detect and correct errors in the alphabetic section of the message. There is no way that we can detect an error in the date (unless it is a number greater than 30), however. The numerical part of message thus contains no redundancy. This is very important because data transmission occurs as a sequence of binary numbers that has no inherent redundancy. It is important to detect occasional errors when they occur, and in data systems a form of redundancy known as 'parity checking' is often employed. The binary digits are divided into groups of 7, e.g., 1000001, and an extra digit is added to make the total number of 1s in the group of 8 either even or odd depending on the system. The receiver then checks each group of 8 digits to determine whether an odd or even parity has been preserved. If the parity check is not valid an error is detected. (This topic is covered in more detail in Section 5.6.)

1.14 INFORMATION TRANSMISSION

The information capacity of a communications channel is specified by Hartley's law [Eqn (1.49)], but this equation does not tell us how to evaluate the number of detectable levels, m. All signals in telecommunications are subject to corruption by noise, and we can make a qualitative statement to the effect that the difference between detectable levels must be greater than the noise present during transmission. If this were not the case, signal plus noise could produce a false level indication. Noise is a random disturbance that may be analysed either in terms of its statistical properties or in terms of its spectral properties. From either description we are able to define a mean square value for the noise that is equivalent to the power developed by the noise voltage in a resistance of $1\,\Omega$.

The relationship between the number of messages that can be transmitted and noise power was obtained by Shannon[2] in 1948 using the mathematics of n-dimensional space. The mathematics of n-dimensional space is a theoretical extension of the familiar mathematics of two- and three-dimensional space. An n-dimensional space is termed a 'hyperspace' and is defined by a set of n mutually perpendicular axes. If q is a point in this hyperspace, its distance from the origin (the point of intersection of the n mutually perpendicular axes) is d where

$$d^2 = x_1^2 + x_2^2 + x_3^2 + \cdots\cdots x_n^2 \tag{1.55}$$

x_n being the perpendicular distance from the point to the nth axis. When dealing wih hyperspace, the 'volume' of an n-dimensional figure is defined as the product of the lengths of its sides. If we are considering an n-dimensional cube (hypercube) in which all the sides have length L units, the volume is given by L^n. (Note that when $n = 2$ the hypercube is actually a square, and the 'volume' is interpreted as an area.)

A circle is a two-dimensional figure whose volume (i.e. area) is πr^2. If two concentric circles are drawn, one with a radius of 1 and the other with a radius of $\frac{1}{2}$, then one-quarter of the total area is enclosed within the inner circle which has half the total radius. The volume of a three-dimensional sphere is $\frac{4}{3}\pi r^3$. This means that a sphere of radius $\frac{1}{2}$ would contain only one-eighth of the volume of a sphere of radius 1. An n-dimensional sphere (hypersphere) has a volume proportional to r^n where r is its radius. The volume of such a hypersphere of radius $\frac{1}{2}$ will thus be equal to $(\frac{1}{2})^n$ of the volume of a hypersphere of radius 1. In other words, the larger the value of n the smaller is the percentage of total volume contained within the hypersphere of radius $\frac{1}{2}$. When $n = 7$ only $1/128$ of the total volume lies within the hypersphere of radius $\frac{1}{2}$. For a hypersphere for which $n = 100$ it turns out that only 0.004% of the total volume of a hypersphere of radius 1 lies within a hypersphere of radius 0.99. This leads to the important conclusion that for a sphere of n dimensions (where $n \gg 1$) practically all of the volume lies very close to the surface of the sphere. This property is fundamental to Shannon's derivation of the information capacity of a channel.

Shannon postulates a signal source that can send out messages of duration

T seconds; the source is limited to a bandwidth of B Hz and has a statistically stationary output (the properties of the output averaged over a long period are constant). We have already indicated in Section 1.10 that any signal band limited to B Hz can be represented by $2B$ samples/s. The energy in each of these samples is proportional to the square of the sample amplitude and the total signal energy will therefore be proportional to the sum of these mean square values. If the signal amplitudes are denoted by x_1, x_2, etc., then the total normalized signal energy is

$$E = x_1^2 + x_2^2 + \cdots + x_n^2 \tag{1.56}$$

This is identical to the expression for the $(distance)^2$ of a point from the origin of the hyperspace. If the average energy/sample is S the total energy in the message is $2BTS$ joules. Hence all messages of energy $2BTS$ joules may be represented as a point in a hyperspace of n dimensions. As $n \to \infty$ virtually all messages will be points very close to the surface of a hypersphere of radius $(2BTS)^{1/2}$.

Received signals are always accompanied by noise, which in an ideal case occupies the same bandwidth as the signal. This means that the noise waveform can also be represented by $2BT$ samples. If the average energy per noise sample is N joules the total noise energy is $2BTN$ joules. We have stated that each message is represented by a point in the hypersphere of radius $(2BTS)^{1/2}$ and therefore the sum of message and noise will be a point whose distance is $(2BTN)^{1/2}$ from each point representing the message alone. This means that all possible message + noise combinations are represented by points that are very close to the surface of a hypersphere of radius $(2BTN)^{1/2}$ centred on the point representing the noise alone. The receiver actually receives message + noise with a total energy of $2BT(S + N)$ and the points representing each combination must lie within a hypersphere of radius $[2BT(S + N)]^{1/2}$.

Because each message must be a distance $(2BTN)^{1/2}$ from points representing message + noise, we can represent the system as a hypersphere of radius $[2BT(S + N)]^{1/2}$ filled with non-overlapping hyperspheres of radius $(2BTN)^{1/2}$. The centres of the small hyperspheres represent a distinguishable message that could have been sent. The total number of distinguishable messages is given by the number of non-overlapping hyperspheres of radius $(2BTN)^{1/2}$ which can exist within a hypersphere of radius $[2BT(S + N)]^{1/2}$. This will be equal to the ratio of the volumes of the two hyperspheres. The physical picture is given in Fig. 1.17. If the number of dimensions is $2BT$, the ratio of the volumes of the two hyperspheres is

$$\left[\frac{(2BT(S + N))^{1/2}}{(2BTN)^{1/2}} \right]^{2BT} = \left[\frac{S + N}{N} \right]^{BT} \tag{1.57}$$

This gives the number of distinguishable messages that can be sent. Assuming each message to be equi-probable, the information transmitted is

$$\log_2 \left[\frac{S + N}{N} \right]^{BT} = BT \log_2 \left(1 + \frac{S}{N} \right) \quad \text{bits} \tag{1.58}$$

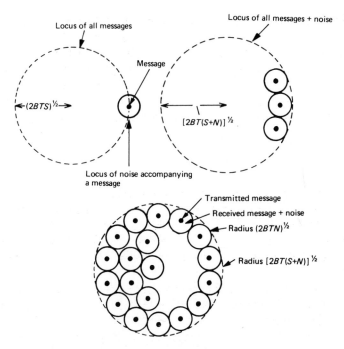

Fig. 1.17 Multidimensional representation of signal + noise.

The channel capacity is thus

$$C = B \log_2 \left(1 + \frac{S}{N} \right) \quad \text{bits/s}$$

The ratio of signal energy to noise energy is identical to the corresponding power ratio so that Shannon's law for channel capacity is

$$C = B \log_2 \left(1 + \frac{S_p}{N_p} \right) \quad \text{bits/s} \qquad (1.59)$$

This law is a fundamental law of telecommunications, and states that if a channel has a bandwidth B and the mean SNR is S_p/N_p the maximum rate at which information may be transmitted is C bits/s. In other words there is a theoretical limit on the amount of information that can be transmitted over any telecommunications channel and, as we shall show in later chapters, all practical systems fall short of this theoretical limit by varying degrees. As an illustration of the theoretical application of Shannon's law we will assume that a link of effective bandwidth 3 kHz is to be used for pulse transmission. This link is to be used for six identical channels, the data rate being 1 kb/s per channel. The SNR on the link is $2/\log_{10} D$ where D is the length of the link in kilometres. We are required to determine the maximum distance over which reliable communication is possible.

The total data rate is $6 \times 1000 = 6$ kb/s. Hence from Shannon's law

$$6 \times 10^3 = 3 \times 10^3 \log_2 \left(1 + \frac{2}{\log_{10} D} \right)$$

which reduces to

$$1 + \frac{2}{\log_{10} D} = 4.0$$

or

$$D = 4.64\,\text{km}$$

This example does not consider the way in which the information is actually transmitted. In order to achieve the theoretical channel capacity Shannon postulated that each message should be represented by a large number of samples (data points). The receiver, in effect, compares the received message, corrupted by noise, with each possible uncorrupted message that could have been transmitted. These uncorrupted messages are the centres of the hyperspheres of radius $(2BTN)^{1/2}$. The receiver then decides that the message which was actually transmitted is the one (in the n-dimensional space) closest to the point representing the received message + noise. As the number of data points describing each message increases, the closer is any individual message to the surface of the hypersphere of radius $(2BTS)^{1/2}$. Alternatively, the larger the number of data points, the more accurate is Shannon's law.

The practical drawback is that the larger the number of data points the longer is the time taken to transmit them over a link of fixed bandwidth. The receiver cannot decide which message was transmitted until all data points have been received. Thus in attempting to realize a communications system that obeys Shannon's law, long delays would be introduced between transmission and reception (the information rate is not affected by a delay).

Shannon's law would not seem to be a practical proposition, but it is extremely useful as a standard for comparing the performance of various telecommunications systems. Of particular importance is the possibility of exchanging bandwidth and SNR in order to achieve a given information transmission rate. We will now examine this possibility in some detail. It will be noted from Eqn (1.59) that the value C can remain fixed when the bandwidth B is changed, provided that the SNR is modified accordingly. This means that SNR and bandwidth can be exchanged without affecting the channel capacity. Figure 1.18 shows bandwidth plotted against SNR for a constant channel capacity of 1 b/s.

Point A on this graph shows that it is possible to transmit 1 b/s in a bandwidth of 0.25 Hz provided that the signal power is 15 times the noise power. If the bandwidth is now doubled (point B on the graph) the same rate of transmission is possible with a SNR of 3. If we assume that the noise power is constant then evidently much less signal power is required at point B. As a simple physical example assume that 1 b/s is being transmitted at point B on the curve using binary pulses. If the bandwidth is decreased then clearly fewer binary pulses can be transmitted per second. To maintain the same information rate, the binary pulses must be replaced by pulses with more than two levels, which consequently increases the mean signal power. For large SNRs, Shannon's law approximates to

$$C = B\log_2(S_p/N_p) \qquad (1.60)$$

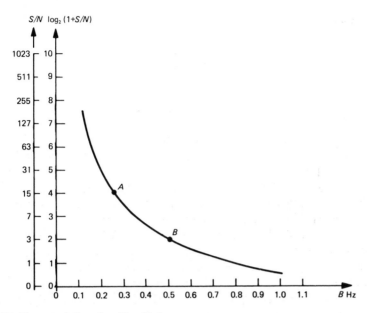

Fig. 1.18 Shannon's law for $C = 1$ b/s.

This equation states that in order to maintain a given rate of transmission of information, a linear change in bandwidth must be accompanied by an exponential change in SNR. Thus very large increases in signal power are required to compensate for relatively small reductions in the channel bandwidth B. The assumption that noise power remains constant is not in fact justified. In telecommunications systems the noise power is linearly related to bandwidth, i.e.

$$N_p = \eta B \tag{1.61}$$

η is known as the noise power spectral density. This means that when the bandwidth is doubled the noise power increases by 3 dB; hence the reduction of required signal power that accompanies an increase in bandwidth is partially offset by the extra noise power. At low SNRs, the saving in received signal power will be cancelled by the extra noise. We can write Shannon's law as

$$C = 1.45\, B \ln(1 + S_p/N_p)$$

If S_p/N_p is very small, $\ln(1 + S_p/N_p) \simeq S_p/N_p$. Hence $C = 1.45\, BS_p/N_p$.
But $N_p = \eta B$; thus

$$C = 1.45\, S_p/\eta \quad \text{bits/s} \tag{1.62}$$

The channel capacity now becomes independent of the bandwidth and a further increase in bandwidth has no effect. Hence there is a limit to the amount of information that can be transmitted with a fixed signal power regardless of the bandwidth. Once again this is a theoretical limit and all practical systems fall short of this limit. In the following chapters we consider both analogue and digital communications systems and compare their relative performance in terms of the theoretical limits imposed by Shannon's law.

1.15 CONCLUSION

In this chapter we have considered, in a general way, methods of describing both signal and channel characteristics met in telecommunications systems. We have introduced the idea that the channel characteristic intimately influences the signal waveform of the transmitted information. In defining information, which is the basic entity to be transmitted, we were able to show once again an intimate relationship between channel characteristics and the rate at which this information can be transmitted.

In the following chapters we will be considering specific telecommunications systems in detail and showing that the relationships outlined in this chapter are common to all systems. Of particular importance to the telecommunications engineer is the effect of the omnipresent noise which corrupts the information-bearing signal, in addition to any distortion produced by channel characteristics. The effect of noise is quite different in analogue and digital systems, and we will show that this is one reason why digital systems have a performance much closer to the theoretical performance of Shannon's law than their analogue counterparts.

REFERENCES

1. Brigham, O., *The Fast Fourier Transform*, Prentice-Hall, London and New Jersey, 1974.
2. Shannon, C. E., 'A mathematical theory of communication', *Bell Systems Technical Journal*, **27**, 379–423 (1948).

PROBLEMS

1.1 Each of the pulse trains shown in the figure represents a voltage across a $1\,\Omega$

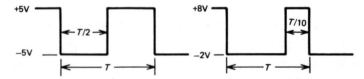

resistance. Find the total average power dissipated by each voltage and also the percentage of the total average power contributed by the first harmonic in each case.

Answer: 25 W; 10 W; 19.2%.

1.2 Assuming the integrator shown in the figure is ideal (the output voltage increases at a rate of $-1/RC$ volts/second when the input is 1 volt), sketch the output voltage when the input is a square wave of amplitude $+A$ volts or $-A$ volts and period T.

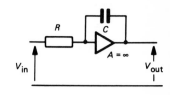

If $1/RC = 2/T$, plot the amplitude spectrum of the input and output signals.

Answer: $(AT/2) \operatorname{sinc}(\pi n f T/2)$; $AT \operatorname{sinc}^2(\pi n f T/2)$.

1.3 Using the principle of superposition, or otherwise, obtain an expression for the amplitude of the nth harmonic of the exponential Fourier series of the waveform shown in the figure.

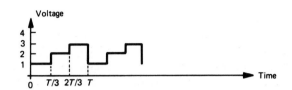

Answer: $[\cos(2n\pi/3) - 1]\pi n f$.

1.4 Plot the single-sided and double-sided power spectral densities for a pulsed waveform with period 0.1 ms and pulse width of 0.01 ms. The amplitude of the narrow pulse is 10 V and the amplitude of the wide pulse is 0 V.

What fraction of the total average power is contained within a bandwidth extending from 0 to 100 kHz?

Answer: 91%.

1.5 A rectangular pulse train with amplitude 0 V or 10 V is applied to the input of the RC network shown in the figure, the pulse repitition rate being 20 000 per second. If the half-power frequency of the network is 500 kHz and the pulse train has a duty cycle of 0.01, sketch the signal power spectrum at the network output. How much power is contained in the second harmonic?

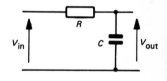

Answer: 127 µW.

1.6 Find the Fourier transform of the function $V(t)$ which is zero for negative values of its argument and is equal to $\exp(-t)$ for positive values. Find the transform when $\exp(-t)$ is replaced by $t * \exp(-t)$.

Answer: $1/(1 + j2\pi f)$; $1/(1 + j2\pi f)^2$

1.7 Show that the response of a linear system with a transfer function $H(jf)$ to a unit impulse function is

$$v(t) = 2 \int_0^\infty m \cos(2\pi f t + \phi) \mathrm{d}f$$

where $m = |H|$ and $\phi = \arg(H)$. (Hint: observe that m is an even function and ϕ is an odd function.)

1.8 Show that the autocorrelation function of a periodic waveform has the same period as the waveform. A voltage waveform is odd with period $2p$ and has the value A volts over half its period and 0 volt over the remainder of its period. Show that for $0 < t < P$ its autocorrelation function is given by

$$R(\tau) = (1 - 2\tau/p) A^2$$

1.9 A voltage waveform $e(t)$ has an arithmetic mean given by

$$\bar{e} = \lim_{T \to \infty} 1/2T \int_{-T}^T e(t) \mathrm{d}t$$

and $R(\tau)$ is its autocorrelation function. Deduce that the autocorrelation function of $e(t) + C$ (where C is a constant) is

$$R(\tau) + 2C\bar{e} + C^2$$

1.10 A system can send out a group of four pulses, each of 1 ms width and with equal probability of having an amplitude of 0, 1, 2 or 3 V. The four pulses are always followed by a pulse of amplitude -1 V to separate the groups. What is the average rate of information transmitted by this system?

Answer: 1600 b/s.

1.11 The probabilities of the previous question are altered such that the 0 V level occurs one-half of the time on average, the 1 V level occurs one-quarter of the time on average, the remaining levels occurring one-eighth of the time each. Find the average rate of transmission of information and determine the redundancy.

Answer: 1400 b/s; 12.5%.

1.12 An alphabet consists of the symbol A, B, C, D. For transmission, each symbol is coded into a sequence of binary pulses. The A is represented by 00, the B by 01, the C by 10, and the D by 11. Each individual pulse interval is 5 ms.

Calculate the average rate of transmission of information if the different symbols have equal probability of occurrence.

Find the average rate of transmission of information when the probability of occurrence of each symbol is $P(A) = 1/5$, $P(B) = 1/4$, $P(C) = 1/4$, $P(D) = 3/10$.

Answer: 200 b/s; 198 b/s.

1.13 A telemetering system can transmit eight different characters which are coded for this purpose into pulses of varying duration. The width of each coded pulse is inversely proportional to the probability of the character it represents. The transmitted pulses have durations of 1, 2, 3, 4, 5, 6, 7 and 8 ms, respectively. Find the average rate of information transmitted by this system.

If the eight characters are coded into 3-digit binary words find the necessary digit rate to maintain the same transmitted information rate.

Answer: 888.8 b/s; 1019.4 digits/s.

1.14 A space vehicle at a distance of 381 000 km from the Earth's surface is equipped with a transmitter with a power of 6 W and a bandwidth of 9 kHz. The attenuation of the signal between the transmitter and a receiver on the Earth is given by $10 + 7\log_{10}(X)$ dB, where X is the distance measured in kilometres. The noise power at the receiver input is $0 \cdot 1$ μW. If the receiver requires an input signal-to-noise ratio 12 dB above the value given by Shannon's law, find the maximum rate of transmission of information.

Answer: 50.41 kb/s.

Analogue modulation theory $\boxed{\mathbf{2}}$

By definition, an information-bearing signal is non-deterministic, i.e. it changes in an unpredictable manner. Such a signal cannot be defined in terms of a specific amplitude and phase spectrum, but it is usually possible to specify its power spectrum.

The characteristics of the channel over which the signal is to be transmitted may be specified in terms of a frequency and phase response. For efficient transmission to occur, the parameters of the signal must match the characteristics of the channel. When this match does not occur, the signal must be modified or processed. The processing is termed modulation, and the need for it may be made clearer by considering two specific examples, viz. frequency multiplexing and electromagnetic radiation from an antenna (aerial).

Frequency multiplexing is commonly used in long-distance telephone transmission, in which many narrowband voice channels are accommodated in a wideband coaxial cable. The bandwidth of such a cable is typically 4 MHz, and the bandwidth of each voice channel is about 3 kHz. The 4 MHz bandwidth is divided up into intervals of 4 kHz, and one voice channel is transmitted in each interval. Hence each voice channel must be processed (modulated) in order to shift its amplitude spectrum into the appropriate frequency slot. This form of processing is termed frequency division multiplexing (FDM), and is discussed in more detail in Section 2.19.

In Chapter 7 we show that, for efficient radiation of electromagnetic energy to occur from an antenna, the wavelength of the radiated signal must be comparable with the physical dimensions of the antenna. For audio-frequency signals, antennas of several hundred kilometres length would be required – clearly a practical impossibility. For convenient antenna dimensions the radiated signal must be of a very high frequency. In this particular example the high-frequency signal would be varied (modulated) in some way to obtain efficient transmission of the low-frequency information.

This chapter is concerned only with continuous wave (i.e. sinusoidal) modulation and assumes a noise-free environment. It should be stressed at this point that real communication is always accompanied by noise, but it is desirable to consider the effects of noise after the basic ideas of modulation have been presented. The general expression for a sinusoidal carrier is

$$v_c(t) = A\cos(2\pi f_c t + \phi) \tag{2.1}$$

The three parameters A, f_c and ϕ may be varied for the purpose of transmitting information giving respectively amplitude, frequency and phase modulation.

2.1 DOUBLE SIDEBAND AMPLITUDE MODULATION (DSB-AM)

With this type of modulation the carrier amplitude is made proportional to the instantaneous amplitude of the modulating signal. Figure 2.1 shows that the original modulating signal is reproduced as an envelope variation of the carrier. Let $A = K + v_m(t)$ where K is the unmodulated carrier amplitude and $v_m(t) = a\cos(2\pi f_m t)$ is the modulating signal. The modulated carrier, assuming $\phi = 0$, is

$$v_c(t) = [K + a\cos(2\pi f_m t)]\cos(2\pi f_c t)$$

i.e.

$$v_c(t) = K[1 + m\cos(2\pi f_m t)]\cos(2\pi f_c t) \tag{2.2}$$

m is the depth of modulation and is defined as

$$m = \frac{\text{modulating signal amplitude}}{\text{unmodulated carrier amplitude}} = \frac{a}{K}$$

For an undistorted envelope, $m \leqslant 1$. If this condition is not observed the envelope becomes distorted, and a carrier phase reversal occurs as shown in Fig. 2.2.

Equation (2.2) may be expanded to yield

$$v_c(t) = K\left[\cos(2\pi f_c t) + \frac{m}{2}\cos\{2\pi(f_c - f_m)t\} + \frac{m}{2}\cos\{2\pi(f_c + f_m)t\}\right] \tag{2.3}$$

The amplitude spectrum of the modulated carrier clearly consists of three components: the carrier frequency f_c and the lower and upper side frequencies $(f_c - f_m)$ and $(f_c + f_m)$. If $v_m(t)$ is a multitone signal, e.g.

$$v_m(t) = a_1\cos(2\pi f_1 t) + a_2\cos(2\pi f_2 t) + a_3\cos(2\pi f_3 t) \tag{2.4}$$

the modulated carrier becomes

$$v_c(t) = K[1 + m_1\cos(2\pi f_1 t) + m_2\cos(2\pi f_2 t)$$
$$+ m_3\cos(2\pi f_3 t)]\cos 2\pi f_c t \tag{2.5}$$

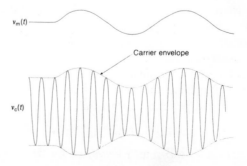

Fig. 2.1 DSB-AM.

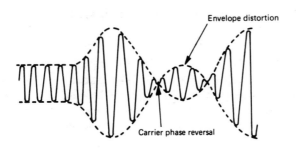

Fig. 2.2 Effect of overmodulation.

The depth of modulation is $m = m_1 + m_2 + m_3$ and, once again for an undistorted envelope, $m \leqslant 1$. The amplitude spectrum of $v_c(t)$ now contains a band of frequencies, termed sidebands, above and below the carrier. If the modulating signal is expressed in terms of a two-sided amplitude spectrum the process of full amplitude modulation (AM) (which is an alternative name for this type of modulation) reproduces the spectrum of $v_m(t)$ centred at frequencies $\pm f_c$. This process is illustrated in Fig. 2.3.

The phasor representation of Eqn (2.3) is a single component of length K rotating with angular velocity $\omega_c (= 2\pi f_c)$ rad/s representing the carrier added to two phasors of length $Km/2$ rotating in opposite directions with angular velocity ω_m rad/s relative to the carrier. This is illustrated in Fig. 2.4. The resultant is a single phasor rotating with angular velocity ω_c rad/s with an amplitude varying between the limits $K(1 - m)$ and $K(1 + m)$.

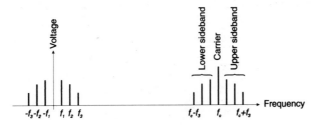

Fig. 2.3 Amplitude spectrum of DSB-AM.

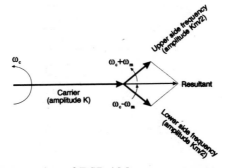

Fig. 2.4 Phasor representation of DSB-AM.

2.2 DOUBLE SIDEBAND SUPPRESSED CARRIER AMPLITUDE MODULATION (DSB-SC-AM)

The total power transmitted in a DSB-AM signal is the sum of the carrier power and the power in the sidebands. The power in a sinusoidal signal is proportional to the square of its amplitude. If the modulating signal is a single tone the total transmitted power is proportional to

$$K^2 + \left(\frac{Km}{2}\right)^2 + \left(\frac{Km}{2}\right)^2 \quad \text{watts}$$

The useful power can be regarded as the power in the sidebands as the carrier component carries no information. The ratio of useful power to total power is therefore

$$\frac{m^2}{2} : \left(1 + \frac{m^2}{2}\right)$$

For peak modulation ($m = 1$) thus ratio has a maximum value of $\frac{1}{3}$. If the carrier can be suppressed, or at least reduced in amplitude, practically all the transmitted power is then useful power. This can be important when the received signal is distorted by noise, as it produces a higher effective SNR than when the full carrier power is transmitted.

The price paid for removing the carrier is an increase in the complexity of the detector. This is discussed in more detail in Section 2.10. The amplitude spectrum of DSB-SC-AM may be derived by assuming a carrier $A \cos(2\pi f_c t)$ and a modulating signal given by $a \cos(2\pi f_m t)$. The modulated signal is simply the product of these two component, i.e.

$$v_c(t) = A \cos(2\pi f_c t)\, a \cos(2\pi f_m t)$$

i.e.

$$v_c(t) = \frac{aA}{2} \cos[2\pi(f_c - f_m)t] + \frac{aA}{2} \cos[2\pi(f_c + f_m)t] \qquad (2.6)$$

When the carrier is suppressed the envelope no longer represents the mod-

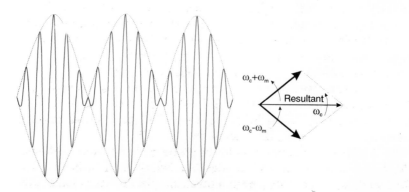

Fig. 2.5 Waveform and phasor representation.

ulating signal. The modulated carrier and the phasor representation of Eqn. (2.6) are illustrated in Fig. 2.5. For this specific case the resultant phasor is a single component with angular velocity ω_c rad/s varying in amplitude between the limits $\pm (aA)$.

2.3 SINGLE SIDEBAND AMPLITUDE MODULATION (SSB-AM)

The reason for suppressing the carrier component was that it contained no information. It can also be observed that the signal information transmitted in the lower sideband is the same as is transmitted in the upper sideband. If either one is suppressed, therefore, it is possible to transmit the same information, but the bandwidth requirement is halved. The SSB signal that is transmitted can be either sideband, e.g.

$$v_c(t) = \frac{aA}{2} \cos\left[2\pi(f_c + f_m)t\right] \qquad (2.7)$$

It should be noted that Eqn. (2.7) seems to infer that a single sinusoidal component is transmitted. This is a special case since it is applicable to a single modulating tone. When the modulating signal is multitone the SSB signal becomes a band of frequencies. The resultant phasor is then the resultant of several phasors each rotating with different angular velocities and lengths. The price paid for reducing the signal bandwidth is an increase in the complexity of the receiver, which is dealt with in Section 2.11.

It becomes clear that, since SSB-AM has half the bandwidth of DSB-AM, twice as many independent information-bearing signals can be transmitted over a channel of fixed bandwidth when SSB-AM is used to produce frequency multiplexing.

2.4 VESTIGIAL SIDEBAND AMPLITUDE MODULATION (VSB-AM)

This is used for wideband modulating signals, such as television, where the bandwidth of the modulating signal can extend up to 5.5 MHz (for a 625 line system). The required bandwidth for a DSB-AM transmission would therefore be 11 MHz. This is regarded as excessive both from the point of view of transmission bandwidth occupation and of cost. It is generally accepted that the wider the bandwidth of a receiver, the greater the cost.

Since the amplitude spectrum of a video waveform has a dc component it would be extremely difficult to produce SSB-AM for a television signal. As a compromise, part (i.e. the vestige) of one sideband and the whole of the other is transmitted. A typical example is shown in Fig. 2.6.

The VSB-AM signal thus has both lower power and less bandwidth than DSB-AM and higher power and greater bandwidth than SSB-AM. The VSB-AM signal, however, does permit a much simpler receiver than a SSB-AM signal.

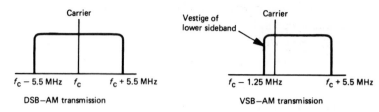

Fig. 2.6 Amplitude spectrum of a television signal.

2.5 DSB-AM MODULATORS

DSB-AM is produced by multiplying together the carrier and the modulating signal. The multiplication is achieved by using a network with a non-linear characteristic. There are basically two types of non-linear networks, one in which the characteristic is continuous and the other in which the characteristic is non-continuous, e.g. a switch.

Non-linear networks are not true multipliers because other components, which have to be filtered off, are produced. The diode modulator is an example of a modulator with a continuous non-linear characteristic of the form shown in Fig. 2.7.

The input/output characteristic of the circuit of Fig. 2.7 can be written in terms of a power series, i.e.

$$V_{\text{out}} = aV_{\text{in}} + bV_{\text{in}}^2 + cV_{\text{in}}^3 + \cdots \tag{2.8}$$

If

$$V_{\text{in}} = [A\cos(2\pi f_c t) + B\cos(2\pi f_m t)]$$

the output is given by

$$V_{\text{out}} = a[A\cos(2\pi f_c t) + B\cos(2\pi f_m t)] + b[A^2\cos^2(2\pi f_c t)$$
$$+ B^2\cos^2(2\pi f_m t) + 2AB\cos(2\pi f_c t)\cos(2\pi f_m t] + \cdots$$

i.e.

$$V_{\text{out}} = aA\cos(2\pi f_c t)[1 + K_1\cos(2\pi f_m t)] + \text{other terms} \tag{2.9}$$

where $K_1 = 2B/a$. If the 'other terms' are filtered off, Eqn. (2.9) has the same form as Eqn. (2.2). Hence the diode produces the required modulation.

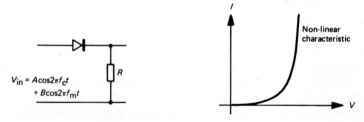

Fig. 2.7 Diode modulator.

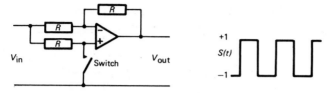

Fig. 2.8 Switching modulator.

Considering the circuit of Fig. 2.8, it can be shown that when the switch is open the circuit has a voltage gain of $+1$ and when the switch is closed the circuit has a voltage gain of -1. If the switch is replaced by a semiconductor device that is pulsed at the carrier frequency, the output of the amplifier is V_{in} multiplied by a square wave of amplitude ± 1. The Fourier series of the square wave is

$$S(t) = C_1 \cos(2\pi f_c t) + C_3 \cos(6\pi f_c t) + C_5 \cos(10\pi f_c t) + \cdots$$

If the input is $V_{in} = V_A + V_B \cos(2\pi f_m t)$ then the output will contain a term

$$C_1 V_A \cos(2\pi f_c t)[1 + K_2 \cos(2\pi f_m t)] \qquad (2.10)$$

where $K_2 = V_B/(C_1 V_A)$, which again has the same form as Eqn. (2.2). Hence the switching modulator also produces DSB-AM. Filtering is required in this case also, to remove the unwanted frequency components.

2.6   DSB-SC-AM MODULATORS

The switching modulator may also be used to produce DSB-SC modulation. If the input of the circuit of Fig. 2.8 is $V_B \cos(2\pi f_m t)$ the output will contain a term

$$C_1 V_B \cos(2\pi f_c t) \cos(2\pi f_m t)$$

i.e.

$$V_{out} = \frac{C_1 V_B}{2} \cos[2\pi(f_c - f_m)t] + \frac{C_1 V_B}{2} \cos[2\pi(f_c + f_m)t]$$

$$+ \text{ higher-frequency terms} \qquad (2.11)$$

This equation, when compared with Eqn. (2.3), will be seen to contain no carrier frequency f_c.

2.7   SSB-AM MODULATORS

SSB-AM can be produced from DSB-SC-AM by filtering off one of the sidebands. The filtering process is relatively difficult to accomplish at high frequency where the sideband separation would be a very small fraction of the filter centre frequency. The problem is eased considerably if the initial modulation takes place at a low carrier frequency. The selected sideband can then be shifted to the required frequency range by a second modulation. The sidebands of the second modulation are widely spaced, and less exact filtering is necessary.

The two-stage modulation process is shown in Fig. 2.9; a typical first modulating frequency would be in the region of 100 kHz. Most modern high-power SSB-AM transmitters working the HF band $(3-30\,\text{MHz})$ generate signals in this way. An interesting alternative method is based on the Hilbert transformation. The Hilbert transform of a signal $v(t)$ is the sum of the individual frequency components that have all been shifted in phase by $90°$. The SSB-AM signal for a single modulating tone is

$$v_c(t) = A\cos[2\pi(f_c + f_m)t]$$

i.e.

$$v_c(t) = A[\cos(2\pi f_c t)\cos(2\pi f_m t) - \sin(2\pi f_c t)\sin(2\pi f_m t)] \qquad (2.12)$$

Equation (2.12) reveals that $v_c(t)$ is obtained by subtracting the outputs of two DSB-SC-AM modulators, often called balanced modulators. There is a quadrature phase relationship between the inputs of the first modulator relative to the inputs of the second modulator. A block diagram of this type of SSB-AM modulator is given in Fig. 2.10, and it should be noted that sideband filters are not required. The modulator does, however, require the provision of wideband phase-shifting networks.

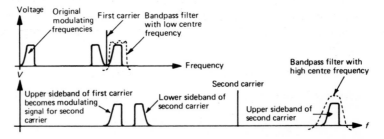

Fig. 2.9 Two-stage production of SSB-AM.

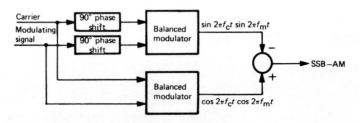

Fig. 2.10 Hilbert transform production of SSB-AM.

2.8   VSB-AM MODULATORS

VSB-AM modulation is produced by filtering the corresponding DSB-AM signal. In a television transmission system the required filtering is usually achieved by a series of cascaded high-frequency tuned amplifiers. The centre frequency of each amplifier in the chain is chosen so that the overall cascaded frequency response is asymmetrically positioned relative to the carrier frequency.

2.9   DSB-AM DETECTION

The detection (i.e. demodulation) of DSB-AM can be considered broadly under two headings; these are non-coherent detection and coherent (synchronous) detection. Traditionally, broadcast receivers have been of the superheterodyne type (see later) and have made use of envelope detection. With the advent of the integrated phase locked loop (PLL), coherent detectors are now attractive. The envelope (non-coherent) detector, as its name suggests, physically reproduces the envelope of the modulated carrier. This detector is basically a half-wave rectifier, and commonly makes use of a silicon diode whose typical current voltage relationship is shown in Fig. 2.11.

The diode acts as a rectifier and its effect on the envelope of a DSB-AM signal is shown in Fig. 2.11. For large carrier amplitudes the envelope is reproduced by the linear part of the characteristic (provided the depth of modulation is less than 100%) and no distortion results. This is termed 'large signal operation'. For small carrier amplitudes the envelope is reproduced by the non-linear part of the characteristic and distortion of the envelope occurs. Assuming the input to the detector is

$$V_{in} = K[1 + m\cos(2\pi f_m t)]\cos(2\pi f_c t)$$

when operating in the large signal mode, the output is given by

$$i = P[1 + m\cos(2\pi f_m t)]V(t) \tag{2.13}$$

where $V(t)$ is a half-wave rectified sinusoid of carrier frequency. Representing $V(t)$ by its Fourier series the output is

$$i = P[1 + m\cos(2\pi f_m t)][C_0 + C_1\cos(2\pi f_c t)$$
$$+ C_2\cos(4\pi f_c t) + C_3\cos(6\pi f_c t) + \cdots] \tag{2.14}$$

i.e.

$$i = PC_0 + PC_0\cos(2\pi f_m t) + \text{unwanted terms} \tag{2.15}$$

The unwanted terms can be filtered off leaving the original modulating signal. To ensure that the envelope remains in the linear region the depth of modulation must be less than 100%. When operating in the small signal mode the diode current is given by

$$i = aV_{in} + bV_{in}^2 + cV_{in}^3 + \cdots \tag{2.16}$$

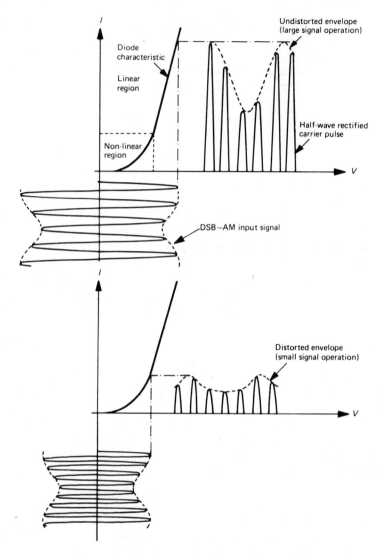

Fig. 2.11 Large and small signal operation of a diode detector.

Letting

$$V_{in} = K[\cos(2\pi f_c t) + \tfrac{1}{2} m \cos\{2\pi(f_c - f_m)t\} + \tfrac{1}{2} m \cos\{2\pi(f_c + f_m)t\}]$$

The square term gives a contribution to the output current of

$$bK^2\{\cos(2\pi f_c t) + \tfrac{1}{2} m \cos[2\pi(f_c - f_m)t] + \tfrac{1}{2} m \cos[2\pi(f_c + f_m)t]\}^2$$

i.e.

$$b\{K^2 m \cos(2\pi f_m t) + \tfrac{1}{4} K^2 m^2 \cos(4\pi f_m t)\} + \text{unwanted terms} \quad (2.17)$$

The unwanted terms are outside the modulating signal bandwidth and can be filtered off. The output current thus contains the required modulating frequency together with a second harmonic distortion term. For the lower modulating frequencies the second harmonic will be within the modulating signal bandwidth and cannot be filtered off. (When the modulating signal is multitone, cross-modulation between the individual components will also occur.) The ratio of fundamental to second harmonic is

$$K^2 m : K^2 \frac{m^2}{4} \quad \text{i.e.} \quad 1 : \frac{m}{4}$$

Thus, the relative distortion is proportional to the depth of modulation, m. (For a depth of modulation of 30%, the second harmonic distortion is 7%.)

To minimize distortion, the receiver should be designed so that the signal voltage is as large as possible. For modulating signals derived from speech or music, the peak modulation is restricted to a depth of 80%; this restricts the average depth of modulation to about 30%, which limits distortion under both large and small signal operation.

The 'unwanted terms' which appear in Eqns (2.15) and (2.17) have frequencies very much higher than the modulating frequencies and can be removed by a simple RC filter as shown in Fig. 2.12. The physical action of this detector is identical to the action of a half-wave rectifier with capacitive smoothing. The RC time constant is chosen such that the capacitor can follow the highest modulating frequencies in the envelope of the rectified carrier without losing excessive charge between carrier pulses.

In the foregoing analysis we assumed that a single modulated carrier appeared at the input of the envelope detector. In a broadcast environment there will be many different modulated carriers present in the antenna circuits of all receivers. A primary function of the receiver, therefore, will be to select one of these carriers and reject all others. The obvious way to achieve this would be to precede the detector with a bandpass filter designed to pass the required carrier and its sidebands and to reject the rest. In modern receivers the necessary frequency response is obtained by use of ceramic resonators. These circuits are available as hybrid units, and a typical example with its frequency response is shown in Fig. 2.13(a).

The circuits can be designed to have a flat response over the required bandwidth and a very large attenuation on either side. One problem, however, is that the centre frequency has a fixed value. It is not possible to vary this centre

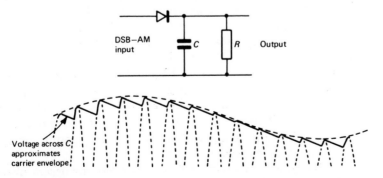

Fig. 2.12 Physical action of envelope detector.

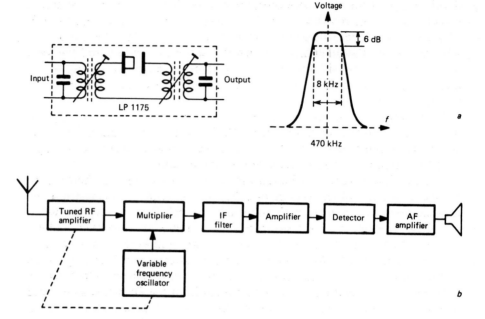

Fig. 2.13 The superheterodyne principle. (a) Ceramic resonator circuit and frequency response; (b) the superheterodyne receiver.

frequency, in order to receive other modulated carriers, and at the same time to maintain the required attenuation on either side of the passband. The superheterodyne receiver was designed specifically to overcome such a problem by transferring the sidebands of the selected carrier frequency to a constant intermediate frequency (IF). A standard IF of 470 kHz has been adopted for receivers working in the medium waveband (540 kHz to 1640 kHz).

A block diagram of the superheterodyne receiver is shown in Fig. 2.13(b) and by examining the functions of each of the blocks the derivation of the term superheterodyne will become apparent. The sidebands of the selected carrier are transferred to a frequency of 470 kHz by multiplying (heterodyning) the carrier with the output of a local oscillator. The output of the multiplier will contain sum and difference frequencies and the difference frequency between the local oscillator and selected carrier is made equal to 470 kHz. To produce this figure, the local oscillator must have a frequency of 470 kHz above or below the required carrier. If the local oscillator is above the carrier frequency the required frequency range of the oscillator is 1010 kHz to 2110 kHz for a receiver operating in the medium waveband. This is a frequency ratio of approximately 2:1 (i.e. one octave). If the local oscillator is below the carrier frequency the corresponding range would be 70 kHz to 1170 kHz, which is a ratio of 17:1. The local oscillator is designed to work above the required carrier frequency as it is then required to have a range of about one octave. Hence the derivation of the term superheterodyne.

It is not the purpose of this text to discuss in detail the design of super-heterodyne receivers,[1] but there is one important point worth noting that is a disadvantage of the superheterodyne principle. When the local oscillator is 470 kHz above the required carrier the sidebands will be transferred to 470 kHz. However, if there is a second carrier at a frequency of 940 kHz above the required carrier the difference between this second frequency and the local oscillator frequency will also be 470 kHz. This is known as the 'image frequency' and is twice the intermediate frequency above the required carrier frequency.

To prevent the sidebands of the image frequency from reaching the detector it must be attenuated; hence the radio frequency amplifier in the superheterodyne receiver also has a bandpass frequency response. A single tuned LC circuit is usually adequate for this purpose.

It is worth noting that although Fig. 2.13(b) shows the superheterodyne receiver as a series of separate blocks, these blocks, with the exception of the ceramic resonator, are available on a single chip.

Coherent detection operates on the principle that is the inverse of modulation, i.e. the frequency of the sidebands is translated back to baseband by multiplying the DSB-AM signal by a sinusoid of the same frequency as the carrier. In the case of DSB-AM the carrier is actually transmitted, so the problem is one of extracting this component from the DSB-AM signal and then applying the extracted component to a multiplier along with the original modulated signal. A block diagram of the coherent detector is shown in Fig. 2.14. The output of the coherent detector may be written

$$V_{out} = K[1 + m\cos(2\pi f_m t)]\cos(2\pi f_c t)\cos(2\pi f_c t)$$

i.e.

$$V_{out} = K[1 + m\cos(2\pi f_m t)]\cos^2(2\pi f_c t)$$

but

$$\cos^2 2\pi f_c t = \tfrac{1}{2}[1 + \cos(4\pi f_c t)]$$

Therefore

$$V_{out} = \tfrac{1}{2}Km\cos(2\pi f_m t) + \text{unwanted terms} \tag{2.18}$$

The unwanted terms will be sidebands of $2f_c$ and are easily filtered off. A

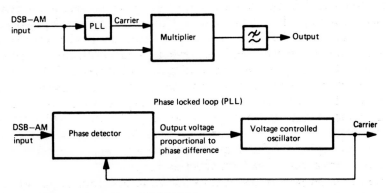

Fig. 2.14 Coherent detection of DSB-AM.

convenient method of extracting the carrier is to use a phase locked loop as shown in Fig. 2.14.

The basic operation of the loop may be described as follows. The input to the phase detector is the DSB-AM signal and a sinusoid from a voltage controlled oscillator (VCO). The output of the multiplier is a signal proportional to the phase difference between the carrier of the DSB-AM signal and the VCO output. This error signal is fed via a low-pass filter to the VCO input. Its effect is to cause the output frequency of the VCO to change in order to minimize the phase error. Hence the PLL is essentially a negative feedback system. When the loop is locked, the frequency of the VCO output is equal in frequency to the incoming carrier.

The PLL is a very important circuit as it also finds use as a frequency modulation (FM) detector. This is described more fully in Section 2.18.

2.10 DSB-SC-AM DETECTION

This type of modulation requires coherent detection, but since there is no carrier transmitted the required component must be generated by a local oscillator. Special circuits are required to ensure that the local oscillator is locked in phase to the incoming signal. If the incoming signal is represented by

$$V_{in} = K \cos(2\pi f_c t) \cos(2\pi f_m t)$$

and the output of the local oscillator is $\cos(2\pi f_c t + \phi)$, the output of the coherent detector will be

$$V_{out} = \tfrac{1}{2} K \cos(2\pi f_m t) [\cos(4\pi f_c t + \phi) + \cos\phi] \qquad (2.19)$$

After filtering off the components centred at $2f_c$ this gives a term $\tfrac{1}{2} K \cos(2\pi f_m t) \cos\phi$. When $\phi = 0$ the output is a maximum and is proportional to the original modulating signal; when $\phi = 90°$ the output is zero. A circuit for demodulating a DSB-SC-AM signal, which makes use of this property, is shown in Fig. 2.15.

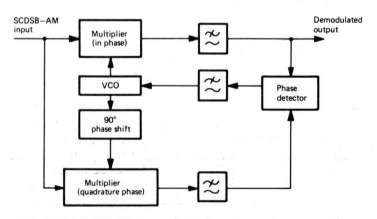

Fig. 2.15 The Costas loop.

If the local oscillator has the same phase as the missing carrier the 'in phase' channel will have the correct output and the 'quadrature channel' will have zero output. The output of the multiplier is a voltage required to maintain the desired value of the VCO phase. If there is a phase error between the VCO output and the missing carrier, the output of the 'in phase' channel will drop and the output of the 'quadrature channel' will become non-zero.

The multiplier now produces an output signal that, when applied to the VCO, will cause the phase error between the missing carrier and the VCO output to tend to zero. Hence the circuit automatically produces a locally generated sinusoid of the correct phase. This circuit was initially developed for data communications and is known as the Costas loop.

2.11   SSB-AM DETECTION

SSB-AM signals require coherent detection and, once again, a local oscillator is required, as no carrier is transmitted.

Assuming the SSB-AM signal is $K \cos 2\pi(f_c + f_m)t$ and the local oscillator signal is $\cos(2\pi f_c t + \phi)$ the output of the coherent detector will be

$$V_{out} = \tfrac{1}{2} K \cos[2\pi(2f_c + f_m)t + \phi] + \tfrac{1}{2} K \cos(2\pi f_m t - \phi) \qquad (2.20)$$

When the term at frequency $(2f_c + f_m)$ is filtered off the remaining components can be written

$$\tfrac{1}{2} K \cos(2\pi f_m t)\cos\phi + \sin(2\pi f_m t)\sin\phi \qquad (2.21)$$

If $\phi = 0$ the output is the $\tfrac{1}{2} K \cos(2\pi f_m t)$ and when $\phi = 90°$ the output is $\tfrac{1}{2} K \sin(2\pi f_m t)$. Hence in contrast to DSB-SC-AM there is also an output (shifted in phase by 90°) when $\phi = 90°$. For human listeners, whose hearing is relatively insensitive to phase distortion, the requirement of phase coherence can be relaxed. It is, however, necessary to have frequency coherence between the missing carrier and the local oscillator output. Modern point-to-point and mobile SSB-AM communications system use crystal-controlled oscillators and frequency synthesizers to achieve the required frequency stability. This is typically 1 part in 10^6 (i.e. a stability of 1 Hz at a frequency of 1 MHz.)

2.12   VSB-AM DETECTION

VSB-AM transmission is virtually exclusive to television systems, and a typical television detection system will therefore be described in this section. Television receivers use a conventional diode detector to demodulate the video carrier but some pre-proccessing of the VSB-AM signal occurs in the amplifiers that precede the detector. To illustrate the detection procedure it will be assumed that a VSB-AM signal has been produced from a two-tone modulating signal, i.e.

$$v_m(t) = q_1 \cos(2\pi f_1 t) + q_2 \cos(2\pi f_2 t)$$

The DSB-AM signal from which the vestigial sideband signal is derived will

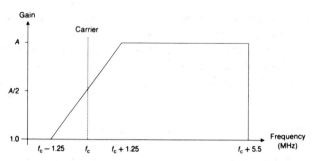

Fig. 2.16 VSB-AM receiver response.

be

$$v_c(t) = K\left[\cos(2\pi f_c t) + \tfrac{1}{2} m_1 \cos\{2\pi(f_c - f_1)t\} + \tfrac{1}{2} m_2 \cos\{2\pi(f_c - f_1)t\}\right.$$
$$\left. + \tfrac{1}{2} m_1 \cos\{2\pi(f_c + f_1)t\} + \tfrac{1}{2} m_2 \cos\{2\pi(f_c + f_2)t\}\right] \qquad (2.22)$$

To produce a VSB-AM signal the component at a frequency of $(f_c - f_2)$ in the lower sideband is assumed suppressed.

It has been shown in Section 2.9 that, if the diode detector is working in the linear region, the input signal is effectively multiplied by the carrier and its harmonics [Eqn (2.14)]. The detector output will therefore contain components

$$K\left[\cos^2(2\pi f_c t) + \tfrac{1}{4} m_1 \cos\{2\pi(2f_c - f_1)t\} + \tfrac{1}{4} m_1 \cos(-2\pi f_1 t)\right.$$
$$+ \tfrac{1}{4} m_1 \cos\{2\pi(2f_c + f_1)t\} + \tfrac{1}{4} m_1 \cos(-\pi f_1 t)$$
$$\left. + \tfrac{1}{4} m_2 \cos 2\pi\{(2f_c + f_2)t\} + \tfrac{1}{4} m_2 \cos(2\pi f_2 t)\right]$$
$$+ \text{unwanted terms} \qquad (2.23)$$

The unwanted terms can be filtered off, and it then becomes clear that the detector output contains components

$$\tfrac{1}{4} K\left[2m_1 \cos(2\pi f_1 t) + m_2 \cos(2\pi f_2 t)\right]$$

i.e.

$$\tfrac{1}{4}\left[2a_1 \cos(2\pi f_1 t) + a_2 \cos(2\pi f_2 t)\right]$$

The amplitude of the component transmitted in both sidebands is doubled relative to the amplitude of the component transmitted in the single sideband. To compensate for this, the gains of the amplifiers preceding the detector are designed to be asymmetric about the carrier frequency. The characteristic is shown in Fig. 2.16. The response is adjusted in such a way that the components present in both sidebands add to give the same output that would occur if only one sideband was transmitted.

2.13 ECONOMIC FACTORS AFFECTING THE CHOICE OF AM SYSTEMS

It is apparent that the special amplifier responses would not be required if the pre-processing was done at the transmitter, which would have the added advantage of reducing the transmitted power. The present standards were chosen on an economic basis. If the processing was done at the transmitter, the television receiver amplifier (of which there are many) would require a flat response over a bandwidth of 6.75 MHz. When the processing is done in the receiver, the required amplifier bandwidths are reduced to 4.25 MHz. This represents a considerable saving in the production costs of a typical television receiver.

The economic argument also decides to a large extent the standards adopted in other forms of telecommunication. A local broadcast transmission system, for example, will have a single high-power transmitter with many receivers. The unit cost of each receiver is lowest when DSB-AM is used, which dictates the use of DSB-AM in this particular situation. The conditions in a mobile communications system will be somewhat different; there will usually be a limited number of receivers, and transmitted power will be at a premium. In this environment SSB-AM with its lower transmitted power becomes attractive. SSB-AM is also used in frequency multiplexed systems over coaxial cables. The obvious advantage here is that twice as many channels may be transmitted as with DSB-AM for a given cable bandwidth.

2.14   ANGLE MODULATION

This is an alternative to amplitude modulation and, as its name suggests, information is transmitted by varying the phase angle of a sinusoidal signal. The sinusoidal signal can be conveniently written as

$$v_c(t) = A \cos \theta(t)$$

where θ is a phase angle that is made proportional to a function of the modulating signal. The time derivative of $\theta(t)$ is defined as the instantaneous frequency of the sinusoid. Strictly speaking, 'frequency' is only defined when $\theta(t)$ is a linear function of time. It is mathematically convenient to write the derivative of $\theta(t)$ as an 'instantaneous frequency', i.e.

$$\dot{\theta}(t) = 2\pi f_i$$

Angle modulation is itself divided into two categories; for example, let

$$\theta(t) = 2\pi f_c t + \phi + K_1 v_m(t) \tag{2.24}$$

This is termed phase modulation because $\theta(t)$ varies linearly with the amplitude of the modulating signal. Frequency modulation is produced if $\dot{\theta}(t)$ varies linearly with the amplitude of the modulating signal, e.g. let

$$\dot{\theta}(t) = 2\pi f_c + 2\pi K_2 v_m(t) = 2\pi f_i \tag{2.25}$$

Therefore

$$\theta(t) = 2\pi f_c t + \phi + 2\pi K_2 \int_0^t v_m(t)\, dt \tag{2.26}$$

In this case $\theta(t)$ is linearly proportional to the amplitude of the integral of the modulating signal.

2.15 PHASE MODULATION (PM)

It is shown in Section 2.16 that angle modulation may be considered as a non-linear process in terms of the relationship between time and frequency domains. This means that it becomes very difficult to derive expressions for angle-modulated waves unless very simple modulating waveforms are considered. In this context a single modulating tone may be regarded as such a signal. The modulating signal is given by

$$v_m(t) = a \cos(2\pi f_m t)$$

The PM carrier produced by this signal has the form

$$v_c(t) = A \cos[2\pi f_c t + \Delta\theta \cos(2\pi f_m t)] \qquad (2.27)$$

where $\Delta\theta = K_1 a$ is the phase shift produced when the modulating signal has its maximum positive value, i.e.

$$v_c(t) = A \cos(2\pi f_c t) \cos[\Delta\theta \cos(2\pi f_m t)]$$
$$- A \sin(2\pi f_c t) \sin[\Delta\theta \cos(2\pi f_m t)]$$

If $\Delta\theta$ is restricted to a low value ($\Delta\theta \ll 1$) then $\cos[\Delta\theta \cos(2\pi f_m t)] \simeq 1$ and $\sin[\Delta\theta \cos(2\pi f_m t)] \simeq \Delta\theta \cos(2\pi f_m t)$. The modulated carrier may then be approximated as

$$v_c(t) = A [\cos(2\pi f_c t) - \Delta\theta \sin(2\pi f_c t) \cos(2\pi f_m t)]$$

This can be expanded to yield

$$v_c(t) = A[\cos(2\pi f_c t) - \tfrac{1}{2}\Delta\theta \sin\{2\pi(f_c - f_m)t\}$$
$$- \tfrac{1}{2}\Delta\theta \sin\{2\pi(f_c + f_m)t\}] \qquad (2.28)$$

When this equation is compared with the expression for DSB-AM [Eqn (2.3)] it will be seen that, provided the maximum phase shift it restricted to low values, the PM signal is equivalent to DSB-AM in which the carrier has been shifted in phase by 90° relative to the sidebands. PM may thus be produced from DSB-SC-AM by reintroducing the suppressed carrier in phase quadrature. This process is illustrated in block schematic form in Fig. 2.17. The phasor representation of narrowband PM ($\Delta\theta \ll 1$) is given in Fig. 2.18. For small values of $\Delta\theta$ the resultant phasor has an almost constant amplitude but the phase angle α is a function of time.

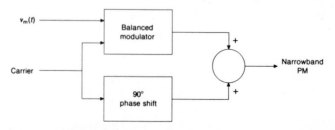

Fig. 2.17 Phase modulation.

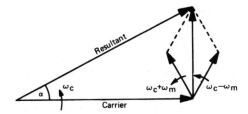

Fig. 2.18 Phasor representation of narrowband PM.

2.16 FREQUENCY MODULATION (FM)

The expression for a frequency modulated carrier is developed in a similar fashion to the expression for a PM carrier. The modulating signal is again assumed to be a single tone, and the phase of the carrier is now related to the time integral of $v_m(t)$, i.e.

$$v_c(t) = A \cos\left[2\pi f_c t + 2\pi K_2 \int_0^t a \cos(2\pi f_m t)\,dt \right] \qquad (2.29)$$

Therefore

$$v_c(t) = A \cos\left[2\pi f_c t + \frac{K_2 a}{f_m} \sin(2\pi f_m t) \right]$$

i.e.

$$v_c(t) = A \cos\left[2\pi f_c t + \beta \sin(2\pi f_m t) \right] \qquad (2.30)$$

The constant β is termed the modulation index. By comparison with Eqn (2.7), it is clear the $K_2 a$ represents the maximum change in carrier frequency that is produced by the modulating signal. Letting $K_2 a = \Delta f_c$ the modulation index is defined as $\beta = \Delta f_c / f_m$. It is important to note that β depends both on the carrier deviation, which is linearly proportional to the

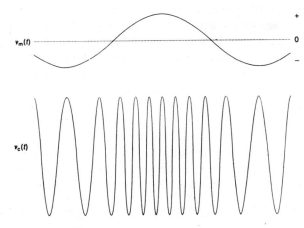

Fig. 2.19 Frequency modulated carrier.

signal amplitude, and also on the frequency of the modulating signal. The phase shift $\Delta\theta$ that occurs in PM is independent of the modulating signal frequency. The FM waveform that results from a single modulating tone is shown in Fig. 2.19.

The amplitude spectrum of the FM carrier is obtained by expanding the trigonometric function of Eqn. (2.30).

$$v_c(t) = A\cos(2\pi f_c t)\cos[\beta\sin(2\pi f_m t)]$$
$$- A\sin(2\pi f_c t)\sin[\beta\sin(2\pi f_m t)] \qquad (2.31)$$

It is convenient to consider Eqn (2.31) for two specific types of FM – narrowband FM and wideband FM. The reason for this distinction is made clearer by considering the series expansion of the trigonometric functions $\cos[\beta\sin(2\pi f_m t)]$ and $\sin[\beta\sin(2\pi f_m t)]$.

$$\cos[\beta\sin(2\pi f_m t)] = 1 - \frac{\beta^2\sin^2(2\pi f_m t)}{2!} + \frac{\beta^4\sin^4(2\pi f_m t)}{4!}$$
$$+ \frac{\beta^6\sin^6(2\pi f_m t)}{6!} + \cdots$$

$$\sin[\beta\sin(2\pi f_m t)] = \beta\sin(2\pi f_m t) - \frac{\beta^3\sin^3(2\pi f_m t)}{3!}$$
$$+ \frac{\beta^5\sin^5(2\pi f_m t)}{5!} + \cdots$$

Narrowband FM is produced when $\beta \ll 1$. The series expansions for this case can be approximated by

$$\cos[\beta\sin(2\pi f_m t)] \simeq 1 \quad \text{and} \quad \sin[\beta\sin(2\pi f_m t)] \simeq \beta\sin(2\pi f_m t)$$

Using these approximations, Eqn (2.31) can be rewritten as

$$v_c(t) = A\left[\cos(2\pi f_c t) - \frac{\beta}{2}\cos\{2\pi(f_c - f_m)t\} + \frac{\beta}{2}\cos 2\pi\{(f_c + f_m)t\}\right] \qquad (2.32)$$

This expression shows that narrowband FM is equivalent to DSB-AM with a phase shift of 180° in the lower sideband. The bandwidth in this case is identical to the bandwidth of a DSB-AM signal and is exactly twice the bandwidth of the modulating signal. The resultant phasor, which is illustrated in Fig. 2.20, is similar to the resultant for PM, i.e. it has almost constant amplitude and a phase angle that is a function of time.

It is evident that as the value of β increases more terms in the series expansions of $\cos[\beta\sin(2\pi f_m t)]$ and $\sin[\beta\sin(2\pi f_m t)]$ become significant and cannot be ignored. This illustrates the non-linear relationship between the time domain and frequency domain, which is a feature of angle-modulated signals. As an example, consider the series expansions of $\cos\beta\sin(2\pi f_m t)$ and $\sin\beta\sin(2\pi f_m t)$ when $\beta = 0.5$. The new approximation are

$$\cos[\beta\sin(2\pi f_m t)] \simeq 1 - \tfrac{1}{2}\beta^2\sin^2(2\pi f_m t) = 1 - \tfrac{1}{4}\beta^2[1 - \cos(4\pi f_m t)]$$
$$\sin[\beta\sin(2\pi f_m t)] \simeq \beta\sin(2\pi f_m t)$$

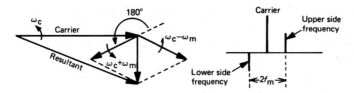

Fig. 2.20 Phasor and spectral representation of narrowband FM.

Equation (2.31) may be rewritten for this example as

$$v_c(t) = A\{(1 - \tfrac{1}{4}\beta^2)\cos 2\pi f_c t$$
$$- \tfrac{1}{2}\beta^2[\cos\{2\pi(f_c - f_m)t\} - \cos\{2\pi(f_c + f_m)t\}]$$
$$+ \tfrac{1}{8}\beta^2[\cos\{2\pi(f_c - 2f_m)t\} + \cos\{2\pi(f_c + 2f_m)t\}]\} \quad (2.33)$$

A simple increase in the value of β (produced by an increase in the amplitude of the modulating signal or a decrease in its frequency) produces a decrease in the amplitude of the component at frequency f_c and results in two additional frequency components spaced at $\pm 2f_m$ relative to f_c. The bandwidth thus effectively doubles, when compared with the previous case, simply as a result of increasing the amplitude of the modulating signal. This non-linear relationship between time and frequency domains becomes apparent when Fig. 2.21 is compared with Fig. 2.20.

Further increases in the value of β will clearly increase the number of significant terms in the series expansion of the two trigonometric functions. Both these functions are, in fact, periodic and can therefore be expressed in terms of a Fourier series, i.e.

$$\cos[\beta\sin(2\pi f_m t)] = C_0 + C_2\cos(4\pi f_m t) + C_4\cos(8\pi f_m t) + \cdots$$
$$\sin[\beta\sin(2\pi f_m t)] = C_1\sin(2\pi f_m t) + C_3\sin(6\pi f_m t) + \cdots$$

The Fourier coefficients C_n are themselves infinite series and may only be evaluated numerically for specific values of n and β. These coefficients are termed Bessel functions of the first kind, i.e. $C_n = J_n(\beta)$. Graphs of $J_n(\beta)$ for several values of n are given in Fig. 2.22. Using the Fourier series representation, Eqn. (2.31) may be rewritten as

$$v_c(t) = A\{J_0(\beta)\cos(2\pi f_c t) - J_1(\beta)[\cos\{2\pi(f_c - f_m)t\} - \cos\{2\pi(f_c + f_m)t\}]$$
$$+ J_2(\beta)[\cos\{2\pi(f_c - 2f_m)t\} + \cos\{2\pi(f_c + 2f_m)t\}]$$
$$- J_3(\beta)[\cos\{2\pi(f_c - 3f_m)t\} - \cos\{2\pi(f_c + 3f_m)t\}]$$
$$+ \cdots\} \quad (2.34)$$

The amplitudes of the carrier and sidebands, of which there is an infinite number, depend on the value of β which fixes the appropriate Bessel function value. Tables of Bessel functions for a range of n and β are given in Appendix A.

The bandwidth of the wideband FM signal is apparently infinite, but in practice some simplifying approximations are possible. With reference to

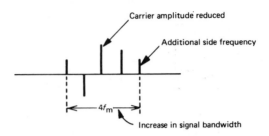

Fig. 2.21 Non-linearity of the FM process.

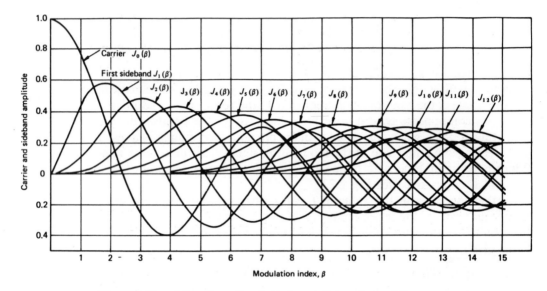

Fig. 2.22 Bessel function of order n for various values of β.

Fig. 2.22, it can be observed that as the value of n is increased then β must be made progressively larger before the appropriate value of $J_n(\beta)$ becomes non-zero. Extrapolating this trend it is found that for large values of n the value of $J_n(\beta)$ is approximately zero when $\beta < n$, e.g. when $\beta = 100$ then $J_{101}(\beta)$, $J_{102}(\beta)$, etc., are approximately zero since $\beta < n$. In this example, therefore, there are 100 values of $J_n(\beta)$ (excluding the carrier $J_0(\beta)$) that have a non-zero value. Hence this leads to the approximation that when β is large there are β values of $J_n(\beta)$ that are non-zero. Under these circumstances there will be β pairs of sidebands, each spaced by f_m, in the amplitude spectrum of the modulated signal.

The signal bandwidth thus becomes

$$B = 2\beta f_m = 2\Delta f_c$$

i.e. for large values of β the signal bandwidth is approximately twice the carrier deviation. An accurate figure for the bandwidth of an FM signal for either very large or very small values of β is given by Carson's rule, which states

$$B = 2f_m(1 + \beta)$$

i.e.

$$\text{when} \quad \beta \gg 1 \quad B \simeq 2f_m\beta = 2\Delta f_c$$

$$\text{when} \quad \beta \ll 1 \quad B \simeq 2f_m$$

For other values of β the signal bandwidth must be computed from Bessel function tables. The spectrum of the FM signal is then defined as containing all sideband components with an amplitude $\geqslant 1\%$ of the unmodulated carrier amplitude.

In the UK, commercial FM broadcasting stations restrict the maximum carrier deviation to $\pm 75\,\text{kHz}$. A single modulating tone of frequency $15\,\text{kHz}$ producing peak frequency deviation of the carrier would yield a modulation index $\beta = 5$. The number of sideband terms with an amplitude of 0.01 or greater is obtained from the Bessel function tables as 8. The bandwidth is then

$$B = 2 \times 8 \times f_m = 240\,\text{kHz}$$

If the peak frequency deviation is produced by a $5\,\text{kHz}$ modulating tone the modulation index will have a value of $\beta = 15$. The number of significant sidebands increases to 19 and the bandwidth becomes

$$B = 2 \times 19 \times f_m = 190\,\text{kHz}$$

Although the larger value of β produces more significant sidebands, the sidebands are actually closer together and the bandwidth of the FM signal is less.

It will be apparent from the foregoing analysis that FM and PM are very closely related, so much so that an 'instantaneous frequency' and 'modulation index' can also be defined for PM. It is instructive to compare PM and FM from the point of view of these parameters, as the comparison gives some insight into why FM is used almost exclusively in preference to PM. For a single modulating tone the instantaneous frequency in a frequency modulated wave has been defined as

$$2\pi f_i = 2\pi f_c + 2\pi K_2 a \cos(2\pi f_m t)$$

i.e.

$$f_i = f_c + \Delta f_c \cos(2\pi f_m t)$$

which gives

$$v_c(t) = A \cos\left[2\pi f_c t + \frac{\Delta f_c}{f_m} \sin(2\pi f_m t) \right]$$

For PM,

$$\theta(t) = 2\pi f_c t + \Delta\theta \cos(2\pi f_m t)$$

Therefore

$$\dot{\theta}(t) = 2\pi f_i = 2\pi f_c - \Delta\theta 2\pi f_m \sin(2\pi f_m t)$$

i.e.

$$\dot{\theta}(t) = 2\pi f_c - 2\pi \Delta f_p \sin(2\pi f_m t)$$

The PM carrier may therefore be written in terms of a frequency deviation $\Delta f_p = -\Delta\theta f_m$, i.e.

$$v_c(t) = A \cos\left[2\pi f_c t + \frac{\Delta f_p}{f_m} \cos(2\pi f_m t) \right] \qquad (2.35)$$

If a modulation index β_p is defined for PM, Eqn. (2.35) can be written

$$v_c(t) = A \cos[2\pi f_c t + \beta_p \cos(2\pi f_m t)]$$

which is similar to Eqn (2.30) except that β_p does not depend on the value of f_m.

Having shown that a PM carrier can be considered in this way, it is then possible to obtain the bandwidth of a wideband PM signal by reference to Bessel function tables. If a frequency deviation $\Delta f_p = \pm 75\,\text{kHz}$ is produced in a PM carrier by a modulating tone of 15 kHz there will be eight significant sidebands and, as with the FM carrier, the bandwidth of the signal will be approximately 240 kHz. However, since β_p does not depend on the value of f_m there will be eight sidebands for all possible values of f_m. A modulating tone of 50 Hz would result in a bandwidth of $2 \times 8 \times 50 = 800\,\text{Hz}$. The value for FM would be approximately 150 kHz. Thus it becomes apparent that PM makes less efficient use of bandwidth than frequency modulation, since the bandwidth of any receiver must be equal to the maximum bandwidth of the received signal. It is shown in Chapter 4 that SNRs in modulated systems are directly related to the bandwidth of the modulated wave. Hence it may be concluded that the signal-to-noise performance of FM will be superior to PM, and for this reason FM is usually chosen in preference to PM. (There is one notable exception to this rule, namely the transmission of wideband data signals, which is dealt with in Chapter 3.)

Because of the inherent non-linearity of the FM process, it is not possible to derive the bandwidth for a multi-tone modulating signal. (The bandwidth of a frequency modulated wave has been derived by Black[2] for a two-tone modulating signal.) One observation that can be made is that if the modulating signal is multitone then no single component can produce peak frequency deviation, or overmodulation would result. Under these circumstances, practical measurements have shown that a reasonably accurate measure of the signal bandwidth is given by a modified form of Carson's rule, i.e. $B = 2(\Delta f_c + D)$, where D is the bandwidth of the modulating signal.

2.17 FREQUENCY MODULATORS

Frequency modulation may be produced directly by varying the frequency of a voltage-controlled oscillator or indirectly from phase modulation. Figure 2.23 illustrates two methods of producing frequency modulation directly. The first method is based on a RC oscillator and the second is based on a parallel resonant LC circuit.

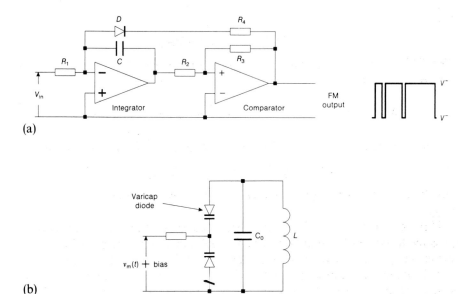

(a)

(b)

Fig. 2.23 Direct FM. (a) RC voltage controlled oscillator, (b) Varicap diode tuning of LC tank circuit.

The RC circuit has the advantage of requiring no inductance and operates as described. When the output of the comparator is V^+ the diode is reversed biased and the integrator charges the capacitor C from a current proportional to V_{in}. When the capacitor charges to the switching voltage of the comparator output goes to V^- which forward biases the diode and produces a rapid discharge of C. This causes the comparator output to return to V^+ reverse biasing the diode and causing C to charge once again from V_{in}. The charging time of the capacitor is inversely proportional to charging current and hence the frequency of oscillation is linearly related to V_{in}. The comparator output will be a variable-frequency pulse train; a sinusoidal frequency modulated waveform is obtained by appropriate filtering.

The LC circuit forms the tank circuit of a LC oscillator. A common method of altering the resonant frequency of such a tuned circuit is to use the dependence of capacitance of a reverse biased p–n junction on the reverse bias voltage. Varactor (varicap) diodes are specifically designed for this purpose and Fig. 2.23 shows a typical arrangement of these diodes in a parallel resonant circuit. The total capacitance in this circuit will be $C = C_0 + \Delta C$ where $\Delta C = K v_m(t)$. The resonant frequency of the circuit is given by

$$2\pi f_r = \frac{1}{(LC)^{1/2}} = \frac{1}{(LC_0)^{1/2}} \times \frac{1}{[1 + (\Delta C/C_0)]^{1/2}}$$

If the change in capacitance is small then $\Delta C \ll C_0$ and the binomial expansion of $[1 + (\Delta C/C_0)]^{-1/2}$ is approximated by $(1 - \Delta C/2C_0)$, i.e.

$$2\pi f_r = \frac{1}{(LC_0)^{1/2}}\left(1 - \frac{\Delta C}{2C_0}\right)$$

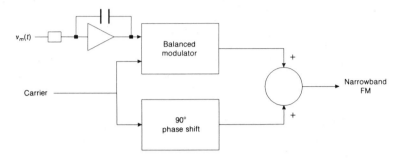

Fig. 2.24 Armstrong frequency modulator.

Therefore

$$2\pi f_r = 2\pi f_c \{1 - [K v_m(t)/2C_0]\}$$

or

$$f_r = f_c[1 - K_c v_m(t)] \tag{2.36}$$

The resonant frequency of the tuned circuit is thus directly related to the amplitude of the modulating signal. The tuned circuit is used as the frequency determining network in a feedback oscillator thus producing FM directly. Such a modulator is restricted to narrowband FM since $\Delta C \ll C_0$ and frequency multiplication is required to produce wideband FM.

The analysis of the previous section has shown that FM is equivalent to PM by the time integral of the modulating signal. Narrowband PM is itself equivalent to DSB-SC-AM with the carrier reinserted in phase quadrature. The Armstrong indirect frequency modulator combines both these properties, and a block diagram is given in Fig. 2.24.

The maximum value of β that can be produced by this modulator is about 0.2. This means that several stages of frequency multiplication are required to produce wideband FM. The advantage of the Armstrong modulator is that the carrier is produced by a stable crystal oscillator. It should be noted that when the frequency of the carrier is multiplied by n (by means of a non-linear device) the frequency deviation, and hence the value of β, is also multiplied by the same factor.

2.18 DEMODULATION OF A FREQUENCY MODULATED WAVE

The primary function of a frequency modulation detector is to produce a voltage proportional to the instantaneous frequency f_i of the modulated wave. The FM waveform is given by

$$v_c(t) = A \cos\left(2\pi f_c t + K \int_0^t v_m(t)\,\mathrm{d}t\right)$$

If this waveform is differentiated, the resultant waveform is

$$\dot{v}_c(t) = -A\left[2\pi f_c + Kv_m(t)\right]\sin\left(2\pi f_c t + K\int_0^t v_m(t)\,\mathrm{d}t\right)$$

This is a frequency modulated wave that now has an envelope of magnitude proportional to the amplitude of the modulating signal $v_m(t)$. The modulating signal may then be recovered by envelope detection. The envelope detector will ignore the frequency variations of the carrier. Traditionally FM detectors have relied upon the properties of tuned circuits to perform the required differentiation. This is demonstrated in Fig. 2.25, where the resonant frequency of the tuned circuit is chosen such that the carrier frequency f_c is on the slope of the circuit response.

The linearity of a single tuned circuit is limited to a relatively small frequency range. This range may be extended by introducing a second tuned circuit with a slightly different resonant frequency. A typical circuit and its response is shown in Fig. 2.26. Each circuit is fed in antiphase by the tuned secondary of the high frequency transformer. Several FM discriminators are based upon this type of circuit; the circuit of Fig. 2.26 is known as a balanced discriminator. Circuits of this type are also sensitive to fluctuations of the amplitude of the FM wave. To avoid this the FM waveform is 'hard limited', as shown in Fig. 2.27, before being applied to the discriminator.

Modern detection techniques are based upon integrated circuit technology, where the emphasis is on inductorless circuits. The hard limited FM signal is in fact a variable frequency pulse waveform. The instantaneous frequency is preserved in the zero crossings of this pulsed signal. A completely digital circuit based upon a zero crossing detector is shown in Fig. 2.28. The number of zero crossings in a fixed interval are gated to the input of a binary counter. The counter outputs are then used as inputs to a digital-to-analogue converter (DAC). The analogue output voltage is thus proportional to the number of zero crossings that occur during the gating interval. Thus the output of the DAC is proportional to the original modulating signal $v_m(t)$.

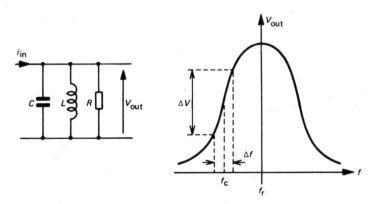

Fig. 2.25 Detection of FM with a single tuned circuit.

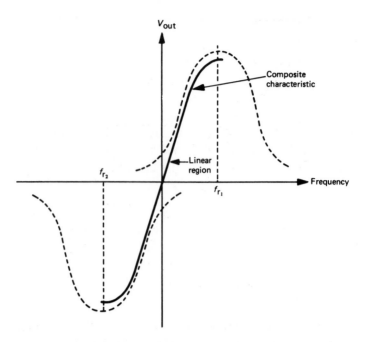

Extended range frequency-to-voltage conversion

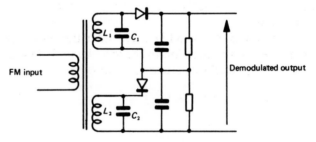

Fig. 2.26 Balanced discriminator.

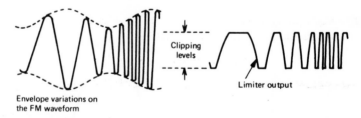

Fig. 2.27 Action of the limiter.

2.18.1 The phase locked loop

This is an important class of inductorless frequency modulation detector which is widely available in integrated form. The circuit is a feedback network with a voltage-controlled oscillator (VCO) in the feedback path. Feedback is arranged so that the output frequency of the VCO is equal to the frequency of

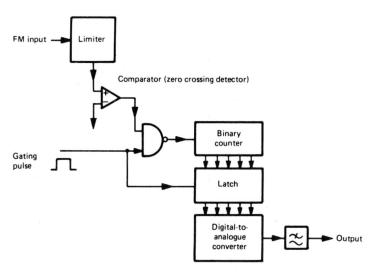

Fig. 2.28 Digital (zero crossing) FM detector.

the input waveform. If the input frequency is modulated by a voltage and the output frequency of the VCO tracks the variation in the input frequency, then the voltage at the VCO input must be equal to the voltage which produced the frequency modulation. The phase locked loop therefore demodulates the frequency modulated input.

The basic phase locked loop is shown in Fig. 2.29, and in the initial analysis it will be assumed that the transfer function of the loop filter $H(f) = 1$. It is convenient to represent the frequency modulated input as

$$v_c(t) = A \cos(2\pi f_c t + \phi_1(t))$$

where

$$\phi_1(t) = k_m \int_0^t v_m(t) \, dt$$

i.e. $\dot{\phi}_1(t) = k_m v_m(t)$, $v_m(t)$ being the modulating voltage.

The output of the VCO may be written as

$$v_r(t) = B \sin(2\pi f_c t + \phi_2(t))$$

The phase comparator is essentially a multiplier followed by a low-pass filter

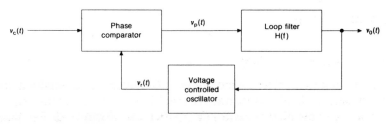

Fig. 2.29 The basic phase locked loop.

and produces an output proportional to the difference in phase between the input waveform and the output of the VCO. The output of the multiplier component of the phase comparator will be the product of $v_c(t)$ and $v_r(t)$ and is given by

$$\tfrac{1}{2} k_d \, AB[\sin\{\phi_2(t) - \phi_1(t)\} + \sin\{4\pi f_c t + \phi_2(t) + \phi_1(t)\}]$$

The high-frequency components are removed by the low-pass filter component and hence the output of the phase comparator may be written

$$v_p(t) = \tfrac{1}{2} k_d \, AB \sin\{\phi_1(t) - \phi_2(t)\}$$

When the loop is locked the phase error will be very small, i.e. $\{\phi_1(t) - \phi_2(t)\} \to 0$.

Thus

$$v_p(t) \approx \tfrac{1}{2} k_d \{\phi_1(t) - \phi_2(t)\}$$

If $H(f) = 1$ then

$$v_o(t) \approx \tfrac{1}{2} k_d \{\phi_1(t) - \phi_2(t)\} \tag{2.37}$$

where k_d is the gain of the phase comparator and has units of volts per radian. It should be noted that when the loop is locked there is actually a phase difference of 90° between $v_c(t)$ and $v_r(t)$. The output frequency of the VCO may be written as $2\pi f_c + \dot{\phi}_2(t)$ where $\dot{\phi}_2(t) = k_0 v_o(t)$, k_0 being the gain of the VCO with units of radians/second/volt, i.e.

$$\phi_2(t) = k_0 \int_0^t v_o(t) \, dt$$

Equation (2.37) can thus be written

$$v_o(t) = \tfrac{1}{2} k_d \, AB \left[k_0 \int_0^t v_o(t) \, dt - \phi_1(t) \right] \tag{2.38}$$

Differentiation of Eqn (2.38) yields

$$\dot{\phi}_1(t) = k_0 \left[\frac{2\dot{v}_o(t)}{k_0 k_d \, AB} - v_o(t) \right] \tag{2.39}$$

It should be noted that the product $k_0 k_d$ has dimensions of seconds^{-1}, which is a frequency. Thus if the **frequency** of $v_o(t) \ll k_0 k_d$ then

$$\frac{2\dot{v}_o(t)}{k_0 k_d \, AB} \to 0 \quad \text{and} \quad \dot{\phi}_1(t) = -k_0 v_o(t)$$

But $\dot{\phi}_1(t) = k_m v_m(t)$, therefore

$$v_o(t) = -\frac{k_m}{k_0} v_m(t)$$

This shows that the PLL demodulates the frequency modulated input waveform.

An insight into the physical action of the PLL can be gained by considering the loop when the input is an unmodulated carrier. The phase of $v_c(t)$ will be a linear function of time, i.e. $2\pi f_c t$, which t can then be regarded as a ramp function. To maintain a small error at the phase detector output the phase of the VCO output must also be a ramp of the same slope. If the frequency of $v_c(t)$ increases the slope of the 'phase ramp' also increases. The error between the phase ramp of $v_c(t)$ and the phase ramp of the VCO output then begins to increase producing an increase in the phase comparator output. This increase modifies the frequency of the VCO output and a new state of equilibrium is reached when the two slopes are again equal, indicating that the two frequencies are also equal. There will, however, be a different output from the phase comparator indicating a different, but constant, phase error.

The analysis presented above deals only with the steady-state response of the PLL the assumption being that the loop is locked. If this is not the case the PLL will behave as a negative feedback system with the usual transient properties. The transient response of the phase locked loop is an important consideration as it determines the range of frequencies over which the loop will acquire lock and also the range of frequencies over which a frequency locked loop will remain in lock. The transient response of the phase locked loop depends on the frequency response of the loop filter and when $H(f) = 1$ the circuit is known as a first-order phase locked loop.

In order to analyse the transient response of the phase locked loop it is convenient to represent the device as a linear system as shown in Fig. 2.30. The frequency domain equivalent is derived in terms of the Laplace transform and is also given in Fig. 2.30.

Considering the frequency domain model

$$V_o(s) = k_d H(s)\left[\phi_1(s) - V_o(s)\frac{k_o}{s} \right]$$

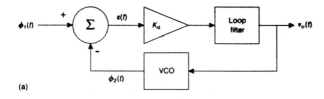

(a)

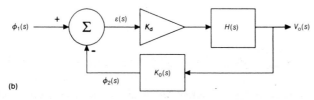

(b)

Fig. 2.30 Alternative time (a) and frequency (b) models of the phase locked loop.

Hence

$$\frac{V_o(s)}{\phi_1(s)} = \frac{sk_dH(s)}{k_ds + k_0k_dH(s)} \tag{2.40}$$

but

$$V_o(s) = \varepsilon(s)k_dH(s)$$

thus

$$\frac{\varepsilon(s)}{\phi_1(s)} = \frac{s}{s + k_0k_dH(s)} \tag{2.41}$$

If $H(s) = 1$ the loop is known as a first-order loop and Eqn (2.40) becomes

$$\frac{V_o(s)}{\phi_1(s)} = \frac{k_ds}{s + k_0k_d} \tag{2.42}$$

The transient response of the loop is examined by applying a unit impulse in frequency at the input, which is equivalent to $\phi_1(s) = 1/s$. Substituting into Eqn (2.42) gives the frequency response of the loop as

$$V_o(s) = \frac{k_d}{s + k_0k_d} \tag{2.43}$$

The 3 dB cut-off frequency is then

$$f_{3\,dB} = \frac{k_0k_d}{2\pi} \tag{2.44}$$

Hence in order to avoid attenuation of high frequency components in the output signal it is necessary that the bandwidth of this signal $< f_{3\,dB}$, which agrees with constraints applied to Eqn 2.39. For a step change in input frequency $\phi_1(s) = 2\pi f/s^2$ which may be substituted in Eqn (2.41) to give the phase error as

$$\varepsilon(s) = \frac{2\pi f}{s(s + k_0k_d)}$$

or

$$\varepsilon(t) = \frac{2\pi f}{k_0k_d}(1 - e^{-k_0k_dt})u(t) \tag{2.45}$$

The steady-state error, after the transient interval, is

$$\varepsilon(\infty) = \frac{2\pi f}{k_0k_d} \tag{2.46}$$

In order to minimize the phase error k_0k_d should be large, which will result in a large value of $f_{3\,dB}$. In practice the bandwidth of the loop should be sufficient only to avoid attenuation of components in the modulating signal. If the bandwidth is significantly greater than this the noise performance of the PLL, as a frequency demodulator deteriorates (see Chapter 4 for noise in frequency modulation transmission). Hence the first-order PLL is seldom used in practice.

The second-order PLL is derived by making the transfer function of the loop filter

$$H(s) = \frac{1 + \tau_2 s}{1 + \tau_1 s} \qquad (2.46)$$

A typical loop filter is shown in Fig. 2.31 and it should be noted that when $R_1 \gg R_2$ the transfer function of the loop filter over the range of frequencies in $V_o(s)$ is approximated by

$$H'(s) = \frac{1 + s\tau_2}{s\tau_1}$$

Substituting for $H'(s)$ in Eqn (2.40) gives

$$\frac{V_o(s)}{\phi_1(s)} = \frac{k_d s(s\tau_2 + 1)}{\tau_1(s^2 + 2\zeta\omega_n s + \omega_n^2)} \qquad (2.47)$$

where

$$\omega_n = 2\pi f_n = \sqrt{\frac{k_o k_d}{\tau_1}}$$

is known as the natural (radian) frequency and

$$\zeta = \frac{\omega_n \tau_2}{2}$$

is known as the damping factor. The bandwidth of the second-order loop depends on the value of ζ and it is usual to operate the loop with $\zeta < 1$. Under

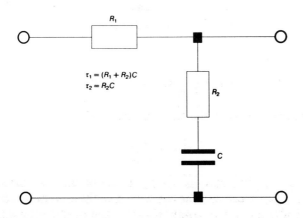

Fig. 2.31 Loop filter for a second-order PLL.

these circumstances the bandwidth of the loop is approximately

$$f_{3\,\text{dB}} = \frac{2\omega_n\zeta}{2\pi} = \frac{k_0 k_d \tau_2}{2\pi\tau_1}$$

The transfer function of the second-order PLL is given by

$$\frac{\varepsilon(s)}{\phi_1(s)} = \frac{s^2}{s^2 + 2\zeta\omega_n s + \omega_n^2} \tag{2.48}$$

For a step change in input frequency $\phi_1(s) = 2\pi f/s^2$ which may be substituted in Eqn (2.48) to give the phase error as

$$\varepsilon(s) = \frac{2\pi f}{s^2 + 2\zeta\omega_n s + \omega_n^2} \tag{2.49}$$

or

$$\varepsilon(t) = \frac{f e^{-\zeta\omega_n t}}{f_n \sqrt{1-\zeta^2}} \sin\left(\omega_n t \sqrt{1-\zeta^2}\right) u(t) \tag{2.50}$$

The steady-state error, after the transient interval, is

$$\varepsilon(\infty) = 0 \tag{2.51}$$

Hence the second-order PLL has zero steady-state error and the transient response and bandwidth can be determined by appropriate choice of natural frequency and damping factor. The second-order PLL is therefore the preferred choice in most practical applications.

2.19 FREQUENCY DIVISION MULTIPLEX (FDM) TRANSMISSION

This is the form of transmission used extensively in telephone systems for the simultaneous transmission of several separate telephone circuits over a wideband link. Each of the telephone signals has a bandwidth limited to

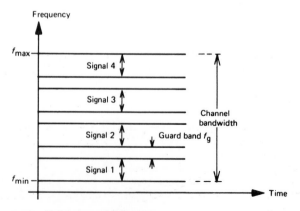

Fig. 2.32 Frequency division multiplexing.

3.4 kHz, and therefore many such signals can be transmitted over, say, a coaxial cable with a bandwidth of several megahertz.

Each signal modulates a sinusoidal carrier of different frequency, single sideband modulation being used. The signals are therefore effectively stacked one above the other throughout the transmission bandwidth. To facilitate separation of the signals at the far end of the link, adjacent signals are separated by a guard band f_g (Fig. 2.32). Each signal in a frequency division multiplex transmission system thus occupies part of transmission bandwidth for the whole of the transmission time.

2.20   CONCLUSION

The economic argument for use of the various amplitude modulated systems was given in Section 2.13. Frequency division multiplexing of telephone circuits is based on SSB-AM because here the emphasis is on packing as many channels as possible into a finite bandwidth.

The arguments for the use of frequency modulation are less well defined. It is shown in Chapter 4 that for a fixed transmitter power frequency modulation has a superior signal-to-noise performance over all types of amplitude modulation. Consequently, frequency modulation has been adopted as the standard for high-quality sound broadcast transmission in many countries. Frequency modulation is also an extremely important technique used in both analogue and digital cellular radio in which the capture effect (covered in Section 4.11) is utilized to increase system capacity.

REFERENCES

1. Scroggie, M. G., *Foundations of Wireless*, 8th edn, Illife, London, 1971.
2. Black, H. S., *Modulation Theory*, Van Nostrand Reinhold, Wokingham, UK, 1953.
3. Stremler, F. G., *Introduction to Communication Systems*, Addison-Wesley, London, 1976.

PROBLEMS

2.1   A transmitter radiates a DSB-AM signals with a total power of 5 kW at a depth of modulation of 60%. Calculate the power transmitted in the carrier and also in each sideband.

Answer: 4.24 kW; 0.38 kW.

2.2   A carrier wave represented by $10\cos(2\pi 10^6 t)$ V is amplitude modulated by a second wave represented by $3\cos(2\pi 10^3 t)$ V. Calculate
(a) the depth of modulation;
(b) the upper and lower side frequencies;
(c) the amplitude of the side frequencies;
(d) the fraction of the power transmitted in the sidebands.

Answer: (a) 30%; (b) 1.001 MHz, 0.999 MHz; (c) 1.5 V; (d) 4.3%.

2.3   A DSB-AM transmitter produces a total output of 24 kW when modulated to a depth of 100%. Determine the power output when
(a) the carrier is unmodulated;
(b) the carrier is modulated to a depth of 60%, one sideband is suppressed and the carrier component is attenuated by 26 dB.

Answer: (a) 16 kW; (b) 1.48 kW.

2.4   A DSB-AM receiver uses a square law detector. What is the maximum depth of modulation that may be used if the second harmonic distortion of the modulating signal, produced by the detector, is restricted to 10% of the fundamental?

Answer: 40%.

2.5   A carrier of 5 V rms and frequency 1 MHz is added to a modulating signal of 2 V rms and frequency 1 kHz. The composite signal is applied to a biased diode rectifier in which the relationship between current and voltage over the range ± 10 V is $i = (5 + v + 0.05 v^2)\ \mu A$ where v is the instantaneous voltage. Find the depth of modulation of the resulting DSB-AM signal and the frequency of each component in the diode current.

Answer: 28.3%; 0 Hz, 1 kHz, 2 kHz, 1 MHz, 2 MHz, 0.999 MHz, 1·001 MHz.

2.6   Narrowband FM is produced indirectly by varying the phase of a carrier of frequency 13 MHz, the maximum phase shift being 0.5 rad. Show that wideband FM may be produced by multiplying the frequency of the modulated carrier. If the modulating signal has a frequency of 1.5 kHz find the frequency deviation of the carrier after a frequency multiplication of 15.

Answer: 11.25 kHz.

2.7   A frequency modulated wave has a total bandwidth of 165 kHz when the modulating signal is a single tone of frequency 10 kHz. Using Bessel function tables, or otherwise, find the maximum carrier frequency deviation produced by the modulating signal.

Answer: 50 kHz.

2.8   In a direct FM transmitter an inductance of 10 μH is tuned by a capacitor whose capacitance is a function of the amplitude of the modulating signal. When the modulating signal is zero the effective capacitance is 1000 pF. An input signal of $4.5 \cos(3\pi 10^3 t)$ V produces a maximum change in capacitance of 6 pF. Assuming the resultant FM signal is multiplied in frequency by 5, calculate the bandwidth of the eventual output.

Answer: 60 kHz.

2.9   A single tone of frequency 7.5 kHz forms the modulating signal for both a DSB-AM and a FM transmission. When modulated the peak frequency deviation of the FM signals is 60 kHz. Assuming that the same total power is transmitted for each of the modulated signals, find the depth of modulation for the DSB-AM signal when the amplitude of the first pair of sidebands of the FM wave equals the amplitude of the sidebands of the DSB-AM wave.

Answer: 50%.

Discrete signals $\boxed{3}$

The signals considered in Chapter 2 were continuous functions of time. There are many advantages that result from the conversion of analogue signals into a binary coded format. Two such advantages that are easily identified are that the transmission and processing of binary signals are generally much easier to achieve than the transmission and processing of analogue signals.

It is not possible to code a continuous analogue signal into binary format because there is an infinite number of values of the continuous signal. Instead the continuous signal is coded at fixed instants of time. These instants are known as sampling instants, and it is important to determine the effect of the sampling action on the properties of the original continuous signal. The rules governing the sampling of continuous signals are specified by the sampling theorem.

3.1 SAMPLING OF CONTINUOUS SIGNALS

The sampling theorem states that if a signal has a maximum frequency of W Hz it is completely defined by samples which occur at intervals of $1/2W$ s. The sampling theorem can be proved by assuming that $h(t)$ is a non-periodic signal band limited to W Hz. The amplitude spectrum is given by

$$H(f) = \int_{-\infty}^{\infty} h(t)\exp(-j2\pi ft)\,dt \qquad (3.1)$$

Since $H(f)$ is band limited to $\pm W$ Hz it is convenient to make $H(f)$ a periodic function of frequency (Fig. 3.1). The value of $H(f)$ in the region $-W$ to $+W$ can be expressed in terms of a Fourier series in the frequency domain.
The time domain Fourier series is

$$h(t) = 1/T \sum_{n=-\infty}^{\infty} C_n(\exp j\,2\pi nt/T)$$

and the corresponding series in the frequency domain is

$$H(f) = 1/2W \sum_{n=-\infty}^{\infty} X_n \exp(j\,2\pi nf/2W) \qquad (3.2)$$

The values of the Fourier coefficients are give by

$$X_n = \int_{-W}^{W} H(f)\exp(-j\pi nf/W)\,df$$

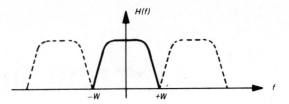

Fig. 3.1 Representation of $H(f)$ as periodic function.

But $H(f)$ is the Fourier transform of $h(t)$, i.e.

$$h(t) = \int_{-W}^{W} H(f) \exp(j2\pi f t)\,df \qquad (3.3)$$

Hence if $t = -n/2W$

$$h(-n/2W) = \int_{-W}^{W} H(f) \exp(-j\pi n f/W)\,df$$

i.e.

$$h(-n/2W) = X_n \qquad (3.4)$$

The values $h(-n/2W)$ are samples of $h(t)$ taken at equally spaced instants of time, the time between samples being $1/2W$ s. These samples define X_n, which in turn completely defines $H(f)$. Since $H(f)$ is the Fourier transform of $h(t)$, then $H(f)$ defines $h(t)$ for all values of t. Hence $h(-n/2W)$ completely defines $h(t)$ for all values of t.

The physical interpretation of the sampling process is shown in Fig. 3.2 The continuous signal $h(t)$ is multiplied by a periodic pulse train $S(t)$ in which the pulse width is much less than the pulse period.

Since the sampling pulse train is periodic it can be expanded in a Fourier series as

$$S(t) = a_0 + a_1 \cos \omega_s t + a_2 \cos 2\omega_s t + \cdots$$

where $\omega_s = 2\pi/T_s$. If the continuous signal $h(t)$ is assumed to be a single tone $\cos \omega_m t$ then the sampled waveform $h_s(t)$ is given by

$$h_s(t) = a_0 \cos \omega_m t + (a_1/2) \cos(\omega_s - \omega_m)t + (a_1/2) \cos(\omega_s + \omega_m)t$$
$$+ (a_2/2) \cos(2\omega_s - \omega_m)t + (a_2/2) \cos(2\omega_s + \omega_m)t + \cdots \qquad (3.5)$$

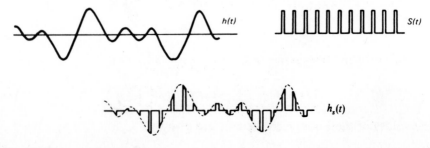

Fig. 3.2 The sampling process.

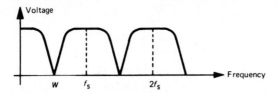

Fig. 3.3 Amplitude spectrum of a sampled signal.

The spectrum of $h_s(t)$ thus contains the original spectrum of $h(t)$ and upper and lower sidebands centred at harmonics of the sampling frequency. The amplitude spectrum for $h_s(t)$ when $h(t)$ is a multitone signal band limited to W Hz is shown in Fig. 3.3. It can be seen from this figure that if the sampling frequency $f_s = 2W$, the sidebands just fail to overlap. If $f_s < 2W$, overlap (aliasing) occurs and distortion of the spectrum of $h(t)$ results.

3.2 RECONSTRUCTION OF THE CONTINUOUS SIGNAL

In order to reproduce a continuous signal from the samples some form of interpolation is required. The output from a sample-and-hold circuit is shown in Fig. 3.4. This circuit holds the level of the last sample until a new sample arrives and then assumes the value of the new sample. It can be seen from the figure that there is a considerable error between the output of the sample-and-hold circuit and the original value of $h(t)$. The exact interpolation function is obtained by considering Eqn (3.3).

If the Fourier series representation of $H(f)$ [Eqn (3.2)] is substituted in Eqn (3.3), then

$$h(t) = \int_{-W}^{W} \left[1/2W \sum_{n=-\infty}^{\infty} X_n \exp(j\pi n f/W) \right] \exp(j2\pi f t)\, df$$

Changing the order of summation and integration gives

$$h(t) = \sum_{n=-\infty}^{\infty} X_n/2W \int_{-W}^{W} \exp[j2\pi n f(t + n/2W)]\, df$$

i.e.

$$h(t) = \sum_{n=-\infty}^{\infty} X_n \operatorname{sinc}[2\pi W(t + n/2W)]$$

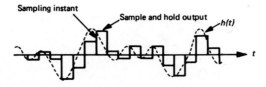

Fig. 3.4 Sample and hold interpolation.

but

$$X_n = h(-n/2W)$$

Therefore

$$h(t) = \sum_{n=-\infty}^{\infty} h(-n/2W) \operatorname{sinc}\left[2\pi W(t + n/2W)\right]$$

or alternatively

$$h(t) = \sum_{n=-\infty}^{\infty} h(n/2W) \operatorname{sinc}\left[2\pi W(t - n/2W)\right] \tag{3.6}$$

The values $h(n/2W)$ are the samples of $h(t)$ and sinc $[2\pi W(t - n/2W)]$ is the required interpolation function. This function is centred (has unity value) at intervals of time spaced at $1/2W$. The function has zeros at instants of time equal to $(n + p)/2W$ where p has integer values, except zero, between $\pm \infty$. The value of $h(t)$ at a sampling instant, therefore, is equal to the amplitude of the sample only, since the weights of the other samples are zero. The value of $h(t)$ between the sampling instants is given by the summation of the corresponding sinc functions and this process is shown in Fig. 3.5.

3.3 LOW-PASS FILTERING OF A SAMPLED SIGNAL

The previous section has shown that the required interpolation function for perfect signal reconstruction is a sinc function. This function is the impulse response of an ideal low-pass filter. If $P(f)$ is the transfer function of a network and $G(f)$ is the spectrum of an input signal, the spectrum of the signal at the network output $L(f)$ is given by

$$L(f) = P(f)G(f)$$

If the input to the network is a very narrow pulse (which approximates to a unit impulse), $G(f) \simeq 1$ and the spectrum of the output is then $L(f) = P(f)$.

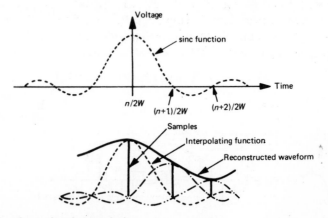

Fig. 3.5 Sinc fuunction interpolation.

The response in the time domain is obtained by taking the Fourier transform of $P(f)$. If the network is an ideal low-pass filter with cut-off frequency f_c the transfer function is given by

$$P(f) = |P(f)| \exp(-j2\pi n f t_0)$$

where

$$|P(f)| = 1 \quad \text{for } |f| \leqslant f_c, \quad = 0 \text{ else}$$

The factor $\exp(-j2\pi f t_0)$ is the linear phase characteristic of the ideal filter. The impulse response $p(t)$ is

$$p(t) = \int_{-f_c}^{f_c} \exp[-j2\pi f(t_0 - t)] \, df$$

i.e.

$$p(t) = 2f_c \frac{\sin[2\pi f_c(t - t_0)]}{2\pi f_c(t - t_0)} \tag{3.7}$$

This function has the required form when $f_c = W$.

In the frequency domain the effect of the ideal low-pass filter is to remove completely all spectral components above W Hz (Fig. 3.6).

In practice the ideal low-pass filter is not physically realizable. This may be seen from the fact that the impulse response exists for $t < 0$ which means that an output exists before the impulse is applied, which is physically impossible. The ideal filter may be approximated by a physically realizable network, the approximation becoming more exact as the complexity of the network is increased. One such approximation with a sharp cut-off characteristic is the Chebyshev approximation. The impulse response of a fifth-order Chebyshev low-pass filter is

$$p(t) = Ae^{-at} + Be^{-bt}(\cos\theta t + \sin\theta t) + Ce^{-ct}(\cos\phi t + \sin\phi t) \tag{3.8}$$

This impulse response will clearly produce some distortion, which is most easily evaluated in the frequency domain as shown in Fig. 3.7.

Since the filter has a finite transmission above W Hz some distortion components outside the signal bandwidth will appear in the filter output. To minimize this distortion the sampling frequency is increased above $2W$ Hz. A practical telephone channel has an upper cut-off frequency of 3.4 kHz and the sampling rate employed is 8 kHz, that is 2.35 times the maximum signal frequency.

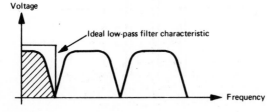

Fig. 3.6 Ideal filtering of a sampled signal.

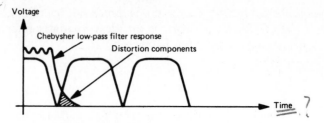

Fig. 3.7 Practical filtering of a sampled signal.

3.4 TIME DIVISION MULTIPLEX (TDM) TRANSMISSION

When a signal is sampled by narrow pulses there are large intervals between the samples in which no signal exists. It is possible during these intervals to transmit the samples of other signals. This process is shown in Fig. 3.8 and is called 'time division multiplex' (TDM) transmission. Since each sampled signal gives rise to a continuous signal after filtering, TDM transmission allows simultaneous transmission of several signals over a single wideband link. It is therefore an alternative to FDM transmission, described in Section 2.19.

The switches at transmitter and receiver (which would be solid-state devices) are synchronized and perform the sampling and interlacing. The samples themselves are very narrow and consequently have a large bandwidth. When transmitted over a link with a fixed bandwidth the samples

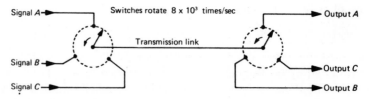

Fig. 3.8 TDM transmission.

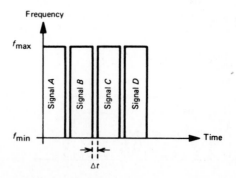

Fig. 3.9 Spectrum/time diagram for TDM.

spread and can overlap adjacent samples. To minimise this, guard intervals Δt are allowed between adjacent signals. In the case of TDM transmission each signal therefore occupies the whole of the transmission bandwidth for part of the transmission time (Fig. 3.9).

3.5   PULSE CODE MODULATION (PCM)

Although it is possible to transmit the samples directly in a TDM system there is a considerable advantage in coding each sample into a binary word before transmission. This is because binary signals have a much greater immunity to noise than analogue signals. A widely used form of digital transmission is pulse code modulation, the essentials of which are shown in Fig 3.10.

In this system, 30 speech channnels are each limited in frequency to 3.4 kHz and sampled at 8 kHz. The sampled signals are converted into binary form for transmission. In addition to the coded signals other signals are sent over the link for synchronization and identification. Once each sample has been converted into a binary code this effectively means that it has been quantized into one of a fixed number of levels. The greater the number of quantization levels the greater is the accuracy of the quantized representation, but also the greater is the number of binary digits (bits) that are required to represent the sample. Since more bits require a higher transmission bandwidth a balance must be struck between accuracy and bandwidth.

The quantization process is shown in Fig 3.11 and it is clear that once quantized the precise amplitude of the original sample can never be restored. This gives rise to an error in the recovered analogue signal, known as quantization error. The quantization error for an eight-level PCM system can be calculated by reference to Fig 3.11. This shows that the peak-to-peak input is I volts. If this value is quantized into M equally spaced levels the spacing between each level is $\delta V = I/M$. The amplitudes reproduced after decoding are usually the mid-point of each quantizing interval, which gives a maximum decoded peak to peak value of $A = (M - 1)\delta V$. If a level V_j is reproduced by the decoder the true amplitude will be anywhere in the range $V_j \pm \partial V/2$; hence

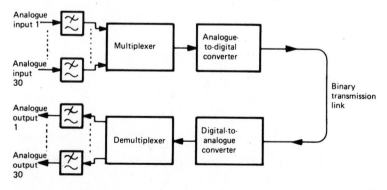

Fig. 3.10 Pulse code modulation system.

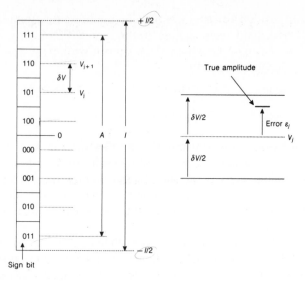

Fig. 3.11 Linear quantization (8 level).

a maximum error of $\partial V/2$ will be present on any decoded output. The error is random in nature and is called quantization noise. In order to calculate the magnitude of this quantization noise it is necessary to know the amplitude probability density function of the signal to be coded. From this knowledge it is possible to calculate the probability that the signal will be in any quantizing interval and the mean square error for each of the quantizing steps.

The simplest case to evaluate is a signal with a uniformly distributed amplitude density function. Such a signal is a triangular wave of amplitude $\pm I/2$. The density function for both a triangular wave and a sine wave are shown in Fig 3.12. A signal with a uniform probability density has equal probability of being in any of the quantizing intervals and also has equal probability of having any particular amplitude within a given quantizing interval. The error produced is shown in Fig 3.11 and has a mean square value given by

$$\bar{\varepsilon}^2 = \frac{1}{\partial V_j} \int_{-\partial V_j/2}^{\partial V_j/2} \varepsilon_j^2 \, \mathrm{d}\varepsilon = \frac{(\partial V_j)^2}{12} \tag{3.9}$$

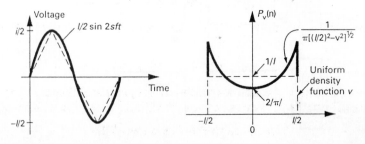

Fig. 3.12 Voltage waveforms and amplitude density functions.

The total mean square error throughout the range of coder is

$$\bar{\sigma}^2 = \sum_j \bar{\varepsilon}_j^2 \, p(V_j)$$

where $p(V_j)$ is the probability that the signal will be in the jth quantization interval, i.e.

$$\bar{\sigma}^2 = \frac{1}{12} \sum_j (\partial V_j)^2 \, p(V_j) \tag{3.10}$$

But $\sum_j (\partial V_j)^2 p(V_j)$ is the mean square value of ∂V_j, i.e.

$$\bar{\sigma}^2 = \frac{\overline{(\partial V_j)^2}}{12} \tag{3.11}$$

If the signal has a uniform amplitude distribution and the quantization is linear, then all values of ∂V_j are equal, i.e.

$$\bar{\sigma}^2 = \frac{(\partial V)^2}{12} \tag{3.12}$$

$\bar{\sigma}^2$ is effectively the quantization noise power. If the uniformly distributed signal has a maximum amplitude of $\pm I/2$, it has a mean square value of

power in signal

$$\bar{S}^2 = \int_{-I/2}^{I/2} V^2 \, p_v(V) \, dV = \int_{-I/2}^{I/2} \frac{V^2}{I} \, dV = \frac{I^2}{12} \tag{3.13}$$

The mean signal to quantization noise power ratio (SQNR) is thus

$$\text{SQNR} = \frac{\bar{S}^2}{\bar{\sigma}^2} = \frac{I^2}{(\partial V)^2} = M^2$$

In a binary system the number of bits m required to code M levels must satisfy the relationship $M = 2^m$; hence

$$\text{SQNR} = 10 \log_{10} 2^{2m} \quad \text{dB}$$

i.e.

$$\text{SQNR} = 20 \, m \log_{10} 2 \quad \text{dB}$$

or

$$\text{mean SQNR} = 6m \quad \text{dB} \tag{3.14}$$

The maximum signal power is $(I/2)^2$, which gives a maximum SQNR of $3M^2$, i.e.

$$\text{maximum SQNR} = (4.8 + 6m) \quad \text{dB} \tag{3.15}$$

If the input signal is a sine wave which occupies the full coder range then $V_{in}(t) = (I/2) \sin 2\pi f t$ and the mean square value is $\bar{S}^2 = I^2/8$, which results in

$$\text{mean SQNR} = (1.8 + 6m) \quad \text{dB} \tag{3.16}$$

Equations (3.14) to (3.16) show that the SQNR of the decoded signal, measured in dB, increases linearly with m which itself is linearly related to transmission bandwidth. This is a similar relationship between SNR and bandwidth to that expressed by Shannon's law. However, in the case of PCM, the noise is quantization noise rather than fluctuation noise. The effect of fluctuation noise on PCM transmissions is covered in Section 5.3.

It is appropriate at this point to investigate the effect of sampling frequency on SQNR by considering the sampling and quantization of the waveform shown in Fig. 3.13. The error between the original waveform and the quantized version is also shown in this figure and is seen to be approximately similar to a sawtooth waveform of period $1/f_s$ where f_s is the sampling frequency. It is clear that as the sampling frequency is increased the frequency of the error waveform also increases. However, the quantization noise power is fixed at $(\partial V)^2/12$ and, as the sampling rate is increased, this power will be distributed over a wider frequency range. Consequently the proportion of quantization noise power occupying the same bandwidth as the original signal, and passing through the reconstruction filter, will decrease as the sampling rate is increased. Thus oversampling a bandlimited waveform will increase the SQNR at the output of the reconstruction filter. This principle is often used to advantage in compact disc players, for example.

In deriving Eqn (3.16) it was assumed that the input waveform fully occupied the range of the coder, which would correspond to the loudest talker in a telephone system. It follows that the SQNR for the quietest talker would be considerably lower than this value. PCM encoders are designed to make the SQNR constant over as wide a range of input amplitudes as possible and this is achieved by making the quantization step size a function of the input

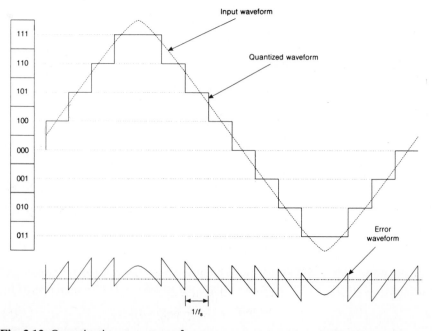

Fig. 3.13 Quantization error waveform.

waveform amplitude. This is known as non-linear (or non-uniform) quantization. Essentially the step size of a non-linear coder is reduced near zero and
increases towards the maximum input level. Non-linear coding is often
achieved in practice by first COMpressing the signal then coding the compressed signal in a linear coder and finally exPANDING the decoded signal
with the inverse of the compression characteristic. The combined process is
known as COMPANDING.

The compression characteristic is required to give a constant signal to
quantization noise ratio for all levels of input. The quantization noise is
calculated by considering the linear quantization of the output of the
compression circuit. This is equivalent to the non-linear quantization of the
input signal. The linear coder and equivalent non-linear coder are shown in
Fig. 3.14. From this figure it is clear that the step size for the linear coder is
$\partial V_{out} = 2/M$ where M is the number of quantization levels. Considering the
jth quantizing interval of the non-linear coder this is assumed to have a
range of $\partial V_{in(j)}$ centred at $V_{in(j)}$. The range of voltages for the jth interval is
$V_{in(j)} \pm V_{in(j)}/2$.

Assuming the M is large, then from Fig. 3.14,

$$\frac{dV_{out}}{dV_{in}} = \frac{2/M}{\partial V_{in(j)}} \text{ or } \partial V_{in(j)} = \frac{2/M}{(dV_{out}/dV_{in})|_j}$$

The mean square error produced by quantizing a voltage in the jth interval is
then

$$\bar{\varepsilon}^2 = \int_{V^-}^{V^+} (V_{in} - V_{in(j)})^2 p_v(V_{in}) \, dV_{in} \qquad (3.17)$$

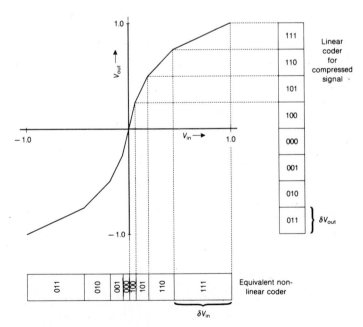

Fig. 3.14 Non-linear quantization.

where

$$V^+ = V_{in(j)} + \frac{\partial V_{in(j)}}{2} \quad \text{and} \quad V^- = V_{in(j)} - \frac{\partial V_{in(j)}}{2}$$

$p_v(V_{in})$ is the amplitude distribution function of the input waveform and when M is large then $p_v(V_{in})$ may be considered constant over the jth interval: i.e.

$$p_v(V_{in}) = p_{vj}(V_{in}) = \text{ constant}$$

The mean square error is then

$$\bar{\varepsilon}^2 = p_{vj}(V_{in}) \frac{\partial V_{in(j)}^3}{12}$$

which may be written

$$\bar{\varepsilon}^2 = p_{vj}(V_{in}) \frac{\partial V_{in(j)}^2}{12} \partial V_{in(j)}$$

but

$$\partial V_{in(j)} = \frac{2/M}{(dV_{out}/dV_{in})|_j}$$

hence the mean square error is

$$\bar{\varepsilon}^2 = p_{vj}(V_{in}) \cdot \frac{(2/M)^2}{(dV_{out}/dV_{in})^2|_j} \cdot \frac{\partial V_{in(j)}}{12} \tag{3.18}$$

The total mean square error is found by summing the contributions from each interval: i.e.

$$\bar{\sigma}_n^2 = 2 \sum_{j=1}^{M/2} p_{vj}(V_{in}) \cdot \frac{(2/M)^2}{(dV_{out}/dV_{in})^2|_j} \cdot \frac{\partial V_{in(j)}}{12}$$

For large values of M the summation may be replaced by an integral over the range ± 1, the total MSE is then

$$\bar{\sigma}_n^2 = 2 \int_0^1 \frac{p_v(V_{in})}{3M^2} \cdot \frac{1}{(dV_{out}/dV_{in})^2} \cdot dV_{in} \tag{3.19}$$

The mean square signal value is

$$\bar{\sigma}_s^2 = 2 \int_0^1 V_{in}^2 p_v(V_{in}) dV_{in}$$

Hence

$$\text{SQNR} = \frac{2 \int_0^1 V_{in}^2 p_v(V_{in}) dV_{in}}{2 \int_0^1 \frac{p_v(V_{in})}{3M^2} \cdot \frac{1}{(dV_{out}/dV_{in})^2} \cdot dV_{in}} \tag{3.20}$$

If $V_{out} = (1 + k \ln V_{in})$ then $dV_{out}/dV_{in} = k/V_{in}$ and the equation for SQNR has a constant value which is independent of the amplitude of V_{in}, which is the required result: i.e.

$$SQNR = 3M^2 \qquad (3.21)$$

Unfortunately such a characteristic cannot be used in practice because when $V_{in} = 0$ then $V_{out} = (1 + k\ln 0) = -\infty$. Under these circumstance a practical approximation to the theoretical compression characteristic is required.

3.6 PRACTICAL COMPRESSION CHARACTERISTICS

Two practical characteristics which overcome this disadvantage are the A law characteristic (Europe) and the μ law characteristic (North America). The A law characteristic is divided into two regions and is given by

$$V_{out} = \frac{AV_{in}}{1 + \ln A} \qquad \text{for} \quad 0 \leqslant |V_{in}| \leqslant \frac{1}{A} \quad \text{(linear)}$$

$$V_{out} = \frac{1 + \ln(AV_{in})}{1 + \ln A} \qquad \text{for} \quad \frac{1}{A} \leqslant |V_{in}| \leqslant 1 \quad \text{(logarithmic)} \qquad (3.22)$$

The linear region ensures that $V_{out} = 0$ when $V_{in} = 0$ and the logarithmic region is specified so that $|V_{out}| = 1$ when $|V_{in}| = 1$, the characteristic is continuous at $|V_{in}| = 1/A$. A is known as the compression coefficient and for large values of A the characteristic is predominantly logarithmic. The characteristic is shown in Fig. 3.15.

The relationship between input coder (non-linear) and output coder (linear) is derived for both regions. For the linear region

$$dV_{out} = \frac{AdV_{in}}{1 + \ln A} \quad \text{or} \quad dV_{in} = \frac{dV_{out}(1 + \ln A)}{A}$$

For the logarithmic region

$$dV_{out} = \frac{1}{1 + \ln A} \times \frac{dV_{in}}{V_{in}} \quad \left(\text{since} \frac{d}{dx}\{\ln(ax + b)\} = \frac{a}{ax + b} \right)$$

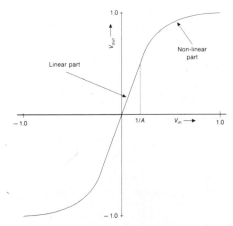

Fig. 3.15 The A law compression characteristic.

or

$$dV_{in} = dV_{out}(1 + \ln A)V_{in}$$

Since the characteristic is normalized with a range of ± 1 then $dV_{out} = M/2$. When the input waveform is in the linear region of the characteristic the quantization noise produced is

$$\sigma^2_{lin} = \frac{(dV_{in})^2}{12} = \frac{\{1 + \ln A\}^2}{3M^2 A^2} = \frac{k^2}{A^2} \tag{3.23}$$

When the input waveform is in the logarithmic region the quantization noise produced is

$$\bar{\sigma}^2_{log} = \frac{(\overline{dV_{in}})^2}{12} = \frac{(1 + \ln A)^2 \bar{V}^2_{log}}{3M^2} = k^2 \bar{V}^2_{log} \tag{3.24}$$

In this case $\bar{V}^2_{log}$ is the mean square value of the input waveform when in the logarithmic region. In practice the input waveform will occupy both linear and logarithmic regions of the characteristic. In order to determine the mean quantization noise it is necessary to know the probability density function of the input waveform. The amplitude probability density function of a typical speech waveform may be approximated by

$$p_v(V) = \frac{1}{\sqrt{2}\sigma_s} \exp\left(\frac{-\sqrt{2}|V|}{\sigma_s}\right) \tag{3.25}$$

where σ_s is the normalized rms voltage of the speech waveform. The probability that the speech waveform is within the linear region is

$$P_{lin} = \int_{-1/A}^{1/A} p_v(V)\,dV$$

or

$$P_{lin} = 2\int_0^{1/A} \frac{1}{\sqrt{2}\sigma_s} \exp\left(\frac{-\sqrt{2}|V|}{\sigma_s}\right)dV = 1 - \exp\left(\frac{-\sqrt{2}}{A\sigma_s}\right)$$

The mean square value when the waveform is in the logarithmic section is

$$\bar{V}^2_{log} = 2\int_{1/A}^1 V^2 p_v(V)\,dV$$

Integrating by parts and assuming that $A \gg 1$ gives

$$\bar{V}^2_{log} = \left(\frac{1}{A^2} + \frac{\sqrt{2}\sigma_s}{A} + \sigma^2_s\right)\exp\left(\frac{-\sqrt{2}}{A\sigma_s}\right)$$

The total quantization noise is

$$\sigma^2_n = \bar{\sigma}^2_{lin} + \bar{\sigma}^2_{log} = k^2\left[\frac{1}{A^2}P_{lin} + \bar{V}^2_{log}\right]$$

and the signal-to-quantization-noise ratio becomes

$$SQNR = \frac{\sigma^2_s}{\sigma^2_n}$$

Note that if $A \gg 1$ then $P_{\text{lin}} \ll 1$ and $\bar{V}_{\text{log}}^2 \approx \sigma_s^2$, i.e.

$$\text{SQNR} \approx \frac{\sigma_s^2}{k\sigma_s^2} = \frac{1}{k^2}$$

Thus for large values of A the SQNR is constant and independent of the value of V_{in}. The full equation for SQNR is

$$\text{SQNR} = \frac{\sigma_s^2}{k^2 \left[\frac{1}{A^2}\left\{1 - \exp\left(\frac{-\sqrt{2}}{A\sigma_s}\right)\right\} + \left\{\frac{1}{A^2} + \frac{-\sqrt{2}\sigma_s}{A} + \sigma_s^2\right\}\exp\left(\frac{-\sqrt{2}}{A\sigma_s}\right)\right]}$$

(3.26)

When $\sigma_s = 1/A$ then

$$\text{SQNR} = \frac{1/A^2}{k^2\left[1/A^2\,(0.76) + 1/A^2\,(1 + \sqrt{2} + 1)\,(0.24)\right]} = \frac{1}{1.58k^2}$$

But $10\log_{10}(1/1.58) = -2\,\text{dB}$, hence the effect of the linear part of the characteristic is to cause a drop in SQNR by 2 dB for input levels below $1/A$. Thus if $A \gg 1$ the input waveform will be in the logarithmic section for most of the time and the SQNR will be approximately constant. The SQNR characteristic for the A law compression characteristic is shown in Figure 3.16.

The system is thus designed so that the rms voltage produced by the quietest talker is equal to $1/A$, which will produce an effective constant SQNR for all users. The SQNR is a function of both A and M which are in turn chosen to give an acceptable performance for a specified dynamic range. The required dynamic range is determined by measurements on a sample of the population. Such measurements reveal that 98% of the population have an rms voice amplitude within $\pm 13\,\text{dB}$ of the median talker, and 99.8% of the population have an rms amplitude within $\pm 17\,\text{dB}$ of the median talker. The ratio of the rms output of the loudest talker to the quietest talker is known as the useful volume range (UVR) and will thus be between 26 dB and 34 dB, depending on which statistic is used.

Further measurements show that talkers have an output amplitude within $\pm 13\,\text{dB}$ of their individual rms values for 99% of the time. The required

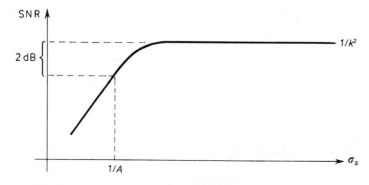

Fig. 3.16 SQNR characteristic for A law compression.

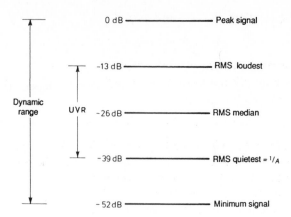

Fig. 3.17 Dynamic range of A law compression.

dynamic range of the coder will therefore exceed the UVR by approximately 26 dB as shown in Fig. 3.17. In this figure the peak signal has been normalized to 0 dB, hence it may be seen that the required value of the compression coefficient is

$$A = \text{UVR} + 13\,\text{dB}$$

Thus if UVR = 30 dB, then $20\log_{10}(1/A) = -43$ i.e. $A = 141$.

The CCITT recommended value of A is 87.6 which gives a UVR of 26 dB. With this value of A and an 8 bit code ($M = 256$) the signal-to-quantization-noise ratio is 38 dB. The UVR may be increased at the expense of SQNR, e.g. for a UVR of 30 dB ($A = 141$) and an 8 bit code then SQNR = 36 dB. In order to give perspective to non-linear compression of voice waveforms it is appropriate to compare the A law compression with linear coding throughout (no compression). To do this it is necessary to define the maximum input voltage which may be handled by the linear coder, in terms of probability. The probability that the input voltage will be within the range of the coder is

$$P_{\text{coder}} = \int_{-I/2}^{I/2} p_v(V)\,\mathrm{d}V$$

i.e.

$$P_{\text{lin}} = 2\int_0^{I/2} \frac{1}{\sqrt{2}\,\sigma_s}\exp\left(\frac{-\sqrt{2}|V|}{\sigma_s}\right)\mathrm{d}V = 1 - \exp\left(\frac{-I}{\sqrt{2}\sigma_s}\right)$$

The maximum input is defined in terms of the probability of being within range of the coder 98% of the time; i.e.

$$0.98 = 1 - \exp\left(\frac{-I}{\sqrt{2}\sigma_s}\right)$$

from which $\sigma_s = 0.18I$. The SQNR for the loudest talker is thus

$$\text{SQNR} = \sigma_s^2 \cdot \frac{12}{\partial v^2} = (0.18P)^2 \frac{12M^2}{I^2}$$

i.e.

$$\text{SQNR} = 0.388M^2$$

To match the performance of the companded system the SQNR for the quietest talker $= 38\,\text{dB}$. It follows that the SQNR for the loudest talker will be $38 + 26 = 64\,\text{dB}$. Hence

$$10\log_{10} 0.388 + 20\log_{10} M = 64$$

but $M = 2m$, hence $\text{SQNR} = -4.1 + 20\,m\log_{10} 2$, from which $m = 11.3$.

Thus 12 bits are required for a linear coder to have the same performance as the 8 bit coder with A law compression. Hence companding produces a saving in transmission bandwidth of the order of 33%.

The actual A law compression characteristic used in practice is shown in Fig. 3.18. In effect the coder is divided into 14 segments, 7 for positive amplitudes and 7 for negative amplitudes. Within these segments linear quantization is employed, the step size varying with segment so that the step size in segment 7, for example, is 56 times the step size in segment 1. This produces a similar quantization noise characteristic to that given in Eqn (3.26). The total number of input levels in this characteristic is 8192 (2^{13}) and it is clear from segment 1 that 64 input levels are transformed into 32 output levels. Hence the low-level input values are quantized at the equivalent of 12 bits linear quantization.

In the USA companding is carried out using the μ law characteristic which is given by

$$V_{\text{out}} = \ln(1 + \mu V_{\text{in}}) \quad \text{for} \quad 0 < |V_{in}| < 1 \tag{3.27}$$

A typical figure for μ is 255. It should be noted that when V_{in} is low $\ln(1 + \mu V_{\text{in}}) \approx \mu V_{\text{in}}$, hence this characteristic is also approximately linear for low input voltages.

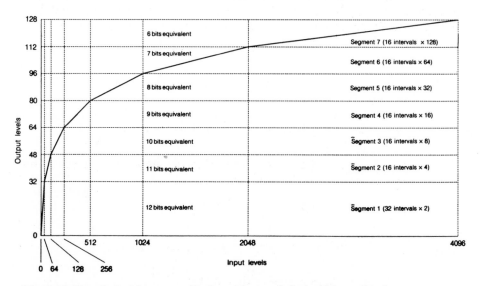

Fig. 3.18 Practical A law quantization characteristic (positive values).

3.7 NICAM

Near instantaneously companded audio multiplex (NICAM) is an alternative compression technique which is used for digital transmission of high-fidelity television stereophonic sound signals. It is essentially a bit reduction technique which is designed to maintain a constant signal-to-quantization-noise ratio over a wide dynamic range. In fact A law compression may also be considered as a bit reduction technique. This is illustrated in Fig. 3.19 which depicts an example of non-linear quantization with 5 to 4 bit compression. This is, in essence, a scaled down version of Fig. 3.18, in which the compression is from 12 to 8 bits.

In Fig. 3.19 the sampled waveform is first quantized linearly into 5 bits and the encoded signal is then processed into a non-linear form to reduce the number of bits to 4. At the receiver the 4 bit codes are converted back into the 5 bit equivalents. This means that the quantization noise for low-level amplitudes is determined by the full resolution of the 5 bit code. It is clearly not possible to reproduce the individual 5 bit code words when several such code words are compressed into a single 4 bit equivalent. In such cases the received 4 bit word is converted into a 5 bit equivalent near the centre of the range. Hence the resolution deteriorates (and quantization noise increases) for the higher signal levels. The advantage of this form of compression is that it is implemented with a simple conversion from 5 to 4 bits, and vice versa.

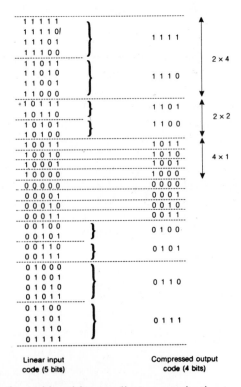

Fig. 3.19 Bit reduction achieved by non-linear quantization.

Table 3.1 Transmitted bits in NICAM

bits	1 MSB	2	3	4	5	6	7	8	9	10	11	12	13	14 LSB
Range 0	*					*	*	*	*	*	*	*	*	*
Range 1	*				*	*	*	*	*	*	*	*	*	
Range 2	*			*	*	*	*	*	*	*	*	*		
Range 3	*		*	*	*	*	*	*	*	*	*			
Range 4	*	*	*	*	*	*	*	*	*	*				

An alternative form of compression can be based on **range coding** in which the input waveform amplitude is divided into a fixed number of ranges. Different groups of bits are then transmitted for different ranges. This is the form of compression used in NICAM transmission and has a superior quantization noise performance, compared to A law, for the higher signal amplitudes. For low input amplitudes only the least significant bits are transmitted, while for high input signal amplitudes only the most significant bits are transmitted. In the case of NICAM, 14 bit resolution is used and this is compressed into 10 bits for transmission. The NICAM signal is divided into, five ranges as shown in Table 3.1. In addition to the bits shown in this table it is also necessary to transmit a 3 bit range code to define the range of the transmitted bits.

If a bit reduction is to be achieved the sum of amplitude bits and range code bits must be less than the uncoded bits. This is achieved by sending a 3 bit range code for a block of amplitude samples, rather than for each individual sample. This has the effect of adding a fraction of a bit per sample. The NICAM system uses a sampling frequency of 32 kHz and transmits one range code for a block of 32 samples. This represents a time interval of 1 ms and may be regarded as **nearly instantaneous** as far as the audio signal is concerned. In order to avoid clipping of the signal the range code corresponds to the largest amplitude in the 1 ms block. A schematic diagram of the NICAM coder is shown in Fig. 3.20.

If one 3 bit range code is transmitted for each 32 samples then this represents a degree of inefficiency as 3 bits can define 8 ranges. This inefficiency is reduced in NICAM by collecting 3×32 sample blocks, which will have $5^3 = 125$ range code combinations and transmitting these combinations as a 7 bit word. This produces an overall delay of 3 ms in the transmitted

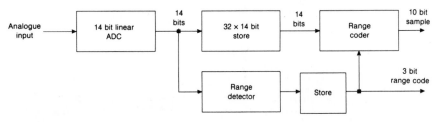

Fig. 3.20 NICAM range coder.

audio signal. Clearly it is important to avoid errors in the range code as this would produce a gross distortion of the reproduced voltage level. To minimize errors 4 parity check digits are added to the 7 bit range code to provide an error correction code with a Hamming distance of 4 (see Section 5.6).

The NICAM system effectively transmits the signal with 10 bits and the relative signal to noise ratio is maintained constant as the quantization noise varies with signal amplitude. It has been found that this system has a SQNR improvement of approximately 12 dB over the equivalent A law system with 10 bits, but requires a more complex coder and decoder. A 10 bit linear quantization would produce a SQNR of approximately 60 dB, but with range coding this is increased to a subjective equivalent of 80 dB.

3.8 PCM LINE SIGNALS

Although it is possible to encode signals at their source this is not in widespread use. TDM transmission with PCM encoding is used extensively between telephone exchanges. This transmission makes use of standard telephone lines that were designed specifically for voice communications. The standards of transmission adopted have been dictated largely by the characteristics of the cables 'already in the ground'. To minimize signal distortion over the audio frequency band these cables are artificially loaded by inductance at intervals of approximately 2 km. This produces very high attenuation of frequencies above about 4 kHz, which makes the line totally unsuitable for digital signals. In addition to the loss of response at high frequency the 2 km lengths of line are transformer coupled and therefore have no dc path. The high-frequency response of the section can be considerably improved by simply removing the loading coil. A typical frequency response for a section of loaded and unloaded cable is shown in Fig. 3.21.

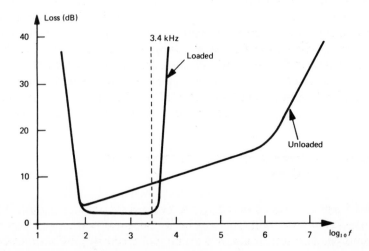

Fig. 3.21 Loss of a loaded and unloaded cable.

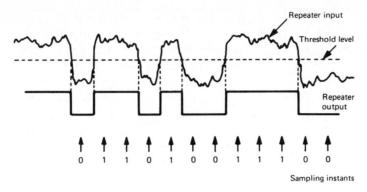

Fig. 3.22 Regenerative repeating of a PCM signal.

In practical PCM systems the loading coils are replaced by regenerative repeaters which effectively isolate each 2 km section of line. The regenerative repeater, in fact, produces a binary signal free from noise at the start of each section. This is shown in Fig. 3.22.

If the repeater input is above the threshold at a timing instant a binary 1 is transmitted to the next section. If the input is below the threshold a binary 0 is transmitted. If the SNR at each repeater input is adequate few decision errors occur and the binary signal is repeated free from noise. This is a distinct advantage of digital transmission over analogue transmission. In the latter case, amplifiers are required at intervals to compensate for signal attenuation. These amplifiers boost the noise as well as the signal. The absence of a dc path presents a serious problem for PCM signals. Its effect on a long sequence of binary 1s is to cause a gradual droop of signal level below the decision threshold, and this will of course produce decision errors. This is illustrated in Fig. 3.23a.

The solution is to remove the dc component in the PCM signal. The first stage in the process is to convert the full-width non-return-to-zero (NRZ) pulses into half-width return-to-zero pulses (RZ). This effectively doubles the signal bandwidth, but is necessary for synchronization purposes (Section 3.8). The dc component in the RZ waveform is removed by inverting alternate binary 1s, the process being called 'alternate mark inversion' (AMI). The original two-level PCM signal is actually transmitted as a three-level signal with zero dc component. The bandwidth of a PCM signal is calculated on the basis of AMI. The processing to produce AMI is illustrated in Fig. 3.23b.

3.9   BANDWIDTH REQUIREMENTS FOR PCM TRANSMISSION

The PCM signal consists of a random sequence of binary 1s and 0s. The transmitted waveform will have maximum bandwidth when the number of transitions per unit time is also a maximum. This corresponds, in the case of NRZ pulses, to an alternating sequence of binary 1s and 0s. The maximum

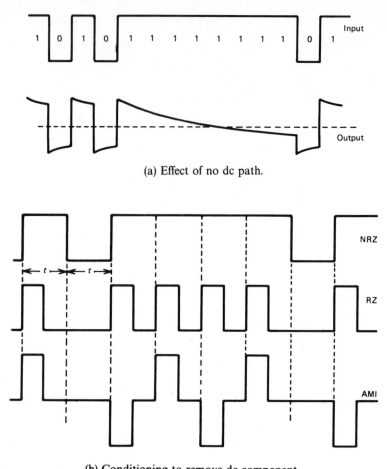

(a) Effect of no dc path.

(b) Conditioning to remove dc component.

Fig. 3.23 Transmission waveforms.

bandwidth in the case of RZ pulses and AMI occurs for a sequence of binary 1s. The transmitted waveforms and their respective amplitude spectra are given in Fig. 3.24.

The CCITT[2] recommendation specifies PCM transmission in terms of time division multiplexing of 32 channels. Each channel is sampled 8×10^3 times per second with each sample represented as an 8 bit code. This gives a total frame bit rate of 2.048 Mb/s. It may be seen from Fig. 3.24 that the minimum bandwidth requirement for AMI is approximately half the bit rate. This is an 'idealized' figure based on the premise that the three levels of the AMI signal could be extracted from a sine wave with a period of twice the bit interval. This bandwidth is insufficient in practice owing to distortion of the pulse waveform, which produces inter-symbol interference. To illustrate this point assume that the line (or channel) has an ideal low-pass response with cut off frequency f_c.

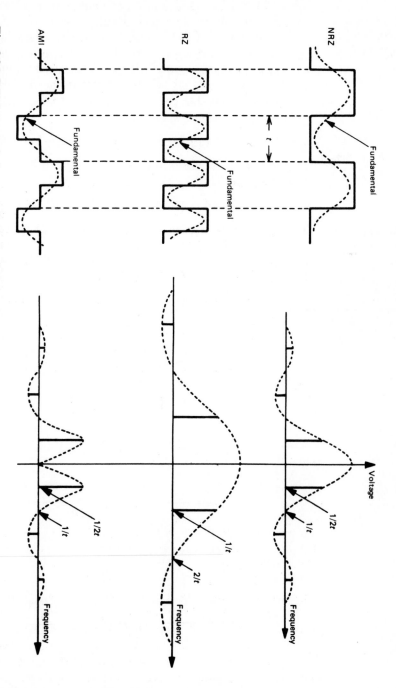

Fig. 3.24 PCM transmission waveforms and their spectra.

The impulse response of such a channel is a sinc function with zeros at intervals $t = n/2f_c$. Hence in theory pulses transmitted at a rate of $2f_c$ per second could be received free of interference from adjacent pulses. It was stated in Chapter 1 that an ideal response cannot be realized in practice; however, the raised cosine response is an approximation to the ideal filter which can be synthesized in practice. The raised cosine response is shown in Fig. 3.25 and it should be observed that the impulse response of such a characteristic also has zeros at intervals of $n/2f_c$.

The raised cosine response is given by

$$H(f) = 0.5\,[1 + \cos(\pi f / 2f_c)] \quad \text{for} \quad |f| < 2f_c \tag{3.28}$$

The impulse response of the network is given by the Fourier transform of Eqn (3.24) and is

$$h(t) = 2f_c\,\text{sinc}\,(2\pi f_c t) \cdot \cos\,(2\pi f_c t)/[1 - (4f_c t)^2] \tag{3.29}$$

Thus in order to transmit pulses at a rate of $2f_c$ per second a raised cosine response with cut-off frequency $2f_c$ is required. (The raised cosine response is a linear phase response which may be approximated by Bessel polynomials[3].) Hence for AMI with a data rate of 2.048 Mb/s the overall response between

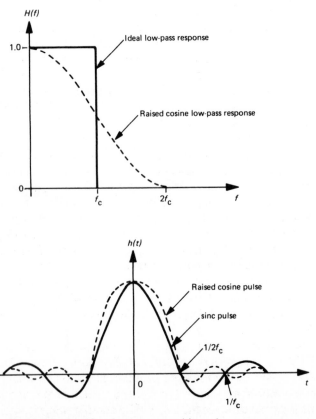

Fig. 3.25 Ideal and raised cosine frequency and impulse response.

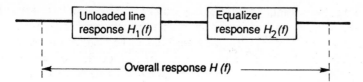

Fig. 3.26 Equalization of line response.

two repeaters should be a raised cosine response with a cut-off frequency of 2.048 MHz. This is achieved by use of equalization as shown in Fig. 3.26.

The overall response $H(f) = H_1(f) H_2(f)$ is designed to be a linear phase approximation of the ideal low-pass filter, i.e. to have a raised cosine amplitude response. The equalizer response $H_2(f)$, which may be realized with digital filters, is thus designed to produce this response.

3.10 SYNCHRONIZATION OF PCM LINKS

A PCM link is essentially made up of many sections, each one being terminated by a regenerative repeater. In such a system it is essential that each repeater operates at the same clock rate as there is no provision for data storage. It is possible to synchronize each repeater by a clock signal inserted at one end of the link, but this would reduce the available bandwidth for data transmission. A more efficient technique is to extract the clock signal for each repeater from the data signal itself. The clock signal is derived from the AMI waveform by full-wave rectification to produce a RZ spectrum. This is necessary because, as illustrated in Fig. 3.24, the AMI waveform has no component at the data rate. The component at t^{-1} is then extracted, possibly by use of a phase locked loop, for repeater timing.

A practical problem arises with timing if a series of binary 0s occurs, as would be the case during pauses in normal speech. To maintain repeater timing during such situations a code known as HDB3 is employed. This code limits the maximum number of successive 0s transmitted in AMI format to three. When four successive 0s occur in the binary (NRZ) signal the AMI waveform, which would be zero, is replaced by a three-level code $(-0+)$. The actual code substituted depends upon the AMI polarity of the previous 1. The receiver must be able to recognize the HDB3 code, and to make this possible the transmitter produces what is termed a bipolar violation. When there is no HDB3 code adjacent 1s in the AMI waveform will have opposite polarity. If adjacent 1s have the same polarity the inclusion of the HDB3 code is detected and hence decoded.

When four successive zeros occur in the binary signal, one of four possible HDB3 line signals is transmitted. The possible three-level codes are $000+, 000-, -00-, +00+$. The actual code transmitted depends on the polarity of the preceding binary 1 and also on whether the number of binary 1s which have occurred since the last HDB3 code (bipolar violation) is odd or even. The substitutions used are given in Table 3.2.

Table 3.2

Polarity of preceding pulse	Number of pulses since last bipolar violation	
	ODD	EVEN
+	000 +	− 00 −
−	000 −	+ 00 +

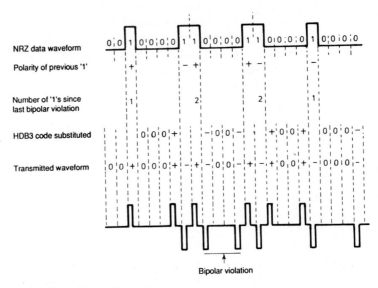

Fig. 3.27 HDB3 coding of line signals.

An example of HDB3 coding is shown in Fig. 3.27 In addition to repeater synchronization it is also necessary to synchronize the multiplexers.

In the CCITT specification two of the 32 PCM channels are reserved for signalling and synchronization (see Section 10.33). The channels are numbered 0 to 31, the 32 channels being called a 'frame'. Frame synchronization is achieved by transmitting a fixed code word in channel 0 on alternate frames. Circuits at the receiver search for this code word and its absence in alternate frames and derive from it a synchronizing signal for the demultiplexer. In this way each 8 bit word in the received frame is routed to its correct destination.

3.11 DELTA MODULATION

This is an alternative binary transmission system using a single digit binary code. Delta modulation does not have the widespread application of standard PCM, it is however used in some rural telephone networks[4] and in digital recording of analogue signals. The fundamental delta modulator, or tracking coder is shown in Fig. 3.28.

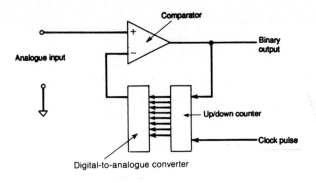

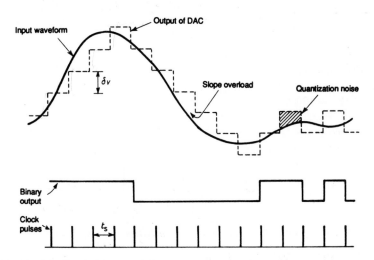

Fig. 3.28 The idealized delta modulator.

The analogue input is compared with the output of a DAC the input of which is derived from an up/down counter. If the amplitude of the analogue input exceeds the output of the DAC the comparator output will be high. This sets the up/down counter to increment on the next clock pulse. If the output of the DAC exceeds the amplitude of the input the comparator output will be low. This sets the up/down counter to decrement on the next clock pulse. The output of the DAC is thus a staircase approximation of the analogue input. The demodulator will consist of the elements in the feedback loop of the modulator. It will be noted from Fig. 3.28 that there are two kinds of distortion produced by this system. Slope overload distortion occurs when the transition from one step to the next fails to cross the input waveform. Quantization distortion occurs due to the finite step size δv.

If t_s is the clocking interval the condition required to prevent slope overload is

$$\dot{h}(t)t_s < \delta v \qquad (3.30)$$

For sinusoidal waveforms $h(t) = A \cos 2\pi f_m t$ and thus $\dot{h}(t)_{max} = A2\pi f_m$, the condition to prevent slope overload is then $A2\pi f_m < \delta v f_s$. Alternatively

$$A_{max} = \frac{\delta v f_s}{2\pi f_m} \tag{3.31}$$

where f_s is the clocking frequency. It may be noted that either δv or f_s can be increased to avoid overload, but with some penalty. Increasing f_s increases the transmitted bit rate (and hence bandwidth requirement), increasing δv increases the quantization error.

In practice a much simpler circuit than that of Fig. 3.28 is used, which has the additional advantage that the effect of any digit errors decreases to zero after a given interval. (This would not be the case for the circuit of Fig. 3.28.) The practical circuit is based on a simple RC integrator and is shown in Fig. 3.29. The circuit will function if the output of the flip-flop is $\pm V$ volts (which may be easily achieved with a CMOS device).

The voltage across the capacitor will be a series of positive and negative exponential decays since the capacitor charges from either $+V$ or $-V$. The capacitor voltage for a typical input waveform is illustrated in Fig. 3.29. The error voltage in this figure is the difference between the input voltage and the capacitor voltage. This is approximately triangular when the input signal is zero and this is known as the idling voltage.

Considering a single RC network in which V is the charging voltage and v is the instaneous voltage across the capacitor, then

$$\frac{V - v}{R} = C\frac{dv}{dt} \text{ and integrating gives } e^{-t/RC} = \frac{V - v}{V - v_i}$$

where v_i is the initial capacitor voltage. Letting $v_i = -V$ (i.e. logic 0) the

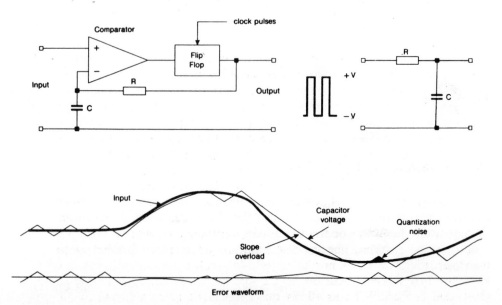

Fig. 3.29 Practical RC delta modulator and waveforms.

capacitor voltage is

$$v = V(1 - 2e^{-t/RC})$$

To avoid overload the slope of the capacitor voltage must be equal to or greater than the slope of the input waveform. If it is assumed that the input $h(t) = A\cos(2\pi f_m t)$ the value of $|\dot{h}(t)|$, when $h(t) = v$, must be less than the slope of the capacitor voltage.

$$|\dot{h}(t)| = A2\pi f_m \sin(2\pi f_m t)$$

or

$$\dot{h}(t) = A2\pi f_m [1 - \cos^2(2\pi f_m t)]^{1/2}$$

but when $h(t) = v$ then $\cos(2\pi f_m t) = v/A$ which means

$$|\dot{h}(t)| = 2\pi f_m (A^2 - v^2)^{1/2}$$

Thus to avoid slope overload

$$\frac{V - v}{RC} > 2\pi f_m (A^2 - v^2)^{1/2} \qquad (3.32)$$

The difference between the slope of the capacitor voltage and the input waveform is

$$D = \frac{(V - v)}{RC} - 2\pi f_m (A^2 - v^2)^{1/2}$$

If overload is to be avoided then D must be positive and in the limit $D \rightarrow 0$. D changes from a positive to negative value when $dD/dv = 0$: i.e.

$$\frac{dD}{dv} = -\frac{1}{RC} + \frac{2\pi f_m}{2}(A^2 - v^2)^{-1/2} 2v = 0$$

Thus

$$\frac{1}{RC} = \frac{2\pi f_m v}{(A^2 - v^2)^{1/2}}$$

From which

$$v = \frac{A}{[1 + (f_m/f_0)^2]^{1/2}} \quad \text{where } f_0 = \frac{1}{2\pi RC}$$

Substituting this into Eqn (3.32) gives

$$A_{max} = \frac{V}{[1 + (f_m/f_0)^2]^{1/2}} \qquad (3.33)$$

The overload characteristic has a frequency response equivalent to that of a single lag characteristic. The optimum value of f_0 is chosen with reference to the amplitude spectrum of normal speech. In practice it is desirable to work as close to overload as possible, and a suitable choice for f_0 is 150 Hz as illustrated in Fig. 3.30. This allows maximum SNR to be achieved at the expense of some overload in the mid-frequency range.

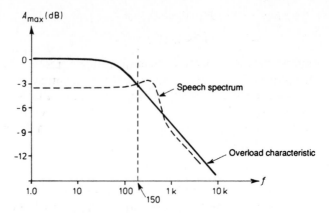

Fig. 3.30 Overload characteristic of RC delta modulator.

3.11.1 Dynamic range and quantization noise

The demodulator in the practical realization is simply the RC network. The quantization noise produced will then be the difference between the input waveform and the voltage across the capacitor of the RC network. When the input signal is zero the error signal is an approximate triangular waveform of period $1/2f_s$ and peak-to-peak amplitude δv. The slope of this waveform at $v = 0$ is $dv/dt = V/RC$ hence $\delta v = V/RC f_s$. This error waveform is shown in Fig. 3.29 and clearly the coder will not deviate from this idling pattern unless the signal exceeds the value $\delta v/2$. The minimum signal level corresponding to the threshold of coding is thus

$$A_{\min} = \frac{V}{2RCf_s} \qquad (3.34)$$

The coding range of the delta modulator is thus

$$\frac{A_{\max}}{A_{\min}} = \frac{V}{[1 + (f_m/f_0)^2]^{1/2}} \frac{2RCf_s}{V}$$

i.e.

$$\text{coding range} = \frac{2RCf_s}{[1 + (f_m/f_0)^2]^{1/2}} \qquad (3.35)$$

The peak-to-peak error is

$$\bar{\varepsilon}^2 = \frac{2}{t_s} \int_0^t \left(\delta v \frac{t}{t_s} \right)^2 dt = \frac{(\delta v)^2}{12}$$

This is the mean square quantization noise output when there is zero input. The error when the input is a sine wave is shown in Fig. 3.29 and it has been shown[5] that the mean square error in this case is numerically equal to

$$\bar{\varepsilon}^2 = \frac{(\delta v)^2}{6} \qquad (3.36)$$

This is constant from the threshold of coding to overload. The error wave-form is actually random but approximately triangular in shape. The power spectrum may thus be approximated by the spectrum of a single triangular pulse (via the autocorrelation function), which is a sinc^2 function with spectral zeros occurring at frequency intervals of $n/2f_s$. The total area under the sinc^2 function is equal to the area under a rectangular figure of the same zero frequency amplitude extending to a frequency of $1/3t_s$, as shown in Fig. 3.31. The total power in the error waveform can thus be represented as $G(f)/3t_s$ where $G(f)$ is a uniform power spectral density, hence

$$\bar{\varepsilon}^2 = G(f)/3t_s \qquad (3.37)$$

If the detector output is low-pass filtered, the cut-off frequency of the filter being f_1, then provided $f_1 < 1/3t_s$ the output quantization noise power will be

$$N_q = G(f)f_1 = \frac{3\bar{\varepsilon}^2 f_1}{f_s} = \frac{(\delta v)^2 f_1}{2f_s}$$

But $\delta v = 2\pi V f_0/f_s$, thus

$$N_q = 2\pi V^2 f_1 (f_0)^2 \qquad (3.38)$$

The maximum signal power which the system can handle before overload is

$$S = \frac{(A_{\max})^2}{2} = \frac{V^2}{2[1 + (f_m/f_0)^2]}$$

hence

$$\text{SQNR}_{\max} = \frac{V^2 (f_0)^2}{2[(f_0)^2 + (f_m)^2]} \frac{(f_s)^3}{2\pi^2 V^2 f_1 (f_0)^2}$$

Assuming, on average, that $f_m \gg f_0$ then

$$\text{SQNR}_{\max} = \frac{(f_s)^3}{4\pi^2 f_1 (f_m)^2} \qquad (3.39)$$

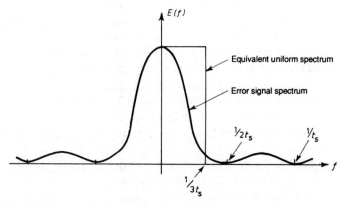

Fig. 3.31 Error signal power spectrum.

This is an approximate result for single tone inputs. With speech waveforms it is found that overload is minimized if the instantaneous speech amplitude is limited to the amplitude of a single tone of frequency 800 Hz, which could be transmitted without overload. Thus if $f_m = 800$ Hz, $f_1 = 3.4 \times 10^3$ and $f_s = 64 \times 10^3$ (which is equivalent to 8 bit linearly quantized PCM) the SQNR is 35 dB. The corresponding figure for 8 bit linearly quantized PCM, obtained from Eqn (3.16) is 49.8 dB. It is usually assumed that the minimum acceptable signal to quantization noise for telephone transmission is about 26 dB. This would give delta modulation a dynamic range of only 9 dB as compared with a figure of 23.8 dB for linearly quantized PCM. (Both of these figures are far short of companded PCM with a UVR of 26 dB at a SQNR of 38 dB.)

The dynamic range of delta modulation may be increased by a form of companding. In the case of the delta modulator the slope of the voltage across the integrating capacitor is varied with input signal amplitude. This gives rise to continuously variable slope delta modulation (CVSDM). Essentially the output of the delta modulator is monitored and when several successive pulses have the value '1' or '0' (indicating that slope overload is occurring) the amplitude of the charging voltage V is increased thereby increasing the slope of the integrator output. When the output returns to 010101 overload is removed and the step size may be progressively reduced. The analysis of several forms of CVSDM is covered in Steele.[6]

3.12 DIFFERENTIAL PCM

A number of analogue waveforms such as speech and video exhibit the property of predictability. This means that the change in value between one sample and the next is small because the rate of change of the analogue waveform is usually low compared with the sampling frequency. Alternatively waveforms such as speech and video have 'instantaneous frequencies' considerably lower than the maximum frequency component on which the sampling frequency is based. The next sample in a waveform can thus be predicted from a knowledge of previous samples. There will be some prediction error, but the peak-to-peak value of the error will be considerably less than the peak-to-peak value of the original waveform. Differential pulse code modulation (DPCM) capitalizes on this fact by coding and transmitting the prediction error. The prediction error requires fewer quantization levels for a given SNR and hence the required transmission bandwidth is less. A simple predictive coder and decoder are shown in Fig. 3.32.

In Fig. 3.32(a), $h(t)$ represents an input sample and $e(t)$ represents the difference between $h(t)$ and the previous sample weighted by the coefficient $a_0 (\leqslant 1)$. T is a delay equal to the sampling period. From this figure

$$e(t) = h(t) - a_0 h(t - T)$$

Considering the circuit of Fig. 3.32(b) it is evident that

$$h(t) = e(t) + a_0 h(t - T) \qquad (3.38)$$

This circuit is known as a predictor because the current sample is predicted

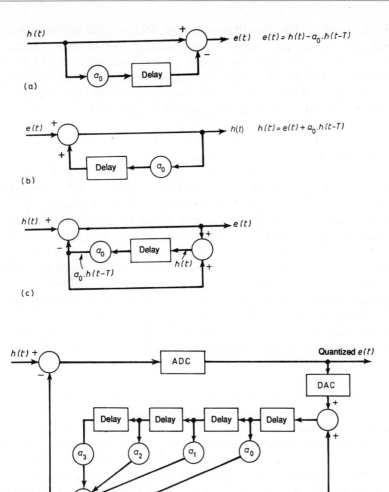

Fig. 3.32 Differential PCM.

from the previous sample and the error signal $e(t)$. The predictor is incorporated into the transmitter as shown in Fig. 3.32(c) and output $e(t)$ is known as the prediction error. There is an advantage in incorporating the predictor in the transmitter as the feedback loop minimizes quantization error when an analogue-to-digital converter is employed. The receiver is simply the feedback loop of the transmitter which, in this case, is the circuit of Fig. 3.32(b).

In practice the estimate of the predictor circuit is based on estimates of the previous four quantized samples as shown in Fig. 3.32(d). This figure also contains an analogue-to-digital converter in the forward path and a digital-to-analogue converter in the feedback path. As stated previously, the feedback action minimizes quantization error. The SQNR for differential PCM is between 5 and 10 dB higher than for PCM without differential coding. This SQNR may be increased further by use of adaptive DPCM (ADPCM). The quantizer step size (i.e. the ADC and DAC) is adapted according to the amplitude of the prediction error. Alternatively, with this type of coder, using a 4 bit quantizer, it is possible to transmit speech with the same quality as 64 kb/s A law compression at a bit rate of 32 kb/s.

3.13   DATA COMMUNICATIONS

PCM and delta modulation have been optimized for the transmission of coded voice signals over trunk routes. Data communications deals primarily with the transmission of digital signals between machines. The bulk of data communications now uses some form of packet switched network, several of which are described in Chapter 13. Traditionally data communications was via the public telephone network and techniques were developed for use on this particular medium and are considered in this section. The most common example of this form of communication is the connection of a terminal to a distant computer via a modem.

When considering data communications, one of the basic parameters that must be defined is the signalling speed. The unit of signalling speed is known as the **baud** after the telegraph engineer Baudot. The signalling speed in baud is in effect the rate at which pulses are transmitted over the communications link. These pulses need not be binary, which means that the data rate, which is usually expressed in bits/s (or b/s), does not necessarily equal the signalling speed.

Unlike PCM, which uses unloaded lines for transmission, data signals which are transmitted over normal telephone circuits must cope with the severely restricted frequency response of such lines. Special problems linked to this response therefore arise in data transmission, and these problems will be considered in some detail. The characteristics of the data signal produced by a VDU keyboard, for example, have two well-defined properties:

(1) low data rate, limited by human typing speed, and
(2) spasmodic output with long periods of no output at all.

To cope with the second of these characteristics, asynchronous communications is used, but it should be noted that dedicated high-speed data links use synchronous communication.

For transmission purposes, each symbol on a keyboard is represented by a unique binary code. The international standard code represents each character by a 7 bit word, i.e.

$$b_6 b_5 b_4 b_3 b_2 b_1 b_0$$

where b_n is a binary digit. Some examples of the 7 bit code are listed in Table 3.3. The 7 bit word has an additional digit, b_7, called a parity check bit, added for error-detection purposes. The parity bit is chosen so that the number of 1s in each 8 bit word is even. If an odd number of 1s occurs at the receiver, the receiver is aware that an error has occurred.

Table 3.3 International 7 bit code

Character	Binary	Octal	Hexadecimal
A	1000001	101	41
B	1000010	102	42
C	1000011	103	43
1	0110001	061	31
2	0110010	062	32
3	0110011	063	33

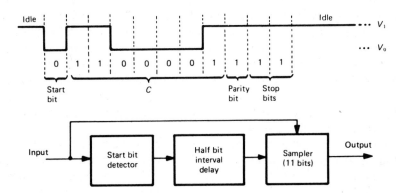

Fig. 3.33 Asynchronous transmission.

The asynchronous transmission system requires extra bits to allow the receiver at either end of the link to determine the beginning and end of each symbol. The transmission of each symbol is preceded by a level change from 1 to 0. The 0 level has a duration of 1 digit interval. The symbol is then transmitted serially, least significant digit (b_0) first, followed by the parity check bit (b_7). The end of each symbol is signalled by a binary 1 which lasts for two digit periods. The idle state (no signal) is thus equivalent to binary 1. A typical digit sequence (10 bits) is illustrated in Fig. 3.33. All timing is initiated by the falling edge of the start bit and the following digits are sampled at their mid-points. This means that timing clocks do not have to be closely matched as synchronization occurs on each start bit. A human operated VDU will produce a maximum of about 10 characters/s, which is, of course, very slow compared with the speed of operation of the computer to which it may be connected. If the transmitted pulses are binary, each pulse has a duration of 9.1 ms, which is equivalent to a signalling speed of 110 baud.

3.14 SPECTRAL PROPERTIES OF DATA SIGNALS

We must know something of the spectral properties of data signals before we can specify the most appropriate form of data transmission over telephone circuits. A typical data signal will consist of a random sequence of pulses of

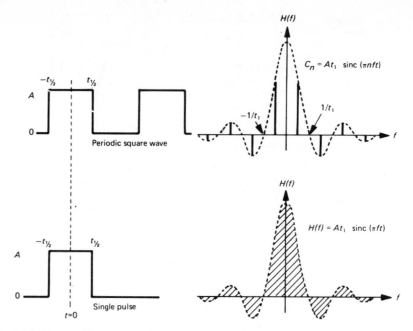

Fig. 3.34 Amplitude spectrum of data signals.

binary 1s and 0s. The power spectral density of such a signal was derived in Section 1.12 from its autocorrelation function. A random binary signal with pulse amplitudes of 0 or A volts and pulse duration t_1 seconds has an amplitude spectrum given by

$$H(f) = At_1 \operatorname{sinc}(\pi f t_1)$$

It can be seen from Fig. 3.34 that most of the energy in the spectral envelope is confined to frequencies below $f = 1/t_1$ hertz. The bandwidth of the data signal is therefore usually approximated by the reciprocal of the pulse width.

The data spectrum, which has a component at zero frequency must be modified for transmission over a telephone circuit which usually has a bandwidth from 300 Hz to 3.4 kHz. Further, since two-way signalling is required over a single circuit, it is necessary to differentiate between transmitted and received data signals. Both these requirements are met by modulating the data signal on to an audio frequency tone. The three possible forms of modulation are AM, FM and PM.

3.15   AMPLITUDE SHIFT KEYING (ASK)

This is the name given to AM when used to transmit data signals. It is not normally used on telephone lines because the large variations in circuit attenuation which can occur make it difficult to fix a threshold for deciding between binary 1 and 0. We shall, however, consider ASK in some detail because it is convenient to represent FM as the sum of two ASK signals.

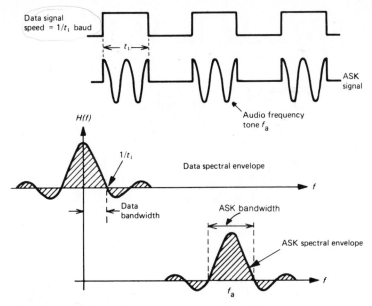

Fig. 3.35 ASK amplitude spectrum.

The ASK signal is generated by multiplying the data signal by an audio tone. This effectively shifts the data spectrum to a centre frequency equal to that of the audio tone. The process is shown in Fig. 3.35. The bandwidth of the modulated signal is twice the bandwidth of the original data signal. This means that the original 110 baud signalling rate requires a transmission bandwidth of 220 Hz using ASK:

$$\text{ASK} \equiv \text{DSBAM} \equiv (\text{carrier} + \text{upper and lower sidebands})$$

3.16 FREQUENCY SHIFT KEYING (FSK)

This is the binary equivalent of FM. In this case a binary 0 is transmitted as an audio frequency tone f_0 and a binary 1 is transmitted as a tone f_1. Hence the binary signal effectively modulates the frequency of a 'carrier'.

Although, strictly speaking, FSK is FM, it is more convenient to consider FSK as the sum of two ASK waveforms with different carrier frequencies. The spectrum of the FSK wave is thus the sum of the spectra of the two ASK waves. This spectrum is shown in Fig. 3.36. Using the FM analogy, it is possible to define a 'carrier frequency' $f_c = f_0 + (f_1 - f_1)/2$ and a 'carrier deviation' $\Delta f = (f_1 - f_0)/2$. The modulation index β is defined as $\beta = \Delta f/B$, where $B = 1/t_1$ is the bandwidth of the data signal. Using these definitions the bandwidth of the FSK signal is

$$B_{\text{FSK}} = 2B(1 + \beta) \tag{3.39}$$

This is similar to Carson's rule for continuous FM. Unlike analogue FM

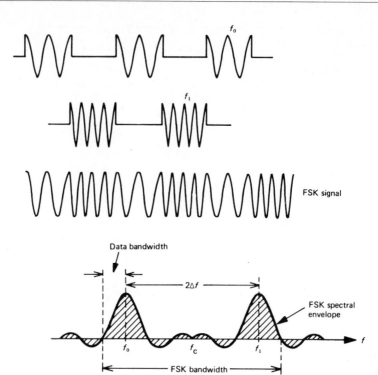

Fig. 3.36 FSK amplitude spectrum.

there is no advantage in increasing Δf beyond the value $\Delta f = B$ since the receiver only needs to differentiate between the two tones f_0 and f_1.

3.17 PHASE SHIFT KEYING (PSK)

This is the binary equivalent of PM, the binary information being transmitted either as zero phase shift or a phase shift of π radians. This is equivalent to multiplying the audio tone by either $+1$ or -1. The bandwidth is thus the same as for ASK. Since there is no dc component in the modulating signal, the carrier in the PSK spectrum will be suppressed. The equivalent modulating signal and the PSK spectral envelope are shown in Fig. 3.37. This form of PSK is sometimes referred to as binary PSK (BPSK) because the phase shift is restricted to two possible values, and it is equivalent to binary DSB-SC-AM.

3.18 PRACTICAL DATA SYSTEMS

We have already indicated the reason for not using ASK for data communications on the public telephone network. The choice between FSK and PSK is determined by the data rate. At low data rates FSK is employed for two-way (duplex) communication. A typical system operating at a signalling rate of

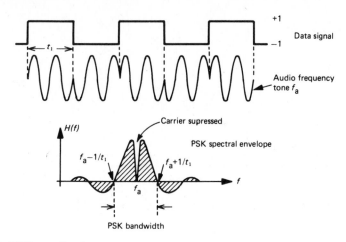

Fig. 3.37 PSK amplitude spectrum.

200 baud uses two tones of 980 Hz and 1180 Hz for binary 1 and 0 in one direction and 1650 Hz and 1850 Hz for binary 1 and 0 in the reverse direction. The incoming FSK is separated into two tones using bandpass filters. Envelope detection is then used to reproduce the binary signal. The combined modulator/demodulator (modem) is illustrated in schematic form in Fig. 3.38.

As the data rate is increased higher carrier frequencies are required; otherwise, each data interval would contain very few cycles of carrier, which would make detection extremely difficult. There is a limit on carrier frequency imposed by the upper cut-off frequency of the telephone line. For this reason

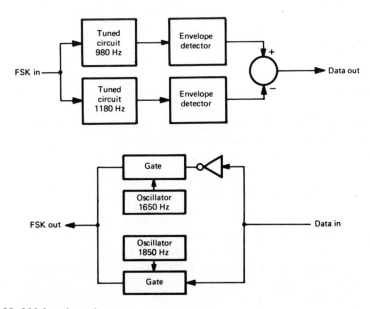

Fig. 3.38 200 baud modem.

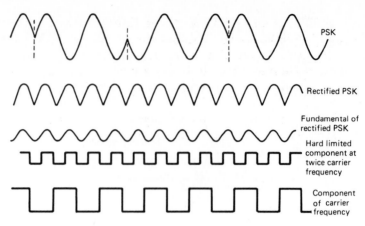

Fig. 3.39 Coherent detection of PSK.

FSK is limited to signalling speeds up to 600 baud. At speeds above this PSK is employed. This type of modulation makes more efficient use of bandwidth but requires more sophisticated coherent detectors. The reference signal for coherent detection is derived from the PSK signal itself. Since the carrier is suppressed in the PSK spectrum the received waveform is first rectified to produce a component at twice carrier frequency. This component is then limited and divided by two to produce the required reference signal. The required signal processing is illustrated in Fig. 3.39.

In the public telephone network any connection between transmitter and receiver will be made via several different paths which will contain several stages of frequency multiplexing and demultiplexing. Imperfections in the various stages of modulation result in random, slowly varying, phase shifts which are introduced into the PSK waveform. This results in phase ambiguity at the receiver and can produce data inversion.[6] The problem is greatly reduced if differential encoding is employed.

3.19 DIFFERENTIAL PHASE SHIFT KEYING (DPSK)

DPSK has the advantage of using the phase of the previous bit interval as the reference for the present bit interval. In order to make this possible, a binary 0

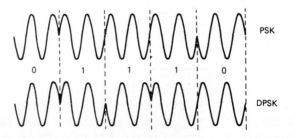

Fig. 3.40 Relationship between PSK and DPSK.

is transmitted as the same phase as the previous digit and a binary 1 is transmitted as a change of phase. The relationship between PSK and DPSK is shown in Fig. 3.40. The receiver compares the phase of the current digit with the phase of the previous digit. If they are the same the current digit is interpreted as a 0; otherwise it is interpreted as a 1. DPSK can be produced by pre-coding the data signal which then modulates the carriers as in standard PSK. If A_n is the present input to the encoder (A_n is binary) and C_{n-1} is the previous output the truth table for the encoder is

A_n	C_{n-1}	C_n
0	0	0
0	1	1
1	0	1
1	1	0

which will be recognized as the exclusive-OR operation

$$C_n = A_n \oplus C_{n-1}$$

We have already noted that pulses can be transmitted at a rate of $1/t_1$ without mutual interference over a channel of cut-off frequency t_1 hertz provided that the channel has a raised cosine frequency response. This applies to the unmodulated signal. A signalling rate of 1200 baud thus requires a raised cosine channel of bandwidth 1200 Hz. The PSK signal will require a band-pass channel with a raised cosine characteristic with a bandwidth of 2400 Hz. A typical PSK signal system would operate at a carrier frequency of 1.8 kHz and a signalling rate of 1200 baud. The bandwidth occupied by this waveform extends from 600 Hz to 3 kHz. Hence a data rate of 1.2 kb/s is an upper limit for BPSK.

3.20   ADVANCED MODULATION METHODS

The data rate of BPSK is sometimes expressed as 1 bit/baud. Since the baud rate is fixed by the channel characteristics the data rate can only be increased by increasing the number of levels per pulse beyond two. If each pulse has four levels the data rate becomes 2 bits/baud and each level can produce a unique phase shift. Thus it is possible to transmit data at a rate of 2.4 kb/s without any increase in bandwidth. It is convenient when considering multiphase PSK to represent the transmitted signal in terms of the sum of two quadrature audi frequency tones, i.e.

$$v_c(t) = a \cos 2\pi f t + b \sin 2\pi f t \tag{3.40}$$

Each of the levels in a four-level pulse can be represented by two binary digits called dibits. Thus it is not actually necessary in practice to produce a four-level pulse; instead the binary signal can be grouped into dibits and each dibit can be used to produce a unique phase shift in multiples of $\pi/2$. When interpreted in this way each dibit represents one pulse, i.e. the signalling rate

Table 3.4

Dibit	Phase shift	In-phase component	Quadrature component
		a	b
00	$\pi/4$	$+1$	$+1$
01	$3\pi/4$	-1	$+1$
11	$-3\pi/4$	-1	-1
10	$-\pi/4$	$+1$	-1

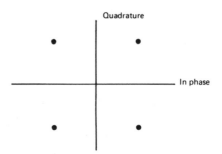

Fig. 3.41 Signal space diagram for QPSK.

in bauds equals half the bit rate. Table 3.4 lists the possible dibits and the values of a and b in Eqn. (3.40) necessary to produce the required phase shifts. The resulting quaternary PSK can be represented on a signal space diagram of the type shown in Fig. 3.41.

The data signal is recovered from the QPSK waveform by using two coherent detectors supplied with locally generated carriers in phase quadrature. The data rate can be increased further by increasing the number of levels of each pulse beyond four. For example, if the number of levels is increased to 16, it is possible to transmit data at a rate of 4 bits/baud. The QPSK signal was characterized by the fact that the coefficients a and b of Eqn. (3.40) always have the same magnitude, thereby producing a resultant of constant amplitude and varying phase. It is also possible for a and b to have different values and the resulting signal is in fact quadrature amplitude modulation. The detection of QAM is covered in Section 11.11 in connection with the transmission of chrominance signals in the PAL colour television system.

The signal space diagram for QAM with 16-1evel pulses is shown in Fig. 3.42. Each individual level is represented by a unique combination of a and b in Eqn (3.40). It is possible with this system to transmit data at a rate of 4.8 kb/s over a raised cosine bandpass channel, with a bandwidth of 2.4 kHz.

At these high data rates, intersymbol interference is a severe problem and elaborate equalization networks (transversal digital filters) are always employed. The lines used are not part of the public telephone network and are maintained to close tolerances in respect of loss and bandwidth. Such lines are often known as leased lines. The sophisticated equalization used on these lines means that signalling rates can approach the Nyquist rate, i.e. pulses can be

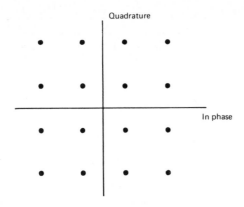

Fig. 3.42 Signal space diagram for 16-level QAM.

transmitted at rates approaching 2400 baud. This means that it is possible to transmit data rates up to 9.6 kb/s using 16-level QAM.

The leased lines referred to in the previous paragraph are basically high grade voice channels. Much higher data rates are possible on specially designed wideband links. These wideband links operate in a synchronous fashion and modern networks are adopting packet switching techniques to maximize the efficiency of usage of these links.

3.21   CONCLUSIONS

This chapter has introduced the basic concepts of digital communications and stressed the advantages of transmission of information in digital format. This is a huge growth area in telecommunications systems engineering and is likely to remain so for the foreseeable future. All-digital telephone networks, including cellular mobile telephones, are progressively being installed. These networks will gradually replace analogue systems and will provide enhanced services and increased reliability. A detailed treatment of packetized transmission is given in Chapter 13.

REFERENCES

1. Schwartz, M., *Information Transmission Modulation and Noise*, 3rd edn, McGraw-Hill, 1980.
2. CCITT, *Orange Book*, Vo. III-2, Recommendations G711, G712, G732.
3. Henderson, K. W. and Kautz, W. H., 'Transient Responses of Conventional Filters', *IRE Transactions on Circuit Theory*, Vol. 5, 1958, pp. 333–47.
4. Johnson, F. B., 'Calculating Delta Modulator Performance', *Trans. IEEE*, Vol. AU-16, No. 1 1968, pp. 121–9.
5. Betts, J. A., *Signal Processing Modulation and Noise*, The English Universities Press, 1972, Ch. 7.

6. Steele, R., *Delta Modulation Systems*, Pentech Press, 1975.
7. Oberst, J. P. and Schilling, D.L., 'Performance of Self Synchronized PSK Systems', *Trans. IEEE*, Vol. COM-17, 1969, pp. 666–9.

PROBLEMS

3.1 The sampling theorem is normally applied to signals with a low-pass spectrum. Show that this theorem can also be applied to signals with a bandpass spectrum. What is the statement of the theorem in this case?

  A signal with a bandwidth extending from 30 to 34 kHz is to be transmitted through an ideal channel with a low-pass characteristic. Determine the minimum theoretical cut-off frequency of the channel. Give a block diagram of the system and describe how the original signal may be reproduced at the receiver.

  Answer: 4 kHz.

3.2 A signal is used to amplitude-modulate a periodic pulse train in which the pulse width is much less than the pulse period. The sampled signal is to be reconstructed by a sample-and-hold circuit. Sketch the output of this circuit and, by considering its frequency response, find the ratio of the signal frequency amplitude to the amplitude of the lowest frequency distortion component. The signal is a single tone of frequency 2.8 kHz sampled at 8 kHz.

  Answer: 1.86:1.

3.3 Explain the difference between the actual bandwidth of a pulse and the bandwidth required for pulse transmission.

  Twenty-four speech signals each with a bandwidth of 0 to 4.5 kHz are to be transmitted over a line with a raised cosine frequency response by TDM. Calculate the minimum theoretical bandwidth of the line. Would this bandwidth be adequate in a practical system?

  Answer: 216 kHz.

3.4 An equalized line with a raised cosine response has an effective bandwidth of 200 kHz. Six speech channels are to be transmitted over this line using time multiplexed PCM. Assuming linear quantization and a sampling frequency of 8 kHz for each signal, what will be the average signal power-to-noise ratio at the decoder output. Assume quantization noise only is to be considered.

  Answer: 26 dB.

3.5 A single information channel carries voice frequencies in the range 50 Hz to 4.3 kHz. The channel is sampled at a 9 kHz rate and the resulting pulses may be transmitted either directly by pulse amplitude modulation (PAM) or by PCM.

  Calculate the minimum bandwidth required for the PAM transmission assuming a line with a raised cosine response.

  If the pulses are linearly quantized into eight levels and are transmitted as binary digits, find the bandwidth required to transmit the digital signal and compare it with the analogue figure. If the number of levels of quantization is increased to 128 what is the new bandwidth required? Calculate the increase in SNR at the decoder output, assuming the peak-to-peak voltage swing at the quantizer is 2 V.

  Answer: 9 kHz; 27 kHz; 63 kHz; 24 dB.

3.6 A PCM system employing uniform quantization and generating a 7 digit code is capable of handling analogue signals of 5 V peak-to-peak. Calculate the mean signal-to-quantizing noise ratio when the analogue waveform has a probability density function given by

$$P(v) = K \exp(-|v|) \quad -2.5 < v < 2.5, \quad = 0 \text{ else}$$

Assume uniform signal distribution within a given quantization interval.

Answer: 38.9 dB.

3.7   Derive an expression for the amplitude spectrum of a single triangular pulse of base width t seconds and amplitude A volts. Hence estimate its bandwidth.

Answer: $2/t$ Hz.

3.8   A data signal consists of a series of binary pulses occurring at a rate of 100 digits/s. This signal is to be transmitted over a telephone line, binary 1 being sent as a 1.5 kHz tone and binary 0 as a 2.8 kHz tone. What is the bandwidth of the transmitted signal?
      If the digit rate is increased to 1000 b/s what are the required upper and lower cut-off frequencies of the line in order that it may transmit this signal?

Answer: 1.5 kHz; 500 Hz; 3.8 kHz.

3.9   If the transmission of question 3.8 is by DPSK, what is the maximum data rate that can be transmitted over the telephone line? What is the optimum carrier frequency in this case?

Answer: 1650 b/s; 2.35 kHz.

3.10 Derive an expression for the amplitude spectrum of a FSK transmission when the digit stream is a series of alternate 1s and 0s.

3.11 A sinusoidal signal is switched periodically from 10 MHz to 11 MHz at a rate of 5000 times/s. Sketch the resulting waveform and identify the modulating signal if the switched sine wave is regarded as FSK.
      Find the approximate transmission bandwidth of the FSK signal and compare this with the bandwidth required if the modulating signal is approximated by a sine wave producing the same carrier deviation.

Answer: 1.02 MHz; 1.01 MHz.

4 Noise in analogue communications systems

4.1 INTRODUCTION

We have considered, in the previous chapters, various ways of transmitting information from one location to another. We concern ourselves now with the performance of these systems in a noisy environment. This comparative analysis will provide the insight required to determine the suitability, or otherwise, of using a particular form of transmission in a specific environment. The relative performance of various systems in noise is, of course, only one of the factors taken into account when choosing a particular method of information transmission. The economic considerations, e.g. cost, complexity, maintenance, are sometimes of paramount importance, but are outside the scope of this text.

Noise is defined as any spurious signal that tends to mask or obscure the information in the transmitted signal. The ratio of signal power to noise power at any point in a telecommunications system is known as the SNR, and the fundamental exchange possible between signal bandwidth and SNR is given by Shannon's law (see Section 1.14). In this chapter we will consider the performance of systems using analogue transmission and then discuss the significance of Shannon's law for the examples chosen.

4.2 PHYSICAL SOURCES OF NOISE

Noise is usually divided into naturally occurring noise and artificial noise. Artificial noise comes from various sources, the most important types being ignition interference, produced whenever sparks occur at electrical contacts, and crosstalk which is produced by inductive or capacitive coupling between one or more communications channels. Artificial noise can, in theory, be eliminated, although the cost of such elimination is often quite uneconomic. Natural noise is produced by many different phenomena, some examples being lightning discharges, thermal radiation and cosmic radiation. Natural noise cannot be eliminated and communication systems must perform efficiently in the presence of this type of noise, and often in the presence of artificial noise also.

The two most important types of natural noise are thermal noise and shot noise. Thermal noise is produced by random motion of charged particles in resistive materials and by thermal radiation from objects surrounding a telecommunications system, particularly one with an antenna. Shot noise is produced in semiconductor devices and results from the fact that currents flow across p–n junctions in finite quanta rather than continuously.

Thermal noise, caused by random motion of charged particles in resistive materials, produces a mean square noise voltage with a magnitude directly related to the temperature of the resistive material, which is given by

$$\bar{v}^2 = 4kTR\Delta f_n \quad (\text{volt})^2 \tag{4.1}$$

In this expression k is Boltzmann's constant $(1.38 \times 10^{-23}\,\text{J/K})$, T is the absolute temperature in kelvins, R is the resistance of the material in ohms, and Δf_n is the bandwidth of the measurement. The derivation of Eqn (4.1) is given by King.[1] Any noisy resistance can be represented by an equivalent circuit consisting of a noise-free resistance in series with a voltage source whose mean square amplitude is given by Eqn (4.1). This allows the determination of the noise properties of networks using conventional network theorems. We will consider one such calculation, of particular importance in telecommunications, to illustrate the procedures. We wish to calculate the noise power delivered by a resistance to a matched load, i.e. a load with the same resistive value. The circuit is shown in Fig. 4.1.

In this circuit R_1 is the noise source and R_2 is the load. Both source and load will produce a mean square noise voltage which will give rise to a mean square current $\bar{i}^2$. The power delivered by R_1 to R_2 is

$$P_n = \frac{\bar{v}_1^2 R_2}{(R_1 + R_2)^2} \tag{4.2}$$

If $R_1 = R_2$ (i.e. the source and the load are matched) the power delivered by R_1 to R_2 is

$$P_n = \frac{\bar{v}_1^2}{4R_2} = \frac{4kT\Delta f_n R_2}{4R_2} = kT\Delta f_n \quad \text{watts}$$

This is often written

$$P_n = \eta \Delta f_n \tag{4.3}$$

The constant η is the single-sided noise power spectral density and is independent of the range of frequencies Δf_n. If Δf_n is measured over both negative and positive frequencies the double-sided noise spectral density is

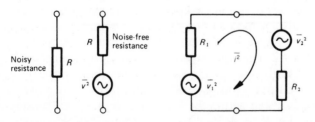

Fig. 4.1 Noise networks. (a) Equivalent circuit of a noisy resistance; (b) noise delivered to a matched load.

$\eta/2$ watts/hertz. This agrees with the Fourier analysis developed in Chapter 1, in which we indicated that half the total power was contributed each by the negative and positive components. When the noise spectral density is independent of frequency the noise is said to be white, by analogy with white light. When white noise is filtered by a frequency-selective network the resulting noise is known as coloured noise.

When thermal noise is derived from an antenna, the temperature of the noise source is not necessarily equal to the antenna temperature. The expression for the noise power delivered by an antenna to a matched load is $P_n = kT_a\Delta f_n$ watts. The effective noise temperature of the antenna T_a is related to the temperature of the radiating bodies surrounding the antenna and is determined by measurement. If the radiating bodies have the same temperature as the antenna, this is the value used for T_a. Antennas pointing into space generally receive far less radiation than antennas directed towards bodies on the Earth's surface and consequently have a much lower effective noise temperature.

4.3 NOISE PROPERTIES OF NETWORKS

All telecommunications systems are characterized by the fact that received signals are always accompanied by noise. The effectiveness of such a system is measured in terms of the ratio of signal power to noise at the system output. The SNR at any point in a telecommunications link is usually expressed in decibels:

$$SNR = 10\log_{10}[S_p/N_p] \qquad (4.4)$$

The minimum acceptable SNR for reliable communication is normally considered to be about 10 dB. Many systems operate at much higher ratios than this: the minimum SNR for telephone circuits is around 26 dB and for high-quality audio transmissions a figure in excess of 60 dB is typical. Some space systems operate with an SNR much less than 10 dB, but such systems require sophisticated techniques, such as correlation detection at the receiver.

All electrical networks generate noise and it will be clear, therefore, that when a signal passes through such a network the SNR at the network output will always be less than at the network input. The amount of extra noise generated by a network is specified by its noise figure. This is given the symbol F and is defined as

$$F = SNR_{in}/SNR_{out} \qquad (4.5)$$

A noiseless network has a noise figure of unity and it therefore follows that real networks always have a noise figure with a numerical value greater than unity. The noise figure of any network is derived in terms of the schematic diagram of Fig. 4.2. The input signal power is S_i and the network power gain is A_p (not necessarily restricted to values greater than unity); the signal power at the network output is thus $S_0 = A_pS_i$. If the input noise power is N_i and the noise power generated within the network is N_a the total noise power at the network output is $N_0 = A_pN_i + N_a$. The output SNR is

$$SNR_0 = A_pS_i/(A_pN_i + N_a)$$

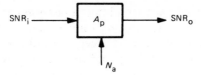

Fig. 4.2 Noise figure of a network.

i.e.

$$F = \frac{S_i}{N_i} \frac{(A_p N_i + N_a)}{A_p S_i} = \frac{A_p N_i + N_a}{A_p N_i} \tag{4.6}$$

The factor $A_p N_i$ is the output noise power of a noise-free network; thus

$$F = \frac{\text{total output noise power}}{\text{output noise power if network was noise free}} \tag{4.7}$$

We see from the definition given as Eqn (4.7) that F is not constant but is related to the noise power at the network input. The value of F is standardized by fixing the input noise power as that produced by a matched source at a standard temperature of 290 K. Use of noise figures in network calculations is thus only valid if the input noise power is $kT_s \Delta f_n$, where $T_s = 290$ K.

The usefulness of noise figures is demonstrated by considering the cascaded networks of Fig. 4.3. Using the definition of Eqn (4.7) we can write the overall noise figure for the cascaded network as

$$F = \frac{(A_{p1} N_i + N_{a1}) A_{p2} + N_{a2}}{A_{p1} A_{p2} N_i}$$

We assume that the networks are matched, i.e. the output resistance of the first network is equal to the input resistance of the second network, and that the input noise N_i is the noise produced by a matched source at 290 K. The noise figure of the first network is therefore

$$F_1 = (A_{p1} N_i + N_{a1})/A_{p1} N_i$$

In defining the noise figure of the second network the input power is also N_i; hence

$$F_2 = (A_{p2} N_i + N_{a2})/A_{p2} N_i$$

Thus the overall noise figure may be written

$$F = \frac{A_{p1} N_i + N_{a1}}{A_{p1} N_i} + \frac{N_{a2}}{A_{p1} A_{p2} N_i}$$

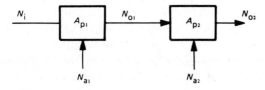

Fig. 4.3 Cascaded noisy networks.

But

$$F_2 - 1 = \left(1 + \frac{N_{a2}}{A_{p2}N_i}\right) - 1$$

Therefore

$$F = F_1 + \frac{F_2 - 1}{A_{p1}} \tag{4.8}$$

We see from Eqn (4.8) that if $A_{p1} \gg 1$, which is usually the case, the major contribution to the overall noise figure is produced by the first network. Evidently it becomes extremely important to ensure that the first network in any cascaded system has as low a noise figure as possible. Equation (4.8) can be expanded to include any number of cascaded networks. The equation for three networks in cascade is

$$F = F_1 + \frac{F_2 - 1}{A_{p1}} + \frac{F_3 - 1}{A_{p1}A_{p2}} \tag{4.9}$$

The effective noise temperature of a network is an alternative method of describing the noise performance of a network. This alternative is especially useful when considering low-noise networks or networks in which the input noise is not produced by a matched source at 290 K. In the latter case the use of noise figure is not valid. The effective noise temperature of a network is determined by replacing the noisy network by a noise-free network with an equivalent noise source at its input. The temperature of the equivalent noise source is chosen to make the noise at the output of the noise-free network equal to the noise at the output of the noisy network. Referring to Fig. 4.2, the noise produced by the network is replaced by an equivalent noise source of

$$N_a = kT_e\Delta f_n A_p$$

The factor $(kT_e\Delta f_n)$ is the noise delivered by an equivalent matched source at a temperature T_e. The temperature T_e is known as the 'effective noise temperature' of the network. If the value of $T_e \ll 290$ K the network itself contributes very little extra noise. The relationship between noise figure and effective noise temperature is

$$F = \frac{A_p N_i + N_a}{A_p N_i} = \frac{A_p k\Delta f_n(T_s + T_e)}{A_p k\Delta f_n T_s}$$

T_s is the standard temperature equal to 290 K. Hence

$$F = 1 + T_e/T_s \quad \text{or} \quad T_e = (F - 1)T_s \tag{4.10}$$

The maser microwave amplifier is an example of a very low noise network; the effective noise temperature of such a device would be between 10 and 30 K, which is equivalent to a noise figure of between 1.03 and 1.11.

It is often convenient to represent cascaded networks in terms of effective noise temperature; substituting Eqn (4.10) into Eqn (4.9) gives

$$T_e = T_1 + \frac{T_2}{A_{p1}} + \frac{T_3}{A_{p1}A_{p2}} \tag{4.11}$$

This equation is particularly useful when considering the noise performance of a cascaded system in which the first element is an antenna with an effective noise temperature not equal to 290 K. The cascaded system, excluding the antenna, is replaced by a noise-free system with an equivalent matched noise source at the input. The antenna noise is then included by adding the effective noise temperature of the antenna to the equivalent noise temperature of the noise source, and using the sum as the noise temperature of the matched source.

We indicated when referring to Fig. 4.2 that the power gain A_p was not necessarily greater than unity; when A_p is less than unity the network is passive and is usually characterized in terms of insertion loss rather than power gain. The insertion loss of a passive network is the reciprocal of power gain:

$$\text{insertion loss } L = \frac{\text{input power}}{\text{output power}} \tag{4.12}$$

When a passive network is matched at both the input and output the insertion loss has the same numerical value as the network noise figure. The noise power delivered to the network by a matched source is $kT_s\Delta f$ watts. The noise delivered by the network to its load will be the sum of the input noise power multiplied by the network power gain (which is less than unity for passive networks) and the noise power generated within the network. This latter component is represented, as in the determination of effective noise temperature, by an equivalent noise source at the input of the network which is then assumed noise free. The total noise power delivered to this noise-free network is $k\Delta f_n(T_s + T_e)$. The noise power delivered to the load is $N_0 = A_p k\Delta f_n(T_s + T_e)$. However, the network is a matched source for its load, and such a source delivers a noise power $N_0 = k\Delta f_n T_s$. Thus these two values must be equal or

$$A_p k\Delta f_n(T_s + T_e) = k\Delta f_n T_s$$

i.e.

$$T_e = T_s(1 - A_p)/A_p = T_s(L - 1) \tag{4.13}$$

But $T_e = (F - 1)T_s$. Thus for a matched passive network the insertion loss L has the same numerical value as the network noise figure, i.e. $F = L$. Both F and L are usually measured in decibels.

EXAMPLE: To illustrate the significance of the analysis presented in this section we will consider a typical system involving a domestic television receiver. The receiver, which has a video bandwidth of 5.5 MHz, is coupled via a 70 Ω coaxial cable with an insertion loss of 6 dB to an antenna with an effective noise temperature of $T_a = 290$ K. The noise figure of the receiver, referred to a matched source of 70 Ω at 290 K, is 6 dB. We are requested to find the SNR at the receiver output when the open-circuit signal voltage at the antenna terminals is 1 mV rms.

Each component in this system will cause a degradation of the SNR. The maximum value of SNR will occur at the input to the coaxial feeder. We begin by determining the SNR at this point and to do this we consider the antenna

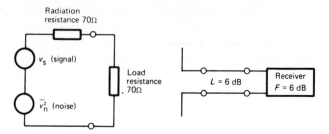

Fig. 4.4 Matched antenna.

as both a matched signal and noise source as shown in Fig. 4.4. We assume that the antenna has a source (see Chapter 7) resistance of $70\,\Omega$, and is therefore matched to the $70\,\Omega$ feeder. We also assume that the receiver has an input resistance of $70\,\Omega$.

Since the antenna open-circuit signal voltage is $1\,mV$ rms the voltage across the $70\,\Omega$ load will be half this value, i.e. $0.5\,mV$ rms. The signal power delivered to the coaxial feeder will be $S_p = (0.5 \times 10^{-3})^2/70$. The noise power delivered by the antenna, which acts as a matched noise source, is $N_p = k\Delta f_n T_a$. The SNR at the feeder input is thus

$$\mathrm{SNR} = 10\log_{10}S_p/N_p = 52.5 \quad \mathrm{dB}$$

We can now calculate the overall noise figure for receiver and feeder. This is valid in this example because the antenna effective temperature is 290 K. The overall noise figure is

$$F = F_1 + (F_2 - 1)/A_{p1}$$

F_1 is the noise figure of the feeder $= L = 6\,dB$, i.e. $F_1 = 3.98$. A_{p1} is the reciprocal of the feeder insertion loss and has a value of 0.251. F_2 is the receiver noise figure $= 6\,dB = 3.98$.

Hence the overall noise figure is $F = 15.85\,(12\,dB)$. The SNR at the receiver output will then be

$$\mathrm{SNR}_{out} = 52.5\,dB - 12\,dB = 40\,dB$$

This figure is actually lower than the minimum acceptable SNR for reasonable picture quality, which is usually considered to be 47 dB. The solution to this problem would be the use of a low-noise pre-amplifier between antenna and receiver. The pre-amplifier would have a typical power gain of 20 dB and a typical noise figure of 3 dB. It is possible to connect such an amplifier either directly to the antenna terminals (i.e. before the feeder) or directly to the receiver input (i.e. after the feeder). We suggest that the reader examines both cases and determines the output SNR when the pre-amplifier is included. It will be found that the output SNR will be 6 dB higher when the pre-amplifier is connected directly to the antenna terminals. This verifies the earlier conclusion that the first network in a cascaded system has the major effect on the overall noise figure.

In this example we represented the domestic television receiver as a

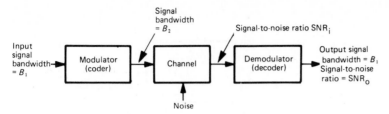

Fig. 4.5 General telecommunications system.

single network with an overall noise figure. In fact the receiver will be divided into several distinct functional blocks including RF and IF amplifiers, vision detector, video frequency amplifier, etc. From our consideration of the super-heterodyne receiver in Chapter 2 it is clear that the signal bandwidth at various points in a receiver (e.g. before and after the detector) is not necessarily constant. This means that possibilities exist for exchanging bandwidth for SNR at various points in a receiver, the theoretical relationship being given by Shannon's law. Thus although it is common practice to specify a single noise figure for a receiver it is more instructive to consider the individual sections of such a system.

A general telecommunications system can be represented in the form shown in Fig. 4.5. In this figure the bandwidth of the signal source is B_1 hertz. The signal enters the modulator and the bandwidth is changed to B_2 hertz. During the transmission noise is added to the signal. It is convenient to show this noise as a single input as in Fig. 4.5, but as we have shown, noise is actually added at every stage of a communications system. The SNR at the detector input is given the symbol SNR_i. At this point the detector transforms the bandwidth of the received signal back to its original value of B_1 hertz and in so doing produces an output signal-to-noise ratio SNR_0. Shannon's law states that whenever there is a change in signal bandwidth there should be an accompanying change in SNR. Shannon's law is, however, a theoretical law in which no allowance is made for any physical constraints that may exist in practical systems. We must therefore examine each transmission system on an individual basis and in order to do this it is necessary to develop an algebraic technique for specifying the effect of noise.

4.4 ALGEBRAIC REPRESENTATION OF BAND-LIMITED NOISE

Equation (4.3) indicates that the noise power delivered by a matched source is $\eta \Delta f_n$ watts. This means that the total noise delivered to the detector in a receiver of bandwidth B hertz is ηB watts. The algebraic representation of this noise is derived by dividing the bandwidth B into small elements Δf and approximating the noise power within each element by a cosine wave. As the element of bandwith $\Delta f_n \rightarrow 0$ this gives a very accurate representation of the noise signal. The technique is illustrated by Fig. 4.6. The noise voltage

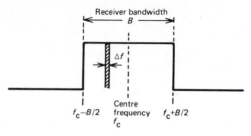

Fig. 4.6 Band-limited white noise.

produced in the elemental bandwidth Δf is represented as

$$n(t) = A_n \cos\left[2\pi f_k t + \theta_k(t)\right] \tag{4.14}$$

where $\frac{1}{2}A_n^2 = \eta \Delta f$ and f_k is the centre frequency of the interval Δf. The phase angle $\theta_k(t)$ is an arbitrary random number. The total noise voltage produced over the entire bandwidth B is calculated by summing the contributions of each element Δf, i.e.

$$V_n(t) = \sum_k A_n \cos\left[2\pi f_k t + \theta_k(t)\right] \tag{4.15}$$

It is convenient at this point to introduce the substitution $f_k = (f_k - f_c) + f_c$ where f_c is the centre frequency of the bandwidth B. Equation (4.14) then becomes

$$n(t) = A_n \cos\left[2\pi(f_k - f_c)t + \theta_k(t)\right] \cos(2\pi f_c t)$$
$$- A_n \sin\left[2\pi(f_k - f_c)t + \theta_k(t)\right] \sin(2\pi f_c t)$$

which means that Eqn (4.15) can be re-written

$$V_n(t) = x(t)\cos(2\pi f_c t) + y(t)\sin(2\pi f_c t) \tag{4.16}$$

where

$$x(t) = \sum_k (2\eta\Delta f)^{1/2} \cos\left[2\pi(f_k - f_c)t + \theta_k(t)\right] \tag{4.16a}$$

and

$$y(t) = -\sum_k (2\eta\Delta f)^{1/2} \sin\left[2\pi(f_k - f_c)t + \theta_k(t)\right] \tag{4.16b}$$

We can thus represent the noise voltage in terms of the sum of two amplitude modulated carriers in phase quadrature. The carrier amplitudes are the random variables $x(t)$ and $y(t)$ that have mean square values $\overline{x^2(t)}$ and $\overline{y^2(t)}$, respectively. The total noise power in the bandwidth B is

$$P = \frac{\overline{x^2(t)}}{2} + \frac{\overline{y^2(t)}}{2} \quad \text{watts} \tag{4.17}$$

but

$$\overline{x^2(t)} = \overline{y^2(t)} = \sum_k \frac{2\eta\Delta f}{2}$$

hence

$$P = \overline{x^2(t)} \quad \text{or} \quad \overline{y^2(t)} \quad \text{i.e.} \quad P = \eta B \quad \text{watts} \tag{4.18}$$

Having derived this representation of band-limited noise we can now examine the effectiveness of various signal transmission systems in the presence of noise. This will be accomplished by comparing the SNR after detection (or decoding) with the value that exists before detection.

4.5 SNR CHARACTERISTICS OF ENVELOPE-DETECTED DSB-AM

The assumptions made in this section are

 (i) the modulating signal is a single tone;
(ii) the envelope detector has an ideal characteristic which infers that its output is directly proportional to instantaneous carrier amplitude.

Some care is required in defining SNR at the detector input; the DSB-AM signal consists of a carrier and two sidebands and we can specify the signal power in terms of combination of these components. The DSB-AM signal plus noise is written as

$$V_{in}(t) = A_c[1 + m\cos(2\pi f_m t)]\cos 2\pi f_c t$$
$$+ x(t)\cos(2\pi f_c t) + y(t)\sin(2\pi f_c t) \tag{4.19}$$

If we normalize this voltage to be the voltage developed across a resistance of $1\,\Omega$ then the carrier power is $\frac{1}{2}A_c^2$ watts, the sideband power is $\frac{1}{4}(mA_c)^2$ watts and the total power is $\frac{1}{2}A_c^2(1 + \frac{1}{2}m^2)$ watts. The noise power is $\frac{1}{2}\overline{x^2(t)} + \frac{1}{2}\overline{y^2(t)} = \overline{x^2(t)}$ watts. There are three commonly used methods of specifying SNR at the detector input: these are

(1) carrier power-to-noise ratio,
(2) sideband power-to-noise ratio, and
(3) total power-to-noise ratio.

The carrier-to-noise ratio is $S_c/N = A_c^2/2\overline{x^2(t)}$, the sideband-to-noise ratio is $S_{sb}/N = (mA_c)^2/4\overline{x^2(t)}$, and the total SNR is

$$S_t/N = A_c^2\left(1 + \frac{m^2}{2}\right)\Big/2\overline{x^2(t)}$$

The resultant input to the detector is given by the vector sum of the amplitude modulated waveform and the noise components. The graphical addition is shown in Fig. 4.7, from which we can see that if the SNR is very large then the phase angle $\phi \to 0$. The input to the detector for large SNR is given approximately by

$$V_{in}(t) = A_c(1 + m\cos 2\pi f_m t) + x(t) \tag{4.20}$$

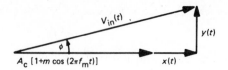

Fig. 4.7 Signal plus noise at the decteor input.

The detector, being ideal, will have an output

$$V_{out}(t) = aA_c(1 + m \cos 2\pi f_m t) + ax(t) \qquad (4.21)$$

The output signal power is $\frac{1}{2}(amA_c)^2$ and the output noise power is $a^2\overline{x^2(t)}$. The SNR at the detector output is thus

$$\text{SNR}_{out} = (mA_c)^2/2\overline{x^2(t)} \qquad (4.22)$$

If we compare this figure with the carrier-to-noise ratio at the detector input then

$$\text{SNR}_{out} = m^2 S_c/N$$

This has its maximum when $m = 1$, i.e.

$$\text{SNR}_{out(max)} = S_c/N \qquad (4.23)$$

This equation states that the SNR at the detector output has the same numerical value as the carrier-to-noise ratio at the detector input. It does not state that the SNR at the detector output is equal to the SNR at the detector input, which is the interpretation sometimes wrongly used in the literature.

If SNR_{out} is compared with the sideband-to-noise ratio at the detector input, then

$$\text{SNR}_{out} = 2S_{sb}/N \qquad (4.24)$$

The output SNR is 3 dB greater than the sideband power-to-noise ratio. It is not correct, however, to say that DSB-AM produces a SNR improvement of 3 dB.

To produce a realistic assessment of DSB-AM we must compare the output SNR with the total SNR at the detector input. When $m = 1$ then

$$\text{SNR}_{out} = \frac{2}{3}\frac{S_t}{N} \qquad (4.25)$$

The SNR at the output of an ideal envelope detector is actually lower than the SNR at the detector input. This is explained by the fact that even when $m = 1$, 66.6% of the total transmitted power is contained within the carrier component, which does not contribute at all to the signal power at the detector output. Thus DSB-AM does not obey Shannon's law in its strict theoretical statement. The situation is improved somewhat if the carrier is suppressed, as this shows an improvement of 3 dB in SNR as given by Eqn (4.24). If the carrier is suppressed, however, it is not possible to employ envelope detection.

4.6   SNR CHARACTERISTICS OF COHERENTLY DETECTED DSB-AM

Equation (4.20) was derived on the assumption that the SNR at the detector input was very large. If this condition is not met the resulting values of SNR_{out} given by Eqns (4.23), (4.24) and (4.25) are not valid. In fact at low values of SNR the performance of the envelope detector deteriorates rapidly. The envelope detector can therefore be employed only in good SNR conditions. This is usually taken to mean that the envelope detector has a signal-to-noise performance that is acceptable only if the SNR at the detector input is greater than 10 dB.

Coherent detection is an alternative demodulation technique for DSB-AM and an essential technique for suppressed carrier and SSB-AM. The performance of the coherent detector is maintained for all values of input SNR. The coherent detector multiplies the received signal by a locally produced referrence signal $E \cos(2\pi f_c t)$. When the received signal is accompanied by noise, the detector output is

$$V_0(t) = EA_c[1 + m\cos(2\pi f_m t)]\cos^2(2\pi f_c t)$$
$$+ Ex(t)\cos^2(2\pi f_c t) + Ey(t)\cos(2\pi f_c t)\sin(2\pi f_c t) \quad (4.26)$$

The frequency terms produced by the multiplication, which are outside the bandwidth occupied by the modulating signal, are removed by the filter that follows the coherent detector, the resulting output being

$$V_{out}(t) = \tfrac{1}{2}EA_c + \tfrac{1}{2}EA_c m\cos 2\pi f_m t + \tfrac{1}{2}Ex(t) \quad (4.27)$$

The signal and noise powers in this expression are respectively $S_p = \tfrac{1}{8}(EA_c m)^2$ watts and $N_p = \tfrac{1}{4}E^2\overline{x^2(t)}$ watts. The SNR at the detector filter output is thus

$$SNR_{out} = (mA_c)^2/2\overline{x^2(t)} \quad (4.28)$$

This is the same result as for the envelope detector but there is no precondition that the SNR should be large. In other words, the coherent detector maintains its SNR performance for all values of input SNR and is therefore superior to the envelope detector in poor SNR conditions.

4.7   SNR CHARACTERISTICS OF DSB-SC-AM

The detection of DSB-SC-AM is discussed fully in Section 2.10. If we assume that such a detector has an ideal characteristic the output SNR will be 3 dB greater than input sideband power-to-noise ratio. In other words, DSB-SC-AM produces a 3 dB SNR improvement.

4.8   SNR CHARACTERISTICS OF SSB-AM

This form of modulation requires coherent detection. In this case the input to the detector will be one sideband only, plus noise. If we assume a single modulating tone of frequency f_m, the signal plus noise at the detector input will be

$$V_{in}(t) = A_c \cos\left[2\pi(f_c + f_m)t\right] + x(t)\cos(2\pi f_c t) + y(t)\sin(2\pi f_c t) \quad (4.30)$$

The input SNR is thus

$$SNR_{in} = A_c^2/2\overline{x^2(t)} \quad (4.31)$$

We should point out here that because the bandwidth of a SSB signal is approximately half that of a DSB signal, the noise power at the detector input will be half the equivalent noise power in a DSB system. The output of the SSB coherent detector after filtering is

$$V_{out}(t) = \tfrac{1}{2}EA_c \cos 2\pi f_m t + \tfrac{1}{2}Ex(t) \quad (4.32)$$

The SNR at the filter output is thus $SNR_{out} = A_c^2/2\overline{x^2(t)}$. Thus for SSB-AM

$$SNR_{out} = SNR_{in} \quad (4.33)$$

It is interesting to compare a SSB transmission with a DSB-SC transmission when the transmitted power is the same in each case. We have shown that there is a 3 dB improvement in the DSB case, but because the noise power in the SSB system is only half that of the DSB system the two are actually equivalent in terms of output SNR when the transmitted power in each case is the same.

It would seem on a purely SNR basis that there is little to choose between DSB-SC-AM and SSB-AM. This is not entirely true: in certain situations (when ionospheric reflections are used, for example) severe distortion can be produced in DSB systems because components in the two sidebands can have differing phase velocities resulting in partial cancellation after detection. This problem is not encountered in SSB systems, when used for audio signal transmission, because the ear is insensitive to phase distortion. In these circumstances there is a considerable advantage in using SSB transmission.

4.9   SNR CHARACTERISTICS OF FM

We will establish the SNR properties of FM by again assuming an ideal detector, i.e. one that has an output voltage directly proportional to the 'instantaneous frequency' of the input signal. The algebraic representation of band-limited noise is identical to the AM case, but it should be borne in mind that the bandwidth of a frequency modulated signal is usually considerably greater than the bandwidth of an AM signal, and that noise power is directly proportional to bandwidth.

The voltage at the FM detector input will be the sum of signal and noise, i.e.

$$V_{in}(t) = A_c \cos\left[2\pi f_c t + \beta \sin(2\pi f_m t)\right]$$
$$+ x(t)\cos(2\pi f_c t) + y(t)\sin(2\pi f_c t) \quad (4.34)$$

The FM signal has constant amplitude and the signal power is also constant and independent of the amplitude of the modulating signal. This is a fundamental difference between FM and AM. The signal power at the detector input has a value of $A_c^2/2$ watts. It is convenient to calculate the signal power at the detector output in the absence of noise, and the noise power in the presence of an unmodulated carrier (i.e. in the absence of signal).

If we assume that the modulating signal is a single tone of frequency f_m the instantaneous frequency of the FM waveform is $f_i = f_c + \Delta f_c \cos(2\pi f_m t)$, and this will produce a voltage at the detector output given by

$$V_s = b 2\pi \Delta f_c \cos(2\pi f_m t) \qquad (4.35)$$

The signal power at the detector output will thus be

$$S_0 = (b 2\pi \Delta f_c)^2/2 \quad \text{watts} \qquad (4.36)$$

We calculate the noise voltage at the detector output using phasor methods. The phasor diagram for the unmodulated carrier and quadrature noise components is shown in Fig. 4.8. The FM detector produces an output proportional to frequency, which is the time differential of the phase angle ϕ. If we assume that the SNR at the detector input is large, then the phase angle is given by

$$\phi(t) \simeq \tan^{-1}\left[\frac{y(t)}{A_c}\right] \simeq \frac{y(t)}{A_c} \qquad (4.37)$$

The 'instantaneous frequency' produced by the noise is

$$\dot{\phi}(t) = \frac{\dot{y}(t)}{A_c} \qquad (4.38)$$

which, if we assume $\dot{y}(t) \ll A_c$, is narrowband FM. The noise voltage at the detector output is

$$V_n = \frac{b}{A_c} \dot{y}(t) \qquad (4.39)$$

where b is the detector constant of proportionality. The noise waveform $y(t)$ is itself the sum of many elemental noise components, i.e.

$$y(t) = \sum_k A_n \sin\left[2\pi(f_k - f_c)t + \theta_k(t)\right]$$

The noise component at a particular frequency f_k is

$$y_k(t) = A_n \sin\left[2\pi f t + \theta_k(t)\right] \qquad (4.40)$$

where $f = (f_k - f_c)$ is the frequency difference between the noise component and the centre frequency, and can have both negative and positive values. The

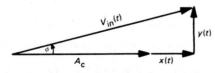

Fig. 4.8 Unmodulated FM carrier plus noise.

noise voltage at the detector output produced by an elemental noise component at the detector input is

$$\delta V_n = \frac{b}{A_c} 2\pi f A_n \cos\left[2\pi ft + \theta_k(t)\right] \qquad (4.41)$$

where we have assumed that $\theta_k(t)$ varies very slowly and does not contribute to the output amplitude. The power at the detector output produced by this elemental component is

$$\Delta N_0 = \frac{(b2\pi A_n f)^2}{2A_c^2} \quad \text{watts} \qquad (4.42)$$

but

$$A_n = (2\eta\Delta f)^{1/2}$$

i.e.

$$\Delta N_0 = \left(\frac{2\pi b}{A_c}\right)^2 \eta f^2 \Delta f$$

Hence

$$\frac{dN_0}{df} = G_0(f) = Kf^2 \quad \text{where} \quad K = \eta\left(\frac{2\pi b}{A_c}\right)^2 \qquad (4.43)$$

In Eqn (4.43) $G_0(f)$ represents the power spectral density of the noise at the detector output and is no longer white but proportional to f^2. This means that the noise power at the output of an ideal FM detector increases with the square of the frequency difference between the centre frequency and the elemental noise frequency. The noise voltage spectral density $[G_0(f)]^{1/2}$ is plotted as a function of f in Fig. 4.9.

The noise at the output of a frequency modulation detector will be produced by the sum of all components within the passband of the filter that follows the detector. If we assume that this filter has an ideal low-pass characteristic with a cut-off frequency of $\pm f_0$ (the negative figure is required for negative values of $f = (f_k - f_c)$), then since the noise produces narrowband FM we can determine the total noise using superposition. The total noise

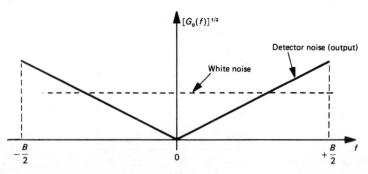

Fig. 4.9 FM detector noise spectral density.

power is

$$N_0 = \int_{-f_0}^{f_0} G_0(f)\,\mathrm{d}f$$

i.e.

$$N_0 = 2K \int_0^{f_0} f^2 \mathrm{d}f$$

which evaluates to

$$N_0 = \eta \left(\frac{2\pi b}{A_c}\right)^2 \tfrac{2}{3} f_0^3 \tag{4.44}$$

The signal power at the detector output is given by Eqn (4.36). Thus the SNR is

$$\mathrm{SNR}_{\mathrm{out}} = \frac{3S_c}{2\eta f_0} \left(\frac{\Delta f_c}{f_0}\right)^2 \tag{4.45}$$

It should be noted that $\Delta f_c / f_0 \neq \beta$ unless $f_0 = f_m$, which is not necessarily the case.

Equation (4.45) is interpreted as stating that the SNR at the output of a frequency modulation detector increases with the square of the carrier deviation, which is independent of the carrier power. It might be argued that the SNR could be increased indefinitely simply by increasing the carrier deviation. This overlooks the fact that increasing Δf_c increases the signal bandwidth with a consequent increase in the noise power, which is itself directly proportional to bandwidth. As Δf_c increases a threshold is reached at which the assumption $y(t) \ll A_c$ is no longer valid. Beyond this threshold a very rapid fall in SNR at the detector output is witnessed. This threshold effect is a characteristic of all wideband systems and is considered in more detail in Section 4.11.

The relative performance of AM and FM systems in the presence of noise can be compared by reference to Eqn (4.45). In this equation, f_0 is the bandwidth of the filter following the detector and will be equal to the bandwidth of the modulating signal. The factor $2\eta f_0$ is thus the noise power that would occur at the output of an AM system with the same modulating signal bandwidth. If we compare FM and AM transmissions in which the AM carrier power equals the FM carrier power, the factor $S_c/2\eta f_0$ of Eqn (4.45) is equal to the output SNR for an AM system when the depth of modulation $m = 100\%$, i.e.

$$\mathrm{SNR}_{\mathrm{out(FM)}} = \mathrm{SNR}_{\mathrm{out(AM)}} 3(\Delta f_c/f_0)^2 \tag{4.46}$$

If we consider a typical FM commercial broadcast system in which $\Delta f_c = 75\,\mathrm{kHz}$ and $f_0 = 15\,\mathrm{kHz}$, then

$$\mathrm{SNR}_{\mathrm{out(FM)}} = 75\,\mathrm{SNR}_{\mathrm{out(AM)}}$$

In fact, in commercial broadcast AM systems the depth of modulation is usually restricted to 30% and the total transmitter power is $(1 + \tfrac{1}{2}m^2)$ times the carrier power. Hence comparing FM and AM on the basis of total transmitted power,

$$\text{SNR}_{\text{out(FM)}} = \frac{3(1 + \frac{1}{2}m^2)}{m^2}\left(\frac{\Delta f_c}{f_0}\right)^2 \text{SNR}_{\text{out(AM)}} \tag{4.47}$$

If

$$m = 0.3, \Delta f = 75\,\text{kHz and } f_0 = 15\,\text{kHz},$$

then

$$\text{SNR}_{\text{out(FM)}} = 870.83\,\text{SNR}_{\text{out(AM)}}$$

Alternatively, to produce the same SNR the transmitted power in the FM case is 29 dB less than the required power in the AM case.

One conclusion we may draw from these figures is that for a given radiated power a FM transmitter will have a greater range than a DSB-AM transmitter, provided that the SNR at the FM detector input is sufficiently high for the noise to produce narrowband frequency modulation of the carrier. In other words, the FM system must be operating above the threshold level. This threshold level is related to both the frequency deviation and the SNR at the detector input. For large values of Δf_c the threshold occurs at a SNR of about 13 dB. The conditions required for FM to exhibit its SNR improvement properties are that $\beta > 1/\sqrt{3}$ [assuming $f_0 = f_m$ in Eqn (4.46)] and the SNR at the detector input must exceed 13 dB. If the SNR at the detector input is below the threshold value, the output SNR decreases rapidly and ultimately becomes poorer than the equivalent AM value.

4.10 PRE-EMPHASIS AND DE-EMPHASIS

The analysis of the previous section was based on a single tone modulating signal and we assumed that irrespective of the frequency of this tone the carrier deviation had its maximum value Δf_c. In other words, we assumed that the FM signal occupied the maximum possible bandwidth for a given modulating signal. In reality the situation is somewhat different; the modulating signal is not a single tone but a complex signal with a particular power spectral density.

The power spectral density of natural speech is closely approximated by the graph shown in Fig. 4.10. We can see from this figure that above a certain frequency f_1 the power spectrum decreases at a rate approaching 6 dB/octave. If such a signal frequency modulates a carrier, the higher-frequency components will produce a lower carrier deviation than the low-frequency components. In other words, the bandwidth occupied by a carrier frequency modulated by a signal of this type will be considerably less than if the power spectrum of the modulating signal were uniformly distributed. In practice, therefore, the FM system would not produce the maximum SNR improvement suggested by Eqn (4.46).

The theoretical SNR can be approached, however, if the power spectrum of

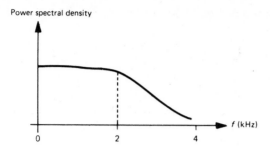

Fig. 4.10 Power spectrum of natural speech.

the modulating signal is made uniform by emphasizing the higher-frequency components before modulation takes place. The spectrum of the modulating signal is restored after detection by applying a corresponding amount of the de-emphasis. The de-emphasis will, of course, operate on the noise produced by the detector as well as the signal, the overall effect being a reduction in the noise power at the detector output. The process of pre-emphasis and de-emphasis is illustrated graphically in Fig. 4.11.

The actual component values used in pre- and de-emphasis networks will vary from circuit to circuit. The amount of pre- and de-emphasis will depend upon the time constant of the filter used, and it is normal to express the pre- and de-emphasis in terms of this time constant. In the UK the value of the time constant used is $RC = 50\,\mu s$ ($f_1 = 3.2\,kHz$); in contrast the value used in the USA is $RC = 75\,\mu s$ ($f_1 = 2.1\,kHz$). The improvement in SNR achieved by this technique is due to the attenuation of the discriminator noise by the de-emphasis network and does not imply an increase in the theoretical SNR given by Eqn (4.46). The attenuation achieved depends on the time constant

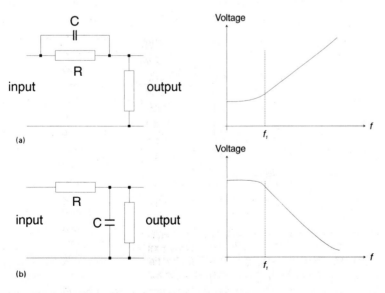

Fig. 4.11 (a) Pre-emphasis and (b) de-emphasis in FM.

used and is approximately 10 dB for a time constant of 50 μs and 13 dB for a time constant of 75 μs.

It is shown by Schwartz[2] that the use of pre- and de-emphasis is not restricted to FM but gives an advantage in all cases where the noise and signal power spectra differ. Hence pre and de-emphasis could also be used in AM systems. It should be noted that pre and de-emphasis is not used in AM systems, in practice, because it is actually advantageous, from the point of view of adjacent channel separation, to have a signal power spectrum that decreases with frequency.

4.11 THE FM CAPTURE (THRESHOLD) EFFECT

Consider the phasor diagram of Fig. 4.12(a) in which X represents the required FM carrier and Y represents an interfering FM carrier. (Y could be due to a second FM station at the receiver image frequency, or Y could represent the frequency modulated carrier produced by noise entering the detector.) It is assumed that the amplitudes X and Y are fixed and that the angle between the phasor ϕ_y is uniformly distributed in the range $-\pi \leqslant \phi_y \leqslant +\pi$. It is required to find the resultant phase angle ϕ_r which will be responsible for any interference generated at the detector output.

From Fig. 4.12(a)

$$\phi_r = \tan^{-1}\left[\frac{Y\sin(\phi_y)}{X + Y\cos(\phi_y)}\right] \tag{4.48}$$

ϕ_r will be a random quantity with mean square value

$$\overline{(\phi_r)^2} = \frac{1}{2\pi}\int_{-\pi}^{+\pi}\left\{\tan^{-1}\left[\frac{Y\sin(\phi_y)}{X + Y\cos(\phi_y)}\right]\right\}^2 d\phi_y \tag{4.49}$$

Equation 4.49 can be solved numerically and the rms value of ϕ_r is plotted against the ratio X/Y in Fig. 4.12(b). It will be noted from this figure that

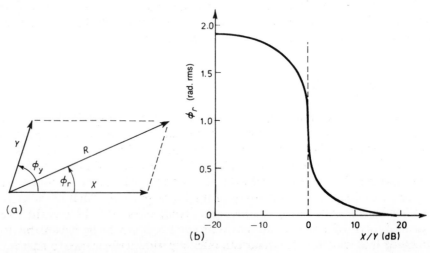

Fig. 4.12 The FM capture effect phasor diagram.

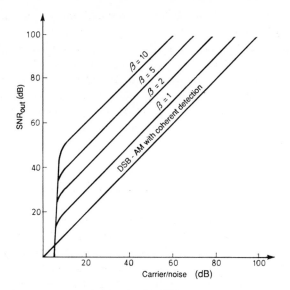

Fig. 4.13 Noise improvement obtained with FM.

a very rapid change in $\sqrt{(\overline{\phi_r})^2}$ occurs in the region $X = Y$. When $X/Y = 3$ the phase angle ϕ_r becomes very small. This means that the interference component produced by Y (which is proportional to $d\phi_r/dt$) also becomes very small. Hence when the ratio $X/Y > 3$ the carrier X takes over, or captures, the system. The FM receiver will thus discriminate in favour of the stronger signal. This is also true when $X/Y < 1$. In the latter case component Y captures the system. Hence if Y is due to noise, a very rapid deterioration in performance is observed. The FM threshold effect is clearly demonstrated in Fig. 4.13 which plots the SNR at the detector output as a function of the SNR at the detector input.

The FM capture effect is actually very useful and is employed in mobile cellular radio (see Fig. 13.21) to suppress interference from base stations in co-channel cells operating on the same frequency.

4.12 CONCLUSION

In this chapter we have compared several analogue transmission systems from a SNR point of view. It may be concluded that wideband systems, such as FM, produce SNR improvements, but not to the degree specified by Shannon's law. (This conclusion is also valid if narrowband pulse amplitude modulation (PAM) is compared with wideband pulse frequency modulation. We have omitted a detailed study of analogue pulse modulation because this has been largely overtaken by digital techniques.) If SNR was the sole figure of merit of a telecommunications system, it would be reasonable to conclude that wide-band systems can offer superior performance to narrow-band systems. There are, however, many other factors that influence the choice of modulation system. For example, the wide bandwidth of FM precludes its use in the medium-wave broadcast band because here the main requirement is to accommodate as many individual stations as possible in a relatively small bandwidth. In any case SNR problems are not usually significant in high-power broadcast transmissions.

There are many instances where the lower bandwidth of SSB is used in preference to DSB-AM. One such instance is the frequency multiplexing of telephone circuits for trunk transmission. In such a system it is possible to transmit a synchronizing signal from which the individual carriers required for coherent demultiplexing can be derived.

Traditionally medium-wave (i.e. local) broadcasts have used DSB-AM although bandwidth is at a premium. Historically DSB receivers were cheaper to produce and more reliable than SSB receivers. Modern technology has completely changed the situation where complexity and reliability are no longer closely related to cost. It is now perfectly possible, both technically and economically, to mass-produce reliable SSB receivers. It is unlikely that any moves will be made in this direction for some time because of the problems of compatibility, i.e. existing DSB receivers, of which there are many millions, would be unable to receive acceptable quality SSB transmission.

Thus, as we have stated, SNR is but one of many factors that influence the choice of a particular transmission system.

REFERENCES

1. King, R., *Electrical Noise*, Chapman & Hall, London, 1966, Chapter 3.
2. Schwartz, M., *Information Transmission Modulation and Noise*, 3rd edn. McGraw-Hill, New York, 1980, p. 412.

PROBLEMS

4.1 Calculate the mean square output noise voltage when a signal generator with an output resistance of $600\,\Omega$ is connected to the input of the two-port network shown.

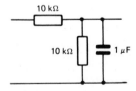

Answer: $4.002 \times 10^{-15}\,\mathrm{V}^2$.

4.2 An amplifier is made up of three identical stages in cascade, each stage having equal input and output resistances. The power gain per stage is 8 dB and the noise figure per stage is 6 dB when the amplifiers are correctly matched. Calculate the overall power gain and noise figure for the cascaded amplifier.

Answer: 24 dB; 6.6 dB.

4.3   The noise figure of a receiver, relative to a matched source at a temperature of 290 K, is 0.9 dB. Calculate the effective noise temperature at the input of the receiver when an antenna of effective noise temperature 200 K is connected.

Answer: 266.6 K.

4.4   The following diagram represents a satellite receiving system coupled by a waveguide to an antenna of effective noise temperature 70 K.

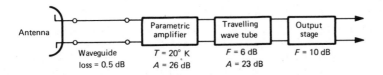

(a) Calculate the equivalent noise temperature of the waveguide and travelling wave tube (a device for amplifying frequencies in the gigahertz range).
(b) Calculate the SNR, at the output of the receiver assuming the antenna radiation resistance to be $50\,\Omega$ and the available received power is 10 pW. The bandwidth of the system is 10 MHz. (Hint: find the equivalent SNR referred to the waveguide input.)

Answer: (a) 35 K, 870 K; (b) 27.5 dB.

4.5   A superheterodyne receiver is connected to an antenna with a noise temperature of 100 K by a coaxial feeder having a loss of 2 dB. The receiver characteristics are

RF bandwidth $= 5\,$MHz
IF $= 20\,$MHz
IF bandwidth $= 1\,$MHz
noise figure $= 4\,$dB

Calculate the total system noise temperature and the required signal power delivered by the antenna to give a SNR of 20 dB at the output of the IF stage. Assume the antenna and receiver are both matched to the co-axial line.

Answer: 964 K; 1.33 pW.

4.6   A single tone of amplitude 2 V rms and frequency 5.8 kHz is used to amplitude modulate a carrier of amplitude 5 V rms, the carrier and both sidebands being transmitted. Given that the noise spectral density at the detector input is 0.1 µW/Hz, find the SNR at the detector output. The audio bandwidth of the receiver is 10 kHz and the carrier amplitude at the detector input is 1 V rms.

Answer: 16 dB.

4.7   A vhf transmitter radiates a DSB-AM signal at a depth of modulation of 45% with an audio bandwidth of 15 kHz. This produces a SNR of 40 dB at the output of a receiver at a distance of 3 km from the transmitter. If the transmitter, is switched to FM radiating the same total power, at a carrier deviation of 60 kHz, find the theoretical distance from the transmitter for the same SNR at the output of a FM receiver. Assume the noise spectral density at the receiver input is the same in each case and that the received power decreases as the square of the distance from the transmitter.

Answer: 48.5 km.

4.8   A radio station transmits a DSB-SC-AM signal with a mean power of 1 kW. If SSB-AM is used instead calculate the mean power for (a) the same signal strength and (b) the same SNR at the detector output.

A single tone of frequency 7.5 kHz forms the modulating signal for both a DSB-AM and a FM system, the power transmitted in each case being the same. When modulated, the peak deviation of the FM carrier is 60 kHz and the amplitude of the first pair of FM sidebands is equal to the sideband amplitude of the AM transmission. Assuming an audio bandwidth of 7.5 kHz for both AM and FM receivers, determine the SNR advantage of the FM receiver. It may be assumed that the noise spectral density has a constant value and is the same for each case.

Answer: (a) 2 kW, (b) 1 kW; 29.4 dB.

4.9 A frequency modulation receiver consists of a tuned amplifier of bandwidth 225 kHz that feeds a limiter and an ideal discriminator followed by a low-pass filter with a bandwidth of 10 kHz. The carrier-to-noise ratio at the discriminator input is 40 dB when the modulating signal is a 10 kHz tone, producing a carrier deviation of 5 kHz. Calculate the SNR at the output of the filter.

If the amplitude of the modulating signal is maintained at the same value when the frequency is changed to 1 kHz, find the new SNR at the filter output. What would be the SNR if the amplitude of the modulating signal is halved?

Answer: 69 dB; 69 dB; 63 dB.

4.10 The SNR at the output of a coherent detector is 25 dB when the input is a SSB-AM wave. If the input to the detector is transferred to DSB-AM with $m = 1$, find the increase in total power at the detector input to maintain an output SNR of 25 dB.

Answer: 4.8 dB.

Noise in digital communications systems | 5

The primary interest in the study of analogue communications systems is the obscuring or masking of the transmitted signal by additive noise. This effect is most conveniently analysed by considering the spectral properties of the noise waveform. The situation in digital systems is quite different since only fixed signal levels are allowed. When these levels are obscured by additive noise the receiver is required to decide which of the allowed levels the noisy signal represents. If the receiver decides correctly the noise has no effect on the received signal whatever. If the receiver makes an incorrect decision the results can be catastrophic. Decision theory is based upon the statistical rather than the spectral properties of noise although, as one might expect, these properties are related. The most important statistical property of white noise, with respect to decision theory, is its amplitude distribution function.

5.1 THE AMPLITUDE DISTRIBUTION FUNCTION OF WHITE NOISE

White noise is a naturally occurring phenomenon produced by the superposition of many randomly occurring events. Consequently it is not possible to specify the instantaneous amplitude of the noise waveform at any given instant. The alternative is to determine the probability that the noise waveform amplitude will exceed a given value. This is possible and requires a knowledge of the amplitude distribution function of the noise waveform. Suppose that n independent samples are taken of a noise waveform. (In order for the samples to be statistically independent, these samples must be spaced in time by an interval not less than $1/B$ seconds, where B is the noise bandwidth.) Suppose also that n_v of these samples have amplitudes between v_1 and v_2, i.e. the total number of samples in the range $\Delta v = (v_1 - v_2)$ is n_v. The probability that any noise sample will be in this range is simply n_v/n. If the interval $\Delta v \to 0$, a continuous distribution results. However, as this limit is approached, the probability that any sample will be in the range Δv also approaches zero. The problem is avoided if we represent probability as an area. The probability that a sample lies within the amplitude range Δv is represented as the area of a rectangle of base width Δv. The area of the rectangle thus has the value n_v/n and its height is $n_v/n\Delta v$ which approaches a finite value as $\Delta v \to 0$ and $n \to \infty$. If the height of the rectangle is plotted for all values of Δv the resulting curve is termed the probability density function.

The probability that a sample will be within the range v_1 to v_2 is the area under the probability density function between these two limits. The probability density function $p_v(v)$ is formally defined by

$$p_v(v) = \lim_{\substack{\Delta v \to 0 \\ n \to \infty}} \left[\frac{n_0}{n\Delta v} \right]$$

Referring to Fig. 5.1, the probability that v has a value between v_1 and v_2 is

$$P = \int_{v_1}^{v_2} p_v(v)\,dv \tag{5.1}$$

The central limit theorem is an important theorem in statistics which states that the probability density function (PDF) of the sum of many random variables approaches a Gaussian distribution regardless of the distributions of the individual variables. The PDF (also known as the amplitude distribution function) of white noise is therefore Gaussian and is given by

$$p_v(v) = \frac{\exp[-(v-a)^2/2\sigma^2]}{\sqrt{(2\pi\sigma^2)}} \tag{5.2}$$

In Eqn (5.2) a is known as the mean or expected value of v and σ is the standard deviation. The Gaussian PDF has a characteristic bell shape with a width proportional to σ. The mean value is given by

$$a = \int_{-\infty}^{\infty} v p_v(v)\,dv \tag{5.3}$$

and the variance (σ^2) is given by

$$\sigma^2 = \int_{-\infty}^{\infty} (v-a)^2\, p_v(v)\,dv \tag{5.4}$$

The total area under the $p_v(v)$ curve must be unity, i.e. the probability that a noise sample has an amplitude between $-\infty$ and $+\infty$ is 1, hence

$$\int_{-\infty}^{\infty} p_v(v)\,dv = 1 \tag{5.5}$$

This gives the Gaussian curve a peak value of $1/\sqrt{(2\pi\sigma^2)}$ at a value of $v = a$ (Fig. 5.2).

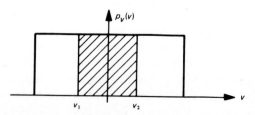

Fig. 5.1 Uniform PDF.

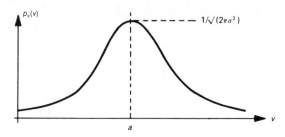

Fig. 5.2 Gaussian PDF.

The mean value of white noise is zero; hence the variance may be interpreted as the mean square value. Alternatively $\sigma = $ rms noise voltage.

A second distribution function, known as the cumulative distribution function, is defined by Eqn (5.6):

$$F_v(v) = \int_{-\infty}^{v} p_v(v)\,dv \qquad (5.6)$$

i.e.

$$F_v(v) = \int_{-\infty}^{v} \frac{\exp[-(v-a)^2/2\sigma^2]}{\sqrt{(2\pi\sigma^2)}}\,dv \qquad (5.7)$$

$F_v(v)$ is the probability that the noise will be less than some value v. Since $p_v(v)$ is symmetrical about $v = a$ it follows that $F_v(v) = 0.5$ when $v = a$. As shown in Fig. 5.3 the cumulative density function for white noise is symmetrical about $v = 0$.

The probability that the noise amplitude is less than some value $K\sigma$ is

$$\text{prob}(-K\sigma \leqslant v \leqslant K\sigma) = \int_{-K\sigma}^{K\sigma} \frac{\exp(-v^2/2\sigma^2)}{\sqrt{(2\pi\sigma^2)}}\,dv \qquad (5.8)$$

Substituting $y = v/\sqrt{(2\sigma^2)}$ and noting that the Gaussian function is symmetrical about $v = 0$ gives

$$\text{prob}(-K\sigma \leqslant v \leqslant K\sigma) = \frac{2}{\sqrt{\pi}} \int_{0}^{K/\sqrt{2}} \exp(-y^2)\,dy \qquad (5.9)$$

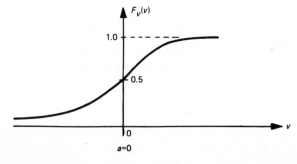

Fig. 5.3 Cumulative density function for white noise with zero mean.

This integral can be evaluated numerically for values of K and is known as the error function, i.e.

$$\text{erf}(P) = \frac{2}{\sqrt{\pi}} \int_0^P \exp(-y^2)\,dy \qquad (5.10)$$

The error function is tabulated in Appendix C. The probability that v will exceed $K\sigma$ is

$$1 - \frac{2}{\sqrt{\pi}} \int_0^{K/\sqrt{2}} \exp(-y^2)\,dy$$

The complementary error function is defined as

$$\text{erfc}(P) = 1 - \frac{2}{\sqrt{\pi}} \int_0^P \exp(-y^2)\,dy \qquad (5.11)$$

Hence the probability that the noise voltage will be less than its rms voltage is, from Eqn (5.9),

$$\text{prob}(-\sigma \leqslant v \leqslant \sigma) = \text{erf}(1/\sqrt{2}) = 0.683$$

Alternatively, the probability that the noise voltage will exceed its rms value is $\text{erfc}(1/\sqrt{2}) = 0.317$. Equations (5.9) and (5.11) are of fundamental importance in determining the probability of decision errors in digital communications systems.

5.2   STATISTICAL DECISION THEORY

This topic deals with the problem of developing statistical tests to determine which of M possible signals was transmitted over a communications link. In the case of binary communications $M = 2$, i.e. it is possible to transmit either a binary 0 or a binary 1. The receiver takes a single sample during each bit interval and then decides that 0 was transmitted or that 1 was transmitted. There are clearly two types of error that can occur, i.e. deciding 1 when 0 was transmitted or deciding 0 when 1 was transmitted. If the total probability of error is to be minimized, the minimization must be based on both types of error.

The decision rule is based on splitting all values of the received voltage v into two regions V_0 and V_1. The boundary between these regions is then chosen to minimize the total error probability. To illustrate this procedure we will assume that P_0 represents the probability of transmitting 0 and P_1 represents the probability of transmitting 1, clearly $P_0 + P_1 = 1$.

The probability that v will fall into region V_0 when a 1 is transmitted is

$$\int_{V_0} p_1(v)\,dv$$

where $p_1(v)$ is the probability density function of the received voltage when a 1 is transmitted. The probability that v will fall into the region V_1 when a 0 is transmitted is

$$\int_{V_1} p_0(v)dv$$

where $p_0(v)$ is the probability density function of the received voltage when a 0 is transmitted. The overall probability of error is then

$$P_e = P_0 \int_{V_1} p_0(v)dv + P_1 \int_{V_0} p_1(v)dv \qquad (5.12)$$

The region $V_0 + V_1$ covers all values of v; hence

$$\int_{V_0+V_1} p_1(v)dv = \int_{V_0} p_1(v)dv + \int_{V_1} p_1(v)dv = 1$$

Hence we can eliminate the integral over V_0 from Eqn (5.12) to give

$$P_e = P_1 + \int_{V_1} [P_0 p_0(v) - P_1 p_1(v)]dv \qquad (5.13)$$

If P_1 is known then P_e can be minimized by making the integral of Eqn (5.13) negative and as large as possible, i.e.

$$P_1 p_1(v) > P_0 p_0(v)$$

The required decision rule is

$$\frac{p_1(v)}{p_0(v)} > \frac{P_0}{P_1} \qquad (5.14)$$

$\lambda = p_1(v)/p_0(v)$ is known as the likelihood ratio. In the binary case, if 0 is represented by zero volts and 1 is represented by A volts and the noise is Gaussian, the two density functions will be

$$p_0(v) = \frac{1}{(2\pi\sigma^2)^{1/2}} \exp[-(v)^2/2\sigma^2] \qquad (5.15)$$

$$p_1(v) = \frac{1}{(2\pi\sigma^2)^{1/2}} \exp[-(v-A)^2/2\sigma^2] \qquad$$

The region V_1 is defined by all values of v for which $\lambda > P_0/P_1$, i.e.

$$\frac{\exp[-(v-A)^2/2\sigma^2]}{\exp[-v^2/2\sigma^2]} > \frac{P_0}{P_1} \qquad (5.16)$$

Taking logarithms of Eqn (5.16)

$$v^2 - (v-A)^2 > 2\sigma^2 \ln P_0/P_1$$

The desired region V_1 is thus defined by all values of v corresponding to

$$v > \frac{A}{2} + \frac{\sigma^2}{A} \ln \frac{P_0}{P_1} \qquad (5.17)$$

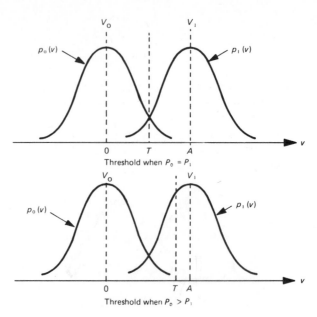

Fig. 5.4 Determination of decision thresholds.

or the boundary between the regions V_0 and V_1 is

$$T = \frac{A}{2} + \frac{\sigma^2}{A} \ln \frac{P_0}{P_1} \qquad (5.18)$$

If binary 1 and 0 are equi-probable the decision threshold becomes $T = A/2$ which is exactly half-way between the voltage levels representing 1 and 0. If 0 is transmitted more often than 1 the threshold moves towards V_1. This is illustrated in Fig. 5.4.

Once the decision threshold has been established the total probability of error can then be calculated. If binary 0 (0 volts) is sent, the probability that it will be received as a 1 is the probability that the noise will exceed $+ A/2$ volts (assuming $P_0 = P_1$) i.e.

$$P_{e0} = \int_{A/2}^{\infty} \frac{\exp(-v^2/2\sigma^2)}{\sqrt{(2\pi\sigma^2)}} \, dv \qquad (5.19)$$

If a 1 is sent (A volts) the probability that it will be received as a 0 is the probability that the noise voltage will be between $- A/2$ and $- \infty$, i.e.

$$P_{e1} = \int_{-\infty}^{-A/2} \frac{\exp(-v^2/2\sigma^2)}{\sqrt{(2\pi\sigma^2)}} \, dv \qquad (5.20)$$

These types of error are mutually exclusive, since sending a 0 precludes sending a 1. Because of the symmetry of the Gaussian function $P_{e0} = P_{e1}$, and the total probability of error, which is $P_0 P_{e0} + P_1 P_{e1}$, can then be written

$$P_e = P_{e1}(P_0 + P_1) = P_{e1}$$

i.e.

$$P_e = \int_{-\infty}^{-A/2} \frac{\exp(-v^2/2\sigma^2)}{\sqrt{(2\pi\sigma^2)}} dv \tag{5.21}$$

This equation can be written in terms of two integrals

$$P_e = \int_{-\infty}^{0} \frac{\exp(-v^2/2\sigma^2)}{\sqrt{(2\pi\sigma^2)}} dv - \int_{-A/2}^{0} \frac{\exp(-v^2/2\sigma^2)}{\sqrt{(2\pi\sigma^2)}} dv \tag{5.22}$$

Using the fact that the Gaussian distribution is symmetrical about its mean value, Eqn (5.22) reduces to

$$P_e = \frac{1}{2} + \int_{0}^{-A/2} \frac{\exp(-v^2/2\sigma^2)}{\sqrt{(2\pi\sigma^2)}} dv$$

If we substitute $y = v/\sqrt{(2\sigma^2)}$ this becomes

$$P_e = \frac{1}{2} + \frac{1}{\sqrt{\pi}} \int_{0}^{A/(2\sqrt{2}\sigma)} \exp(-y^2) dy$$

i.e.

$$P_e = \tfrac{1}{2}\{1 - \text{erf}\,[A/(2\sqrt{2}\sigma)]\} \tag{5.23}$$

The error probability therefore depends solely on the ratio of the peak pulse voltage A to the rms noise voltage σ. A graph of error probability against A/σ is plotted in Fig. 5.5.

This figures shows that when $A/\sigma = 17.4$ dB (a voltage ratio of 7.4:1), $P_e = 10^{-4}$, i.e. on average 1 digit in 10^4 will be in error. If A/σ is increased to 21 dB the error probability drops to $P_e = 10^{-8}$. Hence a very large decrease in error probability occurs for an increase in A/σ of only 3.6 dB. This decrease is much smaller for values of A/σ less than 14 dB. The characteristic thus exhibits a threshold effect for values of A/σ around 18 dB. An error probability of

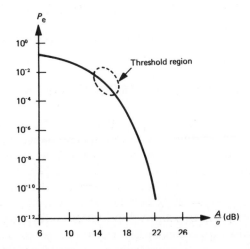

Fig. 5.5 Error probability in binary transmission.

about 10^{-5} is usually considered acceptable for practical data communications systems. This corresponds to a pulse amplitude of about ten times the rms noise voltage.

5.3 DECISION ERRORS IN PCM

In any PCM system decision errors will occur with a low but finite probability at each regenerative repeater. Any errors in transmission will be reflected by voltage errors at the decoder which will be added to the voltage errors produced by the quantisation process. In this section the effect of transmission errors on 8 digit linearly quantised PCM is considered.

If an error occurs in the least significant digit the voltage error produced will be δv. An error in the next significant digit will produce an error of $2\delta v$, and so on. An error in the rth digit will produce a voltage error of $\delta v.2^{r-1}$ volts. If the probability of a digit being in error is $P_e < 10^{-5}$, the probability of more than 1 error in a code word of 8 digits is very low.

If 1 error does occur in 8 digits the mean square voltage error produced by the decoder will be

$$N = \sum_{r=1}^{8} \frac{(\delta v.2^{r-1})^2}{8}$$

N is the "noise power" associated with each error. The probability of 1 error in 8 digits is $8 \times P_e$, thus the mean noise power at the decoder output due to transmission errors is

$$N_e = \delta v^2.P_e \sum_{4=1}^{8} (2^{r-1})^2 = 21845.\delta v^2.P_e \qquad (5.24)$$

If $P_e = 10^{-5}$ this has a value of $0.216 \times \delta v^2$.

This noise power will be added to the quantisation noise power of $\delta v^2/12$. An error in the most significant is clearly much more important than errors in the other digits. In the transmission of high quality sound signals by digital techniques, the most significant digit is usually encoded with additional error protection digits. An error in this digit may thus be corrected before the digital signal is decoded.

5.4 DECISION ERRORS IN CARRIER-MODULATED DATA SIGNALS

In this section we will examine the relative performance of ASK, PSK and FSK in noisy conditions. Unlike the comparison of the analogue modulation systems of Chapter 4, which was done on an SNR basis, we compare modulation methods for digital systems in terms of probability of error. All three modulation systems can, in fact, be detected using coherent detection, although it is more usual to use envelope detection for ASK and FSK. It is convenient, initially, to compare the performance of all three systems assuming coherent detection, because the coherent detector does not alter the PDF of the noise signal.

We consider first of all coherent detection of ASK. The input to the detector will be

$$V_{in}(t) = h(t)\cos(2\pi f_c t) + x(t)\cos(2\pi f_c t) + y(t)\sin(2\pi f_c t) \qquad (5.25)$$

Where $h(t)$ is a binary function with value A or 0 and $[x(t)\cos(2\pi f_c t) + y(t) \sin(2\pi f_c t)]$ represents the band-limited white noise. It was noted in Section 4.4 that both $x(t)$ and $y(t)$ are random variables. We can apply the central limit theorem to Eqn (4.16a) which defines $x(t)$ as the sum of many sinusoidal components with random frequency and phase. The result is that the PDF of both $x(t)$ and $y(t)$ is Gaussian. If the ASK signal plus noise is coherently detected (i.e. multiplied by $\cos[2\pi f_c t]$ and filtered), the baseband output will be

$$V_{out}(t) = \tfrac{1}{2}h(t) + \tfrac{1}{2}x(t) \qquad (5.26)$$

This is similar to Eqn (4.27) which describes the coherent detection of DSB-AM. Ignoring the factor of $\tfrac{1}{2}$ in Eqn (5.26) the output of the coherent detector is

$$V_{out} = (A \text{ or } 0) + x(t) \qquad (5.27)$$

If binary 1 and 0 are equi-probable and σ^2 represents the variance of $x(t)$ the probability of error, assuming a decision threshold of $A/2$, is

$$P_{e(ASK)} = \tfrac{1}{2}\{1 - \text{erf}\,[A/(2\sqrt{2}\sigma)]\} \qquad (5.28)$$

which is identical to Eqn (5.23) for baseband signalling.

For PSK the value of $h(t)$ in Eqn (5.26) will be $\pm A$ and the decision threshold will be 0 volts. The error probability is then

$$P_{e(PSK)} = \tfrac{1}{2}[1 - \text{erf}(A/\sqrt{2}\sigma)] \qquad (5.29)$$

The peak signal-to-noise power ratio for PSK is 6 dB less than for ASK for a given error rate. If we bear in mind that ASK transmits no signal at all for binary 0 (i.e. there is zero signal for half the time) it is evident that to produce the same error rate PSK requires a mean signal-to-noise ratio 3 dB below the value required for ASK. In other words, PSK has a 3 dB advantage, in terms of signal-to-noise power ratio, over ASK and, as is seen from Fig. 5.6, this can represent a significant improvement in terms of error probability.

FSK modems usually employ envelope detection and the block schematic of such a modem is given as Fig. 3.27. A similar schematic for a FSK modem employing coherent detection is presented in Fig. 5.7. The pre-detection filters, which are assumed to have an ideal bandpass characteristic, determine the bandwidth and hence the input noise power for each coherent detector.

The input voltage at each detector will be

$$V_1(t) = h_1(t)\cos(2\pi f_1 t) + x_1(t)\cos(2\pi f_1 t) + y_1(t)\sin(2\pi f_1 t) \qquad (5.30a)$$

and

$$V_0(t) = h_0(t)\cos(2\pi f_0 t) + x_0(t)\cos(2\pi f_0 t) + y_0(t)\sin(2\pi f_0 t) \qquad (5.30b)$$

where $h_1(t)$ is the logical complement of $h_0(t)$ and $x_1(t)$ and $x_0(t)$ are independent random variables, provided that the passband of each predetection filter does not overlap that of the other.

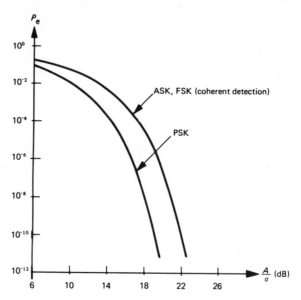

Fig. 5.6 Error probabilities for constant mean signal power.

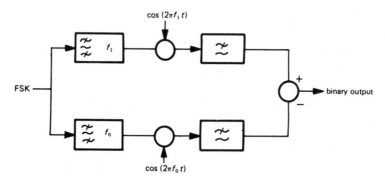

Fig. 5.7 Coherent FSK modem.

The output of the composite detector will be

$$V_{out}(t) = (+A \text{ or } -A) + [x_1(t) - x_0(t)] \qquad (5.31)$$

Since $x_1(t)$ and $x_0(t)$ are independent Gaussian variables, the variable $[x_1(t) - x_0(t)]$ will also be Gaussian with a variance equal to the sum of the original variances. (This might seem to be a surprising result, but remember that if we substract two sinusoids of **different** frequencies the power of the resultant waveform is the sum of the powers in the individual sinusoids.) The noise variance at the detector output will thus be $2\sigma^2$, which gives an effective peak signal-to-rms noise ratio of $2A/\sqrt{2}\sigma$. The probability of error in this case is then

$$P_{e(FSK)} = \tfrac{1}{2}[1 - \text{erf}(A/2\sigma)] \qquad (5.32)$$

This means that if FSK and PSK transmissions are to have the same error probability the FSK transmission requires a SNR which is 3 dB above the value required for PSK.

Thus, compared on a peak power basis, for the same error probability PSK requires an SNR that is 3 dB below that of FSK which in turn requires a SNR which is 3 dB below that of ASK. Compared on a mean power basis the performance of ASK and FSK are identical, but the same error rate can be obtained with PSK with a mean SNR which is 3 dB lower. A comparison of the performance of these three systems is given in Fig. 5.6, where it has been assumed that coherent detection has been used throughout. In practice it is more likely that envelope detection would be used for ASK and FSK. The analysis for envelope detection is slightly different because the envelope detector modifies the statistical properties of the noise. The envelope detector output voltage is proportional to the instantaneous carrier envelope, which is equal to the phasor sum of the inphase and quadrature components of Eqn (5.25). The instantaneous detector output is thus

$$V_{\text{out}} = k\left[\{h(t) + x(t)\}^2 + \{y(t)^2\}\right]^{1/2} \tag{5.33}$$

To calculate the error probability we need to know the PDF of Eqn (5.33) when $h(t) = 0$ (i.e. binary 0) and when $h(t) = A$ (binary 1). In the former case the density function has a Rayleigh distribution given by

$$p_0(v) = (v/\sigma^2)\exp(-v^2/2\sigma^2) \quad \text{for} \quad v \geq 0$$

and in the latter case the density function has a Rician distribution given by

$$p_1(v) = (v/\sigma^2)\exp[-(v^2 + A^2)/2\sigma^2]I_0(vA/\sigma^2) \quad \text{for} \quad v \leq 0$$

$I_0(x)$ is the modified Bessel function of zero order. The Rician distribution approaches the Gaussian distribution for $A/\sigma \gg 1$. The envelope probability density functions are illustrated in Fig. 5.8.

It has been shown[1] that the decision threshold for the envelope detector is

$$T = A^2/2\sigma^2 + \ln(P_0/P_1) \tag{5.34}$$

This reference also shows that for high SNRs envelope detection is marginally poorer for both ASK and FSK than coherent detection. For lower SNRs there is a significant deterioration in the performance of the envelope detection receiver. This, of course, is in agreement with the results for analogue transmission considered in Chapter 4.

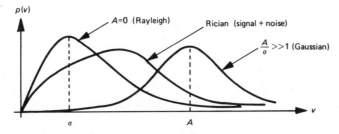

Fig. 5.8 Envelope distribution for signal + noise.

5.5 MATCHED FILTERING AND CORRELATION DETECTION

In all digital communication systems, baseband or carrier modulated, the probability of error ultimately depends upon the SNR. (This, of course, ignores the influence of intersymbol interference, which we considered in Section 3.7.) Thus it is reasonable to consider techniques that may be used to maximize the ratio A/σ. Matched filtering is one such technique, which emphasizes the signal voltage relative to the noise. In so doing, considerable distortion of the signal waveform usually results. Hence where waveform fidelity is important (as in analogue systems) matched-filtering techniques are not applicable. This is not the case with digital transmission, where the receiver is required only to decide between signal and noise.

The matched filter is designed to maximize the SNR at a precise interval t_0 after the signal is applied to its input. We will consider a general voltage waveform $h(t)$ with Fourier transform $H(f)$ which is applied to a filter with frequency response $P(f)$. The amplitude spectrum of the output is given by $G(f) = P(f) H(f)$ and this has a Fourier transform

$$g(t) = \int_{-\infty}^{\infty} P(f) H(f) \exp(j2\pi ft)\, df \tag{5.35}$$

The magnitude of this signal at time t_0 is

$$|g(t_0)| = \left| \int_{-\infty}^{\infty} P(f) H(f) \exp(j2\pi ft_0)\, df \right|$$

Hence the signal power (normalized) at time t_0 is

$$S_{\mathrm{p}} = |g(t_0)|^2 = \left| \int_{-\infty}^{\infty} P(f) H(f) \exp(j2\pi ft_0)\, df \right|^2 \tag{5.36}$$

$P(f)$ is defined for both negative and positive frequencies and it is therefore necessary to use the double-sided power spectral density of $\eta/2$ watts/hertz for the white noise.

The noise at the filter output will no longer be white but will have a power spectrum given by

$$N(f) = \frac{\eta}{2} |P(f)|^2$$

The noise power at the filter output is thus

$$N_{\mathrm{p}} = (\eta/2) \int_{-\infty}^{\infty} |P(f)|^2 df \tag{5.37}$$

The SNR at time t_0 is thus

$$\mathrm{SNR} = \frac{\left| \int_{-\infty}^{\infty} P(f) H(f) \exp(j2\pi ft_0)\, df \right|^2}{(\eta/2) \int_{-\infty}^{\infty} |P(f)|^2 df} \tag{5.38}$$

The filter frequency response is then chosen to maximize the RHS of Eqn (5.38). This is accomplished by applying the Schwartz inequality,[2] which effectively states that the sum of two sides of a triangle must always be greater than or equal to the third side; in integral form this is usually written

$$\int_{-\infty}^{\infty} A^*(x)\,A(x)\,\mathrm{d}x \int_{-\infty}^{\infty} B^*(x)\,B(x)\,\mathrm{d}x \geqslant \left| \int_{-\infty}^{\infty} A^*(x)\,B(x)\,\mathrm{d}x \right|^2 \quad (5.39)$$

The inequality becomes an equality when $A(x)$ and $B(x)$ are co-linear, i.e. $A(x) = KB(x)$. If $A^*(x) = H(f)\exp(j2\pi f t_0)$ then its complex conjugate $A(x) = H^*(f)\exp(-j2\pi f t_0)$, and if $B(x) = P(f)$ then $B^*(x) = P^*(f)$.

Equation (5.39) can then be written

$$\int_{-\infty}^{\infty} |H(f)|^2 \mathrm{d}f \int_{-\infty}^{\infty} |P(f)|^2 \mathrm{d}f \geqslant \left| \int_{-\infty}^{\infty} P(f)H(f)\exp(j2\pi f t_0)\mathrm{d}f \right|^2 \quad (5.40)$$

$$\int_{-\infty}^{\infty} |H(f)|^2 \mathrm{d}f \geqslant \frac{\left| \int_{-\infty}^{\infty} P(f)H(f)\exp(j2\pi f t_0)\mathrm{d}f \right|^2}{\int_{-\infty}^{\infty} |P(f)|^2 \mathrm{d}f}$$

Substituting in Eqn (5.38) we obtain

$$\frac{2}{\eta} \int_{-\infty}^{\infty} |H(f)|^2 \mathrm{d}f \geqslant \mathrm{SNR}$$

But $\int_{-\infty}^{\infty} |H(f)|^2 \mathrm{d}f$ is the energy of the signal, E. The maximum value of SNR is therefore

$$\mathrm{SNR}_{max} = 2E/\eta \quad (5.41)$$

This value depends only on the ratio of signal energy to noise spectral density and is independent of the shape of the signal waveform. The SNR is a maximum when $A(x) = KB(x)$, i.e.

$$H^*(f)\exp(-j2\pi f t_0) = KP(f) \quad (5.42)$$

The impulse response of the matched filter is thus

$$p(t) = \frac{1}{K} \int_{-\infty}^{\infty} H^*(f)\exp(-j2\pi f t_0)\exp(j2\pi f t)\mathrm{d}f$$

i.e.

$$p(t) = \frac{1}{K} h(t_0 - t) \quad (5.43)$$

The impulse response is therefore the time reverse of the input signal waveform $h(t)$ with respect to t_0. Clearly the value of t_0 must be greater than the duration of the signal to which the filter is matched. This is the condition for physical realizability; in other words, the impulse response must be zero for negative values of t.

We will derive, as an example, the matched filter for a rectangular pulse of amplitude A and duration t_1. The Fourier transform of this pulse is

$$H(f) = \int_{-t_{1/2}}^{t_{1/2}} A \exp(-j2\pi f t)\mathrm{d}t$$

Hence from Eqn (5.42) $P(f) = KH^*(f)\exp(-j2\pi f t_0)$ i.e.

$$P(f) = [\text{sinc}(\pi f t_1)\exp(-j2\pi f t_0)]/K \tag{5.44}$$

This is a low-pass filter with a linear phase shift. Matched filters can be specified for any signal shape, e.g. when a FSK waveform is the input to a matched filter the SNR at the output is given by Eqn (5.41).

If we write Eqn (5.35) representing the matched filtering operation in terms of the convolution integral it becomes

$$g(t) = \int_{-\infty}^{\infty} h(t)\, h(t_0 - t)\, dt$$

$h(t_0 - t)$ being the appropriate substitution for $p(t)$. We now compare this equation with the equation for the cross-correlation between two waveforms $h(t)$ and $y(t)$.

$$R_{hy}(\tau) = \int_{-\infty}^{\infty} h(t)\, y(t + \tau)\, dt$$

If we make $y(t + \tau) = h(t_0 - t)$ we see that matched filtering is equivalent to cross-correlating the 'noisy signal' $h(t)$ with a noise-free signal $h(t_0 - t)$, which is the impulse response of the matched filter. The correlation detector is based upon this principle.

The correlation coefficient between the waveforms $h(t)$ and $y(t)$ is defined as

$$R(\tau) = \frac{\displaystyle\int_{-\infty}^{\infty} h(t)y(t + \tau)dt}{\left|\displaystyle\int_{-\infty}^{\infty} h^2(t)dt \int_{-\infty}^{\infty} y^2(t)dt\right|^{1/2}} \tag{5.45}$$

The denominator of this equation is a normalizing factor that makes $R(\tau)$ independent of the actual mean square values of $h(t)$ and $y(t)$ and restricted to the range ± 1. Note when $h(t) = ky(t)$ then $R(\tau) = +1$, and when $h(t) = -ky(t)$ then $R(\tau) = -1$.

$R(\tau)$ is thus a measure of the similarity of the two waveforms $h(t)$ and $y(t)$. Correlation detection is frequently used in situations where the SNR < 1, e.g. space applications. The received waveform is correlated with several noise-free waveforms stored at the receiver. The one that produces the largest correlation coefficient is then assumed to have been the transmitted waveform.

It is found in most practical situations that the improvement obtained using matched filters, rather than conventional low- or band-pass filters, is marginal (usually less than 3 dB). Hence filters are usually designed to minimize intersymbol interference rather than maximize SNR. However, where SNR is of prime importance (e.g. radar signals) matched filtering and correlation techniques are employed.

5.6 ERROR DETECTION AND CORRECTION

The probability of error in a digital transmission system depends ultimately upon the SNR at the receiver input. This ratio is a maximum when matched filtering is used and this produces the lowest probability of error. A significant decrease in this figure can be produced by employing a type of code known as a block code for the digital signal. The purpose of coding the signal in this fashion is to introduce redundancy into the transmitted signal. We observed in Section 1.13 that there are many sources of redundancy in the English language and we showed that the redundancy present reduced the information content but at the same time allowed the receiver to identify and correct errors. Redundancy has the same effect in digital signals. The simplest form of block code is the parity check 8 bit international standard code described in Section 3.11.

The international code represents each character by seven information digits and a single parity check digit. Either an even or an odd parity check can be used; in the former case the parity digit is chosen to make the number of 1s in the 8 bit word even, and in the latter case the parity digit is chosen to make the number of 1s in the 8 bit word odd. The receiver then checks each 8 bit word for odd or even parity depending upon the system in use. If the parity check fails the receiver notes that an error has occurred and requests the transmitter to repeat the 8 bit word. The parity check is carried out simply by exclusive-ORing the eight digits in each code word. If the number of 1s in the code word is odd the result of the exclusive-OR operation is binary 1. If the number of 1s is even, the result of the exclusive-OR operation is 0. A parity check circuit is given in Fig. 5.9.

The error-detecting properties of this code can be demonstrated by comparing the probability of error in the uncoded case with the probability of an undetected error in the coded case. We assume that the probability of error in the uncoded case is P_e. The probability of an undetected error in the coded case will be the probability of an even number of errors. The parity check at

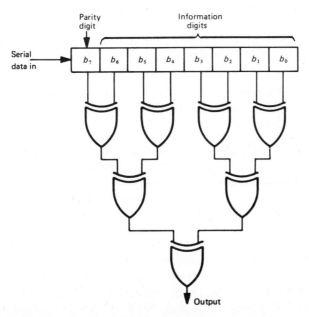

Fig. 5.9 Error detection using a single parity digit.

the receiver will fail, and therefore indicate an error, only if an odd number of errors occurs. We therefore have to compare the probability of error in the uncoded case with the probability of an even number of errors in the coded case.

If there are n digits in each code word the probability of one digit being in error is the joint probability of one incorrect digit and $(n-1)$ correct digits, i.e. the joint probability is

$$P_j = P_e (1 - P_e)^{n-1} \qquad (5.46)$$

There are n digits in the group and Eqn (5.46) gives the probability that any one of these digits will be received incorrectly. The total probability in the n digits is thus

$$P = nP_e (1 - P_e)^{n-1}$$

As an example consider the uncoded case where $n = 7$. If $P_e = 10^{-4}$ the probability of one error in seven digits is

$$P_t = 7 \times 10^{-4} \times (0.9999)^6 \simeq 7 \times 10^{-4}$$

The probability of r errors in a group of n digits is the joint probability that r digits will be received incorrectly and $(n-r)$ will be received correctly, i.e.

$$P_j = P_e^r (1 - P_e)^{n-r}$$

There is a total of ^nC_r possible ways of receiving r digits incorrectly in a total of n digits where

$$^nC_r = \frac{n!}{r!\,(n-r)!}$$

The total probability of error in this case is thus

$$P_t = {}^nC_r P_e^r (1 - P_e)^{n-r} \qquad (5.47)$$

In the coded case, an extra digit is added to the original seven so that we require the probability of an even number of errors in eight digits.

Probability of 2 errors $= {}^8C_2(10^{-4})^2(1 - 10^{-4})^6 = 2.8 \times 10^{-7}$
Probability of 4 errors $= {}^8C_4(10^{-4})^4(1 - 10^{-4})^4 = 7 \times 10^{-15}$
Probability of 6 errors $= {}^8C_6(10^{-4})^6(1 - 10^{-4})^2 = 2.8 \times 10^{-23}$
Probability of 8 errors $= {}^8C_8(10^{-4})^8 \qquad\qquad\quad = 10^{-32}$

Hence the total probability of an undetected error in the coded case is $2.8 \times 10^{-7} + 7.0 \times 10^{-15} + 2.8 \times 10^{-23} \simeq 2.8 \times 10^{-7}$. This is considerably less than the probability of error in the original uncoded group of seven digits. These numerical values show that a significant reduction in error probability is possible even with a very rudimentary error-checking system.

We have assumed in this example that the digit transmission rate is unchanged. This means that the rate of information transmission must necessarily be reduced to $\frac{7}{8}$ths of its value in the uncoded system. It would be more realistic to maintain a constant information rate which means that the transmission bandwidth should be increased to $\frac{8}{7}$ of its original value to accommodate the parity check digit. Such an increase would increase the noise power by the same ratio, i.e. the rms noise would be increased to 1.069 times its previous value. The effect of this increased noise on the error probability must be calculated. Assuming an uncoded probability of error $P_e = 10^{-4}$ for illustration then

$$\text{erf}\,[(A/(2\sqrt{2}\sigma))] = 1 - 2 \times 10^{-4}$$

i.e.

$$A/(2\sqrt{2}\sigma) = 2.63$$

If the rms noise increases by a factor of 1.069, there is a corresponding decrease in this ratio, i.e. the ratio becomes 2.46, which produces an uncoded error probability of $P_e' = 2.51 \times 10^{-4}$. This value must now be a used in Eqn (5.47) to calculate the probability of an undetected error in the coded case, which is $P_t' = 1.75 \times 10^{-6}$. Thus even when the extra noise is taken into account the reduction in undetected errors is still considerable.

This is not always the case, however, if several parity digits are added to the information digits. The extra noise introduced by the increased bandwidth can cancel any advantage obtained by the error-detecting code. In such cases the error probability can sometimes be decreased by increasing the signal power. There are many cases where error-detecting codes do produce a significant advantage, and bearing in mind the threshold region shown by Fig. 5.5 we can only say that each case must be examined separately.

The error-detecting properties of a code can be enhanced to include error correction by increasing the redundancy. This becomes clear by considering an example that codes 16 information digits with 8 parity check digits, i.e. the code redundancy is 33%. In this example the 16 information digits are grouped into a 4×4 matrix. An even or odd parity check digit is transmitted for each row and column of the matrix

```
        Column parity digits              ↓ Column check fails
0  0  1  0↙                      0  0  1  0
 _____                         _____
0  1  1  1 │ 1                   0  0  1  1 │ 1  ←Row check fails
0  1  1  0 │ 0                   0  1  1  0 │ 0
1  0  0  0 │ 1                   1  0  0  0 │ 1
1  0  1  1 │ 1                   1  0  1  1 │ 1
        ↑
   Row parity digits
```

If a single error occurs in the 16 information digits then a row and column parity check will both fail. The point of intersection of row and column then indicates which digit was in error. If a single parity digit is in error then only one column or one row will fail the parity check, and again the incorrect digit can be identified and therefore corrected. This type of code is generally known as a forward error correcting code (FEC).

The above example was chosen to illustrate the possibility of correcting errors, and was not a particularly efficient code. There are many more efficient error-correcting codes in use. The block codes are based upon the work of Hamming.[3] An example of Hamming's single-error-correcting code will now be considered. We assume that the original information digits are split up into blocks of m digits to which are added c parity check digits. For illustration assume that $m = 4$ and $c = 3$. Each coded word therefore contains seven digits and there are $2^7 = 128$ different words. Of these 128 words only $2^4 = 16$ words are required to transmit the original information. The 16 code words are chosen from the possible 128 to give a single-error-correcting capability. A possible choice of code words is shown in Table 5.1.

Each of the code words in Table 5.1 differs from any other in at least three positions. Such a code is said to have a Hamming distance of three. This means that at least three errors are required to convert any code word into one of the others. If a single error occurs in a received code word it will differ from the correct code word in one digit only and from all other code words in at least two digits. Hence the correct code word is chosen as the one with least difference from the received code word. This implies that the receiver requires a list of all allowed code words and then compares each received code word until a match is achieved. In practice this process can be reduced to one of parity checking.

The mathematical relationship between information digits and check digits is based on modulo 2 addition (exclusive-OR) and was devised by Hamming. The rules for a 7 digit block code containing four information digits and three check digits are

$$C_1 = M_1 \oplus M_2 \oplus M_3$$
$$C_2 = M_1 \oplus M_2 \oplus M_4$$
$$C_3 = M_1 \oplus M_3 \oplus M_4 \tag{5.48}$$

Remembering that $C_1 \oplus C_1 = 0$, etc., these equations can be written

$$C_1 \oplus M_1 \oplus M_2 \oplus M_3 = 0$$
$$C_2 \oplus M_1 \oplus M_2 \oplus M_4 = 0$$
$$C_3 \oplus M_1 \oplus M_3 \oplus M_4 = 0 \tag{5.49}$$

Table 5.1

Original message					Transmitted code						
M_1	M_2	M_3	M_4		M_1	M_2	M_3	M_4	C_1	C_2	C_3
0	0	0	0		0	0	0	0	0	0	0
0	0	0	1		0	0	0	1	0	1	1
0	0	1	0		0	0	1	0	1	0	1
0	0	1	1		0	0	1	1	1	1	0
0	1	0	0		0	1	0	0	1	1	0
0	1	0	1		0	1	0	1	1	0	1
0	1	1	0		0	1	1	0	0	1	1
0	1	1	1		0	1	1	1	0	0	0
1	0	0	0		1	0	0	0	1	1	1
1	0	0	1		1	0	0	1	1	0	0
1	0	1	0		1	0	1	0	0	1	0
1	0	1	1		1	0	1	1	0	0	1
1	1	0	0		1	1	0	0	0	0	1
1	1	0	1		1	1	0	1	0	1	0
1	1	1	0		1	1	1	0	1	0	0
1	1	1	1		1	1	1	1	1	1	1

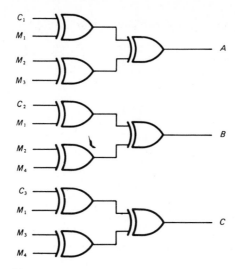

Fig. 5.10 Parity checking network.

If there are no errors at the receiver the modulo 2 additions represented by Eqns (5.49) will all produce a binary 0 result. If there is a single error in any of the information or check digits then one or more of Eqns (5.49) will produce a binary 1. The parity checking circuit for the 7 digit block code is shown in Fig. 5.10.

If we represent a correct digit by Y and an incorrect digit by N the possible values of A, B, C for a single error are listed in Table 5.2.

The outputs A, B, C of the parity check circuits can be regarded as a 3×8 matrix. The number of columns in the matrix equals the number of check digits and each row, except the first, uniquely defines a single error position. This means that c check digits can define $2^c - 1$ error positions assuming only a single error occurs.

If there are m information digits and c check digits then for a single-error correcting code the relationship between m and c is

$$m + c \leqslant 2^c - 1 \tag{5.50}$$

Thus for an ASCII character with $m = 7$ it is necessary to add $c = 4$ check

Table 5.2

M_1	M_2	M_3	M_4	C_1	C_2	C_3	A	B	C
Y	Y	Y	Y	Y	Y	Y	0	0	0
Y	Y	Y	Y	Y	Y	N	1	0	0
Y	Y	Y	Y	Y	N	Y	0	1	0
Y	Y	Y	Y	N	Y	Y	0	0	1
Y	Y	Y	N	Y	Y	Y	1	1	1
Y	Y	N	Y	Y	Y	Y	1	1	0
Y	N	Y	Y	Y	Y	Y	1	0	1
N	Y	Y	Y	Y	Y	Y	0	1	1

Table 5.3 Single-error-correcting codes

m	c	Code type $(m+c, m)$	Efficiency $m/(m+c)$
1	2	$(3,1)$	0.33
4	3	$(7,4)$	0.57
7	4	$(11,7)$	0.64
11	4	$(15,11)$	0.73
26	5	$(31,26)$	0.83
57	6	$(63,57)$	0.90
120	7	$(127,120)$	0.94
247	8	$(255,247)$	0.97

digits for a single-error-correcting capability. Possible single-error-correcting codes are listed in Table 5.3, and it may be seen from this table that the larger the value of c the more efficient the code.

Returning to Table 5.2 it is clear that in a practical system it is necessary only to correct errors in the information digits. This means that for error-correcting purposes only the last four rows of the parity check matrix need to be considered. Noting the following relationship:

$$X \oplus 1 = \bar{X} \quad \text{and} \quad X \oplus 0 = X$$

Then if a received digit is in error, exclusive-ORing it with binary 1 produces the correct value. If a received digit is correct, exclusive-ORing it with binary 0 will preserve the received value. With these conditions it is possible to devise a very simple error-correcting circuit consisting of exclusive-OR gates. This circuit is shown in Fig. 5.11. The binary inputs w, x, y, z are derived from the parity check matrix of Table 5.2 and are given by

$$w = \bar{A}BC$$
$$x = A\bar{B}C$$
$$y = AB\bar{C}$$
$$z = ABC$$

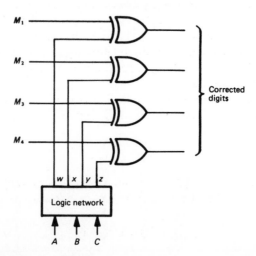

Fig. 5.11 Error-checking network.

The (7, 4) code has a Hamming distance of 3 and can correct single errors. If it is assumed, for illustrative purposes, that an error occurs in both M_1 and M_2 the parity checking network will produce an output of $A = 0, B = 0, C = 1$, which is equivalent to a single error in c_1. The fact that the parity checking network does not give $A = 0, B = 0, C = 0$ means that a double error has been detected, but this will be incorrectly interpreted by the circuit of Fig. 5.11.

Thus, the (7, 4) code can detect a double error provided it is not required to correct a single error also. If it is assumed that the (7, 4) code is designed to correct a single error (using the hardware of Fig. 5.11) a double error will be ignored. In order to evaluate the effectiveness of the (7, 4) code it is necessary to compare the probability of two errors in seven digits with the probability of one error in four digits. If it is desired to maintain the same information rate the bandwidth must be increased in the ratio of 7:4, which increases the rms noise by a factor of 1.322.

Two examples will be considered: in the first the uncoded error probability is assumed to be $P_e = 10^{-4}$ and in the second case $P_e = 10^{-5}$. The probability of a single error in four digits is thus either $P_t = 4 \times 10^{-4}$ or $P_t = 4 \times 10^{-5}$. If the rms noise is increased by a factor of 1.322 the corresponding error probability for an uncoded system is $P'_e = 2.4 \times 10^{-3}$ or $P'_e = 6.31 \times 10^{-3}$. The probability of an undetected error in the coded case is the probability of two or more errors, i.e.

$$P'_t = {}^7C_2 (P'_e)^2 (1 - P'_e)^5 + {}^7C_3 (P'_e)^3 (1 - P'_e)^4 + {}^7C_4 (P'_e)^4 (1 - P'_e)^3$$
$$+ {}^7C_5 (P'_e)^5 (1 - P'_e)^2 + {}^7C_6 (P'_e)^6 (1 - P'_e) + {}^7C_7 (P'_e)^7$$

This evaluates to $P'_t = 1.29 \times 10^{-4}$ or $P'_t = 8.36 \times 10^{-6}$. In the first case the error rate is reduced from 4×10^{-4} to 1.29×10^{-4} (a ratio of 3.1:1) and in the second case from 4×10^{-5} to 8.36×10^{-6} (a greater ratio of 4.7:1). These are fairly modest decreases, the improvement being greater the smaller the initial error probability. If more efficient single-error-detecting codes are used (e.g. 127, 120 code) a very significant reduction in undetected error probability is obtained.

If the code is required to correct more than one error, then clearly the Hamming distance must be increased. The Hamming distance for a double-error-correcting code is 5, which means that each code word must differ from all other code words in at least five digits. If two errors do occur, the received code word will differ by two digits from the correct code and at least three digits from all other code words.

If it is assumed that there are up to y errors in each code word the number of ways that these errors can occur in $(m + c)$ digits is

$$\sum_{t=1}^{y} {}^{m+c}C_t$$

The c parity digits define $(2^c - 1)$ rows in the parity matrix so that the relationship between m and c for a code that corrects up to y errors is

$$\sum_{t=1}^{y} {}^{m+c}C_t \leqslant 2^c - 1 \tag{5.51}$$

which is known as the Hamming bound. Clearly the greater the error-

Table 5.4 Block code properties

Hamming distance	Code properties
1	None
2	Single-error detecting
3	Single-error correcting **or** double-error detecting
4	Single-error correcting **and** double-error detecting
5	Double-error correcting **or** triple-error detecting
6	Double-error correcting **and** triple-error detecting
7	Triple-error detecting

correcting capability of the code the greater is the redundancy. The relationship between the Hamming distance of a code and its error-combating capabilities is listed in Table 5.4 for Hamming distances up to 7.

Table 5.4 indicates that block codes are best suited to correcting a small number of errors in each block. As the number of errors per block rises then block codes become less and less attractive. The interference encountered on most communications links tends to be impulsive rather than Gaussian and this leads to errors occurring in bursts rather than individually. Under such circumstances several errors can occur in a single block and hence simple block codes are not an effective means of error correction. Some care is necessary in defining the duration of an error burst as such a burst often contains some digits which may not be in error. An error burst is bounded by two erroneous digits and must be separated from the next error burst by a number of correct digits greater than or equal to the number of digits in the error burst. This apparently convoluted definition may be clarified by the example shown in Fig. 5.12. In this figure it becomes clear that errors 1 and 3 could not be used to define a single burst of length 13 digits because error 4 would occur within the following 13 digits. Hence it becomes apparent that errors 1 and 2 define an error burst of five digits and this is followed by errors 3 and 4, which define an error burst of six digits.

The unsuitability of the simple block code may be demonstrated by considering the transmission of 7 bit ASCII characters to which have been added four check bits to provide a single error correcting ability. If such characters are transmitted over a channel subject to burst errors of up to 5 bits duration, then clearly the resulting (11,7) block code would not be able to correct the errors even when the bursts occurred infrequently. This problem can be overcome by a process known as interleaving in which the

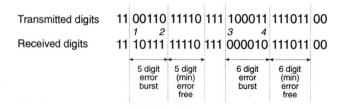

Fig. 5.12 Error bursts in a transmitted sequence.

transmitter assembles five successive 11 bit blocks and transmits the first bit of each block in sequence, followed by the second bit of each block and so on (i.e. the bits of the individual 11 bit blocks are interleaved). The receiver performs the opposite de-interleaving process to reproduce the original 11 bit blocks. In this system if an error burst of up to 5 bits occurs during the transmission of 55 interleaved bits then a maximum of one error per block will occur after the de-interleaving process. Such errors can be corrected by the $(11, 7)$ block code.

5.7 THE CYCLIC REDUNDANCY CHECK MECHANISM

In many instances error coding is restricted to error detection rather than error correction, this is particularly true of packet transmission (see Chapter 11). When an error is detected the transmitter is requested to repeat the data, which clearly requires the provision of a return path between receiver and transmitter. In such circumstances cyclic codes are often employed and these codes are able to detect errors which occur in bursts. Cyclic codes are used to append a cyclic redundancy check (CRC) sequence to a block of information digits and are therefore a form of block code. In general the addition of c parity check digits to a block of m information digits enables any burst of c digits or less to be detected, irrespective of the length of the information block. In addition the fraction of bursts of length $b > c$ which remain undetected by a cyclic code is 2^{-c} if $b > c + 1$. The length of the CRC can be altered to suit the anticipated error statistics but 16 and 32 bits are commonly used. The Ethernet packet shown in Fig. 13.1 appends a CRC of 32 bits to a block of information the length of which varies between 480 and 12 112 bits, which means that if $b > 33$ the fraction of bursts remaining undetected is 2^{-32}, which is a very small number.

The operation of the CRC mechanism can be demonstrated as follows: assume that the information to be transmitted is a binary number M which contains m bits. This is appended by a CRC which may be regarded as a binary number R containing c bits. The CRC is actually the remainder of a division by a another binary number G, known as the generator, which contains $c + 1$ bits. The value of R is obtained from

$$\frac{M \times 2^c}{G} = Q + \frac{R}{G} \tag{5.52}$$

where Q is the quotient. If R/G is added modulo 2 to each side of this equation, then

$$\frac{M \times 2^c + R}{G} = Q \tag{5.53}$$

This indicates that if the number represented by $M \times 2^c + R$ is divided by G, the remainder will be zero. The number $M \times 2^c + R$ is formed by adding the c bit CRC to the original m information bits which are shifted c places to the right. The number so formed has a length $m + c$ bits and is the transmitted code word. If this code word is divided by the same number G at the receiver then, provided that there have been no errors, the remainder will be zero. If errors have occurred the division will result in a non-zero remainder.

It is common practice to represent M, G and R in polynomial form in which case G is known as the generator polynomial. In the case of the Ethernet packet G is 33 bits long and is represented by:

$$G = X^{32} + X^{26} + X^{23} + X^{16} + X^{12} + X^{11} + X^{10} + X^8 + X^7 + X^5$$
$$+ X^4 + X^2 + X + 1 \tag{5.54}$$

In effect the powers of X shown in this polynomial are the powers of 2 in the number G. Thus $G = 100000100100000010001110110110111$.

In theory any polynomial of order $c + 1$ can act as the generator polynomial. However, it is found in practice that not all such polynomials provide good error-detecting characteristics and selection is made by extensive computer simulation.

5.8 CONVOLUTIONAL CODES

Convolutional codes are an alternative to block codes which are more suited to correcting errors in environments in which errors occur in bursts. In a block code the codeword depends only on the m information digits being coded and is therefore memoryless. In convolutional codes a continuous stream of information digits is processed to produce a continuous stream of encoded digits each of which has a value depending on the value of several previous information digits, implying a form of memory. In practice, convolutional codes are generated using a shift register of specified length L, known as the constraint length, and a number of modulo 2 adders (exclusive OR gates). An encoder with a constraint length of 3 and two exclusive OR gates is shown in Fig. 5.13. Clearly, in this example, the output digit rate is double the input digit rate and the code is known as a $\frac{1}{2}$ rate code. (In general the code may be a m/n rate code where m is the number of input digits and n is the number of output digits.)

The constraint length, the number of exclusive OR gates and the way in which they are connected determine the error-correcting properties of the code produced. The goal of the decoder is to determine the most likely input data stream of the encoder given a knowledge of the encoder characteristics

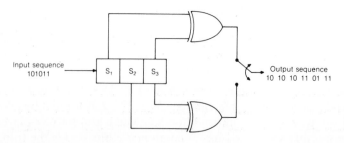

Fig. 5.13 Convolutional encoder.

and the received bit stream. The decoding procedure is equivalent to measuring the distance, in a multi-dimensional space, between the point representing the received sequence and the points representing all possible sequences which could have been produced by the coder. The one with the minimum distance is then chosen as the correct sequence. This minimum distance criterion is similar to that used in the physical interpretation of Shannon's law, presented in Chapter 1. The actual algorithm used in the decoder is known as the Viterbi[4] algorithm and is designed to minimize receiver complexity.

5.9 MULTIPLE DECISIONS

We have assumed in this chapter that the receiver in a system is the only point at which a decision is made. This is not necessarily the case; in a PCM system, for example, a decision is made at each regenerative repeater. Hence when evaluating error probability in a multistage link, the probability of error at each stage must be taken into account. For simplicity we shall assume that the probability of error at each stage in such a link is the same, although this restriction need not apply in a real situation. A typical PCM system is illustrated in Fig. 5.14.

It is evident that errors that may occur at one repeater can be corrected by a second error in the same digit at a subsequent stage. If we consider a particular digit in its passage through a multistage link when it is subject to an even number of incorrect decisions it will be received correctly at the destination.

If there are L decision stages the probability of making U incorrect and $L - U$ correct decision is

$$P = P_e{}^U \times (1 - P_e)^{L-U}$$

but there are ${}^L C_U$ possible ways of making U incorrect decisions in a total of L decisions. Hence for a single digit the probability of making U incorrect decisions in an L stage link is

$$P_t = {}^L C_U P_e \times (1 - P_e)^{L-U} \tag{5.55}$$

The probability of a digit being incorrect at the receiver is found by summing Eqn (5.55) over all odd values of U, i.e.

$$P_m = \sum_{U(\text{odd})} {}^L C_U P_e{}^U (1 - P_e)^{L-U} \tag{5.56}$$

If $P_e \ll 1$ this approximates to $P_m \simeq L P_e$. Thus if P_e is small the probability of a single digit received incorrectly at more than one stage is negligible.

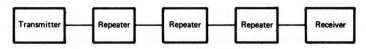

Fig. 5.14 Transmission system with multiple decision points.

However, the use of L decision points makes the individual error L times as likely.

If error coding is employed in a multistage link, equations such as Eqn (5.47) should reflect the higher individual digit error probability, e.g.

$$P_t = {}^nC_r(LP_e)^r(1 - LP_e)^{n-r} \tag{5.57}$$

5.10 CONCLUSION

In this chapter specific attention has been directed towards the effects of Gaussian noise on digital communications and it has been shown that the probability of error is related to the SNR at the receiver. This probability can be reduced either by employing SNR enhancement techniques such as matched filtering or by employing error-coding techniques. The major source of signal impairment on data networks is due to inter-symbol interference, and the noise encountered tends to be impulsive rather than Gaussian. This leads to errors which occur in bursts and the correction of such errors requires the use of specialized codes and techniques of interleaving. This is also the situation in digital cellular radio, described in Chapter 15, where burst errors result from multipath propagation which produces signal fading.

One example where block codes are used is in satellite transmissions, where signal power is at a premium and the received noise is predominantly Gaussian. A second example is covered in detail in Section 11.19, which deals with teletext transmission. In this case page and row address digits are error protected using a code with a Hamming distance of 4 which can correct a single error and detect a double error.

REFERENCES

1. Schwartz, M., Bennett, W. R. and Stein, S., *Communication Systems and Techniques*, McGraw-Hill, New York, 1966.
2. Spiegel, M. R., *Advanced Calculus*, Schaum, 1963.
3. Hamming, R. W., 'Error detecting and correcting codes', *Bell Systems Technical Journal*, **29**, 147 (1950).
4. Virterbi, A. J., 'Convolutional codes and their performance in communications systems', *IEEE Transactions on Communication Technology*, COM **19**(5), (1971).

PROBLEMS

5.1 A binary waveform of amplitude $+2V$ or $-V$ is added to Gaussian noise with zero mean and variance σ^2. The *a priori* signal probabilities are $P(+2V) = \frac{1}{3}$ and $P(-V) = \frac{2}{3}$. If a single sample of signal-plus-noise is taken, determine the decision threshold that minimizes overall error probability.

Answer: $T = V/2 + 0.23\sigma^2/V$.

5.2   A teleprinter system represents each character by a 5 digit binary code that is transmitted as either $A/2$ volts or $-A/2$ volts. The binary signal is received in the presence of Gaussian noise with zero mean at a SNR of 11 dB. Determine the optimum decision level and the probability that the receiver will make an error.

Answer: $T = 0\,V$; $P = 5 \times 0.000\,193$.

5.3   In a binary transmission system, 1 is represented by a raised cosine pulse given by

$$v(t) = V\,[1 + \cos\,(2\pi t/T) \quad \text{for} \quad -T/2 < t < T/2$$

where T is the pulse duration. A binary 0 is represented by the same raised cosine waveform with amplitude $-V$. After transmission over a noisy channel the signal-to-noise power ratio at the detector is 6 dB. If the binary signal is reformed by sampling the received signal at the middle of each pulse interval, find the probability of error. Assume 1 and 0 are equi-probable and that the noise is white.

Answer: 0.000 572.

5.4   Two binary communication links are connected in series, each link having a transmitter and a receiver. If the probability of error in each link is 0.000 01, find the overall probability that:
(a) a 0 is received when a 1 is transmitted;
(b) a 0 is received when a 0 is transmitted;
(c) a 1 is received when a 0 is transmitted;
(d) a 1 is received when a 1 is transmitted.

Answer: (a) 0.000 02; (b) 0.999 98; (c) 0.000 02; (d) 0.999 98.

5.5   If a simple coding scheme is used in the previous question such that each individual digit is repeated three times, what is the probability of deciding incorrectly at the receiver if the following rule is used?
Decide 0 if the received code is 000, 001, 010, 100
Decide 1 if the received code is 111, 110, 101, 011

Answer: 1.2×10^{-9}.

5.6   The noise level on a channel produces an error probability of 0.001 during a binary transmission. In an effort to overcome the effect of noise each digit is repeated five times, and a majority decision is made at the receiver. Find the probability of error in this system if

(a) the digit rate remains constant;
(b) the information rate is half its original value.

Assume white Gaussian noise.

Answer: (a) 9.94×10^{-9}; (b) 1.445×10^{-4}.

5.7   Use Eqns (5.35) and (5.41) to determine the maximum signal and mean square noise voltage at the output of a matched filter.

Answer: E, $E\eta/2$.

5.8   The matched filter is a linear network. This means that when the input to such a filter is Gaussian noise the output noise will also be Gaussian but with a modified variance.
Using this fact, calculate the probability of error in a binary transmission system when the threshold detector is preceded by a matched filter. The transmission

rate is 64 kb/s with 0 and 1 having equal probability. The received waveform has a mean value (normalized) of 0 or 25 mV, and is accompanied by noise with a power spectral density of 1.68×10^{-9} W/Hz.

Answer: 0.000 020 5.

5.9 A signal which is a single pulse of amplitude A volts and duration T seconds is received masked by white Gaussian noise. Calculate the improvement in SNR produced by a matched filter as compared with the SNR at the output of a single-stage RC low-pass filter. Assume the 3 dB cut-off frequency of the RC network is $1/T$ hertz.

Answer: 4.97 dB.

5.10 A FSK receiver consists of two matched filters, one for each of the tone bursts f_0 and f_1. The FSK waveform is fed to both filters, each output being sampled at the instant of maximum signal. The samples are subtracted and then fed to a threshold detector.
Calculate the error probability in terms of the signal energy and noise power spectral density.

Answer: $1 - \mathrm{erf}\sqrt{(E/\eta)}$.

High-frequency transmission lines $\boxed{6}$

The study of transmission lines is the investigation of the properties of the system of conductors used to carry electromagnetic waves from one point to another. Here, however, our attention will be limited to high-frequency applications, i.e. when the length of the transmission line is of at least the same order of magnitude as the wavelength of the signal. In this chapter an idealized model of the line will be used to represent the many different forms found in practice, ranging from twisted pairs to coaxial cables. The theory of transmission lines, which was developed in the early years of the study of electromagnetic propagation, is strictly applicable only to systems of conductors that have a 'go' and 'return' path, or that, in electromagnetic field terms, can support a TEM wave. Hollow-tube waveguides do not fall into this category, although, as we discuss in the chapter on microwaves, many of the concepts of transmission line theory can be applied to them. Transmission line theory is important to communications engineers because it gives them the means for making the most efficient use of the power and equipment at their disposal. By applying their knowledge correctly they can ensure that a transmitting system is designed to transfer as much power as possible from the feeder line into the antenna, or they can take steps to ensure that a receiving antenna is correctly matched to the line that connects it to the receiver itself, so that no power is wasted. The range of systems over which transmission lines are used is as extensive as the subject of communications engineering itself, and the general theory we develop in this chapter may be used to solve a very wide variety of problems.

6.1 VOLTAGE AND CURRENT RELATIONSHIPS ON THE LINE

A transmission line consists of continuous conductors with a cross-sectional configuration that is constant throughout its length. A voltage is placed across the line at the sending end, and it is necessary to be able to determine how it changes with distance so that its value at the load, or at some other point of interest, can be found. Similarly, knowledge of the current on the line may also be required.

To calculate current and voltage values the line must be represented in some way that will allow circuit analysis to be used. However, in such analysis the parameters are discrete and considered to exist at one point, whereas in a transmission line they are evenly distributed throughout its length. This difficulty is overcome by considering a very short length of the line, as we shall see in the next few paragraphs.

The parameters used to describe the line are:

(i)   Resistance (R). The conductors making up the line offer some resistance to the flow of current. R is usually made to include the total resistance of both conductors.

(ii) Inductance (L). The signal on the line is time varying and so there will be an inductive reactance associated with the line. The value of L depends on the cross-sectional geometry of the conductors.

(iii) Conductance (G). The conductors that form the line are held in position by a dielectric material that, because it cannot be a perfect insulator, will allow some leakage current to pass between them.

(iv) Capacitance (C). The conductors, and the dielectric between them, form a capacitor, and there will therefore be a capacitive reactance to a time-varying signal.

These four parameters are usually given per length of line, consequently they must be multiplied by the line length to find the total resistance, inductance, conductance and capacitance of the line.

The continuous distribution of these parameters is approximated by representing the line as a cascaded network of elements, each element having a very short length δz, as shown in Fig. 6.1.

By considering only one of these elements, as shown in Fig. 6.2, the voltage and current dependence on z, the distance from the sending end, can be calculated.

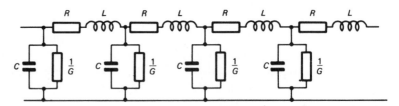

Fig. 6.1 Cascaded network representation of transmission line.

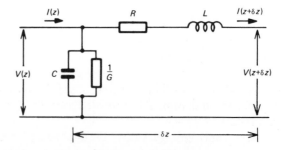

Fig. 6.2 Equivalent circuit for element of length δz.

The current through the shunt circuit is

$$I(z) - I(z + \delta z) = G\delta z V(z) + C\delta z \frac{\partial}{\partial t} V(z) \tag{6.1}$$

and the voltage drop in the series circuit is

$$V(z) - V(z + \delta z) = R\delta z I(z + \delta z) + L\delta z \frac{\partial}{\partial t} I(z + \delta z) \tag{6.2}$$

Note that in these equations the impedance terms have been multiplied by the length of the element, δz. For small δz

$$I(z + \delta z) \simeq I(z) + \frac{\partial I}{\partial z}(z)\delta z$$

Then Eqn (6.1) becomes

$$\frac{\partial I}{\partial z}(z) = -\left(G + C\frac{\partial}{\partial t}\right)V(z) \tag{6.3}$$

Similarly, from Eqn (6.2),

$$\frac{\partial V(z)}{\partial z} = -\left(R + L\frac{\partial}{\partial t}\right)I(z + \delta z) \tag{6.4}$$

i.e.

$$\frac{\partial V(z)}{\partial z} = -\left(R + L\frac{\partial}{\partial t}\right)\left[I(z) + \frac{\partial I(z)}{\partial z}\delta z\right] \tag{6.5}$$

Assuming that the last term, $[\partial I(z)/\partial z]\,\delta z$, is very small compared with the other terms in Eqn (6.5), it can be ignored, and the equation written as

$$\frac{\partial V(z)}{\partial z} = -\left(R + L\frac{\partial}{\partial t}\right)I(z) \tag{6.6}$$

Equations (6.3) and (6.6) may now be combined to give solutions for voltage or current as functions of z. Further simplifications may be made to these equations if it is assumed that the signal on the line is sinusoidal, in which case the time dependence can be expressed by the term $\exp(j\omega t)$, and the time derivative, $\partial/\partial t$, replaced by $j\omega$. If this is done in Eqns (6.3) and (6.6), they become

$$\frac{dI}{dz} = -(G + j\omega C)V \tag{6.7}$$

$$\frac{dV}{dz} = -(R + j\omega L)I \tag{6.8}$$

where the partial differentials have been replaced by total differentials, and V and I are implicitly assumed to be functions of both z and t.
  Differentiating Eqn (6.8) gives

$$\frac{d^2 V}{dz^2} = -(R + j\omega L)\frac{dI}{dz} \tag{6.9}$$

and substituting dI/dz from Eqn (6.7)

$$\frac{d^2V}{dz^2} = (R + j\omega L)(G + j\omega C)V \tag{6.10}$$

More conveniently, Eqn (6.10) is expressed as

$$d^2V/dz^2 = \gamma^2 V \tag{6.11}$$

where

$$\gamma^2 = (R + j\omega L)(G + j\omega C) \tag{6.12}$$

Equation (6.11), which is a standard form of differential equation, has the solution

$$V = V_1 \exp(-\gamma z) + V_2 \exp(\gamma z) \tag{6.13}$$

V_1 and V_2 are determined by the applied voltage and the condition of the line. By a similar analysis, the current is related to z by

$$I = I_1 \exp(-\gamma z) + I_2 \exp(\gamma z) \tag{6.14}$$

Earlier we assumed a time dependence $\exp(j\omega t)$. It is conventional to omit that term from Eqns (6.13) and (6.14), and indeed from the equations in the rest of this chapter, but it is always implicitly assumed because of the time varying signal on the line. We shall return to this later, in Section 6.20.

Equation (6.13) can be expressed in a slightly different form if Eqn (6.12) is examined more closely. γ is a complex quantity and it can therefore be separated into real and imaginary parts, i.e.

$$\gamma = \sqrt{[(R + j\omega L)(G + j\omega C)]} = \alpha + j\beta \tag{6.15}$$

Then, if Eqn (6.15) is used in Eqn (6.13), the total voltage on the line is

$$V = V_1 \exp[-(\alpha + j\beta)z] + V_2 \exp(\alpha + j\beta)z \tag{6.16}$$

$$= V_1 \exp(-\alpha z)\exp(-j\beta z) + V_2 \exp(\alpha z)\exp(j\beta z) \tag{6.17}$$

Thus, the voltage at any point z from the sending end is the sum of two components:

(i) $V_1 \exp(-\alpha z)\exp(-j\beta z)$. The initial voltage V_1 at $z = 0$, is attenuated as it travels down the line, i.e. the amplitude decreases with z as $\exp(-\alpha z)$. The term $\exp(-j\beta z)$ is a phase term and does not influence the amplitude of the voltage. This component is called the forward, or incident, wave.

(ii) $V_2 \exp(\alpha z)\exp(j\beta z)$. Here the amplitude term is of the form $V_2 \exp(\alpha z)$. The component $\exp(\alpha z)$ increases with increasing z, so z must decrease because the voltage must be attenuated as it travels along the line. This component is called the backward, or reflected, wave. It is, as we shall see, produced by a mismatch between the transmission line and the load.

We can therefore conclude that the voltage at a point on the line a distance z from the sending end is the sum of the voltages of the forward and reflected waves at that point.

6.2 LINE PARAMETERS

Referring again to Eqn (6.16), α and β are parameters determined by the characteristics of the line itself, and they have the following interpretation:

α *Attenuation coefficient.* Both the forward and reflected waves are attenuated exponentially at a rate α with distance travelled. α, the real part of Eqn (6.15), is a function of all the line parameters. If R, L, G and C are in their normal units, α is in nepers/m, although it is usually expressed in dB/m.

β *Phase constant.* β, the imaginary part of Eqn (6.15), shows the phase dependence with z of both the forward and backward waves. If z changes from z_1, by a wavelength λ, to $z_1 + \lambda$, the phase of the wave must change by 2π. Therefore

$$\beta(z_1 + \lambda) - \beta z_1 = 2\pi$$

and

$$\beta = 2\pi/\lambda \qquad (6.18)$$

γ *Propagation constant.* The complex sum of the attenuation and phase coefficients, as given by Eqn (6.15), is called the propagation constant because it determines, from Eqn (6.13), how the voltage on the line changes with z.

6.3 CHARACTERISTIC IMPEDANCE

In Eqn (6.14) the current on the line is given as the sum of two component current waves, but by using Eqns (6.8), (6.13) and (6.15) it can be expressed in terms of voltage and the line parameters. From Eqn (6.8)

$$I = -\frac{1}{R + j\omega L}\frac{dV}{dz}$$

From Eqn (6.13)

$$\frac{dV}{dz} = \gamma[V_2 \exp(\gamma z) - V_1 \exp(-\gamma z)]$$

then

$$I = \frac{\gamma}{R + j\omega L}[V_1 \exp(-\gamma z) - V_2 \exp(\gamma z)] \qquad (6.19)$$

Substituting for γ from Eqn (6.15) gives

$$I = \sqrt{\left(\frac{G + j\omega C}{R + j\omega L}\right)}[V_1 \exp(-\gamma z) - V_2 \exp(\gamma z)] \qquad (6.20)$$

This equation is of the form: current $= K \times$ voltage and, by analogy with Ohm's Law, $[(R + j\omega L)/(G + j\omega L)]^{1/2}$ is an impedance. Its value at any particular frequency is determined entirely by the line parameters R, L, G and

C, so it is called the characteristic impedance of the line, Z_0, where

$$Z_0 = \sqrt{[R + j\omega L)/(G + j\omega C)]} \qquad (6.21)$$

It is sometimes helpful to think of Z_0 in one of the following ways:

(i) it is the value which the load impedance must have to match the load to the line;
(ii) it is the impedance seen from the sending end of an infinitely long line;
(iii) it is the impedance seen looking towards the load at any point on a matched line – moving along the line produces no change in the impedance towards the load.

The concept of matching is explained further in the following section.

Having produced the fundamental analysis, we need to look at real problems on transmission lines, and the remainder of this chapter will make considerable use of the equations developed. It is therefore worthwhile to note the assumptions that have been made as we have built up our model of a transmission line:

(i) the line is uniform, homogeneous and straight;
(ii) the line parameters R, L, G and C do not vary with change in ambient conditions, such as temperature and humidity;
(iii) the line parameters are not frequency dependent;
(iv) the analysis applies only between the junctions on the line because the circuit model used in Fig. 6.3 is not valid across a junction.

It is important to be aware that these assumptions have been made and to appreciate that there may be occasions when they should be taken into account.

6.4   REFLECTION FROM THE LOAD

In the last section we noted that a line is matched if it is terminated in a load equal to .its characteristic impedance. When we say the line is matched we mean that the forward wave is totally absorbed by the load; consequently there is no reflected wave, and V_2 is zero.

Usually, however, the load will be different from Z_0, say Z_L, and some of the incident wave will be reflected back down the line. The size of the reflected wave will depend on the difference between Z_0 and Z_L as will be shown in the next section.

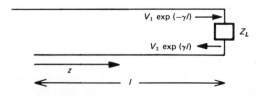

Fig. 6.3 Reflection at the load.

6.5   REFLECTION COEFFICIENT ρ

The amount of reflection caused by the load is expressed in terms of the voltage reflection coefficient ρ. It is defined as the ratio of the reflected voltage to the incident voltage at the load terminals.

The load is at the position $z = l$, as shown in Fig. 6.3, and from Eqn (6.13) the voltage there is

$$V_L = V_1 \exp(-\gamma l) + V_1 \exp(\gamma l) \tag{6.22}$$

The terms $V_1 \exp(-\gamma l)$ and $V_2 \exp(\gamma l)$ are the incident and reflected voltages, respectively, at the load.

So, from the definition of ρ,

$$\rho = V_1 \exp(\gamma l)/V_1 \exp(-\gamma l)$$

$$= (V_2/V_1)\exp(2\gamma l) \tag{6.23}$$

ρ will usually be a complex quantity and it is often convenient to express it in a polar form as

$$\rho = |\rho| \exp(j\psi) \tag{6.24}$$

ψ is referred to as the angle of the reflection coefficient. Now we can find an expression for ρ in terms of the load and the characteristic impedance of the line.

The current at the load, from Eqns (6.20) and (6.21), is

$$I_L = \frac{V_1}{Z_0}\exp(-\gamma l) - \frac{V_2}{Z_0}\exp(\gamma l) \tag{6.25}$$

and the load impedance is

$$Z_L = \frac{V_L}{I_L}$$

Therefore, from Eqns (6.25) and (6.22),

$$Z_L = \frac{V_1 \exp(-\gamma l) + V_2 \exp(\gamma l)}{(V_1/Z_0)\exp(-\gamma l) - (V_2/Z_0)\exp(\gamma l)}$$

Dividing by $V_1 \exp(-\gamma l)$

$$Z_L = Z_0 \left[\frac{1 + (V_2/V_1)\exp(2\gamma l)}{1 - (V_2/V_1)\exp(2\gamma l)} \right]$$

and from Eqn (6.23)

$$Z_L = Z_0 \left(\frac{1 + \rho}{1 - \rho} \right) \tag{6.26}$$

or

$$\rho = (Z_L - Z_0)/(Z_L + Z_0) \tag{6.27}$$

Consider the following three cases:

(i) *Short-circuit load, $Z_L = 0$*
From Eqn. (6.27), $\rho = -1$. Hence, from Eqn (6.24), $|\rho| = 1$ and $\psi = \pi$.

(ii) *Open-circuit load, $Z_L = \infty$*
   From Eqn (6.27), $\rho = 1$, so $|\rho| = 1$ and $\psi = 0$.

(iii) $Z_L = 150 + j\,100\,\Omega$, $Z_0 = 50\,\Omega$
   Then

$$\rho = (100 + j\,100)/(200 + j\,100)$$

from which $\rho = 0.6 + j\,0.2$
which gives $|\rho| = 0.63$ and $\psi = 18.43$.

Each of these cases is shown diagrammatically in Fig. 6.4 in terms of incident and reflected voltages.

So far we have been interested in the voltage on the line and the voltage reflection coefficient. However, we would expect the current wave to be

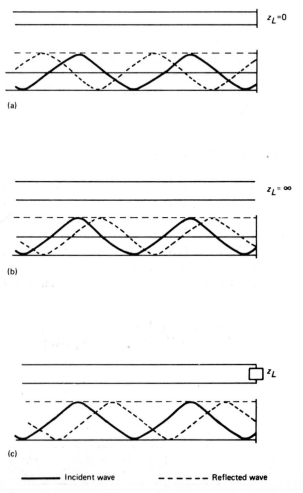

Fig. 6.4 (a) Voltage reflections at a short circuit, $|\rho| = 1$, $\psi = \pi$; (b) voltage reflections at an open circuit, $|\rho| = 1$, $\psi = 0$; (c) voltage reflections at a typical load z_L, $\rho = |\rho| \exp(j\psi)$.

reflected also, and by an analysis similar to that used above the amount of reflection can be seen to depend on the load impedance, as in the voltage case. This time, however, the current reflection coefficient, $\rho_I = |\rho_I| \exp(j\psi_I)$ is given by

$$\rho_I = (Z_0 - Z_L)/(Z_0 + Z_L)$$

For a short-circuit load $\rho_I = 1$ ($|\rho_I| = 1$, $\psi = 0$), which means that there is no change of phase between the incident and reflected waves – as we would expect.

From the discussion about reflection coefficients it is evident that the phase of the reflected wave relative to the incident wave is determined by Z_L and Z_0. However, we have not justified our implicit assumption that the incident current and voltage waves are in phase. Their phase relationship depends on Z_0, as can be seen by considering the case where there are no reflected waves, i.e. $V_2 = I_2 = 0$. Then

$$Z_0 = V_1 \exp(-\gamma z)/I_1 \exp(-\gamma z)$$

and if Z_0 is real, V_1 and I_1 will be in phase. In some applications Z_0 cannot be assumed to be real and then V_1 and I_1 will be out of phase.

In the following part of this chapter, the term reflection coefficient will apply to the voltage reflection coefficient given by Eqns (6.23), (6.24) and (6.27).

6.6 SENDING-END IMPEDANCE

It is often of interest to know the impedance that the combination of transmission line and load presents to the source, so that the degree of mismatch between source and line can be determined. The impedance looking into the line from the source, or generator, is called the sending-end impedance. It can be found in terms of the characteristic impedance Z_0, the load impedance Z_L, and the length of the line, l, as follows.

The impedance at some point A (Fig. 6.5) a distance z from the generator is, from Eqns (6.13), (6.20) and (6.21),

$$Z_A = \frac{V_A}{I_A} = Z_0 \left[\frac{V_1 \exp(-\gamma z) + V_2 \exp(\gamma z)}{V_1 \exp(-\gamma z) - V_2 \exp(\gamma z)} \right]$$

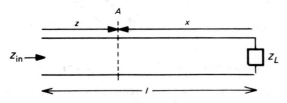

Fig. 6.5 Reference distance to calculate Z_{in}.

Substituting for V_2/V_1 from Eqn (6.23),

$$Z_A = Z_0 \left[\frac{\exp(-\gamma z) + \rho \exp(-2\gamma l)\exp(\gamma z)}{\exp(-\gamma z) - \rho \exp(-2\gamma l)\exp(\gamma z)} \right]$$

or

$$Z_A = Z_0 \left[\frac{\exp \gamma(l - z) + \rho \exp -\gamma(l - z)}{\exp \gamma(l - z) - \rho \exp -\gamma(l - z)} \right]$$

From Fig. 6.5, $x = l - z$; then

$$Z_A = Z_0 \left[\frac{\exp(\gamma x) + \rho \exp(-\gamma x)}{\exp(\gamma x) - \rho \exp(-\gamma x)} \right]$$

Replacing ρ from Eqn (6.27)

$$Z_A = Z_0 \left[\frac{(Z_L + Z_0)\exp(\gamma x) + (Z_L - Z_0)\exp(-\gamma x)}{(Z_L + Z_0)\exp(\gamma x) - (Z_L - Z_0)\exp(-\gamma x)} \right]$$

Rearranging terms

$$Z_A = Z_0 \left[\frac{Z_L[\exp(\gamma x) + \exp(-\gamma x)] + Z_0[\exp(\gamma x) - \exp(-\gamma x)]}{Z_L[\exp(\gamma x) - \exp(-\gamma x)] + Z_0[\exp(\gamma x) + \exp(-\gamma x)]} \right]$$

The exponentials can be replaced by hyperbolic functions to give

$$Z_A = Z_0 \left[\frac{Z_L \cosh \gamma x + Z_0 \sinh \gamma x}{Z_L \sinh \gamma x + Z_0 \cosh \gamma x} \right]$$

or

$$Z_A = Z_0 \left[\frac{Z_L + Z_0 \tanh \gamma x}{Z_L \tanh \gamma x + Z_0} \right] \tag{6.28}$$

The sending-end impedance, Z_{in}, is obtained from Eqn (6.28) by putting $x = l$, i.e.

$$Z_{in} = Z_0 \left[\frac{Z_L + Z_0 \tanh \gamma l}{Z_L \tanh \gamma l + Z_0} \right] \tag{6.29}$$

For general application of this equation it is often more useful if it is normalized to the characteristic impedance of the line. Normalized impedances will be shown as lower-case letters, e.g.

$$z_{in} = Z_{in}/Z_0 \quad \text{and} \quad z_L = Z_L/Z_0$$

From Eqn (6.29),

$$\frac{Z_{in}}{Z_0} = \left[\frac{(Z_L/Z_0) + \tanh \gamma l}{1 + (Z_L/Z_0)\tanh \gamma l} \right]$$

or

$$z_{in} = \frac{z_L + \tanh \gamma l}{1 + z_L \tanh \gamma l} \tag{6.30}$$

6.7   LINES OF LOW LOSS

In Eqn (6.12) we noted that the propagation constant is related to the line parameters by

$$\gamma = (R + j\omega L)^{1/2}(G + j\omega C)^{1/2}$$

$$= j\omega \sqrt{(LC)}\left(1 + \frac{R}{j\omega L}\right)^{1/2}\left(1 + \frac{G}{j\omega C}\right)^{1/2} \qquad (6.31)$$

Using the binomial series, this equation can be expanded to

$$\gamma = j\omega \sqrt{(LC)}\left(1 + \frac{R}{2\,j\omega L} - \frac{1}{4}\frac{R^2}{(j\omega L)^2} + \cdots\right)$$

$$\times \left(1 + \frac{G}{2\,j\omega C} - \frac{G^2}{4(j\omega C)^2} + \cdots\right) \qquad (6.32)$$

Now, in low-loss lines the impedance parameters R and G will be very small, allowing all terms in R^2 and G^2, and higher powers, to be ignored. Then

$$\gamma \simeq j\omega \sqrt{(LC)}\left[1 - \frac{RG}{4\omega^2 LC} - j\frac{R}{2\omega L} - j\frac{G}{2\omega C}\right] \qquad (6.33)$$

and since $\gamma = \alpha + j\beta$ we find that

$$\alpha \simeq \frac{R}{2}\sqrt{\frac{C}{L}} + \frac{G}{2}\sqrt{\frac{L}{C}} \qquad (6.34)$$

and

$$\beta \simeq j\omega \sqrt{(LC)}\left[1 - \frac{RG}{4\omega^2 LC}\right] \qquad (6.35)$$

Since R and G are small, we can reasonably assume that at high frequencies the second term in the bracket is negligible. Then

$$\beta \simeq j\omega \sqrt{(LC)} \qquad (6.36)$$

By a similar argument

$$Z_0 = (R + j\omega L)^{1/2}(G + j\omega C)^{-1/2}$$

$$= \sqrt{\left(\frac{j\omega L}{j\omega C}\right)\left(1 + \frac{R}{j\omega L}\right)^{1/2}\left(1 + \frac{G}{j\omega C}\right)^{-1/2}}$$

$$\simeq \sqrt{\left(\frac{L}{C}\right)\left[\left(1 + \frac{R}{2\,j\omega L}\right)\left(1 - \frac{G}{2\,j\omega C}\right)\right]}$$

if, as before, all terms in R^2, G^2 and above are ignored.

Expanding, and again assuming that the term in RG/ω^2 is negligible,

$$Z_0 \simeq \sqrt{\left(\frac{L}{C}\right)\left[1 + \frac{jR}{2\omega L} - \frac{jG}{2\omega C}\right]}$$

For many applications $R/\omega L$ and $G/\omega C$ will be very small, and a reasonable approximation for Z_0 is

$$Z_0 \simeq \sqrt{(L/C)} \qquad (6.37)$$

Note that when this approximation is valid, Z_0 is real. Using Eqn (6.37) in Eqn (6.34) gives

$$\alpha \simeq \frac{R}{2Z_0} + \frac{GZ_0}{2} \qquad (6.38)$$

6.8 LOSSLESS LINES

In many communications problems concerning transmission lines it is often a reasonable approximation to assume that the line is lossless, i.e. the attenuation is zero. The advantage of making this assumption is that it greatly simplifies the calculations involved, and the disadvantage of a loss in accuracy is not serious when the line is relatively short and operating at very high frequencies.

The effect of zero attenuation is to make the characteristic impedance real, as in Eqn (6.37), and to make the propagation constant imaginary, i.e.

$$\gamma = j\beta$$

We can see immediately the simplification that can result if we replace γ in Eqn (6.29) by $j\beta$. The sending-end impedance is then given by

$$Z_{\text{in}} = Z_0 \left[\frac{Z_L + jZ_0 \tan \beta l}{Z_0 + j Z_L \tan \beta l} \right] \qquad (6.39)$$

using

$$\tanh j\beta l = j \tan \beta l.$$

Similarly, Eqn (6.30) simplifies to

$$z_{\text{in}} = \frac{z_L + j \tan \beta l}{1 + j z_L \tan \beta l} \qquad (6.40)$$

6.9 QUARTER-WAVE TRANSFORMER

The relationship between the input and load impedances of a quarter-wave lossless line can be found easily from Eqn (6.39) expressed as

$$Z_{\text{in}} = Z_0 \left[\frac{Z_L/\tan \beta l + jZ_0}{Z_0/\tan \beta l + jZ_L} \right] \qquad (6.41)$$

When l is a quarter-wavelength long, $\tan \beta l = \tan \pi/2 = \infty$. Then

$$Z_{\text{in}} = Z_0^2/Z_L \qquad (6.42)$$

or

$$Z_{\text{in}} Z_L = Z_0^2 \qquad (6.43)$$

We therefore have a method for matching a known load to a known input (line) impedance by placing between them a quarter-wave section of lossless line of characteristic impedance calculated from Eqn (6.43).

This type of transformer is very frequency sensitive because $\tan \theta$ varies rapidly about $\theta = \pi/2$. Problem 6.10 looks further into this sensitivity.

6.10   STUBS

Froms Eqns (6.30) and (6.40) we can see that the input impedance of a transmission line varies with its length. This property can be used in short lengths of line, known as stubs, to provide an adjustable impedance (or admittance) for use in matching applications, as we shall discuss later. Stubs are terminated in either a short-circuit or open-circuit load. If the load is a short circuit, $z_L = 0$, and from Eqn (6.40), if we assume that the stub is formed from a lossless length of line, the input impedance of the line is

$$z_{in} = j \tan \beta l \qquad (6.44)$$

If the load is an open circuit, $z_L = \infty$, and

$$z_{in} = -j \cot \beta l \qquad (6.45)$$

The variations of the input impedance with the length of the stub is shown in Fig. 6.6. It can be seen that the impedance is always reactive and that its value is repeated at half-wavelength intervals.

The use of stubs as matching devices will be discussed in a later section.

6.11   STANDING WAVES

The total voltage on a lossless line at some point z from the sending end is, from Eqn (6.17),

$$V = V_1 \exp(-j\beta z) + V_2 \exp(j\beta z)$$

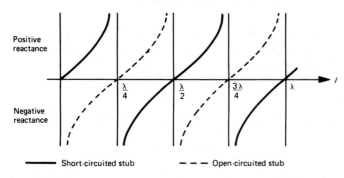

Fig. 6.6 Stub reactance.

and, remembering the time dependence,

$$V = [V_1 \exp(-j\beta z) + V_2 \exp(j\beta z)] \exp(j\omega t)$$

Rerranging the right-hand side, and substituting for V_2/V_1 from Eqn (6.23),

$$V = V_1 \exp(j\omega t) \exp(-j\beta l) [\exp(j\beta x) + \rho \exp(-j\beta x)] \qquad (6.46)$$

where $x = l - z$ as shown in Fig. 6.5.

This equation represents a voltage standing wave, i.e. a stationary wave composed of two component travelling waves, one in the forward direction and the other in the backward direction, reflected from the load.

The precise shape of the voltage standing wave defined by Eqn (6.46) will depend on the value of ρ, which may be complex. In practice, a simple analytical solution of that equation may not be easy to obtain. However, as we noted earlier, when the load is a short or open circuit, ρ has the values -1 or $+1$, respectively, and in each of these cases simple trigonometric solutions of Eqn (6.46) can be found. Without the time dependence, Eqn (6.46) is

$$V = V_1 \exp(-j\beta l) [\exp(j\beta x) + \rho \exp(-j\beta x)] \qquad (6.47)$$

which, for a short-circuit load ($\rho = -1$), becomes

$$V = V_1 \exp(-j\beta l) j^2 [\exp(j\beta x) - \exp(-j\beta x)]/j^2$$
$$= j2V_1 \exp(-j\beta l) \sin \beta x \qquad (6.48)$$

This standing wave could be measured by using a standing-wave detector which displays the real part of the modulus of Eqn (6.48), i.e.

$$|V| = \mathrm{Re}\,[2V_1 \exp(-j\beta l)|\sin \beta x|] \qquad (6.49)$$

as shown in Fig. 6.7(a).

From Eqn (6.49) we can see that minima will occur on the line when $\beta x = (n-1)\pi$, where n is any positive integer. From this expression we can see that adjacent minima will be separated by π/β; that is, by $\lambda/2$, because

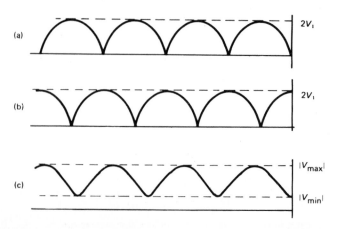

Fig. 6.7 Standing waves: (a) produced by a short circuit; (b) produced by an open circuit; (c) produced by a load z, with reflection coefficient $0.6 + j0.3$.

$\beta = 2\pi/\lambda$. The first voltage minimum will be at the load terminals, $x = 0$, as we have concluded earlier.

In a similar way we can use Eqn (6.47) to show that if there is an open circuit at the load the detected voltage will be

$$|V| = \mathrm{Re}\left[2V_1\exp(-j\beta l)\,|\cos\beta x|\right]$$

which again has minima separated by $\lambda/2$, but with a maximum at the load. The standing wave due to an open circuit is shown in Fig. 6.7(b).

For any other load, ρ may be complex and, as an example, the standing wave for $\rho = 0.6 + j0.3$ is shown in Fig. 6.7(c). It can be seen that the minima do not fall to zero and the maxima do not rise to $2V_1$. However, adjacent minima are still half a wavelength apart, and indeed they will be for all values of ρ.

Current standing waves also exist and they can be discussed in exactly the same way but they will not be considered any further here.

6.12 VOLTAGE STANDING WAVE RATIO (VSWR)

The general shape of the voltage standing wave pattern on a lossless line is shown in Fig. 6.7(c). By using a detector, and moving it along the line, values for V_{max} and V_{min} can be obtained. The ratio of these two is called the voltage standing wave ratio, or VSWR. We shall use the symbol S to denote VSWR and define it as

$$S = \frac{|V_{\mathrm{max}}|}{|V_{\mathrm{min}}|} \tag{6.50}$$

which allows S to take values in the range $1 \leqslant S \leqslant \infty$. The value of S can often be determined experimentally without too much difficulty by using a slotted line.[1] Its value depends on the degree of mismatch at the load, i.e. on the reflection coefficient, as can be seen if the complex form of ρ given in Eqn (6.24) is put into Eqn (6.47):

$$V = V_1\exp(-j\beta l)\left[\exp(j\beta x) + |\rho|\exp(j(\psi - \beta x))\right]$$
$$= V_1\exp(-j\beta(l-x))\left[1 + |\rho|\exp(j(\psi - 2\beta x))\right] \tag{6.51}$$

so that

$$|V_{\mathrm{max}}| = V_1[1 + |\rho|] \tag{6.52}$$

when

$$(\psi - 2\beta x) = 2(m-1)\pi \quad m = 1, 2, 3 \ldots$$

and

$$|V_{\mathrm{min}}| = V_1[1 - |\rho|] \tag{6.53}$$

when

$$(\psi - 2\beta x) = 2(m-1)\pi \quad m = 1, 2, 3 \ldots$$

Therefore, from Eqns (6.52) and (6.53), the standing wave ratio is

$$S = \frac{|V_{min}|}{|V_{min}|} = \frac{1 + |\rho|}{1 - |\rho|} \tag{6.54}$$

We will find it convenient later to use this expression in the form

$$|\rho| = \frac{S - 1}{S + 1} \tag{6.55}$$

6.13 IMPEDANCE AT A VOLTAGE MINIMUM AND AT A VOLTAGE MAXIMUM

The normalized impedance at a voltage minimum can be derived from Eqn (6.51):

$$V = V_1 \exp[-j\beta(l - x)]\{1 + |\rho| \exp[j(\psi - 2\beta x)]\}$$

A voltage minimum occurs when

$$\exp[j(\psi - 2\beta x)] = \exp(j\pi)$$

and this coincides with an impedance minimum.

At the first voltage minimum from the load, when $x = x_{min}$,

$$\psi = \pi + 2\beta x_{min}$$

and at x_{min}

$$Z = Z_{min} = \left(\frac{V}{I}\right)_{x_{min}}$$

Putting $x = x_{min}$ in Eqn (6.51) gives $V_{x_{min}}$ and $I_{x_{min}}$ is found by putting $z = l - x_{min}$ in Eqn (6.20).

Then

$$Z_{min} = \frac{\{V_1 \exp[-j\beta(l - x_{min})]\}\{1 + |\rho| \exp[j(\psi - 2\beta x_{min})]\}}{\{V_1 \exp[-j\beta(l - x_{min})]\}\{1 - |\rho| \exp[j(\psi - 2\beta x_{min})]\}}$$

$$= Z_0 \frac{[1 + |\rho| \exp(j\pi)]}{[1 - |\rho| \exp(j\pi)]} = Z_0 \left[\frac{1 - |\rho|}{1 + |\rho|}\right] = Z_0/S \tag{6.56}$$

or

$$z_{min} = 1/S \tag{6.57}$$

By a similar argument, we can show that

$$Z_{max} = Z_0 S \tag{6.58}$$

and

$$z_{max} = S \tag{6.59}$$

6.14   LOAD IMPEDANCE ON A LOSSLESS LINE

There are often occasions when the load impedance is not known, but its value is required. It can be found, if the line is assumed to be lossless, by measuring the VSWR, the wavelength and the distance from the load to the nearest voltage minimum.

Re-writing Eqn (6.51)

$$V = V_1 \exp[-j\beta(l-x)]\{1 + |\rho|\exp[j(\psi - 2\beta x)]\}$$

and, as we noted in the last section, this is a minimum when

$$\exp[j(\psi - 2\beta x)] = -1$$

i.e. when

$$\psi - 2\beta x = (2m-1)\pi \quad m = 1, 2, 3 \dots$$

The solution of interest is that for the smallest value of x, which we have called x_{min}. It occurs when $m = 1$; then

$$\psi - 2\beta x_{min} = \pi$$

or

$$\psi = \pi + 2\beta x_{min} \tag{6.60}$$

The load impedance and the reflection coefficient are related by

$$Z_L = Z_0 \left(\frac{1+\rho}{1-\rho}\right)$$

i.e. using Eqn (6.24),

$$Z_L = Z_0 \left(\frac{1 + |\rho|\exp(j\psi)}{1 - |\rho|\exp(j\psi)}\right)$$

and substituting for $|\rho|$ from Eqn (6.55),

$$Z_L = Z_0 \left(\frac{1 + [(S-1)/(S+1)]\exp(j\psi)}{1 - [(S-1)/(S+1)]\exp(j\psi)}\right)$$

ψ can be replaced by Eqn (6.60):

$$Z_L = Z_0 \left[\frac{1 + [(S-1)/(S+1)]\exp[j(\pi + 2\beta x_{min})]}{1 - [(S-1)/(S+1)]\exp[j(\pi + 2\beta x_{min})]}\right] \tag{6.61}$$

Rearranging Eqn (6.61), and remembering that $e^{j\pi} = -1$,

$$Z_L = Z_0 \left[\frac{S[1 - \exp(j2\beta x_{min})] + [1 + \exp(j2\beta x_{min})]}{S[1 + \exp(j2\beta x_{min})] + [1 - \exp(j2\beta x_{min})]}\right]$$

Dividing throughout by $\exp(j\beta x_{min})$ gives

$$Z_L = Z_0 \left[\frac{S[\exp(-j\beta x_{min}) - \exp(j\beta x_{min})] + [\exp(-j\beta x_{min}) + \exp(j\beta x_{min})]}{S[\exp(-j\beta x_{min}) + \exp(j\beta x_{min})] + [\exp(-j\beta x_{min}) - \exp(j\beta x_{min})]}\right]$$

or

$$Z_L = Z_0 \left[\frac{-jS \sin \beta x_{min} + \cos \beta x_{min}}{S \cos \beta x_{min} - j \sin \beta x_{min}} \right]$$

$$= Z_0 \left[\frac{1 - jS \tan \beta x_{min}}{S - j \tan \beta x_{min}} \right] \tag{6.62}$$

The normalized load impedance is therefore

$$z_L = \frac{1 - jS \tan \beta x_{min}}{S - j \tan \beta x_{min}} \tag{6.63}$$

6.15 SMITH TRANSMISSION LINE CHART

Transmission line problems can be solved using the equations we have derived in the earlier part of this chapter, but often there is a considerable amount of tedious algebraic manipulation involved that can make calculations rather lengthy. To overcome this problem, several graphical techniques have been developed. We will study only one – the Smith chart, named after P. H. Smith who introduced it in 1944.[2] It is the best known and most comprehensive of all graphical methods, and compared with analysis it has the considerable additional benefit of providing a picture of the conditions on the line. This allows us to assess what sort of adjustments might be necessary to produce the conditions we require.

The derivation of the chart is not difficult to follow, but it is easier to appreciate after some familiarity with its use has been developed, so a discussion on why it has its particular shape will be deferred until a later section. The most commonly used version of the Smith chart is shown in Fig. 6.8. Some time should be spent in reading the various legends inside the chart and on the circumferential scales. You will notice that there are three sets of circular arcs in the chart, and one straight line. The chart is designed to allow any value of impedance or admittance to be represented, and in the examples that we study later it will be used in both modes. The horizontal line dividing the chart into an upper and a lower half is the locus of impedances and admittances having no reactive or susceptive component. Along that line the following equations are satisfied

$$z = r + j0$$

$$y = g + j0$$

Figure 6.9(a) shows one set of circules. They all have their centres on the horizontal line, and they are contiguous at the right-hand end of the chart. These circles give the normalized resistive or conductive component of impedance or admittance, i.e. the r or g in the above equations. The centre of the chart, where the circle representing unity normalized resistance or conductance crosses the horizontal line, has the coordinates $(1, 0)$. It therefore represents the impedance of the line under consideration. When we discuss

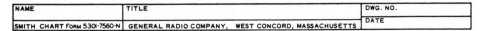

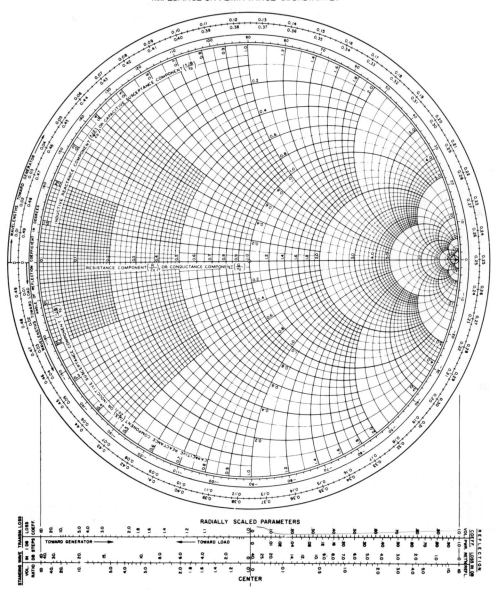

Fig. 6.8 Smith chart. Reproduced by permission of Phillip H. Smith, Analog Instruments Company, New Providence, N.J.

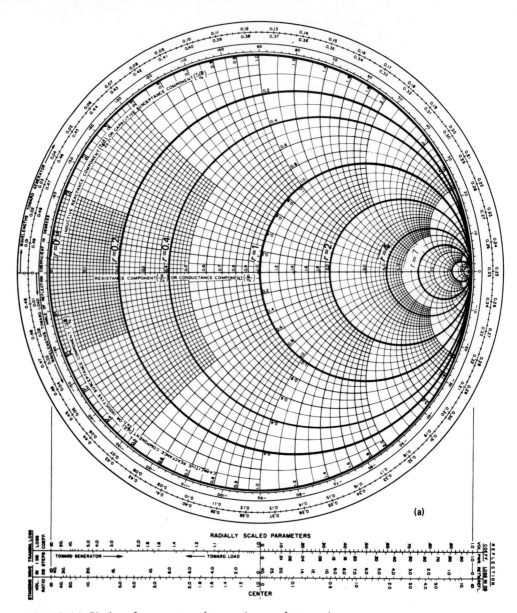

Fig. 6.9 (a) Circles of constant resistance (or conductance).

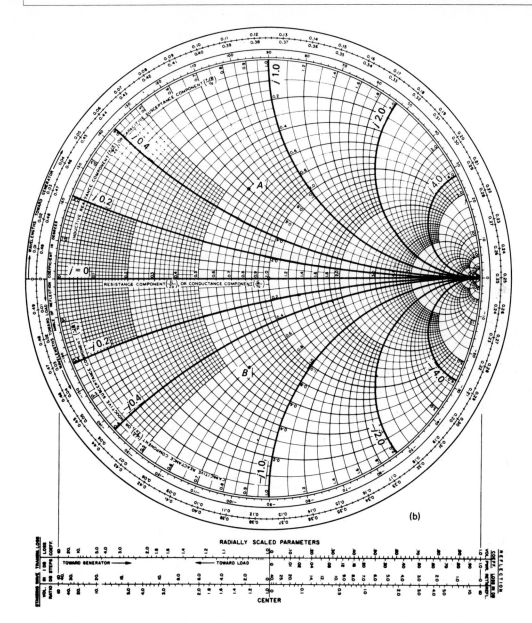

Fig. 6.9 (b) Circles of constant positive reactance (or susceptance), and circles of constant negative reactance (or susceptance).

methods of matching, the object will be to change the impedance at the matching point to the value represented by $(1, 0)$.

The resistance, or conductance, circles range from zero at the outer periphery to infinity at the right-hand side, although the highest value of normalized resistance or conductance shown is 50. Another set of circular arcs is shown in Fig. 6.9(b). These are lines that represent the positive imaginary components of impedance or admittance. For example, the point A has an impedance or admittance of $0.6 + j0.6$.

The third set of arcs, shown in Fig. 6.9(b), is in the lower half of the chart and represents negative imaginary components of impedance or admittance. Point B has an impedance or admittance of $0.6 - j0.6$.

The arcs for the imaginary components are parts of circles that have their centres on a vertical line through the right-hand end of the horizontal line and they all touch at that point (see Section 6.19). Only those parts of the arcs that occur in the range of positive real components are shown.

Another feature that will be used regularly can be seen from Fig. 6.10. Plot $z = 1.6 - j2.0$ on the chart. Then draw a circle, with centre at $(1, 0)$ through that point. From z draw a diameter to cut the circle at the opposite side. The point of intersection has a value $y = 0.25 + j0.31$ and comparing y and z it is found that $z = 1/y$. This relationship applies in general, i.e. the relative impedance and admittance are at opposite ends of a diameter of the circle, with centre $(1,0)$ passing through them. The inner scale round the chart is marked 'angle of reflection coefficient in degrees'. It relates the characteristic impedance of the line, at $(1, 0)$, to any load impedance, e.g. in Fig. 6.10, if point A represents a load of $1.2 + j1.4$, the angle of the reflection coefficient can be read from the intersection of a line from $(1, 0)$ through A, and the inner scale at B, to be $\psi = 49.2°$.

The magnitude of the reflection coefficient, $|\rho|$, is related to the distance between $(1,0)$ and A. Below the chart there are eight radial scales, and the uppermost one on the right is marked 'Voltage Reflection Coefficient'. $|\rho|$ is found by measuring the distance from $(1,0)$ to A along that scale from the centre. In the example given $|\rho| = 0.541$.

The other important radial scale, which is related through Eqn (6.54) to that of $|\rho|$, is the lowest one on the left, marked 'Standing Wave Voltage Ratio'. We know that if $|\rho|$ is constant, S is also represented by a circle. As an example, Fig. 6.10 shows the circle for $S = 4.5$. Its radius is found from the bottom-left radial scale, but we can see from the figure that the circle passes through the point 4.5 on the horizontal line across the chart. The resistance scale to the right of the centre of the chart is identical with the VSWR scale, and either may be used.

It will help if we consider an example of how these various aspects of the Smith chart are used.

EXAMPLE: A lossless transmission line of characteristic impedance $50\,\Omega$ is terminated in a load of $150 + j75\,\Omega$. Find the reflection coefficient, the load admittance, the VSWR, the distance between the load and the nearest voltage minimum to it, and the input impedance, if the line is 92 cm long and the wavelength of the signal on the line is 40 cm.

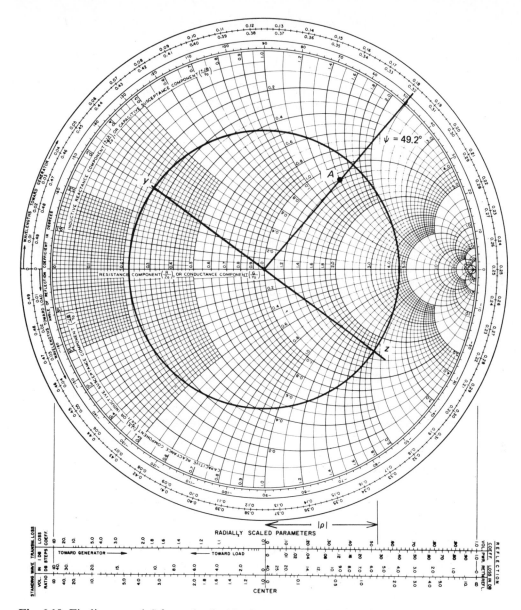

Fig. 6.10 Finding ρ and S from the Smith chart.

The normalized load impedance is z_L where

$$z_L = \frac{150 + j75}{50} = 3 + j1.5$$

and it is plotted on the chart as shown in Fig. 6.11. Then the S circle is drawn through z_L. We can now read the value of VSWR to be $S = 3.8$. By drawing a line from $(1, 0)$ through z_L to the Angle of Reflection Coefficient scale we can read that $\psi = 16°$; $|\rho|$ is measured on the appropriate radial scale to be 0.585. Thus we have found that $\rho = 0.585 \exp(j0.28)$.

The load admittance y_L is on the same S circle as z_L, and at the opposite end of a diameter. Thus, from the chart, $y_L = 0.27 - j0.13$.

The line is 92 cm long which is equivalent to $(92/40)\lambda = 2.3\lambda$. Therefore, to find the input impedance z_{in} we must go from the load a distance 2.3λ towards the generator, on a constant S circle. This we can do by noting where the line from the centre through z cuts the Wavelengths Towards Generator scale. Reading from the chart, this occurs at 0.227λ. The input impedance is therefore at $(2.3 + 0.227)\lambda$ on the Wavelengths Towards Generator scale, i.e. at 2.527λ. One complete revolution on an S circle is 0.5λ; therefore in 2.527λ there are five revolutions $+ 0.027\lambda$. Hence the input impedance is at 0.027λ on the Wavelengths Towards Generator scale and reading from the chart, $z_{in} = 0.27 + j0.16$.

Finally, we want to know the distance of the nearest voltage minimum to the load from the load itself, x_{min}. The normalized impedance at a voltage minimum, which has been shown to be $1/S$ [Eqn (6.57)], must lie on the horizontal line to the left of $(1, 0)$. Then x_{min} is found from the chart by reading the distance from z_L to V_{min} on the Wavelengths Towards Generator scale, and V_{min} will be at point C on the chart which reads 0.0 on the scale. Therefore V_{min} is 0.273λ from the load.

All of these results can be obtained by using the equations which were developed earlier, and to do so would be a very useful exercise.

The decision on whether to use the chart in the admittance or impedance mode depends on the application. If lines are joined in series then, because the impedances at the junction can be added, the impedance chart should be used. Alternatively, if there are parallel junctions on the line, then admittances would be used. The differences in the use of the chart can be summarized as follows (see Fig. 6.12).

1. When the chart is used for impedance
   (a) The intersection of the circle S with the line $r + j0$ gives the maximum impedance on the line at the intersection to the right of (1,0), and the minimum impedance at the intersection to the left.
   (b) Voltage minima are at impedance minima.
2. When the chart is used for admittance
   (a) The intersection of the circle S with the line $g + j0$ gives the maximum admittance at the intersection to the right of $(1, 0)$, and the minimum admittance at the intersection to the left.
   (b) Voltage minima are at admittance maxima.

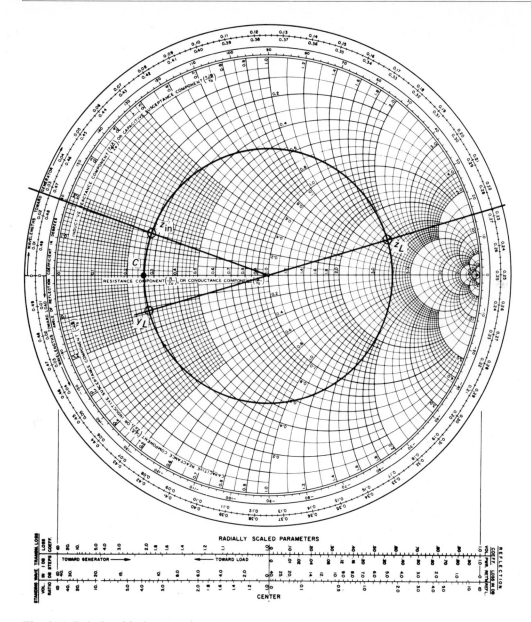

Fig. 6.11 Relationship between impedance and admittance.

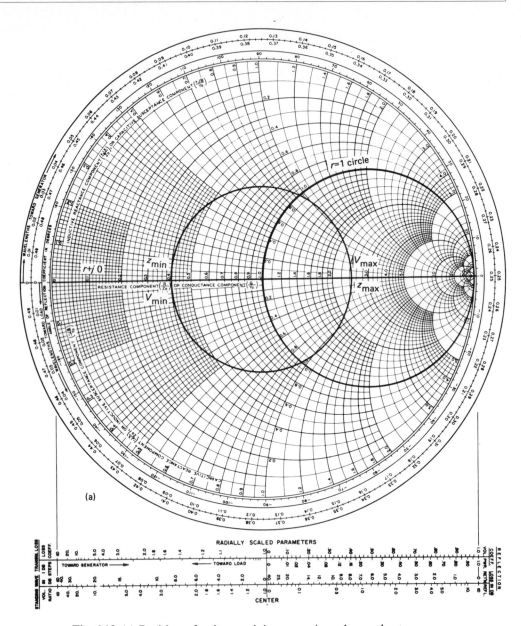

Fig. 6.12 (a) Position of voltage minimum on impedance chart.

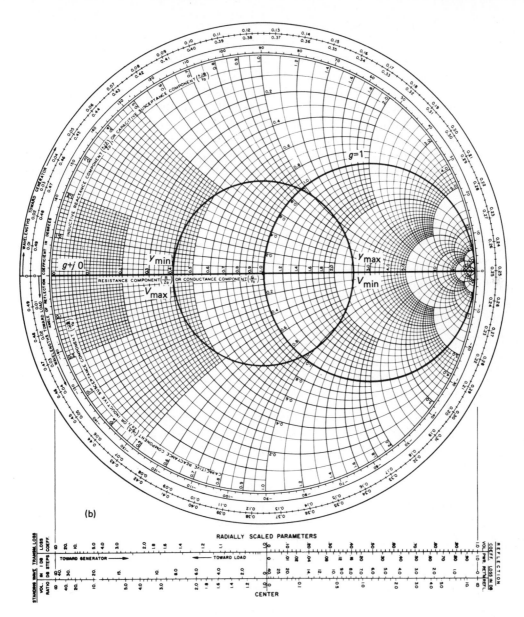

Fig. 6.12 (b) Position of voltage minimum on admittance chart.

6.16 STUB MATCHING

The presence of a reflected wave and a high VSWR, due to a mismatched load, can have undesirable effects on both the generator and the line. The reflected wave may either interfere with the performance of the generator or cause it some damage. The voltage maximum of the standing wave may, in high-voltage applications, produce breakdown stresses in the dielectric. Therefore there is some value in arranging for as much of the line as possible to have a VSWR of unity so that there is no reflected wave at the generator, and this can be done by using stubs to match the line to the load at some point near the load.

6.17 SINGLE-STUB MATCHING

Moving from the load towards the generator the immittance (impedance or admittance) of the line will, at some point, have a normalized real part of unity. If a stub is inserted there and its length adjusted so that it presents an imaginary immittance equal to, and of opposite sign from, that of the line at that point the line will be matched on the generator side of the stub. In this method there are two distances to be determined: first, the position of the stub relative to the load and, second, the length of the stub itself. The stub, as described in Section 6.8, may be either open or short circuited, and may be joined to the line in either series or parallel.

EXAMPLE: Consider a line terminated in an unknown load. Measurements show that the VSWR is 3.4, adjacent minima are 15 cm apart, and $x_{min} = 1.2$ cm. A short-circuit shunt stub is to be used to match the line, as shown in Fig. 6.13(a).

Because a shunt stub is being used, from the information given, calculations should be made in admittances. The VSWR from the load to y_1 is given as 3.5, and this is drawn on the Smith chart as shown in Fig. 6.13(b). y_{in} provides a match to the line, i.e.

$$y_{in} = 1 + j0$$

The stub can only add susceptance, shown as jb, and therefore

$$y_{in} = y_1 + jb$$

or

$$1 + j0 = y_1 + jb$$

and

$$y_1 = 1 - jb$$

Hence y_1 must be on the locus of unit conductance, i.e. the $g = 1$ circle, and as it must also lie on the $S = 3.5$ circle it is situated at the points at which these two intersect, shown in Fig. 6.13(b) as a and b.

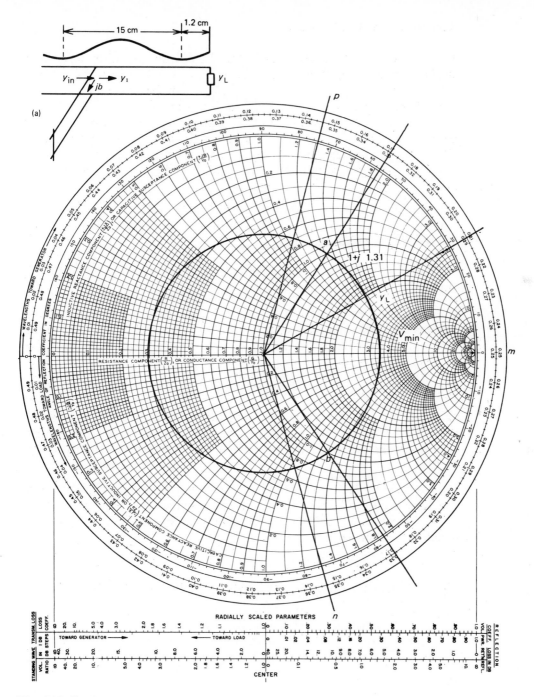

Fig. 6.13 Single-stub matching. (a) Example of shunt stub; (b) Smith chart representation.

Consider point *a*

The admittance of *a* is, from the chart, $1 + j1.31$. Therefore the stub must supply a susceptance of $-j1.31$. The position of V_{min} is known – it is at the point shown on the chart, which is $0.421\,\lambda(= 12.63\,\text{cm})$ towards the load from *a*. The voltage minimum is given as 1.2 cm from the load, making point *a* 13.83 cm from the load.

The length of the stub can be found from the chart with reference to Fig. 6.13(b). The short-circuit load produces a VSWR of ∞ in the stub, which is represented on the chart by the outermost circle of the chart proper. The load admittance of the stub is $y_L = \infty$ which is at the right-hand end of the horizontal line, point *m* in Fig. 6.13(b). The stub length is therefore determined by moving from *m* towards the generator on the $S = \infty$ circle to the point at which susceptance is $-j1.31$, shown as point *n* in the diagram. Thus *mn* is the stub length required.

From the chart, $mn = 0.104\,\lambda$, and since $\lambda = 30\,\text{cm}$, $mn = 3.12\,\text{cm}$

Hence one solution to the problem is to make the short-circuited stub 3.12 cm long and place it 13.83 cm from the load.

Alternatively, consider point *b*

The susceptance at *b* is $-j1.31$ so the stub susceptance must be $j1.31$.

Then, following the same procedure as before, the stub length is given by the distance from *m* to the point on the $S = \infty$ circle where susceptance is $j1.31$, i.e. point *p*.

From the chart, $mp = 0.396\,\lambda = 11.88\,\text{cm}$

The stub positions, from V_{min} to *b*, is the same distance as *a* from V_{min}, but towards the generator. V_{min} to *b* is $0.079\,\lambda$, making the total distance from the load to the stub $(0.079 + 0.04)\,\lambda = 3.57\,\text{cm}$.

The other solution, therefore, is to make the stub 11.88 cm long, and place it 3.57 cm from the load.

6.18 DOUBLE-STUB MATCHING

The use of a single stub may not be convenient for some applications because of the need to be able to adjust the position of the stub relative to the load if the load impedance is changed. By using two stubs the two variables necessary to produce a match are the lengths of the stubs, and the stub positions can be fixed. Their position is not entirely arbitrary, as will be seen later. This method of matching is referred to as double stub, and the basic idea behind it can be understood with the help of Fig. 6.14. Admittances are used again, because short-circuited shunt stubs are shown. If series stubs were being used, it would be necessary to work in impedances.

The input admittance of the stub nearest to the generator, y_{in}, must match the line, so

$$y_{in} = 1 + j0$$

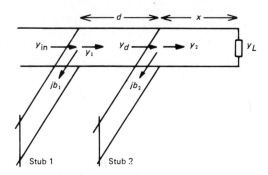

Fig. 6.14 Diagram of two-stub matching.

which is at the centre of the chart. The admittance on the load side of the first stub, y_1, must be given by

$$y_{\text{in}} = y_1 + jb_1$$

or

$$y_1 = 1 - jb_1$$

which lies somewhere on the unity conductance $g = 1$ circle. y_d, the admittance on the generator side of stub 2, must lie on the same VSWR circle as y_1, since it belongs to the same section of line, and it can be found by moving from y_1 on a constant S circle a distance d towards the load. However, the precise position of y_1 is not known. All that we know is that it lies on the $g = 1$ circle, so y_d must lie on the $g = 1$ circle translated through distance d towards the load. This circle we will call, after Everitt and Anner,[3] the L circle.

If y_d is on the L circle, y_1 lies on the $g = 1$ circle, as required, because each point on the L circle has a corresponding point, distance d towards the generator, on the $g = 1$ circle. The position of the L circle is determined by the stub separation d, and it can be drawn on the chart, once a value for d has been fixed, by drawing a circle of the same radius as the $g = 1$ circle on a centre equal to the $g = 1$ circle moved a distance d, on a constant S circle, towards the load.

The function of the first stub is to add the susceptance or reactance required to cancel that of y_1, in much the same way as was done with the single stub in the previous method.

EXAMPLE: A lossless transmission line, of characteristic impedance $50\,\Omega$, is terminated in a load of $100 + j25\,\Omega$. Two short-circuited series stubs are to be used to match the line. One stub is 8 cm and the other 32 cm from the load. The signal on the line is at 750 MHz. Find the lengths of the stubs.

Figure 6.15(a) shows the arrangement. Because series stubs are used, the problem will be solved by using impedances.

First, the stub positions must be fixed in terms of wavelength. 750 MHz is equivalent to a wavelength of 40 cm. So the first stub is $0.2\,\lambda$ from the load, and the distance between the stubs is $0.6\,\lambda$.

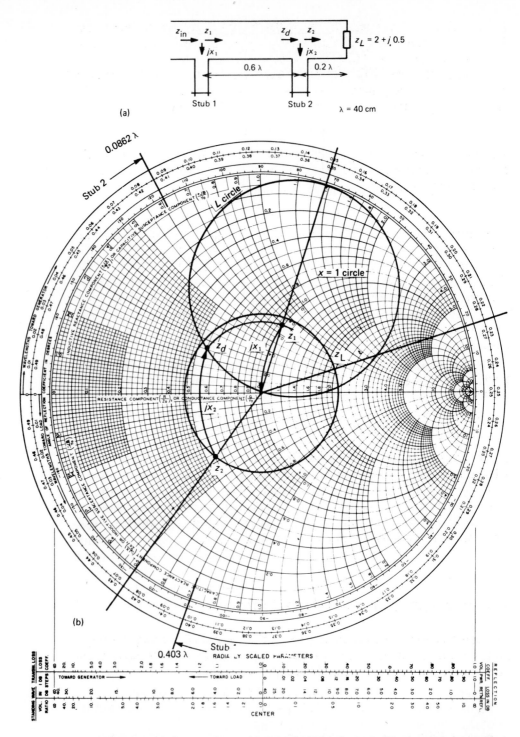

Fig. 6.15 Example of two-stub matching. (a) Series stubs; (b) Smith chart representation.

The L circle is drawn with its centre 0.6 $\lambda(\equiv 0.1\,\lambda)$ towards the load from the $x = 1$ circle, as shown in Fig. 6.15(b). The normalized load, $2 + j0.5$, is fixed on the chart, and the VSWR circle is drawn.

The value of z_2 can be found by moving on the constant S circle a distance of 0.2λ towards the generator from z_L as shown in Fig. 6.15(b). Thus the impedance z_2, at the load side of the terminals of stub 2 is

$$z_2 = 0.55 - j0.36$$

The stub must add a reactance to z_2 which is sufficient to put z_d on the L circle. Moving on a constant resistance circle from z_2, z_d is on the L circle at the point shown. From the chart

$$z_d = 0.55 + j0.25$$

The difference in reactance between z_d and z_2 is that provided by stub 2. From the chart, then, stub 2 adds reactance of $j0.61$ to z_2 to give z_d on the L circle.

z_1 is found by moving towards the generator a distance d on a constant S circle from z_d, i.e. by moving on a constant S circle from z_d, on the L circle, to the corresponding point, d wavelengths towards the generator, on the $x = 1$ circle. Note that once z_d is fixed there is only one point on the $x = 1$ circle that corresponds to z_1. From Fig. 6.15, $z_1 = 1 + j0.7$ and stub 1 must add the reactance required to reduce the reactance of z_1 to zero, i.e. $-j0.7$.

The stubs are short-circuited, so to find their length it is necessary to start at point $z = 0$, and go round the $S = \infty$ circle towards the generator to the required reactance.

Stub 1 is $0.8\,\lambda$ from the load, and 16.12 cm long.
Stub 2 is $0.2\,\lambda$ from the load, and 3.45 cm long.

Earlier, it was stated that the choice of stub separation cannot be entirely arbitrary. In the above example, if the first stub were $0.08\,\lambda$ and the second $0.29\,\lambda$, from the load the L circle would be drawn as shown in Fig. 6.16. In that case it would not be possible to place z_d on the L circle by adding positive or negative reactance to z_2, and a match could not be obtained. In that event, different stub positions would have to be used.

6.19 DERIVATION OF THE SMITH CHART

The construction of the chart is based on a conformal transformation of impedance from the z plane to the ρ plane. Earlier, we derived the relationship between the load impedance and the reflection coefficient.

$$z_L = \frac{1 + \rho}{1 - \rho} \qquad (6.26)$$

In general, both z_L and ρ are complex.

Let $z_L = r + jx \quad \rho = u + jv$

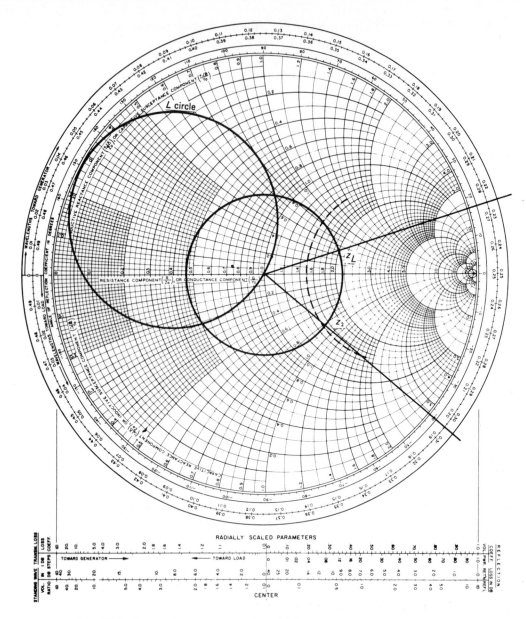

Fig. 6.16 Limitation on two-stub matching arrangement.

Substituting these values into Eqn (6.26) gives

$$r + jx = \frac{1 + u + jv}{1 - (u + jv)} = \frac{1 - u^2 - v^2 + 2jv}{(1 - u)^2 + v^2}$$

and from this expression

$$r = \frac{1 - u^2 - v^2}{(1 - u)^2 + v^2} \quad x = \frac{2v}{(1 - u)^2 + v^2}$$

These equations can be put in the form

$$\left(u - \frac{r}{1 + r}\right)^2 + v^2 = \frac{1}{(1 + r)^2} \qquad (6.64)$$

$$(u - 1)^2 + \left(v - \frac{1}{x}\right)^2 = \frac{1}{x^2} \qquad (6.65)$$

Equation (6.64) is the equation of a circle in the ρ plane if r is constant, with the centre of the circle at

$$u = \frac{r}{1 + r} \quad v = 0$$

r cannot be negative, so the centre of the circle must lie on the horizontal axis between $u = 0$ and $+1$. The radius of the circle is $1/(1 + r)$.

Similarly, Eqn (6.65) is also the equation of a circle in the ρ plane, when x is constant. Now the centre of the circle is at

$$u = 1, \quad v = 1/x$$

which lies on a vertical line, one unit to the right of the origin. If x is positive, the centre points will be above $v = 0$, and if x is negative they will lie below that point.

The Smith chart therefore consists of plots of Eqns (6.64) and (6.65) for various values of r and x. The result is a series of orthogonal circles, with resistance circles having their centres on the horizontal line, and all passing through the point $u = 1$, $v = 0$, at the right-hand edge of the chart; and reactance circles having their centres on the vertical line $u = 1$, and their perimeters also tangential at the point $u = 1$, $v = 0$. As we noted earlier, all $r =$ constant circles are shown, but only those segments of the $x =$ constant circles situated in the region $r \geqslant 0$. On the chart, apart from $v = 0$, which corresponds to the line $x = 0$, there are no u or v values shown. The numbers given on the circles are for r and x, but if it is remembered that the radius of the circle $r = 0$, is one unit of u, then Eqns (6.64) and (6.65) are consistent with the normalized values on the chart.

The assumption that $\rho = u + jv$ means that $\tan\psi = v/u$. Therefore, any line from the centre of the chart to the outer circumference will be of constant ψ. Angular movement round the chart therefore represents a movement with ψ, and so the inner circumferential scale is marked 'Angle of Reflection Coefficient in Degrees'.

Any circle with centre at the origin must satisfy the equation

$$u^2 + v^2 = |\rho|^2$$

and therefore it has radius $|\rho|$. From Eqn (6.54) the VSWR and $|\rho|$ are directly related, so constant $|\rho|$ means constant S.

6.20 TRAVELLING WAVES ON A LOSSLESS LINE

In Section 6.1 we assumed that forward and backward travelling waves would be produced by the time-varying input signal, and in deriving the voltage equations we ignored the effect of time. Now we will consider again our earlier ideas, and show that we do indeed have waves travelling along the line.

Referring to Eqns (6.3) and (6.6), i.e.

$$\frac{\partial I}{\partial z} = -\left(G + C\frac{\partial}{\partial t}\right)V \tag{6.3}$$

and

$$\frac{\partial V}{\partial z} = -\left(R + L\frac{\partial}{\partial t}\right)I \tag{6.6}$$

Differentiating Eqn (6.3) with respect to t, and Eqn (6.6) with respect to z,

$$\frac{\partial^2 I}{\partial t \partial z} = -\left(G + C\frac{\partial}{\partial t}\right)\frac{\partial V}{\partial t} \tag{6.66}$$

and

$$\frac{\partial^2 V}{\partial z^2} = -\left(R + L\frac{\partial}{\partial t}\right)\frac{\partial I}{\partial z} \tag{6.67}$$

6.21 WAVE EQUATION FOR A LOSSLESS TRANSMISSION LINE

In a lossless line it is assumed that R and G can be ignored. Then Eqns (6.66) and (6.67) become

$$\frac{\partial^2 I}{\partial t \partial z} = -C\frac{\partial^2 V}{\partial t^2} \tag{6.68}$$

and

$$\frac{\partial^2 V}{\partial z^2} = -L\frac{\partial^2 I}{\partial t \partial z} \tag{6.69}$$

Substituting in Eqn (6.69) for $\partial^2 I/(\partial t \partial z)$ from Eqn (6.68),

$$\frac{\partial^2 V}{\partial z^2} = LC\frac{\partial^2 V}{\partial t^2} \tag{6.70}$$

This is the usual form of what is known as the wave equation. It is satisfied by any function of the form

$$f\left(t - \frac{z}{v}\right)$$

so that, if

$$V = V_1 f\left(t - \frac{z}{v}\right)$$

then

$$\frac{\partial^2 V}{\partial z^2} = \frac{V_1}{v^2} \quad \text{and} \quad \frac{\partial^2 V}{\partial t^2} = V_1$$

and Eqn (6.70) is satisfied if

$$1/v^2 = LC$$

A similar argument shows that with the same condition on V_2, the function $f[t + (z/v)]$ is also a solution of Eqn (6.70), so that the total voltage on the line is

$$V = V_1 f\left(t - \frac{z}{v}\right) + V_2 f\left(t + \frac{z}{v}\right) \tag{6.71}$$

The function $f[t - (z/v)]$ can be seen to represent a travelling wave if it is considered that in order to retain a position of constant phase on the line it is necessary that $[t - (z/v)]$ is constant, i.e. that a movement in the z direction of δz must be related to the time it takes to make the movement δt, in such a way that

$$t - \frac{z}{v} = t + \delta t - \frac{z + \delta z}{v}$$

or

$$\delta t = \frac{\delta z}{v}$$

Hence the velocity of propagation, defined as the change in z in unit time, is

$$v = \frac{\delta z}{\delta t}$$

Thus v in Eqn (6.71) is the velocity of the wave and is given by

$$v = 1/\sqrt{(LC)} \tag{6.72}$$

v is referred to as the **phase velocity** of the wave.

If a sinusoidal signal is placed on the line, the functions $f[t \pm (z/v)]$ are written

$$\exp[j\omega(t \pm \beta z)] \tag{6.73}$$

for a lossless line, as noted earlier. Then,

$$V = V_1 \exp[j\omega(t - \beta z)] + V_2 \exp[j\omega(t + \beta z)] \tag{6.74}$$

Comparing Eqns (6.71) and (6.74),

$$\frac{\beta}{\omega} = \frac{1}{v} \quad \text{i.e.} \quad v = \frac{\omega}{\beta} \tag{6.75}$$

so, from Eqn (6.72),

$$\beta = \omega \sqrt{(LC)} \tag{6.76}$$

6.22 CONCLUSION

In this chapter we have introduced and examined several concepts such as reflection, standing waves, line impedance, characteristic impedance, matching and travelling waves. We shall see in some of the following chapters that these same ideas occur in different contexts, but with similar meaning, and it is this wide applicability of the main features of transmission line theory that makes it so important to the study and design of communications transmission systems in general.

REFERENCES

1. Chipman, R. A., *Transmission Lines*, Schaum – McGraw-Hill, New York, 1968.
2. Smith, P. H., 'An improved transmission line calculator', *Electronics*, Jan 1944, 130.
3. Everitt, W. E. and Anner, G. E., *Communication Engineering*, McGraw-Hill, New York, 1956.
4. Davidson, C. W., *Transmission Lines for Communications*, Macmillan, London, 1978.

PROBLEMS

Note: In all problems assume velocity of wave $= c$.

6.1   A lossless transmission line of $Z_0 = 100\,\Omega$ is terminated by an unknown impedance. The termination is found to be at a maximum of the voltage standing wave, and the VSWR is 5. What is the value of the terminating impedance?

Answer: $500\,\Omega$.

6.2   A load of $90 - j120\,\Omega$ terminates a $50\,\Omega$ lossless air-spaced transmission line. Find, both analytically and by using the Smith chart,

(a) the reflection coefficient;
(b) the VSWR;
(c) the distance from the load to the first voltage minimum
if the operating frequency is 750 MHz.

Answer: (a) $0.69e^{-j0.54}$; (b) 5.4; (c) 8.28 cm.

6.3   Calculate the impedance at terminals AA of the lossless transmission line system shown. All three lines have a characteristic impedance of $50\,\Omega$. The reflection coefficients, at an operating frequency of 750 MHz, are $\rho_1 = 0.6 - j0.5$ and $\rho_2 = 0.8 + j0.6$.

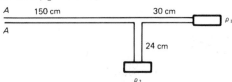

Answer: $0.9 + j17.75\,\Omega$.

6.4　A section of line $\lambda/4$ long at 100 MHz is to be used to match a $50\,\Omega$ resistive load to a transmission line having $Z_0 = 200\,\Omega$.
(a) Find the characteristic impedance of the matching section.
(b) What is the input impedance of the matching section at 12 MHz?
(c) Repeat part (b) for a frequency of 80 MHz.

Answer:　(a) $100\,\Omega$; (b) $(155 - j68)\,\Omega$; (c) $67.5 + j47.3\,\Omega$.

6.5　A load Z produces a VSWR of 5 on a $50\,\Omega$ lossless transmission line. The first minimum is 0.2λ from the load and the frequency of operation is 750 MHz. Without using a Smith chart, find the position and length of a short-circuit shunt stub that will match the load to the line.

Answer: Either 3.252 cm long, 5.608 cm from load, or 16.478 cm long, 10.948 cm from load.

6.6　A $50\,\Omega$ lossless transmission line is terminated by a load $60 + j90\,\Omega$. Find the position and length of an open-circuit series stub that will provide a match at 800 MHz.

Answer:　Either 15.4 cm long, 4.95 cm from load or 3.26 cm long, 18.4 cm from load.

6.7　A lossless transmission line is carrying a signal of 47 cm wavelength, $S = 4.5$ and x_{min} is 5 cm from the load. Find the lengths of the two open-cricuited series stubs that will match the line if one is placed $0.43\,\lambda$ and the other $0.22\,\lambda$ from the load.

Answer: 11.49 cm at 0.22λ, 19.27 cm at 0.43λ.

6.8　A lossless transmission line with $Z_0 = 200\,\Omega$ is terminated by a load Z_L. If the VSWR is 6.5, and x_{min} is $0.168\,\lambda$, find Z_L and the length of the two short-circuited shunt stubs required to match the line to the load if one stub is placed at the load and the other one-quarter of a wavelength away.

Answer: $Z_L = 120 - j320\,\Omega$, $l_1 = 0.423\,\lambda$, $l_2 = 0.226\,\lambda$.

6.9　A double-stub matching system is to be used on a lossless transmission line of load impedance z_L. If the distance from the load to the stubs is s and $d + s$, show, using a Smith chart if necessary, that there are values of s for which the system cannot be used.

6.10　A transmission line is matched by two stubs, one open cirucuited and placed in parallel with the line $0.15\,\lambda$ from the load, and the other short circuited, $0.3\,\lambda$ long, and placed in series with the line a distance d towards the generator from the first stub. Find the length of the parallel stub, and the distance d, if the line is assumed to be lossless and the wavelength of the signal is 40 cm. The first voltage minimum is $0.07\,\lambda$ from the load, and the VSWR at the load is 3.

Answer: Stub length $= 12.78$ cm, $d = 6.47$ cm.

7 | Antennas

In the last chapter we discussed transmission lines, assuming that each line was terminated in a load. That load was drawn as a lumped circuit, and we did not enquire about its purpose, or its form. In practical systems, a transmission line is often used to connect either a power source to a transmitting antenna, or a receiving antenna to a detecting circuit. Antennas are reciprocal devices, so the ideas we will develop in this chapter apply to both transmitting and receiving systems. However, to avoid unnecessary repetition we will only discuss antennas in the transmitting mode.

When a transmission line feeds an antenna, it 'sees' the impedance of the antenna, which will be frequency dependent. If the antenna is matched to the transmission line, all the energy transmitted along the line is absorbed by the antenna and, apart from losses within the antenna itself, is radiated.

From an antenna designer's point of view, there are several parameters that contribute to a design specification. Those of particular interest are polar diagram, beamwidth, bandwidth and impedance. Other quantities relating to the mechanical specification of the antenna will also be important, such as its size, weight, shape and material.

Mechanisms involved in the propagation of radio waves have an important effect on the thinking of the antenna designer. This chapter concludes with some of the fundamental features of radio propagation and indicates the extent to which they are dependent on the frequency of operation.

7.1 RADIATION PATTERN (POLAR DIAGRAM)

The radiation from an antenna is an electromagnetic wave, and for most applications our interest is in the strength of this wave at a distant point. In terms of metres, what constitutes a distant point depends on the size of the antenna and the wavelength of the radiation, but in principle it refers to the region in space in which the wave can be considered to be plane, and normal to the direction of the antenna. There will be just two field components, an E field vector and an H field vector, as shown in Fig. 7.1.

The radiation pattern is the variation of this distant electric field as a function of angle; an example is shown in Fig. 7.2. A detector is placed at a fixed distance R from the antenna, and the E field is measured [Fig. 7.2(a)]. The value of this E field varies with the angle, θ, relative to some reference line, shown as $\theta = 0$. The variation of E with θ is the radiation pattern of the antenna in the particular plane in which the measurement is taken. The

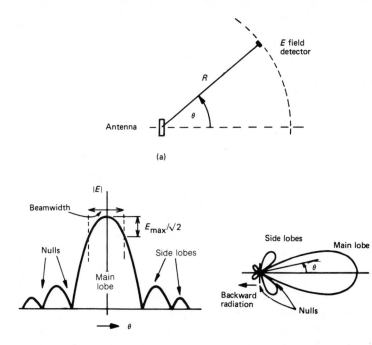

(a)

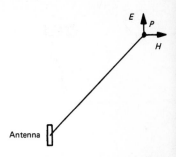

Fig. 7.1 Field components due to an antenna at a distant point.

(b)

(c)

Fig. 7.2 (a) Reference for radiation patterns; (b) parameters at a radiation pattern; (c) polar diagram.

radiation pattern can be represented in either Cartesian coordinates [Fig. 7.2(b)], or in polar coordinates, when it is usually referred to as a polar diagram [Fig. 7.2(c)]. Clearly, the radiation pattern of any real antenna is three-dimensional and therefore, to represent it fully, measurements must be made in at least two orthogonal planes. The traditional planes used are the vertical and the horizontal (which are sometimes, especially in radar applications, called elevation and azimuth, respectively), and these are related to the orientation in which the antenna is to be used. The reference direction, $\theta = 0$, is generally assumed to be normal to the plane of the antenna, as shown. For symmetrical antennas this is the direction of the main lobe. The various terms associated with the radiation pattern are shown in Fig. 7.2(b) and (c).

7.2 LOBES

We are sometimes interested in the absolute power radiated in any particular direction; for example, the actual power output from a broadcast transmitter will determine the range over which the transmission can be detected. More usually, however, we are more interested in the shape of the radiation pattern. Then we use a normalized pattern, with magnitudes plotted in relation to the main lobe maximum, which is given the value of unity. The finite size of the wavelength, relative to the dimensions of the antenna, produces diffraction effects, causing side lobes. These side lobes, separated from the main lobe and from each other by field minima, generally represent inefficiencies in the antenna. For most applications the designer wishes to keep side lobe levels to a minimum.

7.3   BEAMWIDTH

For most applications we are interested to know the width of the main lobe, i.e. over what angular distance the main lobe is spread. A definition of the beamwidth could be the distance between the nulls on either side of the main lobe. Although appealing, because it should be easy to read from a radiation pattern, in practice there are three disadvantages to its use:

(i) there may not be a null;
(ii) if a null exists, its precise location may be difficult to determine accurately;
(iii) the detectable level for any practical measurement is well above zero field.

However, some convention must be agreed upon, and that normally used is to define the beamwidth as the angular separation, in a given plane, between the points on either side of the main lobe that are 3 dB in power below that of the maximum. On an E field radiation plot, which has a normalized maximum of unity, the beamwidth is between those points on each side of the main lobe which have field values of $1/\sqrt{2}$.

7.4   ANTENNA IMPEDANCE AND BANDWIDTH

Antennas can be considered to belong to one of two categories, resonant or non-resonant. Resonant antennas have a dimension that is near to a half-wavelength, or one of its multiples, whereas non-resonant systems do not. The difference between the two is particularly noticeable in respect of antenna

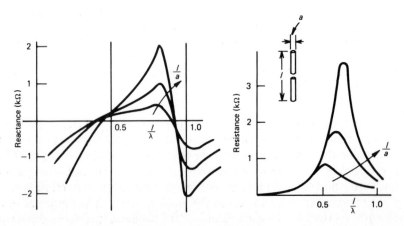

Fig. 7.3 Dipole impedance variation with dipole length.

impedance and bandwidth. The bandwidth is the frequency range over which the input impedance is substantially constant. In resonant systems, the impedance and bandwidth vary significantly with small changes in operating frequency; the impedance changes from negative reactance, through pure resistance, to positive reactance, as the frequency changes from below to above resonance. Figure 7.3 shows the impedance as a function of wavelength for a dipole. From it we can see that at about resonance, $l \simeq \lambda/2$, there is a rapid change of impedance with frequency resulting in a small bandwidth, which means that the characteristics of the antenna are strongly frequency dependent. In non-resonant antennas the radiation pattern and the input impedance are not so variable with frequency, and these antennas are sometimes referred to as broadband.

7.5 GAIN

The final parameter of interest to the systems designer is antenna gain. In any physically realisable antenna there will be one direction in which the power density is highest. To give a quantitative value to this maximum, it is compared with some reference, the most common being the power density from an isotropic antenna. An isotropic antenna has a spherical radiation pattern. There is no practical device that can produce such a pattern, but it is a useful notional tool for gain comparisons. There is a distinction to be made between two types of gain. Understandably, the power at both the isotropic antenna and the antenna under test must be the same. However, that may refer to the input power, or to the total radiated power. For the isotropic antenna, which has no loss mechanisms, the input and radiated powers are identical, but for the test antenna there will be losses that reduce its efficiency by several per cent, resulting in a difference between these two values of gain. In some of the literature they are distinguished by the terms 'directive gain' (D) and 'power gain' (G), where D assumes equal radiated powers from, and G assumes equal input powers to, the test and isotropic antennas. Thus, we can establish the relationship

$$G = kD \tag{7.1}$$

where k, an efficiency factor, is the ratio of the total radiated power to the total input power for the antenna under test.

For applications in which the power transmitted in a specified direction is of particular interest (e.g. in a microwave link, or a satellite earth station), the gain is an indicator of the antenna's effectiveness.

7.6 FIELD DUE TO A FILAMENT OF CURRENT

By considering the simplest possible antenna, we can obtain some insight into radiated fields, and derive some general ideas that are applicable to all antennas. In Fig. 7.4 we see a short, very thin element of length l, carrying a sinusoidal current $I \sin \omega t$. The current flowing along the filament induces a

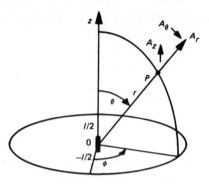

Fig. 7.4 Coordinate system relating fields at distant point P to antenna current.

magnetic field in the surrounding air. This magnetic field has an associated electric field and the result is a movement of electromagnetic energy away from the filament. We will examine the field at some point P, a distance r from the centre of the filament at O. Imagine that P is situated on a sphere centre O. Then P is related to O by r, and ϕ. In spherical coordinates, the field components at P are E_r, E_θ, E_ϕ and H_r, H_θ, H_ϕ. Allowing the current through the filament to be $I \sin \omega t$, the field at P will be produced by the current flowing through the filament at time $t - r/c$, where r/c is the time taken for the field to travel from the filament to P at the velocity of light, c.

To find the field at P we use the concept of a retarded magnetic vector potential, $\bar{A}$, which is related to the current in the filament by[1]

$$\bar{A} = \frac{1}{4\pi} \int_V \frac{i\,[t - (r/c)]}{r} \, \mathrm{d}V \tag{7.2}$$

In this example, the current has only one component, in the z direction (see Fig. 7.4), which is constant throughout the length of the filament. We can therefore replace $\bar{A}$ by A_z, and the volume integral by the linear integral over the range $-l/2 \leqslant z \leqslant l/2$, giving

$$A_z = \frac{1}{4\pi} \int_{-l/2}^{l/2} \frac{I \sin \omega [t - (r/c)]}{r} \, \mathrm{d}z \tag{7.3}$$

$$= \frac{Il}{4\pi r} \sin \omega \, [t - (r/c)] \tag{7.4}$$

The vector potential is a mathematical device. It allows us, in some cases, to find the electromagnetic field components more easily than would be possible if we were to try to relate the fields directly to the electric and magnetic charge on the antenna. The vector potential we are using here is related to the current in the antenna (the thin filament) through Eqn (7.4), and also to the magnetic field $\bar{H}$. The quantities $\bar{A}$ and $\bar{H}$ are, by definition, related by

$$\nabla \times \bar{A} = \bar{H} \tag{7.5}$$

Hence, knowing the components of $\bar{A}$, those of $\bar{H}$ can be found. From

Eqn (7.4) and Fig. 7.4, the spherical coordinate components of $\bar{A}$ are

$$A_r = A_z \cos \theta = \frac{Il}{4\pi r} \sin \omega \left(t - \frac{r}{c} \right) \cos \theta$$

$$A_\theta = -A_z \sin \theta = \frac{-Il}{4\pi r} \sin \omega \left(t - \frac{r}{c} \right) \sin \theta$$

$$A_\phi = 0$$

(7.6)

By expanding Eqn (7.5), we can see that the only component of $\bar{H}$ is H_ϕ. Both H_θ and H_r are zero. Appendix 7.1 gives a fuller derivation of the electric and magnetic field components at P. We can see from there that

$$H_\phi = \frac{Il \sin \theta}{4\pi} \left[\frac{\omega}{cr} \cos \omega \left(t - \frac{r}{c} \right) + \frac{1}{r^2} \sin \omega \left(t - \frac{r}{c} \right) \right]$$

(7.7)

$$E_\theta = \frac{Il \sin \theta}{4\pi\varepsilon} \left[\frac{\omega}{c^2 r} \cos \omega \left(t - \frac{r}{c} \right) + \frac{1}{cr^2} \sin \omega \left(t - \frac{r}{c} \right) - \frac{1}{\omega r^3} \cos \omega \left(t - \frac{r}{c} \right) \right]$$

(7.8)

$$E_r = \frac{Il \cos \theta}{2\pi\varepsilon} \left[\frac{1}{cr^2} \sin \omega \left(t - \frac{r}{c} \right) - \frac{1}{\omega r^3} \cos \omega \left(t - \frac{r}{c} \right) \right]$$

(7.9)

We are not really interested in the magnitudes of these expressions, but in their form. Within each of the brackets there are terms related to powers of $1/r$, and these allow us to divide the space surrounding the current-carrying element into two parts. In the region near the antenna, terms in $1/r^3$ and $1/r^2$ will predominate, but at large distances these terms will be negligible, leaving only the terms in $1/r$. The fields in these two regions are referred to as the induction and radiation fields, respectively.

7.7 INDUCTION FIELD

The near-field region, or Fresnel zone, is occupied chiefly by the induction field, which does not radiate. Energy is stored in the field terms having $1/r^2$ coefficients for one half-cycle, and passes back to the antenna for the other half-cycle. The $1/r^3$ terms are confined to a region very close to the antenna before becoming negligible as r increases.

When considered in detail, the induction field can be subdivided into two parts. That nearest the antenna, known as the reactive near field, consists of strong reactive terms and is confined to a radius of less than $0.62\sqrt{D^2/\lambda}$, where D is the largest dimension of the antenna and is much greater than the wavelength λ.

Beyond the reactive near field is the radiating near field, or Fresnel zone, and that extends to the boundary with the far field. In this radiation near field region the distribution of the field as a function of angle from the antenna varies with radius.

7.8 RADIATION FIELD

In the far field, or Fraunhofer region, only the $1/r$ terms are significant, and, as we shall see below, it is those that provide the energy outflow from the antenna.

We noted earlier that in most applications it is the far field that is of interest, and radiation patterns refer to that area. The distance from the antenna at which the radiation field predominates can be found from the above equations by comparing the coefficients of the $1/r$ and $1/r^2$ terms for various values of r at the wavelength of interest.

A figure often used as the near/far field boundary is a radius of $2D^2/\lambda$.

Clearly, since the field is a continuous function of radius, the boundary expressions given above are merely guides, and in a particular application the transition into the far field will be at a radius determined by the dimensions of the antenna, and the wavelength of the signal.

7.9 POWER RADIATED FROM A CURRENT ELEMENT

We have called the fields in the Fraunhofer region the radiation fields, thus suggesting that it is the $1/r$ terms in Eqns (7.7) and (7.9) that are the components of power flow. We can justify this claim by considering the Poynting vector which, by taking the vector cross-product of the electric and magnetic fields, gives the value and direction of the net power flow from the antenna.

In vector notation, the Poynting vector $\bar{P}$ gives the power density in terms of the electric and magnetic field vectors:

$$\bar{P} = \bar{E} \times \bar{H} \qquad (7.10)$$

For the short current element, the fields that exist at a point outside the antenna are E_θ, E_r, and H_ϕ; therefore

$$P_\phi = 0$$

and

$$P_\theta = -E_r H_\phi$$

i.e. from Eqns (7.7) and (7.9),

$$P_\theta = -\frac{Il\cos\theta}{2\pi\varepsilon}\left[\frac{\sin\omega[t-(r/c)]}{r^2 c} - \frac{\cos\omega[t-(r/c)]}{\omega r^3}\right]$$
$$\times \frac{Il\sin\theta}{4\pi}\left[\frac{\omega\cos\omega[t-(r/c)]}{rc} + \frac{\sin\omega[t-(r/c)]}{r^2}\right]$$

Since the time average of terms in $\cos 2\omega[t-(r/c)]$ and $\sin 2\omega[t-(r/c)]$ is zero, P_θ is zero. This leaves the radial power density component

$$P_r = E_\theta H_\phi \qquad (7.11)$$

i.e. from Eqns (7.7) and (7.8),

$$P_r = \frac{I^2 l^2 \sin^2 \theta}{16\pi^2 \varepsilon} \left\{ \left[\frac{\omega}{rc} \cos \omega \left[t - (r/c) \right] + \frac{1}{r^2} \sin \omega \left[t - (r/c) \right] \right] \right.$$

$$\left. \times \left[\frac{\omega}{rc^2} \cos \left[t - (r/c) \right] - \frac{1}{\omega r^3} \cos \omega \left[t - (r/c) \right] + \frac{1}{cr^2} \sin \omega \left[t - (r/c) \right] \right] \right\}$$

(7.12)

Again, the terms in $\sin 2\omega \left[t - (r/c) \right]$ and $\cos 2\omega[t - (r/c)]$ have zero time average, leaving

$$P_r = \frac{\omega^2 I^2 l^2 \sin^2 \theta}{16\pi^2 \varepsilon r^2 c^3} \frac{1}{2}$$

giving

$$P_r = \frac{\eta I^2 l^2 \sin^2 \theta}{8\lambda^2 r^2}$$

(7.13)

where η = the space impedance, $\sqrt{(\mu_0/\varepsilon_0)}, f$ = the frequency of the radiation = $\omega/2\pi = c/\lambda$ and c = the velocity of the radiated wave = $1/\sqrt{(\mu_0\varepsilon_0)}$.

This remaining component of the Poynting vector gives the power density flowing from the antenna. Before finding the actual power from the filament, it is worth nothing that P_r in Eqn (7.13) is formed from the $1/r$ terms in E_θ and H_ϕ, which is why they are called the radiation terms. The positive sign of P_r indicates a flow of power outwards from the antenna.

We can find the total radiated power from the current element by integrating the power density at a distant point P over the surface of a sphere through P. Referring to Fig. 7.5 the total power P_T is given by

$$P_T = \int P_r \, da$$

integrated over the area of the sphere.

From the diagram, $da = 2\pi r^3 \sin \theta \, d\theta$. Hence, from Eqn (7.13),

$$P_T = \int_0^\pi \frac{\eta I^2 l^2 \sin^2 \theta}{8\lambda^2 r^2} 2\pi r^2 \sin \theta \, d\theta$$

$$= \frac{\eta I^2 l^2 \pi}{4\lambda^2} \int_0^\pi \sin^3 \theta \, d\theta$$

$$= \frac{\eta I^2 l^2 \pi}{8\lambda^2} \left[-\frac{\cos \theta}{3} (\sin^2 \theta + 2) \right]_0^\pi$$

$$= \frac{\eta I^2 l^2 \pi}{3\lambda^2}$$

(7.14)

I is the peak value of current in the filament. If, instead, the rms value is used,

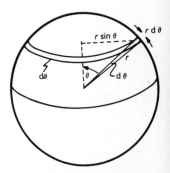

Fig. 7.5 Power radiated from an antenna at the centre of a sphere.

the total power from the filament is

$$P = \frac{2\eta I_{\mathrm{rms}}^2 l^2 \pi}{3\lambda^2} \quad \text{watts} \tag{7.15}$$

In free space, the impedance η has the value $120\pi = 377\,\Omega$, and using this value

$$P = I_{\mathrm{rms}}^2 \, 80\pi^2 \, (l/\lambda)^2 \quad \text{watts} \tag{7.16}$$

We can see immediately that the quantity $80\pi^2(l/\lambda)^2$ has the dimensions of resistance. It is known usually as the radiation resistance of the antenna, and for this particular antenna it is related to the electrical length of the element, l/λ.

The idea of radiation resistance is applicable to any antenna. Its value for a particular antenna can be found if the current distribution over the antenna surface is known.

In the short element considered above, we assumed that the current was constant over the length of the antenna. Usually, however, this assumption is not valid, and in order to determine the amount of power radiated, the current distribution must be known. There are, then, two difficulties associated with finding the power radiated from a practical antenna.

(i) The current distribution cannot be determined very readily. Experimental measurements are not easy to obtain, and theoretical analysis, in which Maxwell's equations are solved for the antenna surface, is very complex. Current distributions have been calculated for simple, idealized shapes, but even then the solutions are complicated and difficult to use. In any practical analysis, unless lengthy numerical methods are employed, the usual approach is to build up an approximate model from simpler shapes, but even then the analysis is not easy.

(ii) Once the current distribution is known, the radiation fields must be calculated. That, too, may be a very difficult analytical process which must sometimes be approached by making realistic simplifications.

We will not concern ourselves with the solution of such problems, but we will accept that any antenna has

(i) a radiation field at some distance from the antenna, appearing as a plane wave moving radially outwards,
(ii) an induction field near to the antenna,
(iii) a radiation resistance that gives a measure of the amount of power radiated by the antenna,

and that the distance from the antenna at which the near field merges into the far field depends on the electrical length of the antenna.

7.10   RADIATION PATTERN OF A SHORT DIPOLE

The short current element we have just considered is like a short dipole. Equation (7.41) gives E_θ, the electric field component at a distant point. If the

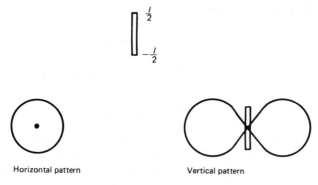

Fig. 7.6 Polar diagram of short dipole.

range from the antenna to the point of interest is in the far field, only the radiation term will be significant, i.e.

$$E_\theta = \frac{\eta I l}{2\lambda\pi} \sin\theta \cos\omega\,[t - (r/c)]$$

which has the magnitude

$$\frac{60\pi I}{r}\frac{l}{\lambda}\sin\theta \qquad (7.17)$$

where, as before, η has been replaced by 120π.

Referring again to Fig. 7.4 we can see that for fixed θ the magnitude will be constant for all values of ϕ. As θ is varied, the magnitude will vary from zero at $\theta = 0$ and π, to a maximum at $\theta = \pi/2$ from Eqn (7.17). If the dipole is placed vertically, the horizontal radiation pattern, with θ constant, is a circle, and the vertical radiation pattern, with ϕ constant, is a figure of eight. We can see that the three-dimensional radiation pattern is a solid of revolution around the dipole axis (Fig. 7.6). From a practical point of view this type of antenna could be used where no preference was to be given to any direction, or where nulls were required, as in some direction-finding methods. The basic shape of the radiation pattern in Fig. 7.6 is maintained as the value of l increases into the resonant region of $\lambda/2$, the half-wave dipole. The impedance of the antenna changes quite dramatically as l is increased, as we saw in Fig. 7.3.

7.11   ANTENNA ARRAYS

The circular horizontal radiation pattern of a dipole antenna may not be satisfactory for some applications. By placing several dipoles in line, the radiation pattern can be modified to give a large variety of radiation patterns depending on the parameters of the array. Here we shall start with the simplest array of two elements, and then consider arrays of several elements, equally spaced and in line. Other, more complex, arrays have been studied, and details can be found in, for example, Kraus.[2]

7.12 TWO-ELEMENT ARRAY

In the arrangement of Fig. 7.7, the elements are fed with currents I_0 and I_1, which we will assume to be of equal magnitude, but out of phase, i.e.

$$I_1 = I_0 \underline{/\alpha}$$

If P, the point of observation, is well into the far field, the path-length difference to P from the elements will be approximately equal to $d \cos \phi$, where d is the element spacing.

In the position shown, the phase of the radiation at P from element 1 will lead that from element 0 by ψ, where

$$\psi = \beta d \cos \phi + \alpha$$

$\beta (= 2\pi/\lambda)$ is the phase constant of the transmitted wave.

Given that the field at P due to element 0 is E_0, then the total field E_p is given by

$$E_p = E_0 [1 + \exp(j\psi)]$$
$$= E_0 [\exp(-\tfrac{1}{2} j\psi) + \exp(\tfrac{1}{2} j\psi)] \exp(j\psi/2)$$

which has a magnitude

$$E = 2E_0 \cos \tfrac{1}{2}\psi \qquad (7.18)$$

i.e.

$$E = 2E_0 \cos \tfrac{1}{2}(\beta d \cos \phi + \alpha)$$
$$= 2 E_0 \cos \left(\frac{\pi d}{\lambda} \cos \phi + \frac{\alpha}{2} \right) \qquad (7.19)$$

Thus, for any given separation and phase difference, the radiation pattern, which is the variation of E with ϕ, can be found. Major changes in the polar diagram can be produced by varying d/λ and α. Problem 7.1 clearly demonstrates these changes, and it is a useful exercise to plot the radiation pattern for each value, and then to compare the results. Jordan,[1] Kraus[2] and Jasik[3] show many of these patterns.

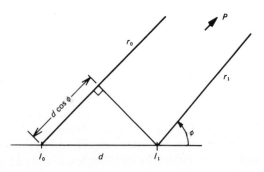

Fig. 7.7 Two-element array.

7.13   LINEAR ARRAYS

Several elements in line create a linear array. In this section we restrict our attention to the simplest form of array, in which n equally spaced elements are fed with currents of equal magnitude, and the phase difference between adjacent elements is constant.

From Fig. 7.8 the field at some distant point P is

$$E_p = E_0[1 + \exp(j\psi) + \exp(j2\psi) + \cdots \exp[j(n-1)\psi]] \qquad (7.20)$$

where, as before, E_0 is the field at P due to radiation from element 0, and ψ is the phase difference at P due to radiation from element 1 compared with that from element 0, i.e.

$$\psi = \beta d \cos \phi + \alpha \qquad (7.21)$$

To find the magnitude of E_p we can use the relationship

$$1 + \exp(j\psi) + \cdots \exp[j(n+1)\psi] = \frac{1 - \exp(jn\psi)}{1 - \exp(j\psi)} \qquad (7.22)$$

Then, the magnitude of Eqn (7.22) is

$$E = E_0 \left| \frac{1 - \exp(jn\psi)}{1 - \exp(j\psi)} \right|$$

$$= E_0 \left| \frac{\sin n\psi/2}{\sin \psi/2} \right| \qquad (7.23)$$

The quantity $|(\sin n\psi/2)/(\sin \psi/2)|$ is known as the array factor because it determines the shape of the radiation pattern. It has a minimum, as we could verify by using L'Hospital's Rule, when $\psi = 0$, i.e. when

$$\beta d \cos \phi = -\alpha$$

By choosing α correctly we can place the maximum where required. Consider two examples.

Broadside array. The maximum is normal to the line of the elements, in the direction $\phi = \pi/2$. The radiation from the elements must be in phase in this

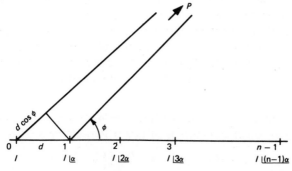

Fig. 7.8 Uniform linear array of n elements.

direction, and from Eqn (7.21) this requires, as we would expect, that the elements are fed in phase, i.e. $\alpha = 0$.

End-fire array. In this case the maximum is placed along the line of the elements, i.e. in the direction $\phi = 0$. Here the radiation from adjacent elements will add in one direction, and subtract in the reverse, backward, direction. Again, we are not surprised to find from Eqn (7.21) that the phase relationship between adjacent elements is

$$\alpha = -\beta d$$

so that the phase difference in the currents between, for example, elements 0 and 1 is cancelled by their separation.

In both these cases, and for any linear array, the width of the main lobe is reduced by increasing the number of elements, which is similar to making the array physically longer.

To find the shape of the radiation pattern of a particular array it is often sufficient to establish the direction of the maximum, from Eqn (7.23), and to find the position of the nulls. These null directions, where the field at P is zero, can be determined much more easily than the directions of the maxima of the secondary lobes. From Eqn (7.23) we can see that a zero field occurs when the array factor is zero, i.e. when

$$\frac{n\psi}{2} = \pm k\pi$$

or

$$\tfrac{1}{2}n(\beta d \cos \phi + \alpha) = \pm k\pi \qquad (7.24)$$

where $k = 1, 2, 3 \cdots$. Further insight into the behaviour of these arrays can be found in Problems 7.3 and 7.4 at the end of the chapter.

A detailed evaluation of the array factor is necessary if a complete radiation pattern is required. Often, however, it is sufficient to estimate the minor lobes by calculating one or two values for sample angles between adjacent nulls.

7.14 PATTERN MULTIPLICATION

There are some more complex linear arrangements of elements that can be analysed by decomposing the array into components that have a known, or easily found, radiation pattern, and then recombining these results to find the overall pattern. This method of pattern multiplication can be described most easily by using an example. Consider the array of elements shown in Fig. 7.9(a). Assuming that the elements are isotropic, we can decompose the array into

(i) a three-element linear array with $d = 3\lambda/4$, and $\alpha = 30°$ [Fig. 7.9(b)];
(ii) each of these three elements is in fact a two-element array with

$$d = 0.6\lambda \quad \text{and} \quad \alpha = 0 \quad [\text{Fig. 7.9(c)}].$$

In this example the six elements of the initial array, $A\,B\,C\,D\,E\,F$ have been

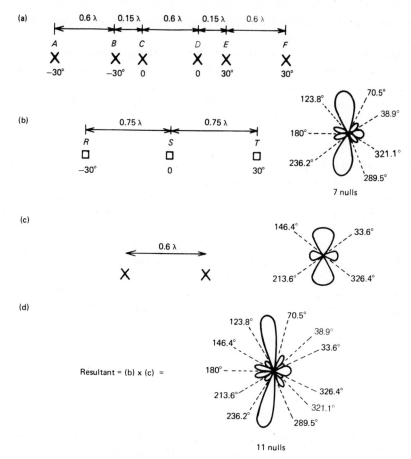

Fig. 7.9 Application of pattern multiplication.

replaced by three elements, $R \, S \, T$, where R has replaced A and B, S has replaced C and D, and T has replaced E and F.

The radiation pattern of $R \, S \, T$, assuming each element to be isotropic , is derived from Section 7.13. The result is shown in Fig. 7.9(b). The radiation pattern of R (or S or T) is found from Eqn (7.19) and it has the shape given in Fig. 7.9(c).

The resultant radiation pattern for the six-element array is therefore the product of the array pattern [Fig. 7.9(b)] and the group pattern [Fig. 7.9(c)]. We can see that the number of nulls in the resultant pattern is the sum of the nulls in the component patterns, i.e. four nulls from the group pattern and seven nulls form the array pattern, giving eleven in all [Fig. 7.9(d)].

7.15   ANTENNA MATCHING

Efficient connection between the feeder transmission line and the antenna depends on matching the impedance of one to the other. For narrowband

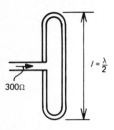

$l = \frac{\lambda}{2}$

300Ω

Fig. 7.10 Folded dipole.

applications, the antenna impedance can be calculated or measured, and an appropriate transmission line, with matching unit if necessary, can be used. In wideband applications, however, such matching may not be possible. Non-resonant antennas have a wide bandwidth capability because their impedance variation with frequency is less marked than that of resonant structures. We can see the effect of resonance on antenna terminal impedance by considering a simple half-wave dipole. Figure 7.3 shows the typical form of the variation for several thicknesses of antenna rod.

The resonant length is that which produces a purely resistive impedance, and we can see from the diagram that the reactance is zero at length slightly less than $\lambda/2$, depending on the thickness of the dipole rod. As the rod diameter falls to zero the resonant length approaches $\lambda/2$. This is due in part to the capacitive effect of the rod ends.

The half-wave dipole impedance at resonance is about $73\,\Omega$, and therefore a cable having a nominal characteristic impedance of $75\,\Omega$ will provide a good match. For some applications, however, a $300\,\Omega$ balanced cable is more appropriate, and this does not match with the $73\,\Omega$ dipole. If the dipole is folded, as in Fig. 7.10, the impedance is approximately $300\,\Omega$, which will match the characteristic impedance of the line. There are other advantages of using a folded dipole; it is mechanically more robust, it can be attached more easily to a support, and the feed area is free from supporting brackets.

7.16 PARASITIC ELEMENTS

In discussing the dipole, we established that its horizontal polar diagram is circular, and therefore has no directive properties. By using several dipoles in line a more directive antenna can be produced, but that is not always convenient. The feeding arrangements, requiring specified magnitude and phase of the current at each element, add to the complexity and hence the expense of the antenna.

For domestic broadcasting, in the frequency ranges where a half-wave dipole is a manageable size, parasitic elements are used to distort the dipole field to produce a directive radiation pattern. A parasitic element does not have any feeder cable. It is placed near to the driven element, and picks up current from it by mutual inductance. The parasitic then acts as a separate, weaker, radiator and the resulting radiation pattern is a combination of the radiation from both the driven and parasitic dipoles. The position of the parasitic element, relative to the direction of the main beam, determines its function. In Fig. 7.11 we see the parasitic element acting as a director, being placed at such a distance in front of the driven element that the fields from the two add along the line joining them, in the direction of the required maximum. Alternatively, a parasitic element placed behind the driven element acts as a reflector, with the reradiated energy adding to that of the driven element, again in the forward direction (Fig. 7.12).

By using many director elements, as is common in a VHF antenna (Fig. 7.13), a highly directional beam results. When used as a receiver, this type of antenna is most sensitive in the forward direction, and therefore

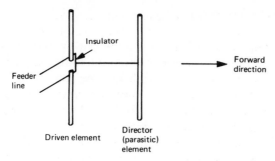

Fig. 7.11 Dipole with passive director element.

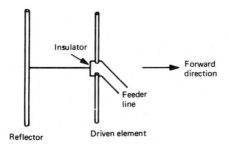

Fig. 7.12 Dipole with passive reflector element.

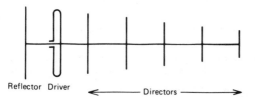

Fig. 7.13 Yagi array.

discriminates against signals from other directions. The beamwidth decreases as the number of elements increases.

The antenna is named after Hidetsugu Yagi, who was active in the general field of high-frequency communications. A description of the antenna he developed was published in the Proceedings of the Institute of Radio Engineers in 1925. Emphasizing the high directivity of the array, he referred to it as a wave projector, or wave canal.

We are not attempting to analyse these antennas. The mutual coupling is complex, depending on the spacing, length, diameter, and material of each element, and it exists not only between the driven and parasitic elements, but between the parasitic elements themselves. The design of multi-element antennas is based on empirical experience, rather than analysis, because of the difficulty in modelling the interaction between the elements.

7.17   MICROWAVE ANTENNAS

At microwave frequencies the antennas are generally significantly larger than a wavelength, and this allows techniques akin to those of optical devices, such as reflectors and lenses, to be used, thus making some microwave antennas different in kind from those possible at lower frequencies. The range of applications of microwaves in radar and communications is very large, and growing, and hence the demand for efficient, robust and inexpensive antennas, particularly in the millimetre-wave range, will increase.

Section 14.5 discusses some of the design problems related to satellite antennas.

7.18   PARABOLIC REFLECTOR

Probably the most commonly used microwave antenna is the parabolic dish. It reflects energy from a primary source placed at its focus. The principle is the same as that used in an optical mirror; when the source is at the focus, on the axis of the parabola, the reflected beam is parallel to the axis. Figure 7.14 shows the main features. Assuming that ray theory applies, in order to produce a parallel beam, all rays leaving the focus F and striking the reflecting surface must be in phase in some plane normal to the axis of the reflector, such as AA in the diagram. Typical rays would be FCD, FGH, FJK and FLM as shown. The requirement that DHK and M are in phase means that

$$FC + CD = FG + GH = FJ + JK = FL + LM$$

and this relationship is satisfied by a parabolic reflecting surface. The usual shape of antenna is the paraboloid, formed by revolving a parabola on its main axis, and designed to give a symmetrical radiation pattern. Two main factors prevent the reflected beam from being parallel with the axis: first, the feed cannot be a point source; and second, the wavelength of the radiation is not so small that diffraction effects can be ignored.

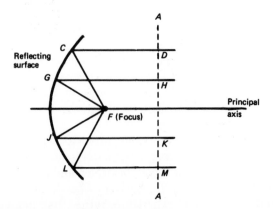

Fig. 7.14 Geometry of a parabolic dish antenna.

The feed introduces some problems. It has a comparatively large size and therefore blocks energy reflected from the paraboloid, and its own radiation pattern may cause destructive interference with energy from the reflector.

The comparatively large wavelength of microwave radiation produces diffraction effects at the edge of the reflector, allowing energy to spill round the back of the dish, and be wasted. One way of limiting such loss is to ensure, by design of the primary feed, that the reflector is not illuminated uniformly. A typical design figure is that the energy from the source, incident at the edge of the reflector, is only 10% of that along the axis.

The primary feeds normally used are shown in Fig. 7.15. The dipole, with parasitic reflector to prevent radiation in the forward direction, has a less directive radiation pattern than the horn antenna, and therefore illuminates the edge of the reflector too strongly, but it has a much smaller supporting structure than that required for the horn, and so does not block as much of the reflected wave. To reduce this blocking by the support, a Cassegrain antenna is sometimes used, particularly in receiving applications. The Cassegrain antenna (Fig. 7.16) consists of a primary horn feed, set in the surface of the reflector, and a sub-reflector of hyperbolic shape placed just in front of the focus of the parabola. Energy from the horn strikes the sub-reflector and is

Dipole feed
(with reflector)

Horn feed

Reflected feed

Fig. 7.15 Parabolic dish feeds.

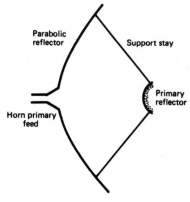

Parabolic
reflector

Support stay

Primary
reflector

Horn primary
feed

Fig. 7.16 Cassegrain antenna.

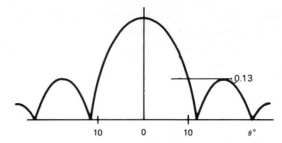

Fig. 7.17 Radiation pattern of parabolic dish with uniform illumination.

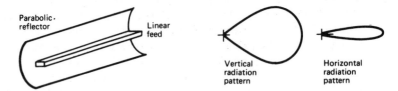

Fig. 7.18 Parabolic cylinder antenna and its polar diagrams.

then redirected to the surface of the parabolic dish. The supports used for the sub-reflector are much less robust than the waveguide feed necessary for a horn-type primary source.

The beamwidth of the parabolic reflector is related to the wavelength of the signal, and the diameter of the dish at its rim, as well as to the illumination variation across the reflector surface. A typical radiation pattern, assuming uniform illumination, is shown in Fig. 7.17. The first side lobes rise to only 13% of the main lobe intensity, and the beamwidth is a few degrees. As the diameter is increased, the beamwidth decreases, according to the relationship

$$\theta = k\lambda/D \tag{7.25}$$

where θ is the beamwidth, D is the diameter of the reflector at the rim and λ is the radiation wavelength. k is a constant depending on the illumination from the primary feed, with a value of about 60. Thus, for $\theta = 10°$, D will need to be about 6λ.

Other shapes of reflector are used, usually to produce beamshapes that are not equal in the horizontal and vertical planes. For example, the parabolic cylinder shown in Fig. 7.18 is fed by a line source, such as an array of dipoles, and produces a polar diagram that has a very narrow horizontal beamwidth and a broad vertical beamwidth. Used as a radar antenna it can therefore identify the horizontal direction of a target with reasonable accuracy, but it will have no discrimination in elevation.

7.19 HORN ANTENNAS

If a waveguide is unterminated it will radiate energy, producing a broad radiation pattern, as in Fig. 7.19. By flaring the end of the guide a more directional pattern is produced.

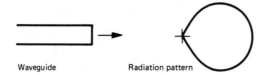

Fig. 7.19 Radiation pattern from waveguide end.

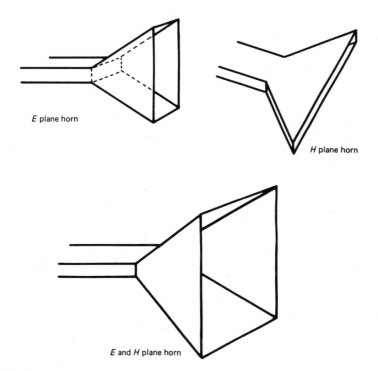

Fig. 7.20 Horn antennas.

The flare can be in the E plane [Fig. 7.20(a)], the H plane [Fig. 7.20(b)] or both [Fig. 7.20(c)], producing a beam which is narrow in the direction of the flare. It is worth noting that radiation from a horn antenna is polarized, that is the direction of the E field is related to the orientation of the waveguide, hence the examples shown in Fig. 7.20 are all vertically polarized.

The shape of the polar diagram from a waveguide horn will depend on two factors–the flare angle θ and the length of the horn L (Fig. 7.21). By using a large flare angle, the horn aperture can be made large in a short distance from the throat. However, a large θ will result in a large phase difference across the wavefront at the horn aperture, shown as δ in Fig. 7.21. A reasonable limit for δ is $\lambda/4$, in which case from the geometry of the horn the relationship between L, D and the wavelength λ is

$$L = D^2/2\lambda \qquad (7.26)$$

At small wavelengths, in the millimetre range for example, Eqn (7.26) does not

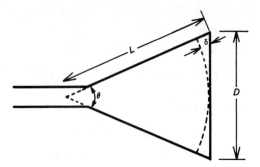

Fig. 7.21 Horn antenna dimensions.

impose difficult conditions on the size of the horn, but at longer wavelengths the value of L is so large that any practical horn would be too long, too bulky and too heavy. The dependence of beamwidth on horn aperture is approximately the same as that given in Eqn (7.25) for the paraboloidal reflector.

7.20 DIELECTRIC LENS

The length required to produce a horn with an aperture large enough to give a narrow main beam can be reduced by placing a perspex lens in the mouth of the horn.

The lens works in exactly the same way as an optical lens. It is made from a dielectric with a permittivity greater than that of air and therefore slows a wave travelling through it. If we refer to Fig. 7.22 we can see a ray theory approximation of the lens. Consider the ray that starts at the point F on the convex side of a convex-planar lens, strikes the lens at A, and passes through the lens to leave at B. We can compare it with the ray that travels along the axis from F to T, leaving the lens at P.

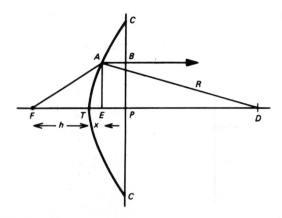

Fig. 7.22 Plano-convex lens antenna.

Given the dimensions in the diagram, for the two rays to have equal travelling times from F to the plane CC,

$$FA/c + AB/v = FT/c + TP/v \qquad (7.27)$$

where c is the velocity of the wave in air and v is the velocity of the wave in the dielectric.

Equation (7.27) can be written

$$FA/c = FT/c + (TP - AB)/v$$

From Fig. 7.22, FA, FT, and $TP - AB$ can be replaced by r, h and x, giving

$$r/c = h/c + x/v \qquad (7.28)$$

The velocity v is related to c by

$$v = c/\sqrt{(\varepsilon_r)}$$

where ε_r is the relative permittivity of the lens material. The quantity $\sqrt{(\varepsilon_r)}$ is called the refractive index, n, so $v = c/n$. Using this expression in Eqn (7.28) allows the velocity of the wave in air, c, to be cancelled, leaving

$$r = h + nx \qquad (7.29)$$

From the triangle TAF, x is related to r, θ and h by

$$x = r \cos \theta - h$$

Hence Eqn (7.29) becomes

$$r = h + n(r \cos \theta - h)$$

Rearranging terms

$$r(n \cos \theta - 1) = h(n - 1)$$

or

$$r = h(n - 1)/(n \cos \theta - 1)$$

This equation is the polar coordinate expression for a hyperbola, and it defines the curve of the convex surface of the lens.

If the focal length h is short, the thickness of the lens at the axis (TP) will be substantial, resulting in a device which is both bulky and heavy. Provided that the application is at a fixed frequency, the weight and structure of the lens

Fig. 7.23 Stepped lens antennas.

can be improved by taking away zones that are a wavelength deep. The zones can be removed from either the convex or plane side of the lens, as shown in Fig. 7.23.

7.21 MICROSTRIP ANTENNAS

Antennas that are physically robust, that can be mounted in the surface of a body such as an aeroplane, that are relatively cheap to produce and that can be built into steerable one- or two-dimensional arrays, are extremely attractive to the antenna designer. Such antennas can be made using the microstrip approach; in this case a conducting patch is mounted on to a dielectric ground plane (Fig. 7.24).

The patch can have any geometry, although rectangular and circular are the most common. The feed is either directly from a stripline, or via a coaxial probe in which the centre conductor is connected to the patch and the outer to the ground on which the dielectric sits.

Although the microstrip feed would appear to be the obvious choice, in fact the function of a microstrip line is rather different from that of a microstrip antenna, in that the first is designed to constrain and guide the transmitted power, whereas the second is designed to radiate it. This results in different design parameters for the dielectric in each case. In the guide the dielectric is thick and made from a material with a low permittivity; in the antenna the dielectric is thin and has a high permittivity.

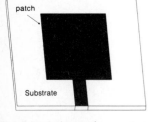

Fig. 7.24 Microstrip rectangular patch antenna.

7.21.1 Modelling

The designer is keen to be able to correlate the parameters of the antenna with its electrical properties such as bandwidth, impedance, radiation pattern and efficiency, so that the best balance between the parameters can be chosen to meet the specifications of a particular application. To relate properties to physical parameters, some analysis of the behaviour of microstrip antennas is required, and if possible the analysis should provide a guide to appropriate CAD tools.

Two approaches to modelling have been used, in much the same way as happens in many of the other systems considered in this book. An approximate approach is used to provide intuitive insights into the operation of the antennas, with the hope that the results obtained are reasonably accurate, and sufficient for a design in some cases. The other is to develop a rigorous, accurate analysis that gives a detailed assessment of the performance of the system, taking into account as many elements and parameters as possible. The cost of this approach is long and complex computation, following some difficult analysis related to the particular system being studied.

For accurate results, and as a basis for building CAD packages that are not dependent on too many assumptions, the second approach is the best, but it requires some powerful computing resource. For many initial designs, the more approximate method will be adequate, provided its limitations are clearly understood.

There can be an interpretative problem with rigorous analysis. Because of its complexity it is difficult to sustain an awareness of the effects of each of the parameters of the system on the final result, and there are many applications where the designer needs to assess what trade-offs are available. Intuition is more easily guided by an approximate analysis in which the main elements of the system involved in determining a particular characteristic are explicitly available, and others are supressed. A helpful initial review is contained in reference 8.

7.21.2   Types of antenna

The polar diagram of a patch antenna is generally broadside and symmetrical in relation to the plane of the patch; most of the energy is radiated normally to the patch, and the beamwidth itself is broad. The particular arrangement of the feed, and the thickness of the antenna, will influence the radiation pattern and the input impedance at the feed point. In the case of a coaxial feed, this impedance is a function of the position of the probe relative to the patch. Towards the edge the impedance is high and it falls towards the centre. Some degree of matching is therefore possible by choosing the position of the feed point.

The patch is frequency sensitive. It can be modelled as a microwave cavity. To a first approximation, there is zero tangential E-field along the patch surface, and the edge of the patch can be considered as a magnetic wall from which the magnetic field is reflected. This approach leads to expressions similar to those in Section 9.19. Patch antennas have a high Q and hence a narrow bandwidth.

In addition, the resonant frequency is high for small patches. For example, for use as a mobile cellular car antenna, a reasonably sized device, fitted in the car roof, would resonate at about 2.2 GHz, whereas cellular radio frequencies are presently below 1 GHz. An antenna with a natural resonant frequency as low as that would be too large for the application, so one approach is to modify the patch itself by cutting a series of overlapping concentric slots into it. These oblige the current to go from the probe, at the centre, to the patch edge via a longer path since it has to travel round the slots. In this way the resonant frequency is moved into the region required.

There are many applications, particularly in moving vehicles, or in structures that are exposed to extreme conditions, where conformal surface antennas are extremely attractive, and as communication technology is increasingly related to mobile use, microstrip antennas will continue to attract considerable attention from theorists and designers. This is one of several areas in antenna technology that has yet to realize its full potential.

7.22   RADIO WAVE PROPAGATION

The mechanism whereby a radio signal transmits through the air between transmitters depends chiefly on the frequency of the wave. In the electromag-

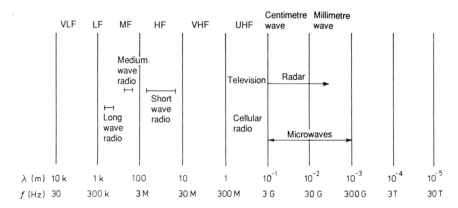

Fig. 7.25 Radio spectrum.

netic spectrum shown in Fig. 7.25 the bands of frequencies used for various applications are indicated. For domestic radio broadcasts the medium and long wave bands are used, and broadcast television frequencies are normally in the VHF and UHF bands. Between these two ranges there is the commonly used shortwave radio or HF band between 1 and 30 MHz. Other bands of common interest are also indicated in the diagram. It is noteworthy that there is comparatively little interest at present in the enormous range between millimetre radar (100 GHz) and optical fibre transmission frequencies. This is due mainly to a lack of suitable sources and detectors, and the problems of transmission.

The propagation mechanisms involved in radio transmission are (a) groundwave (b) skywave (c) troposphere scatter and (d) line of sight. At low frequencies the groundwave is the most important; at HF frequencies it is the skywave that is predominant and at higher frequencies still, line of sight becomes the most important mechanism.

At very low frequencies the atmosphere between the earth's surface and the ionosphere acts as a waveguide which allows the low-frequency energy to be guided around the surface of the earth. This mechanism supports propagation distances of thousands of kilometres and is therefore greatly used for long distance navigation.

The groundwave becomes increasingly attenuated as the frequency rises. It is still predominant at the radio broadcast frequencies but at the upper end of the MF band the groundwave has become a surface wave which attenuates rapidly if the surface has high conductivity. In practice the conductivity of the surface of the earth is complicated by, first, the large range of materials of which the earth's surface is constituted, and secondly the presence of water contaminated by salt solutions. Both the dielectric constant and the conductivity of the earth's surface are much dependent on the moisture content and the nature of the moisture. This has the effect that propagation over the sea is better than over the land and the depth of penetration of the radiowave over the earth is highly related to the dryness of the ground, its constituent materials, and the presence of water below its surface. A particulary useful discussion of groundwaves is given in reference 1 to which the reader is referred for further information.

7.22.1 Skywaves

In the HF frequency band the main mode of propagation is by skywave. Radiation from the earth is propagated outwards into the sky and reflected or refracted back to earth where it can be detected. The mechanism which causes the refraction or reflection is the ionosphere. The ionosphere consists of a number of layers of ionized gas which are produced at various heights due to the absorption of radiation from the sun. The predominant layers are known as D, E, Fl and F2, with the D layer at the lowest altitude, of round about 50 km, extending to approximately 300 km for the F2 layer. Because these layers are generated by the sun's activities the lower layers are in place only during the day; at night the ionization disperses and there are no free electrons. At the upper reaches of the atmosphere the sun is effective throughout the 24 hours. The ionized gases in the various layers act like a refractive medium which, depending on the frequency and angle of presentation of the radiowave, will cause reflection or refraction to a degree dependent on both the density of the free electrons and the frequency of the wave.

We can see from Fig. 7.26 that there is a distance measured along the ground known as the skip distance; this is the distance between the transmitting antenna and the first position at which the wave can be received. To some extent the area around the transmitter will be covered by the groundwave but there will be a significant region which does not receive any signal.

A simple argument for the refraction mechanism, based on geometric optics and a very simplified model of the ionospheric layers is as follows.

To a first approximation the refractive index of an ionized medium, relative to that of free space, is related to the electron density, the electron mass and charge, and the frequency of the incident wave by

$$\varepsilon_r \simeq 1 - \frac{Ne^2}{\varepsilon_0 m \omega^2}$$

$$\simeq 1 - \frac{81N}{f^2}$$

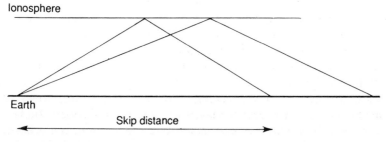

Fig. 7.26 Skip effect in skywave.

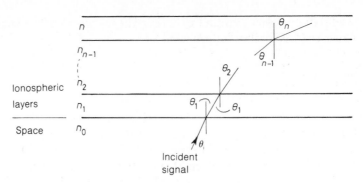

Fig. 7.27 Simple ionospheric model.

using $e = 1.59 \times 10^{-19}$ C, $m = 9 \times 10^{-31}$ kg, $\varepsilon_0 = 10^{-9}/36\pi$ farads/m and $\omega = 2\pi f$. N is measured in electron/m^3.

We can model the ionosphere as a series of layers, each with an increasing electron density, as shown in Fig. 7.27. The thickness of each layer can be made arbitrarily small. The angle of incidence θ_i at the ionosphere is related to the refractive index of the first layer by Snell's Law of Refraction:

$$n_0 \sin \theta_i = n_1 \sin \theta_1$$

Assuming $n_0 = 1$

$$\sin \theta_i = n_1 \sin \theta_1$$

Repeating this equation for each boundary interface,

$$\sin \theta_i = n_n \sin \theta_n$$

The refractive index $= \sqrt{\varepsilon_r}$.

Hence, if N is the electron density in the nth layer,

$$\sin \theta_i = \left[1 - \frac{81\,N}{f^2} \right]^{1/2} \sin \theta_n$$

As the electron density increases there will be some layer for which the above relationship cannot be satisfied, i.e. when $\sin \theta_n \geqslant 1$ and the wave no longer passes through the ionosphere but is reflected back to earth. For a given value of incident angle θ_i, this occurs, from the last equation, when N satisfies the relationship

$$N = \frac{f^2}{81} \cos^2 \theta_i$$

The electron density is a function of altitude, so the altitude at which reflection occurs varies with the angle of the incident wave θ_i and the frequency of the signal, f.

There is a very significant variation in the density and height of the ionospheric layers with time of day and season of the year. In fact the prediction of radiowave propagation through the ionosphere is very much akin to the prediction of weather forecasts and it has the same level of success.

It is possible to forecast reasonably well how the ionosphere is likely to behave and therefore to choose those frequencies which will produce the most effective reception. This generally means that through the 24-hour period the transmission frequency has to vary significantly; it would not be unusual for the frequency to have to change by a factor of 2 over a very small period of time to respond to, say, the end of darkness. Nowadays these considerations of ionospheric condition are studied by computer, and graphs of recommended transmission frequency against time of day are plotted.

Apart from the skip distance problem, which can be overcome to some extent by having a broad elevation in the radiated pattern, there is another problem with multiple reflections. Under certain circumstances a wave can be reflected by the ionosphere and then reflected once at the ground and again at one of the ionospheric layers so that a multipath signal is eventually received. In that case interference will result between the various signals.

A useful discussion on the general approach to coping with ionospheric problems is given in reference 7.

7.22.2 Line of sight

At frequencies above the VHF range the most important propagation mechanism is line of sight. At these frequencies the ionosphere ceases to refract the wave which goes straight through the various layers; and the groundwave is attenuated very rapidly. This means that from a practical point of view transmitter and receiver must be in visible contact, although there will be a small degree of defraction around objects at the lower frequencies. This results in relatively poor over-the-horizon communication and it raises a problem for coverage in the higher television broadcast bands where transmitter and customers are in hilly regions.

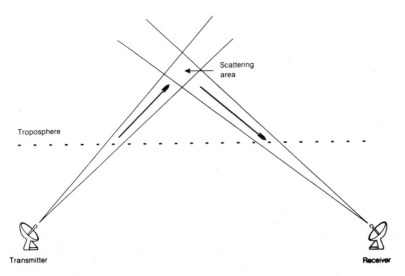

Fig. 7.28 Tropospheric scatter.

7.22.3 Tropospheric scatter

At frequencies above about 400 MHz there is significant scatter in the troposphere (below the ionosphere). This provides the means for extending the range of a communication link, as shown in Fig. 7.28. The signal level may be low, and the link is subject to fading due to variations in the tropospheric height, but for some applications it is the most practical method of providing a high capacity communications channel.

REFERENCES

1. Jordan, E. C., *Electromagnetic Waves and Radiating Systems*, Prentice-Hall, London, 1968.
2. Kraus, J. D., *Antennas*, McGraw-Hill, Maidenhead, 1950.
3. Jasik, H., *Antenna Engineering Handbook*, McGraw-Hill, Maidenhead, 1961.
4. Fradin, A. Z., *Microwave Antennas*, Pergamon, Oxford, 1961.
5. Weeks, W. L., *Antennas Engineering*, McGraw-Hill, Maidenhead, 1968.
6. Rudge, A. W., Olver, A. D., Milne, K. and Knight, P., *The Handbook of Antenna Design* (Vol. I), Peter Peregrinus, London, 1982.
7. Rudge, A. W., Olver, A. D., Milne, K. and Knight P., *The Handbook of Antenna Design* (Vol. II), Peter Peregrinus, London, 1983.
8. Pozar, D. M., Microstrip Antennas, *Proc. IEEE*, Jan. 1992, 79–91.

APPENDIX 7.1   FIELD COMPONENTS PRODUCED BY A SHORT CURRENT ELEMENT

In Section 7.3 we used Eqn (7.5) to represent the relationship between the retarded vector potential and the magnetic field $\bar{H}$, viz.

$$\nabla \times \bar{A} = \bar{H} \tag{7.5}$$

and the components of $\bar{A}$ for the current filament shown in Fig. 7.4 are

$$
\left.
\begin{aligned}
A_r &= A_z \cos \theta = \frac{Il}{4\pi r} \sin \omega \left[t - (r/c) \right] \cos \theta \\[2mm]
A_\theta &= - A_z \sin \theta = - \frac{Il}{4\pi r} \sin \omega \left[t - (r/c) \right] \sin \theta \\[2mm]
A_\phi &= 0
\end{aligned}
\right\} \tag{7.6}
$$

In spherical coordinates, $\nabla \times \bar{A}$ is expressed as

$$
\nabla \times \bar{A} = \bar{a}_r \frac{1}{r \sin \theta} \left[\frac{\partial}{\partial \theta} (A_\phi \sin \theta) - \frac{\partial A_\theta}{\partial \phi} \right] + \bar{a}_\theta \frac{1}{r} \left[\frac{1}{\sin \theta} \frac{\partial A_r}{\partial \phi} - \frac{\partial}{\partial r} (r A_\phi) \right]
$$
$$
+ \bar{a}_\phi \frac{1}{r} \left[\frac{\partial}{\partial r} (r A_\theta) - \frac{\partial A_r}{\partial \theta} \right] \tag{7.30}
$$

where $\bar{a}_r, \bar{a}_\theta, \bar{a}_\phi$ are the unit vectors in the r, θ and ϕ directions, respectively.

Substituting from Eqn (7.6) into Eqn (7.30), the $\bar{a}_r$ and $\bar{a}_\theta$ components are zero, leaving

$$\nabla \times \bar{A} = \bar{a}_\phi \frac{1}{r}\left[\frac{\partial}{\partial r}\left(-\frac{Il}{4\pi}\sin \omega\,[t-(r/c)]\sin \theta\right)\right.$$

$$\left. - \frac{\partial}{\partial \theta}\left(\frac{Il}{4\pi r}\sin \omega\,[t-(r/c)]\cos \theta\right)\right]$$

$$= \bar{a}_\phi \frac{1}{r}\left[\frac{Il}{4\pi}\frac{\omega}{c}\cos \omega\,[t-(r/c)]\sin \theta + \frac{Il}{4\pi r}\sin \omega\,[t-(r/c)]\sin \theta\right] \qquad (7.31)$$

and, knowing that $\bar{H} = \bar{a}_r H_r + \bar{a}_\theta H_\theta + \bar{a}_\phi H_\phi$, we have, from Eqns (7.5) and (7.31),

$$H_\phi = \frac{Il}{4\pi}\sin \theta \left[\frac{\omega}{cr}\cos \omega\,[t-(r/c)] + \frac{1}{r^2}\sin \omega\,[t-(r/c)]\right] \qquad (7.32)$$

For the space outside the current element, the relationship between the H field and the E field is given by one of Maxwell's equations [Eqn. (9.10) with $J = 0$], i.e.

$$\nabla \times \bar{H} = \varepsilon \frac{\partial \bar{E}}{\partial r} = \varepsilon \left[\bar{a}_r \frac{\partial E_r}{\partial t} + \bar{a}_\theta \frac{\partial E_\theta}{\partial t} + \bar{a}_\phi \frac{\partial E_\phi}{\partial t}\right] \qquad (7.33)$$

Thus, if we expand the left-hand side of this equation, then with the help of Eqn (7.32) we can find the electric field components. Since there is only one component to the $\bar{H}$ field, H_ϕ, we have by analogy with Eqn (7.30)

$$\nabla \times \bar{H} = \bar{a}_r \frac{1}{r\sin \theta}\frac{\partial}{\partial \theta}(H_\phi \sin \theta) - \bar{a}_\theta \frac{1}{r}\cdot\frac{\partial}{\partial r}(rH_\phi)$$

Substituting from Eqn (7.32), this becomes

$$\nabla \times \bar{H} = \bar{a}_r \frac{1}{r\sin \theta}\frac{\partial}{\partial \theta}\left\{\frac{Il\sin^2\theta}{4\pi}\left[\frac{\omega}{cr}\cos \omega\,[t-(r/c)] + \frac{1}{r^2}\sin \omega\,[t-(r/c)]\right]\right\}$$

$$- \bar{a}_\theta \frac{1}{r}\frac{\partial}{\partial r}\left\{\frac{Il\sin \theta}{4\pi}\left[\frac{\omega}{c}\cos \omega\,[t-(r/c)] + \frac{1}{r}\sin \omega\,[t-(r/c)]\right]\right\}$$

After performing the differentiations, and rearranging the terms,

$$\nabla \times \bar{H} = \bar{a}_r \frac{Il\cos \theta}{2\pi}\left[\frac{\omega}{cr^2}\cos \omega\,[t-(r/c)] + \frac{1}{r^3}\sin \omega\,[t-(r/c)]\right]$$

$$+ \bar{a}_\theta \frac{Il\sin \theta}{4\pi}\left[-\frac{\omega^2}{c^2 r}\sin \omega\,[t-(r/c)] + \frac{\omega}{cr^2}\cos \omega\,[t-(r/c)]\right.$$

$$\left. + \frac{1}{r^3}\sin \omega\,[t-(r/c)]\right] \qquad (7.34)$$

Equating the $\bar{a}_\theta$ from the right-hand sides of Eqns (7.33) and (7.34),

$$E_\theta = \frac{1}{\varepsilon}\frac{Il\sin \theta}{4\pi r}\int\left[-\frac{\omega^2}{c^2}\sin \omega\,[t-(r/c)] + \frac{\omega}{cr}\cos \omega\,[t-(r/c)]\right.$$

$$\left. + \frac{1}{r^2}\sin \omega\,[t-(r/c)]\right]dt$$

which, after integrating, becomes

$$E_\theta = \frac{Il\sin\theta}{4\pi\varepsilon}\left[\frac{\omega}{rc^2}\cos\omega\left[t-(r/c)\right] + \frac{1}{cr^2}\sin\omega\left[t-(r/c)\right]\right.$$
$$\left. - \frac{1}{\omega r^3}\cos\omega\left[t-(r/c)\right]\right] \tag{7.35}$$

By a similar argument, the $\bar{a}_r$ terms on the right-hand sides of Eqns (7.33) and (7.34) are used to find the radial component of the E field:

$$E_r = \frac{Il\cos\theta}{2\pi\varepsilon}\left[\frac{1}{cr^2}\sin\omega\left[t-(r/c)\right] - \frac{1}{\omega r^3}\cos\omega\left[t-(r/c)\right]\right] \tag{7.36}$$

We have shown, therefore, that the non-zero field components in the space surrounding the current element are

$$H_\phi = \frac{Il\sin\theta}{4\pi}\left[\frac{\omega}{cr}\cos\omega\left[t-(r/c)\right] + \frac{1}{r^2}\sin\omega\left[t-(r/c)\right]\right] \tag{7.37}$$

$$E_\theta = \frac{Il\sin\theta}{4\pi\varepsilon}\left[\frac{\omega}{c^2r}\cos\omega\left[t-(r/c)\right] + \frac{1}{cr^2}\sin\omega\left[t-(r/c)\right]\right.$$
$$\left. - \frac{1}{\omega r^3}\cos\omega\left[t-(r/c)\right]\right] \tag{7.38}$$

$$E_r = \frac{Il\cos\theta}{2\pi\varepsilon}\left[\frac{1}{cr^2}\sin\omega\left[t-(r/c)\right] - \frac{1}{\omega r^3}\cos\omega\left[t-(r/c)\right]\right] \tag{7.39}$$

corresponding with Eqns (7.7)–(7.9) in Section 7.3.

By substituting $\omega = 2\pi f, c = f\lambda$ and $\eta = \sqrt{(\mu/\varepsilon)}$ (where η is the free-space impedance), these equations can be expressed in a form which explicitly involves λ:

$$H_\phi = \frac{Il\sin\theta}{2\lambda}\left[\frac{1}{r}\cos\omega\left[t-(r/c)\right] + \frac{\lambda}{2\pi r^2}\sin\omega\left[t-(r/c)\right]\right] \tag{7.40}$$

$$E_\theta = \frac{\eta Il\sin\theta}{2\lambda}\left[\frac{1}{r}\cos\omega\left[t-(r/c)\right] + \frac{\lambda}{2\pi r^2}\sin\omega\left[t-(r/c)\right]\right.$$
$$\left. - \frac{\lambda^2}{4\pi^2 r^3}\cos\omega\left[t-(r/c)\right]\right] \tag{7.41}$$

$$E_r = \frac{\eta Il\cos\theta}{\lambda}\left[\frac{\lambda}{2\pi r^2}\sin\omega\left[t-(r/c)\right] - \frac{\lambda^2}{4\pi^2 r^3}\cos\omega\left[t-(r/c)\right]\right] \tag{7.42}$$

PROBLEMS

7.1 Draw the polar diagram produced by two isotropic antennas, which are fed with equal currents if:

(a) the separation between the antennas is (i) 0.25λ; (ii) 0.7λ; (iii) 1.1λ;
(b) the phase difference between the currents is (i) $0°$; (ii) $30°$; (iii) $90°$.

7.2 Two isotropic elements, 0.4λ apart, are fed with currents of equal magnitude and a phase difference of 90°. Find the number of nulls in the polar diagram, and their direction, and the magnitude of the field normal to the line joining the elements.

Answer: 2 nulls, $\pm 51.2°$, 0.7.

7.3 Find the position of the main beam, relative to the line of the array, if five elements, 0.5λ apart, are fed with currents of equal magnitude and phase.

Answer: Main beam normal to the line of elements.

7.4 Repeat problem 7.3 for separations between the elements of (i) 0.25λ and (ii) 0.625λ, and sketch the radiation pattern in all three cases, noting the effect of increasing the spacing between the elements.

7.5 Plot the radiation pattern for an array of ten identical isotropic sources in line, and spaced at intervals of 0.375λ, if the phase difference between each source is $-135°$.

7.6 Use pattern multiplication to find the polar diagram of the array of isotropic radiators shown, assuming that the current magnitude is the same in each radiator.

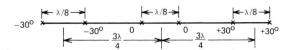

7.7 Use pattern multiplication to find the radiation pattern of an array of three isotropic elements in line, spaced at half-wavelength intervals, if the currents are in phase, but the magnitude of the current in the middle element is twice that in the outer ones.

7.8 Sketch the polar diagram of a six-element linear array in which the elements are fed with currents of equal magnitude, and a constant phase difference of 30°. The distance between adjacent elements is 0.6λ.
Compare this polar diagram with that produced by the array shown:

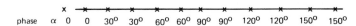

Spacing between adjacent elements is 0.3λ.

8 Active microwave devices

We are considering here systems that operate at frequencies above about 1 GHz, covering the bands used for industrial and domestic heating, microwave communications, satellite communications and radar.

A notable feature of this range of frequencies is that some of the active devices are vacuum tubes. The power levels required for many systems cannot be achieved with solid-state techniques and this has meant a continuous development and refinement of the vacuum tubes that were first introduced many years ago.

Whether or not a solid-state device can be used in a particular application will depend almost entirely on the power levels involved. If these levels do not exceed 100 W, solid-state will be a natural choice. For higher powers, however, vacuum tubes must be used; which type will depend on other considerations such as bandwidth and noise level.

Here we will not look at a comprehensive range of devices, but we will study the operation of the most common, starting with the oldest of the vacuum tubes currently in use, the klystron.

8.1 KLYSTRON

The origins of this tube are not clear. In the late 1930s, workers in several countries wrote about the possibility of using velocity modulation to generate microwaves; the space-charge modulation approach, which formed the basis of operation of lower frequency valves such as the triode, could not operate at such high frequencies. Which group should be credited with actually making the first prototype klystron is not known. The onset of war veiled such things in secrecy. However, we do know that much early work was done in the USA and in the UK, where there was intense pressure to develop satisfactory oscillators and amplifiers for use in the new detection system called radar.

8.2 TWO-CAVITY KLYSTRON

The basic form of the klystron amplifier is shown in Fig. 8.1. As we can see, it consists of a cathode, electron gun, two cavities and a collector. Electrons are generated at the cathode, directed into a fine stream along the axis of the tube by the electron gun, and, after passing through two cavities, they leave the

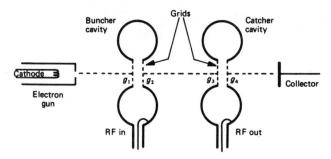

Fig. 8.1 Diagram of a two-cavity klystron.

tube via the collector. The cavity nearer to the cathode is called the buncher, and the other is called the catcher. The microwave signal to be amplified is coupled into the buncher cavity by either a waveguide or a coaxial loop. A similar device is used to abstract the amplified signal from the catcher cavity. The shape of the cavities will depend on the particular design; it could be inside the vacuum tube surrounding the electron gun, electron beam and collector, with the coupling by waveguide designed to launch the signal into the cavity, or it could be partly outside the tube, which would facilitate easier coupling of the microwaves, but would also require glass-to-metal seals between the cavities and the vacuum tube. Whatever the method used, the basic shape of the cavities is toroidal with two fine mesh grids across the axis to allow the electron beam to penetrate. The grids contain the cavity fields between them. There are some klystrons in which grids are not used; then there is no impairment to the passage of the electrons, but there are fringe fields around the axial gaps in the cavities.

The term 'velocity modulation' describes the action of the microwave signal on the electron beam. In our discussion we are talking about the steady-state operation of the valve; that is, the way in which the valve works after it has warmed up and any transient effects have died away.

In all microwave tubes there is a transfer-of-energy mechanism that we will accept without attempting to explain how it takes place. If an electron passes through an electric field it experiences a force that causes it to change velocity. If the electron is speeded up, its kinetic energy is increased, and that increase must be taken from the field. If the electron is decelerated it loses energy, and the field gains an equivalent amount, so that energy in the system is conserved.

Now we consider the passage of an electron between the grids of a cavity (Fig. 8.2). In the steady state, the potential across the grids will change at the microwave frequency of the field within the cavity itself. On one half-cycle [Fig. 8.2(a)] grid 1 will be positive with respect to grid 2. The electron passing through the grids in the direction shown will experience a decelerating force and will therefore slow down, giving up energy to the field in the cavity. The degree of deceleration will depend on the field across the grids at the time of the passage of the electron. We assume in our discussion that the grids are close together and the field remains constant during the transit of the

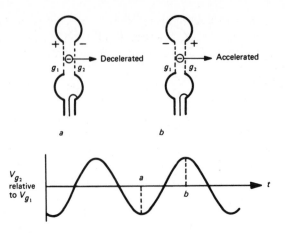

Fig. 8.2 Effect of grids on the electron beam in a two-cavity klystron.

electron. On the other half-cycle [Fig. 8.2(b)] the electron sees an accelerating force and is therefore speeded up, absorbing energy from the cavity field. When there is no potential difference between the grids the electron will pass through with velocity unchanged.

To clarify this velocity modulation mechanism further we can represent the movement of electrons, relative to the microwave field across the cavity grids, on a distance–time diagram, sometimes called the Applegate diagram after an early American worker on klystron characteristics (Fig. 8.3).

The sinusoid at the bottom of the diagram represents the fields across the grids. The lines above it represent the passage of an electron; the angle of each line is a measure of its velocity. Consequently, an electron passing through the grids when the field is maximum accelerating (point A) will have a larger slope (further distance in a given time) than one that leaves when there is no potential difference between the grids (point B); and that in turn has a larger slope than an electron suffering maximum deceleration (point C). As we might

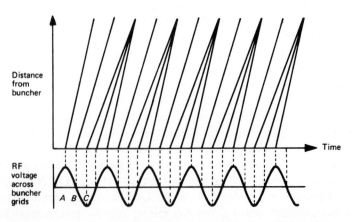

Fig. 8.3 Distance–time diagram for electrons passing through buncher.

expect the faster electrons catch up the slower ones, and bunches of electrons are formed. The diagram shows that these bunches form around the electrons that pass through with unchanged velocity as the field goes from decelerating to accelerating.

In the two-cavity klystron it is the function of the first cavity to create these bunches – hence its name. The cavity is tuned, by either adjusting a screw in the wall or deforming the wall itself in some way, to the incoming microwave frequency. When the electron stream, which we assume to be approaching the buncher at uniform velocity, passes through the grids, the velocity modulation causes bunches to form at some point along the axis of the tube in the direction of the collector. The catcher is placed at this point, allowing the tight bunches to arrive between the grids. The distance between the buncher and catcher will be fixed once the valve is constructed. However, the distance taken to form the bunches will depend on their initial velocity, and this allows the collector voltage to control the bunching distance. In the steady state, the bunches arrive between the catcher grids when grid 3 is positive with respect to grid 4 (Fig. 8.1). Then the bunch is decelerated, and the cavity field absorbs the loss in kinetic energy, thus enhancing the strength of the field in the catcher cavity.

We can see from Fig. 8.3 that, although bunches are formed, if there is a continuous stream of electrons from the cathode, not all electrons are involved in the bunching process. Those that are not will not add to the catcher field and may in fact take energy from it. This mechanism reduces the efficiency of the valve.

The electron gun, consisting of a complex arrangement of electrodes, is designed to focus the electron stream into a filament along the axis of the tube. There will be some blurring at the edges of the beam, because of the space-charge forces that tend to spread the beam out.

The collector is designed to perform two functions. It must collect the electrons from the beam, without allowing any reflections or secondary emissions, and it must conduct away the heat generated by the impact of the electron beam. The first is achieved by shaping the collector in such a way that any secondary electrons do not pass back down the tube. The second requires that either a large structure of good conducting material is used to allow natural convection cooling, or that forced-air blowing be included in the overall system design.

It is possible to feed some of the catcher field back into the buncher by providing a suitable coupling mechanism between them, and so create a two-cavity klystron oscillator, but it is not much used in practice.

8.3 REFLEX KLYSTRON

The commonest klystron oscillator, particularly for laboratory work or for use as a local oscillator, is the reflex klystron. It uses only one cavity. A schematic representation is shown in Fig. 8.4. An electron gun directs electrons from the cathode through the grids of the cavity. The velocity of the electron beam is determined by the resonator – cathode voltage. As the beam

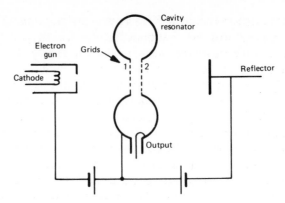

Fig. 8.4 Diagram of a reflex klystron.

passes through the resonator cavity grids, it is velocity-modulated, as we discussed in the last section with the aid of Fig. 8.2. Passing from the resonator grids the beam travels towards the reflector which is at a large negative potential. The beam is therefore in a retarding field, which it is eventually unable to resist, and it is forced to reverse direction and travel back down the tube towards the cathode. It therefore passes through the grids a second time and, on emerging from the grids, the beam is collected on the body of the resonator. The technical problem of providing adequate heat dissipation limits the power output of this tube.

The influence of the field in the resonator on the electron beam is similar to that of the buncher and catcher in the two-cavity klystron. The modulated beam travelling through the cavity for the first time gradually breaks up. The faster electrons travel further towards the reflector, before having to turn back, than those electrons that have already been retarded by the action of the resonator field. The depth of penetration of the electrons into the reflector region is governed by the reflector voltage. To make the valve oscillate, the returning beam must arrive back between the grids in such a way

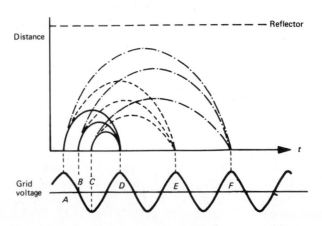

Fig. 8.5 Distance–time diagram for reflex klystron.

that (a) the bunches are completely formed, and (b) the field is such that it provides maximum retardation, i.e. grid 1 is maximum negative with respect to grid 2. We can see this condition by referring to Fig. 8.5, which shows a distance–time diagram for the reflex klystron. The bunches form around electrons that pass through the grids for the first time as the field is moving from accelerating to decelerating (point B), and to achieve maximum transfer of energy from the beam to the resonator field the bunch must arrive back between the grids when the field is at D, E, F, etc. The transit time for the electrons in the resonator–reflector region, to give maximum output, is $(n - \frac{1}{4})T$, where T is one cycle of the field and $n = 1, 2, 3 \ldots$.

Evidently, from the diagram, point D is at $n = 1$, E at $n = 2$, etc. Two other conditions are of interest. If the bunches arrive back between the grids when grid 1 is positive with respect to grid 2, they will be in an accelerating field and will therefore take energy from the field in the cavity, and there will be no output. Second, the bunches could arrive between the grids when there is a retarding field, but it is not at a maximum. Then the amount of energy taken from the bunches will be reduced, and the output from the klystron will fall.

When using a reflex klystron it is normal practice to adjust the resonator and reflector voltages to achieve maximum output. The frequency can normally be varied by a tuning screw, which slightly deforms the shape of the resonator.

The variation of output power with reflector voltage is shown in Fig. 8.6. The maxima are clearly defined. Each hump is called a mode and, as we can see, there may be several. In general, the amplitude of successive maxima increases as the reflector voltage increases; thus the largest output is achieved in mode 1, when $n = 1$ in the above formula, corresponding to a high voltage on the reflector. Occasionally, however, mode 1 is not the largest; second-order effects, such as multiple transits through the grids, reduce its level below that of mode 2.

Although the resonant frequency of the cavity itself is determined entirely by its geometry and size, when the valve is operating other factors affect the output frequency. There are three main components that combine to form the total impedance of the system; in their simplest form they can be considered as three parallel admittances, as shown in Fig. 8.7.

Y_e is the admittance of the electron beam, Y_c is the admittance of the cavity, and Y_L is the load, or output, admittance. Variation of the reflector or

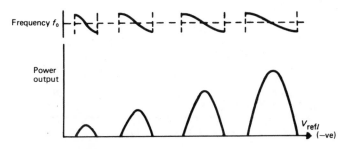

Fig. 8.6 Output characteristics of reflex klystron.

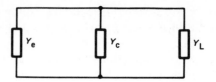

Fig. 8.7 Simplified equivalent circuit for reflex klystron.

resonator voltage will change the bunching and velocity of the electron beam, thus varying the beam current and hence Y_e. The output frequency will therefore be varied. Figure 8.6 shows a typical example of the variation of frequency with reflector voltage. We can see that the output frequency varies across each mode; the frequency is the same at each mode maximum, but the frequency range of the variation across the mode may not be constant. This frequency dependence on the reflector voltage is often called electronic tuning, and it is this characteristic that made the reflex klystron attractive as a local oscillator in a radar receiver. By means of an automatic frequency-control circuit, the frequency of the received microwave signal was compared with that from the klystron, and if the difference was not equal to the required intermediate frequency (typically 45 or 60 MHz) the reflector voltage could be adjusted to bring the klystron output frequency to an appropriate value.

We have already noted that the cavity can be tuned mechanically by adjusting a screw or deforming a flexible wall. This mechanical change results in a change in Y_c, the cavity admittance. Since Y_L is included in the equivalent admittance of the tube, the load can have an effect on the output; if the load changes, Y_L will change, and hence the output frequency will alter. This effect can cause difficulties in systems where the load impedance is not constant, and to overcome it some form of device, such as an isolator, is used to block any reflections coming back into the generator.

The reflex klystron can exhibit a hysteresis effect, which sometimes causes problems in applications where large frequency variations are required. Figure 8.8 shows two representations of this effect. In Fig. 8.8(a) we see that as the reflector voltage is increased the output follows the normal mode shape, but as the reflector voltage is decreased the characteristic follows a different path. Figure 8.8(b) shows how hysteresis affects the curve of resonator voltage against reflector voltage for a fixed output power. Instead of being a single curve it is double valued. The characteristic is

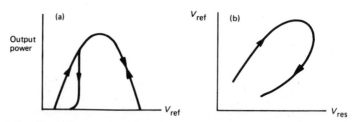

Fig. 8.8 Hysteresis effect (a) across a mode; (b) constant power curve.

produced by following one side of the curve, by increasing both resonator and reflector voltages, and then following the other side as those voltages are decreased.

Hysteresis is caused by second-order effects that we have not taken into account.

8.4 MAGNETRON

The reflex klystron is not able to deliver large quantities of power. For applications such as industrial heating, domestic cooking and radar, from 1 kW up to several megawatts of pulsed power may be required. To generate the high powers required for radar and microwave heating a different approach from that of the reflex klystron is necessary. The cavity magnetron developed by Randell and Boot in 1940 is the most common means of generating high powers in the low frequencies of the microwave range. In fact, as has been pointed out by Mr Selby Lounds, the magnetron principle was invented some 20 years earlier, being the subject of a patent by A W Hull of the General Electric Company of America in 1921, and Yagi and his group were working on similar devices in the 1920s. It was the shaping of the anode block into a series of cavities that changed a mechanism for generating low powers into one that could produce several megawatts. Although its design has been refined since then, there have been very few significant changes from the crude prototypes that demonstrated that the first magnetrons would work.

Figure 8.9 shows the essential features of the valve. It has five major components:

(i) an axial magnetic field, usually produced by a permanent magnet;
(ii) a cathode that is heated to generate electrons;
(iii) an anode block that has several slots cut into it (slow-wave structure);
(iv) a coupling device, either a loop or a hole, to couple energy from a slot in the anode block to external circuitry;
(v) the space between the cathode and the anode.

The operation of the magnetron is dependent on the interactions between

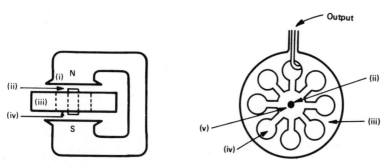

Fig. 8.9 Diagram of magnetron.

the electrons travelling in the anode–cathode space and the static and RF electromagnetic fields in that region. The analysis is complex and will not be considered here.

We will consider briefly the action of the dc fields on electrons leaving the cathode, and we shall also look at the anode block and consider it as a slow-wave structure, whose purpose is to slow down a wave circulating in the cathode–anode space to a velocity approximately the same as that of the electrons moving in that space. Energy can then be transferred from the electron beam to the electromagnetic wave. The frequency of the output depends on the design of the slow-wave structure, as we shall see.

Between the anode and the cathode there is a strong electric field, produced by a high negative dc voltage on the cathode. In the absence of a magnetic field, an electron leaving the cathode moves radially to the anode (path 1 in Fig. 8.10). In this diagram we assume that the anode block has no indentations. An axial magnetic field applied to the structure will create an additional, rotational force on an electron in the anode–cathode space. The force is a function of both the electron velocity (determined by the anode–cathode voltage) and the strength of the magnetic field. The combination of radial and rotational forces on the electron can produce different trajectories, as shown in Fig. 8.10. If the electron velocity is not up to some critical value, the electron will follow a curved path on to the anode – path 2 in the diagram. If the velocity of the electron is very high, the rotational force of the magnetic field will draw the electron back into the cathode (path 4). Between these two extremes an electron, by a suitable combination of fields, will follow a curved trajectory almost to the anode, and then follow a wide loop back to the cathode (path 3). These paths are those that would occur in the concentric arrangement shown in Fig. 8.10. However, the function of the magnetron is to produce microwave oscillations of an electromagnetic field. To do that, the field must be able to exist inside the tube, and it must also be capable of interacting with the electron beam, from which it will receive its energy. The basic mechanism involved is of an electromagnetic wave travelling around the space between cathode and anode and, during its progress, absorbing energy from the electron cloud in that space. For satisfactory interaction between electrons and the beam to occur, the electron velocity and the electromagnetic field travelling-wave velocity must be substantially the same, as we noted earlier. The slow-wave structure used to support a travelling

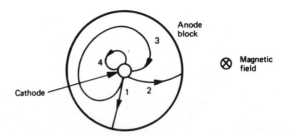

Fig. 8.10 Electron-paths in anode-cathode space.

wave, whose phase velocity is about the same as that of the electron beam (i.e. about one tenth of the velocity of light), consists of a series of cavities cut into the anode block. The simplest shape is an even number of circular cavities evenly distributed around the block, and connected to the cathode–anode space by a short coupling slot. The cavities are as similar as they can be made, and we can assume that they resonate at identical frequencies. They behave in the same way as the cylindrical cavity described in Section 9.21. Normally they would be operated in the fundamental TE_{101} mode.

The fields at the coupling holes will fringe into the cathode–anode space and influence any electron passing by. If the polarity of the field is such that an electron is accelerated it will take up energy from the cavity. The consequence of the increase in electron velocity is to move the electron back towards the cathode, and there it will create secondary electrons. (In the design of the cathode, this secondary emission, due to the return of high-velocity primary electrons, is a significant source of electrons, and allows the valve to run at a lower cathode temperature than would otherwise be possible.)

On the other hand, if the polarity of the field is in the opposite direction, the electrons passing through it will be slowed down and lose energy to the field. This is the condition required for successful operation. The loss of energy causes the electron to move into a wider orbit, and as it leaves the influence of that cavity it is affected by the field from the next. If by the time the electron has moved from the field due to the first cavity to that due to the second, the field polarities have changed, then it will be decelerated by that field, move out into a wider orbit, and continue to the next cavity where the field will again be a retarding one, and so on. Eventually the electron orbit will take it on to the anode. The trajectory just described is shown in Fig. 8.11.

In fact, the structure of the anode block can support a large number of travelling wave modes. The one we have described, in which the fields at alternate slots are 180° out of phase, is called the π mode. It is the one most commonly used in practice. Because a large number of modes are possible there is an inherent instability as the travelling wave can slip from one mode to another. Early in the development of the magnetron it was appreciated that better stability could be achieved if the π relationship could be maintained. A

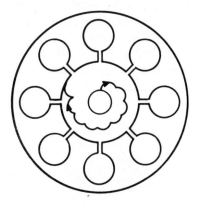

Fig. 8.11 Path of electron undergoing deceleration at successive slots.

Fig. 8.12 Strapping between alternate slots.

simple, but effective, method was 'strapping', in which alternate 'pole pieces' are interconnected by metal straps over the anode block, as shown in Fig. 8.12. This device was very successful in reducing the valve instabilities.

The coupling loop or waveguide aperture used to take energy from the valve (Fig. 8.9) can be placed in any one of the cavities. Work performed by some of the original designers showed that there was no advantage in coupling from more than one cavity; that chosen will depend on which is the most convenient mechanically.

Many different configurations have been used for the anode block, and some are shown in Fig. 8.13. Of particular significance is the 'rising sun' arrangement of Fig. 8.13(a), which is designed for use at high frequencies, where the slot and cavity shape are more difficult to manufacture. The alternate short and long slots support field patterns that stabilize the π mode without the need to use strapping.

8.5 TRAVELLING-WAVE TUBE (TWT)

The klystron amplifier has the drawback of being a narrowband device. For some applications, particularly in communications, wideband operation is essential but the resonant cavities of the klystron limit its response to a few megahertz. The purpose of the klystron cavities is to provide an RF field that can velocity-modulate an electron beam, in the case of the buncher, or respond to such modulation in the case of the catcher. However, Kompfner[1] reported that another method of allowing the RF field to interact with the electron beam could be used without relying on resonant cavities. He introduced the travelling-wave tube in which the electromagnetic signal to be amplified is transmitted close to an electron beam for a distance of several wavelengths; their proximity allows an interaction to take place in which the beam loses energy to the RF wave. We have noted several times in the preceding sections that, before a constructive interaction can occur, the electron beam must be travelling slightly faster than the electromagnetic wave. This means that, for the TWT to function as an amplifier, the wave must be slowed down to about the same velocity as the beam, which for normal operating conditions is about one tenth of the velocity of light. The wave velocity of interest is that along the axis of the tube. To achieve this reduction, a slow-wave structure is used. For low-power tubes, the slow-wave structure is a helix. A diagram of a helix TWT is shown in Fig. 8.14. The axial velocity component of the wave propagating along the helix is $c \sin \alpha$, where α is the helix

Fig. 8.13 Various anode block shapes. (a) Rising sun, (b) hole and slot, (c) slot.

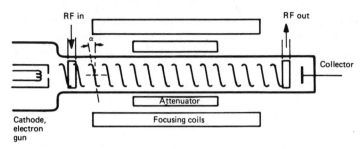

Fig. 8.14 Diagram of travelling-wave tube.

pitch and c is the velocity of light. For small α this velocity is approximately $c\alpha$, where α is in radians.

Other components of the TWT can be seen in the diagram. The electron beam is generated by a cathode, and focused by an electron gun. Additional focusing is necessary to confine the beam to the axis over the length of the tube, so it is provided by an axial magnetic field, produced by a coil surrounding the tube. The beam passes through the helix and then strikes a collector, where it is absorbed. The collector is a robust piece of copper, shaped to reduce secondary emission, and attached to a large heat sink.

The helix is made from fine wire and supported by ceramic rods. Its diameter is substantially greater than that of the beam. At each end of the helix some coupling device must be used to interconnect the electromagnetic wave with the external circuitry.

There are two interrelated mechanisms involved in the operation of the TWT:

(i) the creation of travelling waves on the electron beam, due to the action of the axial field produced by the signal on the helix;
(ii) the effect of these electron beam travelling waves on the propagation characteristics of the helix.

The slow-wave on the helix produces an axial electric field that, for a single-frequency signal, is sinusoidal in the z direction. Some sections of the beam will be accelerated by this axial field, and some decelerated. The natural charge repulsion inside the beam will cause slight bunching to occur, indicative of velocity modulation as we discussed in earlier sections. Figure 8.15 shows schematically the relationship between the axial field and the beam. The diagram represents one instant of time. However, the electric field will travel along the helix, and that induces two travelling waves on the beam. A theoretical description of this mechanism of setting up travelling waves on the beam by slight perturbations of the electron flow is given in several texts, e.g. Atwater[2] and Collin.[3]

The existence of the two travelling waves on the beam has an influence on the characteristics of the slow-wave structure. In Fig. 8.16 the beam current, and a transmission line representation of the slow-wave structure, are shown. There will also be a small leakage current δi between the beam and the structure. This leakage current, which is directly proportional to the current

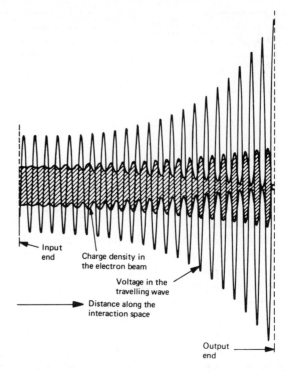

Fig. 8.15 Transfer of energy from beam to signal. Reproduced from Reich, Shalnik, Ordung and Kraus (1957), *Microwave Principles*, Van Nostrand Reinhold, New York, by permission.

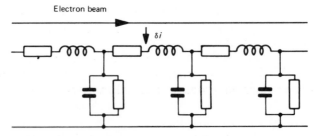

Fig. 8.16 Transmission line representation of travelling-wave tube.

on the beam, will contain components of the two beam travelling waves. The combination of the normal transmission line travelling waves on the helix when no beam is present and the travelling waves on the electron beam, is to produce four travelling waves on the helix in the presence of the beam. One is a backward wave and the other three are forward waves. The backward wave, if allowed to travel to the input coupler, would be reflected there and then travel forward, thus causing interference with the forward waves and possibly generating oscillations. An attenuator (or an isolator) is placed at some point along the tube to prevent such interference. Of the three forward waves, one

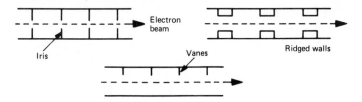

Fig. 8.17 Waveguide slow-wave structures.

has no amplitude variation (the propagation constant is imaginary), another is attenuated, but the third grows exponentially with z, the distance from the input end of the helix. It is this last wave that gives the amplification required. The actual gain achieved will depend on how efficiently the energy in the beam can be transferred to the field on the helix. The growth wave is continuously amplified as it passes along the tube, and at the output end it is coupled to the external circuit.

Typically, a helix type TWT will operate over an octave bandwidth, and have a gain of 30 dB or more. Uniform gain across the band may be essential for many applications, and this will limit the bandwidth available.

The helix slow-wave structure is convenient, and easy to manufacture, but it can be used only for low-power tubes. The description above assumes that the electron beam consists of electrons moving in straight lines from cathode to anode. In practice the electron flow is much more complicated because of the mutual repulsion between the electrons, and this causes the beam to be blurred at the edges. For high power, the electron beam is of large diameter and it could easily damage the helix if it came in contact with it. The helix itself is not able to withstand bombardment by stray electrons; hence other structures are used in high-power systems. Figure 8.17 shows examples of some slow-wave structures based on waveguide propagation.

One of the most important civil applications of the TWT is as a high power amplifier (HPA) in a satellite transponder (see Section 14.4).

8.6 SOLID-STATE DEVICES

The vacuum tubes described earlier, although capable of operating at high power levels, suffer from several major disadvantages; they require stable, high voltage, power supplies; they are mechanically fragile, and they have relatively short lives. Solid-state devices, on the other hand, are extremely robust, have long working lives, and operate at a few tens of volts. They are inherently smaller and more convenient to use, and cheaper to buy. Consequently they are used for all applications not requiring a continuous power greater than several tens of watts. As has happened at lower frequencies, there have been several different devices proposed and developed, but here we shall look only at the three most important; the Gunn diode, the IMPATT diode and the GaAsFET.

8.7   GUNN DIODE

The excitement following the invention of the transistor in 1948 led to an upsurge in active device research for microwave applications. Although many modifications were made to bipolar transistor designs, in an attempt to make them suitable for operation at microwave frequencies, there were clearly some intrinsic limitations that prevented lower-frequency techniques from being applied successfully. From 1961 to 1963, however, there was a major research breakthrough. In 1961 Ridley and Watkins,[4] and in 1962 Hilsum,[5] wrote theoretical papers predicting that, given certain conditions that are satisfied by a few compounds, microwave oscillations could be produced directly from a thin sample of such a material. In 1963, Gunn,[6] while examining the noise properties of a slice of GaAs, observed a regular pulse of current from his sample when the applied voltage exceeded some level; the Gunn device was discovered. Gunn followed up this work with a set of very precise and difficult measurements in which he investigated the variations of electric field across the device when it was operating in an oscillatory mode, and as a result he confirmed many of the earlier theoretical predictions.

We can see the requirements of a suitable material from the energy level diagram of GaAs shown in Fig. 8.18. The conduction band is separated from the valence band by what is called the 'forbidden gap', which is 1.4 eV wide. When there is no field across the diode (so called because it has two terminals, a cathode and an anode, and not because it exhibits any rectifying properties) the free conduction electrons reside in the lowest conduction state, shown as the lower valley in the diagram. As the electric field across the device increases, the drift velocity also increases; the straight line OA in Fig. 8.19 shows the relationship between them. The gradient of OA is known as the electron mobility in the lower valley, say μ_L. As the electric field is increased, it eventually achieves a value E_{Th} at which the linear relationship between electron velocity and electric field breaks down, and the velocity begins to decrease, along CD. There is now a negative gradient until the field reaches E_v at point D, after which the velocity again increases linearly with electric field. Beyond D the characteristic follows the line OB which represents an electron mobility of μ_u. Clearly, μ_u is smaller than μ_L.

Along DB the conduction electrons, which were in the lower valley, have now moved to the upper valley where their mobility is reduced, and their

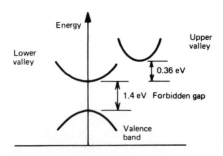

Fig. 8.18 Energy-level diagram for GaAs.

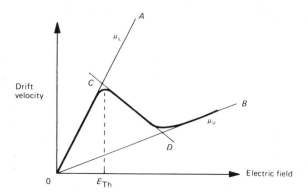

Fig. 8.19 Variation of drift velocity with electric field in a Gunn device.

mass is increased. Between C and D the transition over the 0.36 eV gap between the valleys is taking place. To avoid breakdown of the diode material, which must have suitable conduction band energy levels, the difference between them must be much smaller than the forbidden gap. There are some III–V compounds that satisfy this requirement. GaAs is the best understood because, since the Gunn experiments, it has been used widely as a diode material. Another compound of interest is InP which may eventually prove to be even more appropriate. Intensive work is being done in several countries to overcome some of the technical problems associated with its production.

The distribution of electric field across the diode is constant until E_{Th} is exceeded. Thereafter the diode is operating in the unstable CD region. Gunn showed that in this region a domain of high-intensity field and low-mobility electrons forms at the cathode, and gradually drifts across to the anode, where it leaves the device and produces a pulse of current in the external circuit. The variation of output current with time is shown in Fig. 8.20. Once a domain forms at the cathode no further domain can be sustained, so there is only one domain present at any time. Some properties of the cathode region ensure that the domain begins there consistently, and therefore the periodicity of the pulses is determined by the transit time of the domain across the device, which in turn is related to the device thickness. For operation in the range 5–10 GHz the material would be about 10 μm across. To sustain oscillations in a circuit, the circuit resistance and the device resistance must cancel,

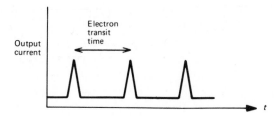

Fig. 8.20 Output pulses from Gunn device..

implying that at the frequency of interest the device resistance must be negative. That this is so can be seen from Fig. 8.19. The electron drift velocity is related directly to current flow through the device, and the electric field is proportional to the terminal voltage; hence the line CD represents a negative resistance. In practice it will have a value of no more than $10\,\Omega$. Apart from this negative resistance, there will be reactive components associated with the device, its leads and the packaging, all of which will affect the operating frequency. A simplified equivalent circuit is shown in Fig. 8.21.

The mode of operation described above, in which the applied voltage is maintained above E_{Th}, is called the transit-time mode. To produce oscilla-

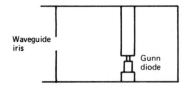

Fig. 8.21 Simplified equivalent circuit of Gunn device.

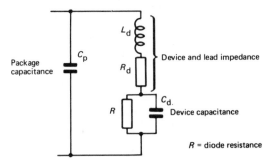

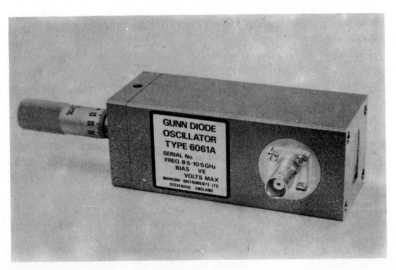

Fig. 8.22 Waveguide mounting for Gunn device.

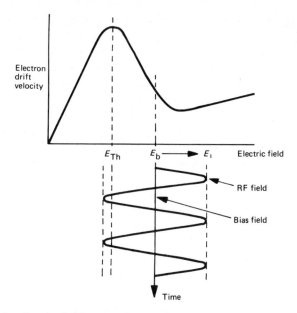

Fig. 8.23 Biasing for the LSA operation mode.

tions, the diode is usually placed in a resonant microwave circuit which acts as a high Q tuned system, giving a sinusoidal output. A schematic of a mounting for the Gunn diode is shown in Fig. 8.22. The electrodes are produced by crystal growth on to a conducting substrate, or by ion implantation. One of the factors that limits the power obtainable is the heat dissipation that can be achieved. To maximize the conductivity of heat from the device a large integral heat sink, forming part of the substrate, is used.

In the transit-time mode we saw that the natural oscillating frequency is determined by the length of the sample: the thinner the sample, the shorter the transit time, and hence the higher the frequency. However, the transit time mode is not the only one that is possible.[7] By applying a dc bias which takes the electric field just below E_{Th}, the diode can be made to switch into the negative resistance region as the alternating voltage takes the field value above E_{Th} (Fig. 8.23). Then the frequency of oscillation can be determined entirely by the external circuit in which the device is mounted. This mode of operation is known as the limited space-charge accumulation (LSA) mode.[8,9]

8.8 IMPATT DIODE

The IMPATT diode takes its name from IMPact Avalanche and Transit Time diode. It is one of several microwave devices based on the avalanche growth of carriers through a diode junction placed in a very high electric field. Sometimes called the Read diode after the man who proposed it, this device was first produced experimentally in 1965 by Johnston, de Loach and Cohen.[10] Since then IMPATT oscillators have been made for frequencies throughout the microwave spectrum, from 300 MHz to 300 GHz. They are more effficient (typical value about 15%), and can produce significantly more power than the Gunn device, but at the expense of greater noise and higher power supply voltage.

As the name implies, there are two factors that are particularly important in the operation of IMPATT diodes. First is avalanche multiplication. This phenomenon can occur at electric fields above about 10^5 V/cm. Carriers receive sufficient energy to produce electron–hole pairs by impact ionization. The ionization rate is strongly, but not linearly, related to the electric field; a five-fold increase in ionization rate can be produced by doubling the electric field.

The other important factor is the finite carrier drift velocity, and its variation with electric field strength. The electric fields used in IMPATT devices are much higher than the threshold fields discussed in the previous section on Gunn devices. Operating fields are above 10^5 V/cm and at such levels the electron drift velocity variation with electric field suffers a saturation effect and remains substantially constant. The value of the drift velocity depends particularly on the device material; 6×10^6 cm/s and 8×10^6 cm/s are typical for Si and GaAs, respectively, when used at an operating temperature of about 200 °C.

The basic form of the IMPATT is shown in Fig. 8.24; electrons are released at the p–n junction and then drift through an intrinsic n region. There are two significant widths associated with the device; W_A, the avalanche region, in which ionization takes place and a concentration of electrons is formed, and W_D, the drift region through which the electron bunch moves at velocity v_s.

The field across the p^+–n junction is well in excess of the breakdown value, with the maximum field E_p being approximately given by

$$E_p = \frac{q}{\varepsilon} N W$$

where W is the depletion width $= W_A + W_D$, N is the doping concentration in the n-type material, q is the electron charge and ε is the permittivity of the material.

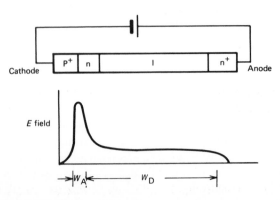

Fig. 8.24 IMPACT diode structure.

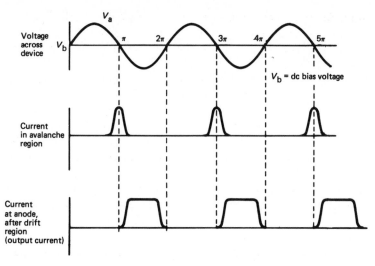

Fig. 8.25 Delay between output current and applied voltage in IMPATT diode.

When an IMPATT is used in an oscillator circuit it is reverse biased by a fixed voltage V_b, and superimposed on V_b is the voltage due to the microwave field in the oscillator, V_a. If V_b is chosen such as to just support avalanche breakdown, we can see from Fig. 8.25 how the diode sustains oscillations in the circuit.

The ionization mechanism has an inherent inertia; consequently the ionization rate does not respond instantaneously to changes in junction voltage. As V_a increases, the ionization rate increases, and continues to increase when V_a falls, until V_a reaches zero. The ionization then decreases also. The resulting electron density is shown as a function of time, relative to V_a in Fig. 8.25. The avalanche current peak is approximately $90°$ behind the maximum of V_a. The electrons leave the avalanche region after distance W_A and enter the drift region which they traverse at velocity v_s. The current at the terminals of the IMPATT, in response to V_a, is shown by the bottom curve of Fig. 8.25. The fundamental will be approximately $180°$ behind the V_a maximum. The angular displacement through the drift region is θ where

$$\theta = \omega \frac{W_D}{v_s} \qquad (8.1)$$

For complete antiphase between V_a and the output current, the oscillation frequency will be such as to make $\theta = \pi$. This occurs at the Read frequency, f_r where

$$f_r = \frac{v_s}{2W} \qquad (8.2)$$

The width of a Si-based device, which has $v_s \simeq 6 \times 10^6$ cm/s for an operating frequency of 10 GHz, would therefore be about 3 μm.

Read's equation

The theoretical basis for the IMPATT device was provided by Read in 1958. Following his work, several refinements have been made to his analysis to account for the manner of realizing the device. The variation of avalanche current density J_a with time is given by the Read equation:

$$\frac{\mathrm{d}J_a(t)}{\mathrm{d}t} = \frac{3J_a(t)}{\tau_a}\left[\int_0^W \alpha(E)\,\mathrm{d}x - 1\right] \qquad (8.3)$$

where x is the distance through the device from the junction and α is the ionization coefficient, which is the probability that a carrier will experience an ionizing collision in a unit length. τ_a is the time taken for an electron travelling at the saturation velocity v_s to travel over the avalanche region, i.e.

$$\tau_a = W_A/v_s$$

Associated with the avalanche region is an inductance L_A and a capacitance C_A:

$$L_A = \frac{W_A}{AJ_{DC}k} \qquad (8.4)$$

$$C_A = \frac{\varepsilon A}{W_A} \qquad (8.5)$$

A is the device area, J_{DC} the mean current density and k is a factor relating the characteristics of the semiconductor material to the ionization coefficient, at maximum field:

$$k = \frac{3}{\tau_a}\frac{\varepsilon}{qN}\alpha(E_p)$$

From Eqns (8.4) and (8.5) the resonant frequency of the avalanche region is f_{ra} where

$$f_{ra}^2 = \frac{1}{(2\pi)^2 L_A C_A} = \frac{J_{DC}k}{\varepsilon} \qquad (8.6)$$

At frequencies above f_{ra} the device resistance becomes negative, and the reactance changes from inductive to capacitive.

Power frequency limitations

There is a maximum voltage V_m, which may be applied to the diode, corresponding to the maximum field E_m which the diode can withstand without breakdown over the depletion layer W. When the electron velocity is at its saturation level v_s, the maximum current I_m is

$$I_m = \frac{E_m \varepsilon v_s A}{W} \qquad (8.7)$$

where A is the cross-sectional area of the device and ε the permittivity of the semiconductor material.

The maximum power input

$$P_m = I_m V_m \qquad (8.8)$$

$$= \frac{E_m \varepsilon v_s A}{W} E_m W$$

$$= E_m^2 \varepsilon v_s A$$

The depletion layer capacitance $C = \varepsilon A/W$. Therefore

$$P_m = \frac{E_m^2 v_s^2}{\Delta \pi f^2 X_c} \qquad (8.9)$$

where

$$X_c = \frac{1}{2\pi f C} \quad \text{and} \quad f = \frac{v_s}{2W}$$

Hence if X_c is limited, P_m is proportional to $1/f^2$ in the millimetre wave range.

Thermal limitation

The major limiting factor other than the intrinsic limit implied above is the amount of heat that can be conducted away from the junction. The dissipation of heat can be improved by mounting the device on a high conductivity material such as copper or diamond. Most of the heat is generated in the avalanche region where, for GaAs or Si, the operating temperature may be about 200 °C. The thermal conductivity of the diode depends on the thermal conductivity of the semiconductor material, the thermal resistance between the semiconductor and the heat sink, and the thermal conductivity of the sink material. For a particular system, the power dissipation decreases linearly with frequency.

8.9 FIELD-EFFECT TRANSISTORS

Probably the most promising type of microwave solid-state device is the field-effect transistor in one or other of its many forms. Based on GaAs, it is used for various purposes: high power generation, low noise and high frequency amplification in satellite and microwave terrestrial links. GaAs has a higher electron mobility than Si, and therefore a shorter transit time, allowing better high-frequency performance.

The basic form of the device is shown in Fig. 8.26. The buffer layer isolates the n-type active layer from the semi-insulating substrate. The active layer, which is less than half a micrometre deep, has a doping level of some 10^{17} donors/cm^3. Source and drain ohmic contacts are placed either directly on to the active layer, or on to another, more heavily doped, layer in the contact area. Electrons travel from the source to the drain, which is positive relative to the source, via the active layer. The path available for the electrons is called the channel. With this arrangement the current from source to drain, I_{DS},

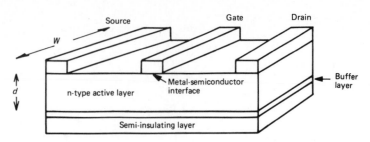

Fig. 8.26 Microwave FET.

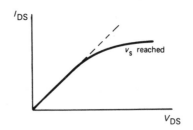

Fig. 8.27 Source–drain current–voltage characteristic for FET.

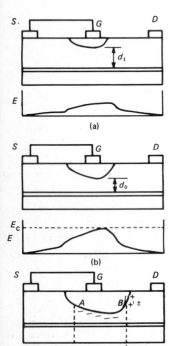

Fig. 8.28 Effect of applied voltage on current channel in a FET

increases as the applied voltage V_{DS} increases. The relationship between I_{DS} and V_{DS} is linear at low voltage as the electron velocity v varies linearly with the electric field produced by V_{DS}. As the voltage increases, v increases less quickly than the electric field, and eventually when a critical value of E is reached, $E_c \simeq 3 \times 10^3\,\text{V/cm}$, the electron drift velocity saturates, at v_s. This characteristic is shown in Fig. 8.27.

The current I_{DS} is given by

$$I_{DS} = wdvqn \qquad (8.10)$$

where wd = width × depth of the current channel, and $v.q.n.$ = velocity, unit charge and density of the electron flow, respectively.

In the case just discussed, where the field just reaches E_c, the minimum depth of the channel is d_0. Clearly I_{DS} varies directly with v and therefore reaches a saturation value when $v = v_s$.

By introducing an electrode (the gate) between source and drain, an additional control on I_{DS} is possible. The gate, which is reverse biased, creates a depletion region in the adjacent active layer, which has a depth proportional to the gate voltage. This depletion region has no conduction electrons, and therefore the depth of the current channel is now reduced to d_1, shown in Fig. 8.28(a).

For low values of V_{GS} the current I_{DS} varies linearly with V_{DS}, and, as V_{DS} increases, eventually saturates. As we would expect, the saturation current and the value of V_{DS} at which it is reached decrease as VGS becomes more negative. A typical family of curves is shown in Fig. 8.29.

When V_{GS} is zero the gate still has an effect: the field beneath the gate produces a depletion layer. The reduction in channel depth produced by the

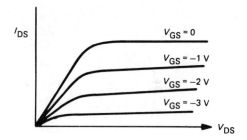

Fig. 8.29 Variation of output current with applied voltage.

depletion layer induces an increase in electron drift velocity as the field beneath the active layer increases [Fig. 8.28(b)]. When the field reaches the critical value E_c the velocity saturates, the depth of the channel is d_0, and the current through the device is given by

$$I_{DS} = wqnv_s d_0 \qquad (8.11)$$

Any further increase in V_{DS} deepens the depletion layer, hence reducing d still further. For I_{DS} to remain constant in the area of saturated velocity, the electron density increases. Figure 8.28(c) shows the effect of this deeper depletion layer. The point at which the field becomes equal to E_c moves towards the source, and therefore is reduced in voltage (point A), producing a deeper channel than d_0, and therefore a larger current. This suggests that the current characteristic, after the saturation effect in Fig. 8.28(c), increases again with V_{DS}. GaAs demonstrates such a gentle positive slope beyond the initial saturation.

Another feature of the depletion layer on the drain side of A, in Fig. 8.28(c), is that an accumulation of electrons develops over the distance AB where the field is greater than E_c; point B is that position where the channel depth reaches the value it had at A. The value of n increases to maintain I_{DS} since d is reduced, and the electron velocity is saturated. Beyond B, a positive charge layer is formed which helps to maintain constant I_{DS}.

The bottom edge of the depletion layer is not parallel to the surface of the device, but is deeper on the drain side. This is due to the voltage gradient over the width of the gate producing a wedge-shaped layer, with rounded edges. As we can see from the diagrams, the precise shape varies with V_{DS}.

This qualitative description shows the effect of the gate electrode when it is connected directly to the source. By its influence, the I_{DS}/V_{DS} characteristic is modified. However, the device is designed to be used as a transistor with a control voltage on the gate. As we noted in Fig. 8.29, by negatively biasing the gate the saturation effect occurs at a lower V_{DS} than for $V_{GS} = 0$. The reverse biased gate junction deepens the depletion layer and hence restricts I_{DS}. Eventually the depletion layer reaches the buffer layer and the channel is closed. The gate voltage at which this occurs, known as the pinch-off voltage V_p, is determined by the electrical properties, and the depth, of the active

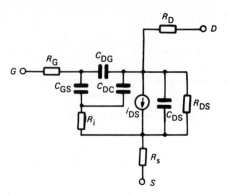

Fig. 8.30 FET equivalent circuit.

layer:

$$V_p = \frac{qnd^2}{2\varepsilon\varepsilon_0}$$ (8.12)

where q, n and d are as we defined them earlier, and ε, ε_0 are the relative permittivity of GaAs and the permittivity of free space, respectively.

A lumped component equivalent circuit for the device is shown in Fig. 8.30. From this circuit, the maximum frequency of oscillation is

$$f_m = \frac{g_m}{4\pi(C_{GS} + C_{DG})}\left[\frac{R_i + R_s + R_G}{R_{DS}} + \frac{g_m}{(C_{GS} + C_{DG})}C_{DG}R_G\right]^{1/2}$$ (8.13)

The performance of the device is related to the bias circuit used to establish and maintain the dc gate voltage; variations in this voltage induce noise and therefore low values of source and gate resistance are desirable. To minimize the variation in bias point, a two-source biasing configuration is preferred, such as that shown in Fig. 8.31.

This brief description refers to the MESFET (metal semiconductor field-effect transistor) which has a metal gate (Schottky) contact. It has a very low noise figure, and is capable of operating at high millimetre wave frequencies. Many other FETs exist, or are being developed; by varying the geometry of the device and the nature of the gate, particular features are enhanced. For further information, see Lamming[11] and Ha.[12]

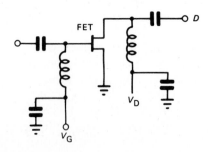

Fig. 8.31 Biasing network for an FET.

8.10 CONCLUSIONS

FET devices are the most common of the semiconductor devices currently in use. There are various forms, and several versions of each. One of the most promising recent developments has been the HEMT (high electron mobility transistor), which, although invented in the 1970s has now generated considerable interest. Previously it offered increased gain when operated at low temperatures, being dependent on a thin layer of AlGaAs for which the mobility increased considerably as the temperature was reduced to 77 K. New combinations of III–V materials have been used, based on InAs, and these promise significant gain improvements at room temperatures.

As we noted either, there are many applications that require power levels above those which can be achieved from solid-state devices. Consequently, there continues to be significant development in microwave vacuum tube design, in an attempt to improve the efficiency and performance of those tubes that have been established for nearly half a century, and to discover entirely new ways of generating high power.

REFERENCES

1. Kompfner, R., "The travelling-wave tube", *Wireless World*, **52**, 349 (1946).
2. Atwater, H. A., *Introduction to Microwave Theory*, McGraw-Hill, Maidenhead, 1962.
3. Collin, R. E., *Foundation for Microwave Engineering*, McGraw-Hill, Maidenhead, 1966.
4. Ridley, B. K. and Watkins, T. B., "The possibility of negative resistance effects in semiconductors", *Proceedings of the Physical Society*, **78**, 293–304 (Aug 1961).
5. Hilsum, C., "Transferred electron amplifiers and oscillators", *Proceedings of the Institute of Radio Engineers*, **50**(2), 185–9 (Feb 1962).
6. Gunn, J. B., "Microwave oscillations of current in III–V semiconductors", *Solid-state Communications*, **1**, 88–91 (Sept 1963).
7. Pattison, J., "Active microwave devices", *Microwave Solid-state Devices and Applications* (ed. Morgan, D. V. and Howe, M. J.), Pergamon, Oxford, 1980, Chapter 2.
8. Hobson, G. S., *The Gunn Effect*, Clarendon, Oxford, 1974.
9. Bosch, B. G. and Engelmann, R. W., *Gunn-effect Electronics*, Pitman, London, 1975.
10. Johnston, R. L., de Loach, B. C. and Cohen, B. G., "A silicon diode microwave oscillator", *Bell System Technical Journal*, **44**, 2 (Feb 1965).
11. Lamming, J., "Active microwave devices – FET's and BJT's", *Microwave Solid-state Devices and Applications* (ed. Morgan, D. V. and Howe, M. J.), Pergamon, Oxford, 1980, Chapter 4.
12. Ha, Tri T., *Solid-state Microwave Amplifier Design*, Wiley, Chichester, 1981.

9 Passive microwave devices

In the last chapter we considered some common microwave active devices. When microwave energy is generated it is used by either radiating it into space via an antenna or passing it along a transmission system. In either case some circuitry will be involved at both the transmitting and receiving ends. This circuitry, and the transmission system, have features that are special to signals of microwave frequencies. In this chapter we will consider first the commonest microwave component, the rectangular waveguide, and go on to look at several of the passive devices that comprise microwave circuitry.

9.1 WAVEGUIDES

Waveguides are distinguished by having only one guiding surface. They are of two main types: (a) metal tubes of any cross-section, and (b) dielectric rods. In the first case the wave travels down the inside of the tube and there is no leakage to the outside. In the second, the wave is propagated over the outer surface of the rod and there is a considerable field in the surrounding medium.

In this chapter we shall be looking at the first type only.

For best transmission, tubular waveguides should be made of a low-loss, conducting material. If runs are short, and cost is important, aluminium or brass will be used, but good quality waveguides having low attenuation are made from copper which, for some applications where low loss is paramount, might be plated with silver or gold. The energy is propagated along the inside of the tube so the inner surface should be smooth and the cross-section must be uniform throughout the length of the guide.

9.2 RECTANGULAR WAVEGUIDE

Power is transmitted down a waveguide by electric and magnetic fields which travel in a manner similar to that of travelling voltage waves in a transmission line.

Figure 9.1 shows a rectangular waveguide of internal cross-section $a \times b$, in relation to a right-handed system of coordinates. This means that the broad face of the guide is in the x direction, the narrow face in the y direction, and the wave is propagated in the z direction.

The wave inside the guide is described in terms of a field pattern. Figure 9.2 shows the pattern for the fundamental rectangular waveguide mode, TE_{10}

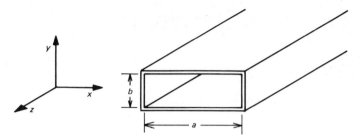

Fig. 9.1 Rectangular waveguide dimensions.

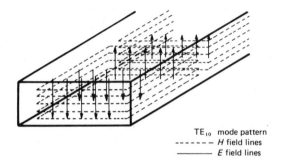

TE$_{10}$ mode pattern
------ *H* field lines
———— *E* field lines

Fig. 9.2 Field patterns of TE$_{10}$ rectangular waveguide mode.

(defined below). The way in which this field pattern is set up can be considered from two points of view, which will be considered in turn in the following sections.

9.3 INTERFERENCE OF TWO PLANE WAVES

In Figure 9.3 two uniform plane waves of identical amplitude and phase are shown travelling in different directions. In each, the electric field vector is normal to the plane of the paper. As the waves travel across each other there will be constructive and destructive intereference, resulting in a field pattern that is the algebraic sum of the component fields at each point. Where a maximum of one of the waves crosses a maximum of the other there will be a maximum in the resultant field, and conversely there will be a minimum where the minima of the two component fields are coincident. Examples of these points are at E for the minimum and F for the maximum in Fig. 9.3.

At points where there is a maximum of one field coincident with a minimum of the other there will be zero resultant field. In Fig. 9.3 a line is drawn joining such points, shown as AA. Since there is zero electric field along this line, a conducting surface could be inserted parallel to the E field lines along AA without perturbing the field. Similarly, a conducting surface could be placed along BB. These two conducting planes could thus form the side

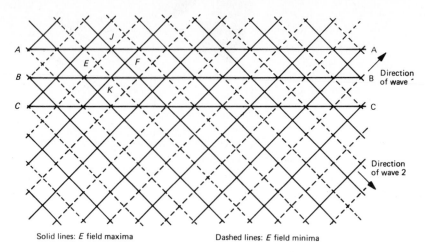

Solid lines: *E* field maxima   Dashed lines: *E* field minima

Fig. 9.3 Waveguide field pattern from two plane waves.

walls of a rectangular waveguide. Top and bottom conducting walls, parallel with the plane of the paper, can be used without altering the field pattern since the *E* field lines will meet them normally.

The resultant field pattern between *AA* and *BB* in Fig. 9.3 is therefore a representation of one that can exist inside a rectangular waveguide. As we shall see later the pattern between *AA* and *BB* represents the fundamental rectangular waveguide mode. If, instead of the side walls being along *AA* and *BB*, they were placed along *AA* and *CC*, a different pattern, representing a second-order mode, would exist. The question of modes is discussed in Section 9.7.

9.4   CUT-OFF WAVELENGTH

The pattern shown in Fig. 9.3 between lines *AA* and *BB* can be used to quantify the cut-off wavelength in terms of the free space wavelength and the guide width, i.e. the distance between *AA* and *BB*. An expanded representation of the pattern between a maximum and a minimum is given in Fig. 9.4. We can see that:

λ_g = wavelength of the field pattern inside the guide, i.e. the distance between two adjacent points of similar phase in the direction of propagation

λ_0 = wavelength of the wave in the medium filling the guide, say air

$$= \frac{\text{velocity of light in air}}{\text{frequency of the wave}} = \frac{c}{f}$$

From the diagram, if the angle between the direction of propagation of the two component waves is 2α (i.e. each wave propagates at an angle α to the axis

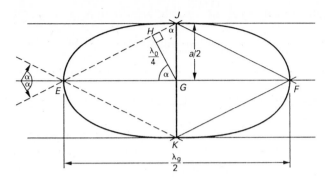

Fig. 9.4 Details of field pattern.

of the guide),

$$\sin\alpha = \frac{\lambda_0}{2}\frac{1}{a} \quad \text{and} \quad \cos\alpha = \frac{\lambda_0}{\lambda_g}$$

Using the identity $\cos^2\alpha + \sin^2\alpha = 1$

$$\frac{\lambda_0^2}{(2a)^2} + \frac{\lambda_0^2}{\lambda_g^2} = 1$$

or

$$\frac{1}{\lambda_0^2} = \frac{1}{\lambda_g^2} + \frac{1}{(2a)^2} \tag{9.1}$$

where a is the width of the guide. $2a$ is called the cut-off wavelength λ_c for the TE$_{10}$ mode; hence Eqn (9.1) becomes

$$\frac{1}{\lambda_0^2} = \frac{1}{\lambda_g^2} + \frac{1}{\lambda_c^2} \tag{9.2}$$

The cut-off wavelength is so called because it is the maximum value that λ_0 can have. If λ_0 becomes greater than λ_c the wave will not propagate down the guide. Thus the waveguide behaves as a high-pass filter: only waves of frequency greater than the cut-off value $f_c (= c/\lambda_c)$ can be carried. This aspect of waveguide behaviour will be considered in further detail in later sections.

9.5   PHASE VELOCITY

The rate at which a particular point in the field pattern travels down the guide is known as the phase velocity. From Fig. 9.4 consider the movement of point E towards F. In order to maintain the phase of E constant it must arrive at G at the same time as point H on one of the wavefronts. H is moving with velocity c, and covers a distance $\lambda_0/4$, whereas E travels $\lambda_g/4$ in the same time.

Therefore the velocity of E, i.e. the phase velocity v_{ph}, is given by

$$v_{ph} = c \frac{\lambda_g}{\lambda_0} \tag{9.3}$$

9.6   GROUP VELOCITY

It would violate the theory of relativity if energy were transported down the guide at the phase velocity, and in fact that does not happen. The velocity of energy movement is known as the group velocity, and in terms of our plane wave model in Fig. 9.4 it is equal to the component of the plane wave velocity in the z direction. The plane wave travels at velocity c in air so the component in the z direction is

$$v_g = c \cos \alpha = c \frac{\lambda_0}{\lambda_g} \tag{9.4}$$

It is worth noting from the last two equations that

$$v_{ph} \cdot v_g = c^2 \tag{9.5}$$

9.7   RECTANGULAR WAVEGUIDE MODES

The above discussion has been concerned with a particular field pattern inside the waveguide – that given by Figs. 9.2 and 9.3 and the section between AA and BB of Fig. 9.4. As we noted earlier, it is the TE_{10} mode, often called the fundamental rectangular waveguide mode because it has the lowest cut-off frequency.

In the designation TE_{10} the letters stand for transverse electric, indicating that the field pattern has no electric component in the z direction, it being contained entirely in the $x-y$, transverse, plane. The numerals 1, 0 are used to indicate the number of half-sinusoidal variations of the E field in the x and y directions, respectively. For any value of y the electric field goes through one half-sinusoidal variation with x, i.e. from zero at one wall, through a maximum in the guide-centre, to zero at the other wall (Fig. 9.2). For any value of x there is no variation of field with y, i.e. the field is constant in the y direction. Thus TE_{10} designates a transverse electric mode having one half-sinusoidal variation in the x direction (across the broad side of the guide) and no variation in the y direction.

In fact, there are many possible TE modes. The general designation is TE_{mn} where m is the number of half-sinusoidal variations in the x direction, and n the number of similar variations in the y direction. We shall see later that the cut-off wavelength for the TE_{mn} mode is given by

$$\frac{1}{(\lambda_{c_{mn}})^2} = \left(\frac{m}{2a}\right)^2 + \left(\frac{n}{2b}\right)^2 \tag{9.6}$$

where a and b are the dimensions shown in Fig. 9.1.

There is another set of modes possible in a rectangular waveguide; the transverse magnetic, or TM, modes, in which there is no component of magnetic field in the direction of propagation. The general mode is designated TM_{mn}, and it too has a cut-off wavelength given by Eqn (9.6). The suffices mn have the same meaning as in TE_{mn}.

Unlike coaxial transmission line, there is no TEM mode in a waveguide: a mode cannot exist in which both the electric and magnetic vectors have no component in the z direction.

9.8   FIELD THEORY OF PROPAGATION ALONG A RECTANGULAR WAVEGUIDE

Although the above discussion, in which the fields inside the waveguide are considered to result from the intereference of two plane waves, gives a very useful insight into the mechanism of waveguide propagation, it has some disadvantages. In particular, it does not allow an evaluation of the field components inside the guide. A deeper appreciation can be obtained by using Maxwell's equations. These equations state in generalized form the relationship that must exist between the electric and magnetic fields in any medium, provided the characteristics of the medium and the initial conditions of the fields are known. It would be inappropriate to develop Maxwell's equations here, but several texts discuss their derivation in some detail, and reference should be made to one of these, e.g. Bleaney and Bleaney[1] or Paul and Nasar,[2] for background information.

We start by stating the equations in generalized vector form:

$$\nabla \cdot \bar{E} = \frac{\rho}{\varepsilon} \tag{9.7}$$

$$\nabla \cdot \bar{B} = 0 \tag{9.8}$$

$$\nabla \times \bar{E} = -\frac{\partial \bar{B}}{\partial t} \tag{9.9}$$

$$\nabla \times \bar{H} = \bar{J} + \varepsilon \frac{\partial \bar{E}}{\partial t} \tag{9.10}$$

$$\bar{B} = \mu \bar{H} \tag{9.11}$$

$\bar{E}$ and $\bar{H}$ are the electric and magnetic field vectors, respectively, $\bar{B}$ is the magnetic flux density, and $\bar{J}$ is the conduction current density. All four vectors are related to the space inside the guide; for air the value of $\bar{J}$ is zero.

The other symbols represent the electric characteristics of the material filling the guide. Normally the charge density (ρ) is zero, and for air the permittivity (ε_0) and the permeability (μ_0) are those for free space, i.e.

$$\varepsilon_0 = \frac{1}{36\pi \times 10^9} \quad \text{farad/m}$$

$$\mu_0 = 4\pi \times 10^7 \quad \text{H/m}$$

Although the following analysis is lengthy, it allows us to build up an awareness of an analytical description of waveguide propagation. By considering the argument in some detail, rather than merely isolating the important equations, we can appreciate more fully the underlying theory, and that will be of value in discussing several of the devices considered later in the chapter.

If we assume immediately that there is a sinusoidal signal being transmitted down the guide, then the time derivatives in Eqns (9.9) and (9.10) can be replaced by $j\omega$, and they become, when $\bar{B}$ is replaced by μH,

$$\nabla \times \bar{E} = -j\omega\mu\bar{H} \tag{9.12}$$

$$\nabla \times \bar{H} = j\omega\varepsilon\bar{E} \tag{9.13}$$

and, since $\rho = 0$,

$$\nabla \cdot \bar{E} = 0 \tag{9.14}$$

Now we must solve these equations for waveguide having a rectangular cross-section.

Taking the curl of Eqn (9.12), and substituting from Eqn (9.13),

$$\nabla \times \nabla \times \bar{E} = \omega^2 \varepsilon \mu \bar{E} \tag{9.15}$$

Using Eqn (9.15) in the vector identity,

$$\nabla \times \nabla \times \bar{E} = -\nabla^2 \bar{E} + \nabla(\nabla \cdot \bar{E})$$

and combining the result with Eqn (9.14), we have

$$\nabla^2 \bar{E} + \mu\varepsilon\omega^2 \bar{E} = 0 \tag{9.16}$$

This is a vector representation of the E field wave equation, and to solve it for the waveguide it must be put into an appropriate coordinate system. In rectangular coordinates

$$\nabla^2 \bar{E} = \bar{x}\,\nabla^2 E_x + \bar{y}\,\nabla^2 E_y + \bar{z}\,\nabla^2 E_z \tag{9.17}$$

From Eqns (9.16) and (9.17) there are three equations to be solved, one for each coordinate direction:

$$\nabla^2 E_x = -\mu\varepsilon\omega^2 E_x \tag{9.18}$$

$$\nabla^2 E_y = -\mu\varepsilon\omega^2 E_y \tag{9.19}$$

$$\nabla^2 E_z = -\mu\varepsilon\omega^2 E_z \tag{9.20}$$

By developing Eqns (9.16) and (9.17) in terms of $\bar{H}$ rather than $\bar{E}$, a similar set of equations is produced for the $\bar{H}$ components. The six equations resulting are related by the boundary conditions at the waveguide walls, and these conditions must be used when the equations are solved.

We will now solve Eqn (9.19), which can be expanded to

$$\frac{\partial^2 E_y}{\partial x^2} + \frac{\partial^2 E_y}{\partial y^2} + \frac{\partial^2 E_y}{\partial z^2} = -\mu\varepsilon\omega^2 E_y \tag{9.21}$$

In principle, E_y may vary with x, y and z, and variation in one direction will be independent of variation in another, so the technique of separation of variables can be used.

Let the solution of E_y be

$$E_y = XYZ \qquad (9.22)$$

where XYZ are functions of only x, y, and z, respectively.

Then, substituting Eqn (9.22) into Eqn (9.21),

$$\frac{\partial^2 X}{\partial x^2} YZ + X \frac{\partial^2 Y}{\partial y^2} Z + XY \frac{\partial^2 Z}{\partial z^2} = -\mu\varepsilon\omega^2 XYZ$$

or

$$\frac{1}{X}\frac{\partial^2 X}{\partial x^2} + \frac{1}{Y}\frac{\partial^2 Y}{\partial y^2} + \frac{1}{Z}\frac{\partial^2 Z}{\partial z^2} = -\mu\varepsilon\omega^2 \qquad (9.23)$$

At any specific frequency the right-hand side of Eqn (9.23) is constant, so for this expression to be valid each of the terms on the left-hand side must also be constant; hence we can write

$$\frac{1}{X}\frac{\partial^2 X}{\partial x^2} = k_x^2 \qquad (9.24)$$

$$\frac{1}{Y}\frac{\partial^2 Y}{\partial y^2} = k_y^2 \qquad (9.25)$$

$$\frac{1}{Z}\frac{\partial^2 Z}{\partial z^2} = k_z^2 \qquad (9.26)$$

The significance of k_x, k_y and k_z will be clear later.

Introducing $k^2 = \mu\varepsilon\omega^2$, Eqn (9.23) gives

$$k_x^2 + k_y^2 + k_z^2 = -k^2 = -\mu\varepsilon\omega^2 \qquad (9.27)$$

We can distinguish between the transverse and z direction components; letting

$$k_x^2 + k_y^2 = -k_c^2 \qquad (9.28)$$

we have

$$k_z^2 = -k^2 + k_c^2 \qquad (9.29)$$

or

$$k_z = \pm j\sqrt{(k^2 - k_c^2)} \qquad (9.30)$$

The solution of Eqn (9.26), which gives the variation in the z direction, is, by analogy with voltage waves on a transmission line (following Eqn (6.11))

$$Z = \exp(\pm k_z z) \qquad (9.31)$$

and from Eqn (9.30), when $k^2 > k_c^2$,

$$k_z = \pm j\beta_g \qquad (9.32)$$

where

$$\beta_g = \sqrt{(k^2 - k_c^2)} \qquad (9.33)$$

hence

$$Z = \exp(\pm j\beta_\mathrm{g} z) \qquad (9.34)$$

Then E_y is given by

$$E_y = XY \exp[j(\omega t \pm \beta_\mathrm{g} z)] \qquad (9.35)$$

where here the time dependence implicit in assuming a sinusoidal signal is included in the term $\exp(j\omega t)$.

Equation (9.35) represents two travelling waves, one in the forward and one in the backward direction, in a similar way to the waves on a transmission line (see Section 6.1).

This method of solution could be used to develop expressions for the other components in the E and H fields.

For normal propagation $k^2 > k_\mathrm{c}^2$, making β_g real and k_z imaginary, and therefore $\exp(\pm j\beta_\mathrm{g} z)$ is a phase term. However, if $k_\mathrm{c}^2 > k^2$, β_g is imaginary and k_z is real, indicating attenuation. In that case there is rapid attenuation. The transition between propagation and attenuation occurs at that value of $\omega (=\omega_\mathrm{c})$ for which

$$k_\mathrm{c}^2 = k^2$$

i.e.

$$k_\mathrm{c}^2 = \mu\varepsilon\omega_\mathrm{c}^2$$

or

$$\omega_\mathrm{c} = k_\mathrm{c}/\sqrt{(\mu\varepsilon)} \qquad (9.36)$$

k_c has dimensions of 1/length and it can be defined by

$$k_\mathrm{c} = 2\pi/\lambda_\mathrm{c} \qquad (9.37)$$

We can rewrite Eqn (9.29), i.e.

$$k_z^2 = -k^2 + k_\mathrm{c}^2$$

as

$$-\left(\frac{2\pi}{\lambda_\mathrm{g}}\right)^2 = -\left(\frac{2\pi}{\lambda_0}\right)^2 + \left(\frac{2\pi}{\lambda_\mathrm{c}}\right)^2$$

or, rearranging terms,

$$\frac{1}{\lambda_0^2} = \frac{1}{\lambda_\mathrm{g}^2} + \frac{1}{\lambda_\mathrm{c}^2} \qquad (9.2)$$

where

$$\lambda_\mathrm{g} = \frac{2\pi}{\beta_\mathrm{g}} \quad \text{(cf. Eqn (6.18))}$$

and, using $\omega = 2\pi f_0 = c/\lambda_0$ in Eqn (9.27)

$$\lambda_0 = \frac{2\pi}{k} = \frac{2\pi}{\mu\varepsilon\omega^2}$$

These wavelengths λ_0, λ_g and λ_c have been discussed in Section 9.4. Here we are not yet able to specify λ_c, because we have not considered a particular mode, but that will be done shortly.

9.9   TRANSVERSE ELECTRIC (TE) MODES

In the last section we derived some general expressions for modes in a rectangular waveguide, but we cannot go further without specifying the class of mode in which we are interested. As we saw in Section 9.7 there are two classes, TE and TM. In this section we consider TE modes in a rectangular guide, and we can use our earlier equations to set up the field equations in the waveguide; we then introduce the effect of having no component of the E field in the z direction, thus allowing a solution to Maxwell's equations to be obtained.

Returning to Eqn (9.12),

$$\nabla \times \bar{E} = -j\omega\mu\bar{H}$$

and expressing it in Cartesian coordinates,

$$\bar{x}\left(\frac{\partial E_z}{\partial y} - \frac{\partial E_y}{\partial z}\right) + \bar{y}\left(\frac{\partial E_x}{\partial z} - \frac{\partial E_z}{\partial x}\right) + \bar{z}\left(\frac{\partial E_y}{\partial x} - \frac{\partial E_x}{\partial y}\right)$$

$$= -j\omega\mu(\bar{x}H_x + \bar{y}H_y + \bar{z}H_z) \tag{9.38}$$

For TE modes there is no E field in the z direction, and hence $E_z = 0$. From Eqn (9.31) the dependence of the other fields in the z direction is $\exp(\pm k_z)$, giving $\partial/\partial z = k_z$.

Using these two conditions in Eqn (9.38),

$$\bar{x}(-k_z E_y) + \bar{y}(k_z E_x) + \bar{z}\left(\frac{\partial E_y}{\partial x} - \frac{\partial E_x}{\partial y}\right)$$

$$= -j\omega\mu(\bar{x}H_x + \bar{y}H_y + \bar{z}H_z)$$

Equating x, y and z components,

$$k_z E_y = j\omega\mu H_x \tag{9.39}$$

$$k_z E_x = -j\omega\mu H_y \tag{9.40}$$

$$\frac{\partial E_y}{\partial x} - \frac{\partial E_x}{\partial y} = -j\omega\mu H_z \tag{9.41}$$

By a similar development from Eqn (9.13), additional relationships between the E and H fields can be found, viz.

$$\frac{\partial H_z}{\partial y} - k_z H_y = j\omega\varepsilon E_x \tag{9.42}$$

$$k_z H_x - \frac{\partial H_z}{\partial x} = j\omega\varepsilon E_y \tag{9.43}$$

$$\frac{\partial H_y}{\partial x} - \frac{\partial H_x}{\partial y} = 0 \tag{9.44}$$

To find the field components we can first use these equations to express the transverse field components E_x, E_y, H_x, H_y in terms of the single component in the z direction, H_z.

From Eqns (9.40) and (9.42),

$$\frac{\partial H_z}{\partial y} + \frac{k_z^2}{j\omega\mu} E_x = j\omega\varepsilon E_x$$

$$j\omega\mu \frac{\partial H_z}{\partial y} + k_z^2 E_x = -\omega^2\mu\varepsilon E_x$$

or

$$j\omega\mu \frac{\partial H_z}{\partial y} = -(k_z^2 + \omega^2\mu\varepsilon) E_x$$

i.e.

$$
\begin{aligned}
E_x &= -\frac{j\omega\mu}{k_z^2 + \omega^2\mu\varepsilon} \frac{\partial H_z}{\partial y} \\
&= -\frac{j\omega\mu}{k_c^2} \frac{\partial H_z}{\partial y}
\end{aligned}
\tag{9.45}
$$

where

$$k_c^2 = k_z^2 + \omega^2\mu\varepsilon$$

Substituting Eqn (9.45) back into Eqn (9.40), and rearranging terms gives

$$H_y = \frac{k_z}{k_c^2} \frac{\partial H_z}{\partial y} \tag{9.46}$$

In a similar way we can find expressions for the other transverse components:

$$E_y = \frac{j\omega\mu}{k_c^2} \frac{\partial H_z}{\partial x} \tag{9.47}$$

and

$$H_x = \frac{k_z \partial H_z}{k_c^2 \partial x} \tag{9.48}$$

Thus in order to find the transverse field components we only need to determine H_z. It can be found as follows.

By analogy with the discussion on E_y in Section 9.8, H_z must be a solution of

$$\nabla^2 H_z + \mu\varepsilon\omega^2 H_z = 0 \tag{9.49}$$

and in the same way it will have the form

$$H_z = XY \exp\left[j(\omega t \pm \beta_g z)\right] \tag{9.50}$$

We remember that X and Y must each satisfy second-order differential equations, so they must each have two solutions. Two satisfactory solutions are

$$X = A \sin jk_x x + B \cos jk_x x \qquad (9.51)$$

$$Y = C \sin jk_y y + D \cos jk_y y \qquad (9.52)$$

Then the full solution for H_z is

$$H_z = (A \sin jk_x x + B \cos jk_x x)(C \sin jk_y y + D \cos jk_y y) \exp [j(\omega t \pm \beta_g z)] \qquad (9.53)$$

The unknowns in this expression (A, B, C, D, k_x, k_y) are found from either the boundary or the initial conditions. The boundary conditions imposed by the walls of the guide are as follows:

(i) there can be no normal component of magnetic field at the guide walls;
(ii) there can be no tangential component of electric field at the guide walls.

Condition (i) means that

$$\frac{\partial H_z}{\partial x} = 0 \quad \text{at} \quad x = 0, a$$

and

$$\frac{\partial H_z}{\partial y} = 0 \quad \text{at} \quad y = 0, b$$

thus making $A = C = 0$ in Eqn (9.53), leaving

$$H_z = H_1 \cos jk_x x \cos jk_y y \exp (\pm j\beta_g z) \qquad (9.54)$$

(omitting, but not forgetting, the time-dependent term $\exp [j\omega t]$), $H_1 = BD$, and is determined by the initial field applied to the guide.

If we use this value of H_z in Eqn (9.47) we have

$$E_y = \frac{\omega \mu k_x H_1}{k_c^2} \sin jk_x x \cos jk_y y \exp (\pm j\beta_g z)$$

and from condition (ii) $E_y = 0$ at $x = 0, a$, which is satified by making

$$jk_x = m\pi/a \qquad (9.55)$$

By a similar argument, consideration of E_x imposes the relationship

$$jk_y = n\pi/b \qquad (9.56)$$

Then, substituting for jk_x and jk_y from Eqns (9.55) and (9.56) into Eqn (9.54) gives

$$H_z = H_1 \cos \frac{m\pi}{a} x \cos \frac{n\pi}{b} y \exp (\pm j\beta_g z) \qquad (9.57)$$

where m and n can be any integer and indicate the number of half-sinusoidal variations of field in the x and y directions, respectively, i.e. they refer to the TE_{mn} mode.

By inserting Eqn (9.57) into Eqns (9.45), (9.46) and (9.48), the full set of field components for the TE$_{mn}$ mode are obtained, i.e.

$$E_x = \frac{j\omega\mu n\pi H_1}{bk_c^2} \cos\frac{m\pi x}{a} \sin\frac{n\pi y}{b} \exp(\pm j\beta_g z) \qquad (9.58)$$

$$E_y = -\frac{j\omega\mu m\pi H_1}{ak_c^2} \sin\frac{m\pi x}{a} \cos\frac{n\pi y}{b} \exp(\pm j\beta_g z) \qquad (9.59)$$

$$E_z = 0 \qquad (9.60)$$

$$H_x = -\frac{j\beta_g m\pi H_1}{ak_c^2} \sin\frac{m\pi x}{a} \cos\frac{n\pi y}{b} \exp(\pm j\beta_g z) \qquad (9.61)$$

$$H_y = -\frac{j\beta_g n\pi H_1}{bk_c^2} \cos\frac{m\pi x}{a} \sin\frac{n\pi y}{b} \exp(\pm j\beta_g z) \qquad (9.62)$$

$$H_z = H_1 \cos\frac{m\pi x}{a} \cos\frac{n\pi y}{b} \exp(\pm j\beta_g z) \qquad (9.63)$$

where, from Eqns (9.27) and (9.28) as before,

$$k_c^2 = \mu\varepsilon\omega^2 + k_z^2$$

From Eqn (9.28)

$$k_c^2 = -(k_x^2 + k_y^2)$$

Substituting for k_x and k_y from Eqns (9.55) and (9.56), respectively,

$$k_c^2 = \left[\left(\frac{m\pi}{a}\right)^2 + \left(\frac{n\pi}{b}\right)^2\right]$$

therefore

$$\mu\varepsilon\omega^2 + k_z^2 = \left(\frac{m\pi}{a}\right)^2 + \left(\frac{n\pi}{b}\right)^2$$

From $\omega = 2\pi f$, $\qquad \omega^2 = \left(\frac{2\pi c}{\lambda_0}\right)^2 \qquad$ and $\qquad c = 1/\sqrt{(\mu\varepsilon)}$

therefore

$$\mu\varepsilon\omega^2 = \left(\frac{2\pi}{\lambda_0}\right)^2$$

and

$$k_z^2 = (j\beta_g)^2 = -\left(\frac{2\pi}{\lambda_g}\right)^2$$

hence

$$\left(\frac{2\pi}{\lambda_0}\right)^2 - \left(\frac{2\pi}{\lambda_g}\right)^2 = \left(\frac{m\pi}{a}\right)^2 + \left(\frac{n\pi}{b}\right)^2$$

Comparing this equation with Eqn (9.6)

$$\left(\frac{2\pi}{\lambda_c}\right)^2 = \left(\frac{m\pi}{a}\right)^2 + \left(\frac{n\pi}{b}\right)^2$$

therefore

$$\frac{1}{\lambda_c^2} = \left(\frac{m}{2a}\right)^2 + \left(\frac{n}{2b}\right)^2 \tag{9.64}$$

This expression, which for the TE_{10} mode gives $\lambda_c = 2a$ as in Section 9.4, allows us to find the cut-off wavelength for any TE mode, given the cross-sectional dimensions of the waveguide. We can see that λ_c will vary with the size of guide, and therefore with its aspect ratio a/b.

9.10 TRANSVERSE MAGNETIC (TM) MODES

As we noted in Section 9.7, the other class of modes in a rectangular waveguide is that in which the magnetic field has no component in the z direction (i.e. $H_z = 0$). Modes of this type are called transverse magnetic because the magnetic field is contained entirely in the xy plane. The general mode designation is TM_{mn} where m and n have the same meaning as in Section 9.7.

By an analysis similar to that used in considering TE modes, the six field components of the TM_{mn} mode are obtained:

$$E_x = \frac{j\beta_g m\pi E_1}{ak_c^2} \cos\frac{m\pi x}{a} \sin\frac{n\pi y}{b} \exp(\pm j\beta_g z) \tag{9.65}$$

$$E_y = \frac{j\beta_g n\pi E_1}{bk_c^2} \sin\frac{m\pi x}{a} \cos\frac{n\pi y}{b} \exp(\pm j\beta_g z) \tag{9.66}$$

$$E_z = E_1 \sin\frac{m\pi x}{a} \sin\frac{n\pi y}{b} \exp(\pm j\beta_g z) \tag{9.67}$$

$$H_x = \frac{j\omega\varepsilon n\pi E_1}{bk_c^2} \sin\frac{m\pi x}{a} \cos\frac{n\pi y}{b} \exp(\pm j\beta_g z) \tag{9.68}$$

$$H_y = -\frac{j\omega\varepsilon m\pi E_1}{ak_c^2} \cos\frac{m\pi x}{a} \sin\frac{n\pi y}{b} \exp(\pm j\beta_g z) \tag{9.69}$$

$$H_z = 0 \tag{9.70}$$

Equations (9.64) and (9.33) for the cut-off wavelength and the phase constant are also valid for TM modes. If we examine Eqn (9.64) again, and express it in terms of cut-off frequency rather than wavelength, it becomes

$$f_c = c\left[\left(\frac{m}{2a}\right)^2 + \left(\frac{n}{2b}\right)^2\right]^{1/2} \tag{9.71}$$

and it is clear that there is a dependence on a and b as well as on m and n. Taking as an example $b = 2a$ we can show the relationship between the cut-off frequencies of several modes on the simple diagram of Fig. 9.5. The cut-off values are given relative to that for the TE_{10} mode, and it can now be seen why it is called the fundamental. Some modes (e.g. TE_{11}, TM_{11}) have the same value of f_c; they are called degenerate because propagation that starts in one mode can easily change (i.e. degenerate) into the other. Such things as

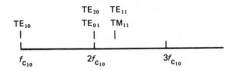

$f_{c_{10}}$ − cut-off frequency for TE$_{10}$ mode

Fig. 9.5 Cut-off frequencies for low-order rectangular modes.

bends, roughness in the guide or poorly fitted flanges can cause this mode conversion and because waveguide detectors are mode dependent it results in a loss of power.

9.11 FIELD EQUATIONS FOR THE FUNDAMENTAL TE$_{10}$ MODE

Putting $m = 1$, $n = 0$ into Eqns (9.58) to (9.63) and adjusting the coefficients, we have the following field components:

$$E_x = 0 \tag{9.72}$$

$$E_y = E_1 \sin \frac{\pi x}{a} \exp\left(\pm j\beta_g z\right) \tag{9.73}$$

$$E_z = 0 \tag{9.74}$$

$$H_x = \frac{\beta_g}{\omega\mu} E_1 \sin \frac{\pi x}{a} \exp\left(\pm j\beta_g z\right) \tag{9.75}$$

$$H_y = 0 \tag{9.76}$$

$$H_z = \frac{jak_c^2 E_1}{\omega\mu\pi} \cos \frac{\pi x}{a} \exp\left(\pm j\beta_g z\right) \tag{9.77}$$

These are the equations of the field patterns considered in Section 9.9 and which are shown in Fig. 9.2. As well as being the fundamental, this is also the simplest rectangular waveguide mode; there is only one half-sinusoidal variation of electric field across the guide in the x direction and none in the y direction. Similar points of phase down the guide are separated by λ_g.

From Eqns (9.72) to (9.77) the field patterns for the TE$_{10}$ mode can be drawn, giving the diagram we discussed earlier (Fig. 9.2). The introduction of a cut-off frequency implies that the phase constant β_g is frequency dependent, unlike the phase constant β discussed in Chapter 6 on transmission lines. The variation of β_g with frequency can be shown on an ω–β diagram. Figure 9.6 shows a typical characteristic for rectangular waveguide with β_g falling to zero (no propagation) when $\omega = \omega_c$, the cut-off frequency.

At high frequencies the ω–β diagram is linear. For comparison, the ω–β characteristic of a transmission line is also shown. The equations relating ω and β are

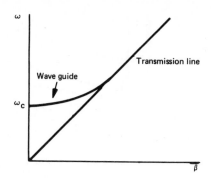

Fig. 9.6 Rectangular waveguide $\omega - \beta$ diagram.

for waveguide $\qquad\qquad \omega \sqrt{(\mu\varepsilon)} = \sqrt{(k_c^2 + \beta_g^2)}$ $\qquad\qquad$ (9.78)

where, for a given guide, μ, ε and k_c are constant, and

for transmission line $\qquad\qquad \omega \sqrt{(\mu\varepsilon)} = \beta$ $\qquad\qquad$ (9.79)

9.12   ATTENUATION IN RECTANGULAR WAVEGUIDE

In the above analysis we have assumed implicitly that the walls of the guide are perfect conductors. Although the conductivity may be very high, there will be some loss at the waveguide walls caused by two factors. First, the finite conductivity of the wall material gives rise to ohmic loss and, second, surface roughness, which becomes more important as the frequency increases. The second cause can be reduced at the cost of a high-quality polished finish on the inside wall surface. The ohmic loss is a function of the wall itself and therefore it is intrinsic once the wall material has been chosen.

An expression for the attenuation along a rectangular air-filled waveguide, due to ohmic loss, can be obtained by finding the wall currents and then calculating the power loss into the guide wall. In this way the attenuation is shown[5,7] to be given by

$$\alpha = \frac{R_s}{120\pi b} \frac{2\pi}{\lambda_0 \beta_g} \left[1 + 2\frac{b}{a}\left(\frac{f_c}{f}\right)^2 \right] \quad \text{nepers/m} \qquad (9.80)$$

where R_s is the surface resistivity of the waveguide inside wall.

Figure 9.7 shows how this attenuation varies with frequency. Each curve has a characteristic shape; as the frequency increases, there is a rapid fall in attenuation as the cut-off value is exceeded. Thereafter, the attenuation falls gently to a minimum, a short distance from cut-off, and then gradually increases with frequency.

It can be seen that the value of attenuation varies with the size and aspect ratio of the waveguide cross-section and with the mode number, as well as with the guide conductivity.

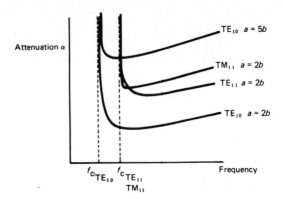

Fig. 9.7 Rectangular waveguide attenuation characteristics.

9.13 EVANESCENT MODES

Although we have discussed the cut-off mechanism as an effect that occurs suddenly as the frequency is reduced, in fact there is some propagation below cut-off, but very little. Below cut-off the propagation coefficient is real, and therefore the wave attenuates exponentially with distance from the source. The modes that are present in the region of the source end of the waveguide when it is operating below cut-off are called evanescent modes.

For transmission purposes, evanescent modes cannot be used, but there are some applications where the rapid attenuation can be useful. Because the forward wave attenuates very rapidly, there is little or no reflection from a short circuit or other junction placed in the guide, and any reflected wave is also heavily attenuated. This means that the reflected wave can be ignored allowing the guide input impedance to appear to be purely reactive, a characteristic that finds application in a number of measuring techniques.

9.14 RECTANGULAR MODE PATTERNS

We discussed the existence of higher-order modes in an earlier section, but the variation of fields in both the x and y directions makes the shapes of the field patterns for the general mn mode difficult to visualize.

In Fig. 9.8 we present patterns for some of the lower-order modes alongside diagrams showing methods of exciting them. Notice that the E and H field lines cross at right-angles, that the magnetic field lines do not terminate on the walls of the guide but form closed loops, and that the positions of maximum field in the cross-section of the guide vary with the mode number.

9.15 CIRCULAR WAVEGUIDES

Although the rectangular cross-section is that most commonly used in waveguides, there are several others that are used from time to time, some of

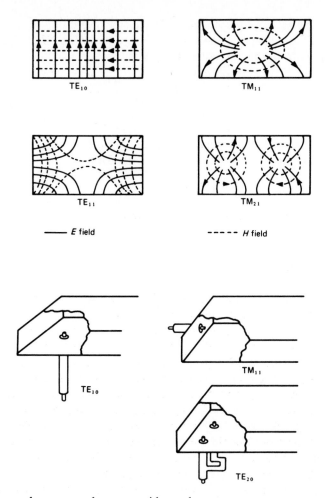

Fig. 9.8 Low-order rectangular waveguide modes.

which are shown in Fig. 9.9. The field patterns can be found by solving Maxwell's equations and applying the appropriate boundary conditions, which usually means that a coordinate system allied to the shape of the cross-section must be introduced. As an example, we shall look briefly at the solution for a circular waveguide, which is used in several special devices that need the mode-pattern symmetry which is characteristic of some of the low-order modes.

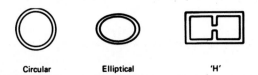

Fig. 9.9 Waveguide cross-sections.

It is not necessary to go through all the steps we followed earlier; we can assume that the space inside the guide has the same effect on Maxwell's equations as in the rectangular waveguide case.

Starting with Eqns (9.12) and (9.13),

$$\nabla \times \bar{E} = -j\omega\mu\bar{H} \tag{9.12}$$

$$\nabla \times \bar{H} = j\omega\varepsilon\bar{E} \tag{9.13}$$

we can use a similar argument to that of Section 9.7. Equation (9.12), expressed in cylindrical coordinates, has the form

$$\bar{r}\frac{1}{r}\left[\frac{\partial E_z}{\partial \theta} - \frac{\partial(rE_\theta)}{\partial z}\right] + \bar{\theta}\left[\frac{\partial E_r}{\partial z} - \frac{\partial E_z}{\partial r}\right] + \bar{z}\frac{1}{r}\left[\frac{\partial(rE_\theta)}{\partial r} - \frac{\partial E_r}{\partial \theta}\right]$$
$$= -j\omega\mu[\bar{r}H_r + \bar{\theta}H_\theta + \bar{z}H_z] \tag{9.81}$$

Separating the three components, and equating left and right sides of Eqn (9.81),

$$\frac{1}{r}\left[\frac{\partial E_z}{\partial \theta} + \frac{\partial(rE_\theta)}{\partial z}\right] = -j\omega\mu H_r \tag{9.82}$$

$$\frac{\partial E_r}{\partial z} - \frac{\partial E_z}{\partial r} = -j\omega\mu H_\theta \tag{9.83}$$

$$\frac{1}{r}\left[\frac{\partial(rE_\theta)}{\partial r} - \frac{\partial E_r}{\partial \theta}\right] = -j\omega\mu H_z \tag{9.84}$$

As before, we will assume that the z dependence of the fields is $\exp(k_z)$, and therefore we can replace $\partial/\partial z$ in Eqns (9.82) and (9.83) by k_z. Then

$$\frac{1}{r}\frac{\partial E_z}{\partial \theta} - k_z E_\theta = -j\omega\mu H_r \tag{9.85}$$

$$k_z E_r - \frac{\partial E_z}{\partial r} = -j\omega\mu H_\theta \tag{9.86}$$

$$\frac{1}{r}\left[\frac{\partial(rE_\theta)}{\partial r} - \frac{\partial E_r}{\partial \theta}\right] = -j\omega\mu H_z \tag{9.87}$$

Similarly, starting with Eqn (9.13), the following additional equations are developed:

$$\frac{1}{r}\frac{\partial H_z}{\partial \theta} - k_z H_\theta = j\omega\varepsilon E_r \tag{9.88}$$

$$k_z H_r - \frac{\partial H_z}{\partial r} = j\omega\varepsilon E_\theta \tag{9.89}$$

$$\frac{1}{r}\frac{\partial}{\partial r}(rH_\theta) - \frac{1}{r}\frac{\partial H_r}{\partial \theta} = j\omega\varepsilon E_z \tag{9.90}$$

Again, two families of modes are possible, transverse electric (TE) and transverse magnetic (TM).

9.16 CIRCULAR WAVEGUIDE (TE) MODES

By definition, the z component of electric field is zero, and putting $E_z = 0$ into Eqns (9.85) to (9.86) reduces them to

$$k_z E_\theta = j\omega\mu H_r \tag{9.91}$$

$$k_z E_r = -j\omega\mu H_\theta \tag{9.92}$$

$$\frac{1}{r}\frac{\partial}{\partial r}(rE_\theta) - \frac{1}{r}\frac{\partial E_r}{\partial \theta} = -j\omega\mu H_z \tag{9.93}$$

$$\frac{1}{r}\frac{\partial H_z}{\partial \theta} - k_z H_\theta = j\omega\varepsilon E_r \tag{9.94}$$

$$k_z H_r - \frac{\partial H_z}{\partial r} = j\omega\varepsilon E_\theta \tag{9.95}$$

$$\frac{1}{r}\frac{\partial}{\partial r}(rH_\theta) - \frac{1}{r}\frac{\partial H_r}{\partial \theta} = 0 \tag{9.96}$$

The only z component is H_z, and by algebraic manipulation of the above equations the other components can be obtained in relation to it. For example, from Eqn (9.88),

$$H_\theta = -\frac{1}{k_z}\left[j\omega\varepsilon E_r - \frac{1}{r}\frac{\partial H_z}{\partial \theta}\right]$$

and putting that into Eqn (9.92) gives

$$E_r = \frac{j\omega\mu}{k_z^2}\left[j\omega\varepsilon E_r - \frac{1}{r}\frac{\partial H_z}{\partial \theta}\right]$$

Hence

$$E_r = \frac{-j\omega\mu}{r(k_z^2 + \omega^2\varepsilon\mu)}\frac{\partial H_z}{\partial \theta}$$

or, if k_c is introduced from Eqn (9.45),

$$E_r = \frac{-j\omega\mu}{k_c^2}\frac{1}{r}\frac{\partial H_z}{\partial \theta} \tag{9.97}$$

By a similar sort of argument the other three components are

$$E_\theta = \frac{j\omega\mu}{k_c^2}\frac{\partial H_z}{\partial r} \tag{9.98}$$

$$H_r = \frac{k_z}{k_c^2}\frac{\partial H_z}{\partial r} \tag{9.99}$$

$$H_\theta = \frac{k_z}{k_c^2}\frac{1}{r}\frac{\partial H_z}{\partial \theta} \tag{9.100}$$

Thus if we can find an expression for H_z we can then find the values of the other components.

By analogy with the arguments leading to Eqn (9.20), H_z must satisfy the vector wave equation

$$\nabla^2 H_z = -\omega^2 \varepsilon \mu H_z \tag{9.101}$$

which has a solution of the form

$$H_z = R\theta Z \tag{9.102}$$

where R, θ and Z refer to the three cylindrical coordinates of Fig. 9.10.
In cylindrical coordinates

$$\nabla^2 H_z = \frac{\partial^2 H_z}{\partial r^2} + \frac{1}{r}\frac{\partial H_z}{\partial r} + \frac{1}{r^2}\frac{\partial^2 H_z}{\partial \theta^2} + \frac{\partial^2 H_z}{\partial z^2} \tag{9.103}$$

Putting Eqns (9.103) and (9.102) into Eqn (9.101), and remembering that the z component solution is

$$Z = H_z \exp(k_z z) \tag{9.104}$$

Eqn (9.101) becomes

$$\theta Z \frac{\partial^2 Z}{\partial r^2} + \frac{\theta Z}{r}\frac{\partial R}{\partial r} + \frac{RZ}{r^2}\frac{\partial^2 \theta}{\partial \theta^2} + R\theta k_z^2 Z = k^2 R\theta Z$$

and dividing by $R\theta Z$

$$\frac{1}{R}\frac{\partial^2 R}{\partial r^2} + \frac{1}{Rr}\frac{\partial R}{\partial r} + \frac{1}{\theta r^2}\frac{\partial^2 \theta}{\partial \theta^2} + k_z^2 = -k^2$$

or

$$\frac{1}{R}\frac{\partial^2 R}{\partial r^2} + \frac{1}{r}\frac{1}{R}\frac{\partial R}{\partial r} + \frac{1}{r^2}\frac{1}{\theta}\frac{\partial^2 \theta}{\partial \theta^2} + k_c^2 = 0 \tag{9.105}$$

where, as before [Eqn (9.29)],

$$k^2 + k_z^2 = k_c^2$$

From this we can establish two independent expressions:

$$\frac{r^2}{R}\frac{\partial^2 R}{\partial r^2} + \frac{r}{R}\frac{\partial R}{\partial r} + k_c^2 = n^2 \tag{9.106}$$

and

$$\frac{1}{\theta}\frac{\partial^2 \theta}{\partial \theta^2} = -n^2 \tag{9.107}$$

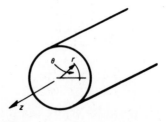

Fig. 9.10 Circular waveguide.

where n^2 is a constant. The solution for θ is straightforward:

$$\frac{\partial^2 \theta}{\partial \theta^2} + n^2 \theta = 0 \qquad (9.108)$$

Hence

$$\theta = A_n \sin n\theta + B_n \cos n\theta \qquad (9.109)$$

However, the solution for R is less simple. Equation (9.106),

$$\frac{\partial^2 R}{\partial r^2} + \frac{1}{r}\frac{\partial R}{\partial r} + \left(k_c^2 - \frac{n^2}{v^2}\right)R = 0$$

does not have a simple closed-form solution. It appears in several problems and it is called the Bessel equation. Solutions are given in tabular form and designated by the symbols $J_n(k_c r)$ and $N_n(k_c r)$. Thus the solution Eqn (9.105) is written

$$R = C_n J_n(k_c r) + D_n N_n(k_c r) \qquad (9.110)$$

$J_n(k_c r)$ is called the nth-order Bessel function of the first kind, of argument $k_c r$, and $N_n(k_c r)$ is the nth-order Bessel function of the second kind.

Combining Eqns (9.104), (9.109), and (9.110) into Eqn (9.102),

$$H_z = [C_n J_n(k_c r) + D_n N_n(k_c r)][A_n \sin n\theta + B_n \cos n\theta]\exp(k_z z) \quad (9.111)$$

The coefficients, C_n, D_n, A_n and B_n are obtained from the boundary conditions or initial fields in the guide.

The nth-order Bessel function of the second kind, $N_n(k_c r)$, increases to infinity as r approaches zero. The effect of having this term in the expression for H_z would therefore be to have a field of infinite magnitude along the axis of the waveguide. This is not possible and hence, on physical grounds, $D_n = 0$, leaving

$$H_z = C_n J_n(k_c r)[A_n \sin n\theta + B_n \cos n\theta]\exp(k_z z) \qquad (9.112)$$

The terms giving the variation of field with θ depend on the positioning of the $\theta = 0$ reference. By choosing it correctly, A_n can be made zero without imposing additional constraints on the field. Then

$$H_z = B_n \cos n\theta \cdot C_n J_n(k_c r)\exp(k_z z)$$

or

$$H_z = H_1 \cos n\theta \cdot J_n(k_c r)\exp(k_z z) \qquad (9.113)$$

where H_1 is determined by the initial field applied to the guide. This solution is not yet complete: H_z must satisfy the conditions at the internal surface of the waveguide. At the guide wall, radius a, $J_n(k_c a)$ must be zero since H_z must be zero there. In addition, also at the guide walls, radius a, $\partial H_z/\partial r = 0$ using boundary conditions (ii) following Eqn (9.53), which implies that E_θ is zero from Eqn (9.113), [Eqn (9.98)]. For $\partial H_z/\partial r = 0$,

$$\frac{\partial}{\partial r}[H_1 J_n(k_c a)\cos n\theta \exp(\pm j\beta_g z)] = 0$$

giving

$$\frac{\partial}{\partial r} J_n(k_c a) = 0 \tag{9.114}$$

$J_n(k_c a)$ does not have a regular periodicity so there is no simple expression for k_c. Instead, tabulated values of the roots of Eqn (9.114) must be used. These depend on n and they are denoted by p'_{nm}. The prime indicates the roots of the first derivative of $J_n(k_c a)$, n is the Bessel function order, and m is the number of the root or the number of zeros of the field [i.e. of $J'_n(k_c a)$] in the range 0 to a.

So now we have developed a full solution of H_z, i.e.

$$H_z = H_1 J_n\left(\frac{p'_{nm}}{a}, r\right) \cos n\theta \exp(\pm j\beta_g z) \tag{9.115}$$

Returning to the other field components, Eqn (9.115) can be used in Eqns (9.97), (9.99) and (9.100), giving

$$E_r = -\frac{j\omega\mu}{k_c^2}\frac{1}{r}\frac{\partial}{\partial\theta}\left[H_1 J_n\left(\frac{p'_{nm}}{a}, r\right)\cos n\theta \exp(\pm j\beta_g z)\right]$$

$$= \frac{j\omega\mu n}{rk_c^2} H_1 j_n\left(\frac{p'_{nm}}{a}, r\right)\sin n\theta \exp(\pm j\beta_g z) \tag{9.116}$$

$$E_\theta = \frac{j\omega\mu}{k_c^2}\frac{\partial}{\partial r}\left[H_1 J_n\left(\frac{p'_{nm}}{a}, r\right)\cos n\theta \exp(\pm j\beta_g z)\right]$$

$$= \frac{j\omega\mu}{k_c^2} H_1\left(\frac{\partial}{\partial r}\left[J_n\left(\frac{p'_{nm}}{a}, r\right)\right]\right)\cos n\theta \exp(\pm j\beta_g z) \tag{9.117}$$

$$H_r = \frac{k_z}{k_c^2} H_1\left[\frac{\partial}{\partial r}J_n\left(\frac{p'_{nm}}{a}, r\right)\right]\cos n\theta \exp(\pm j\beta_g z) \tag{9.118}$$

$$H_\theta = \frac{k_z}{k_c^2}\frac{1}{r}nH_1 J_n\left(\frac{p'_{nm}}{a}, r\right)\sin n\theta \exp(\pm j\beta_g z) \tag{9.119}$$

As before, we have obtained the transverse field components in terms of the single longitudinal component H_z.

9.17 CIRCULAR MODE CUT-OFF FREQUENCY

The nomenclature used to describe the transverse electric fields in a circular waveguide is similar to that in a rectangular waveguide. TE_{nm} designates the nm mode. The cut-off wavelength in the circular case is given by

$$\lambda_{c_{nm}} = \frac{2\pi a}{p'_{nm}} \tag{9.120}$$

and the cut-off frequency by

$$f_{c_{nm}} = \frac{p'_{nm} c}{2\pi a} \tag{9.121}$$

Table 9.1 First ten TE modes in circular waveguide, and corresponding values of p'_{nm}

Mode	p'_{nm}
TE_{11}	1.841
TE_{21}	3.054
TE_{01}	3.832
TE_{31}	4.201
TE_{41}	5.318
TE_{12}	5.331
TE_{51}	6.416
TE_{22}	6.706
TE_{02}	7.016
TE_{32}	8.015

Values of p'_{nm} are given in Table 9.1. We can see from the table that the lowest cut-off frequency is for the TE_{11} mode, when $p'_{11} = 1.841$. This is called the fundamental or dominant mode. The field patterns across the guide for some of the lowest-order modes are shown in Fig. 9.11.

9.18 ATTENUATION IN CIRCULAR WAVEGUIDE

The waveguide walls have finite conductivity and so give rise to some ohmic loss. The attenuation of a TE_{nm} wave caused by the wall losses is given by

$$\alpha = \frac{R_s}{a\eta} \frac{1}{(1 - (f_c/f)^2)^{1/2}} \left[\left(\frac{f_c}{f}\right)^2 + \frac{n^2}{(p'_{nm})^2 - n^2} \right] \tag{9.122}$$

We can see that $n = 0$ is a special case, for then

$$\alpha = \frac{R_s}{a\eta} \frac{1}{\sqrt{[1 - (f_{c01}/f)^2]}} \tag{9.123}$$

and the variation of α with frequency is negative, i.e. the attenuation decreases as the frequency increases.

Typical curves of TE mode attenuation are shown in Fig. 9.12 and the difference between the TE_{0m} modes and the others is clear.

Another peculiar property is that the attenuation of these modes decreases as a, the guide radius, is increased. These two properties, although appreciated for some time, led to a worldwide interest in the mid-1950s when communication systems were required which could transmit high-frequency carriers over long distances with low loss.[3] The TE_{01} circular mode was

TE_{11} TM_{01} TE_{01} TM_{11}

Fig. 9.11 Low-order circular waveguide modes.

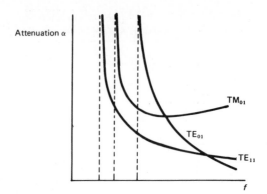

Fig. 9.12 Circular waveguide attenuation characteristics.

chosen, and laboratories in the UK, USA and Japan spent the next 20 years in developing production systems. Several were used successfully. From our discussion there would appear to be few problems, but in practice there are several which produced major difficulties. The TE_{01} mode is not the fundamental, and the waveguide is overmoded because of its large diameter. These factors mean that several modes can propagate down the guide, and at bends and corners, which are essential parts of a realistic system, the TE_{01} mode can be easily converted into other modes, with a consequent loss of power. These other spurious modes may have different propagation velocities from the TE_{01}, and that can have the effect of degrading the shape of the signal if there is a double mode conversion – from TE_{01} to spurious mode at one point along the guide and then back from spurious mode to TE_{01} at a further point. The performance of the guide can be improved by using something more complex than a simple circular cylindrical tube. By putting a helix of fine wire, set in resin, on the inner surface, many of the spurious modes can be filtered out since the helix will attenuate all non-TE_{0m} modes. Of course, the helix adds significantly to the cost of making the guide, but if the spacing between repeaters along the guide can be reduced thereby, a net saving may result. The precision quality of the waveguide requires that it is inserted into some protective jacket and laid in such a way that there are no tight bends. The latter considerations again add to the cost. Until the early 1970s it seemed that, on balance, circular waveguide would be an attractive long-haul transmission medium. Then optical fibres came along, and although for a few years both systems were developed in parallel in the large telecommunications laboratories, it soon became clear that fibre systems (see Chapter 12) could be better, and now their performance is such that the waveguide approach is no longer viable.

9.19 RECTANGULAR CAVITY

Figure 9.13(a) shows a rectangular waveguide cavity, and we can recognize it as a length of waveguide, terminated at each end by a short-circuiting face.

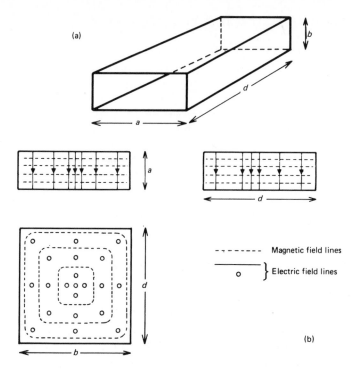

Fig. 9.13 (a) Rectangular waveguide cavity; (b) field pattern for TE_{101} mode.

Waves are reflected back and forth from these end faces. The cavity is highly resonant; oscillations occur over a very narrow band of frequencies centred on the resonant frequency. Tuning is either fixed, when all the cavity dimensions can have just one value, or variable, when one end face is attached to a plunger, allowing the resonant length of the cavity to be varied.

Any of the rectangular waveguide modes of Section 9.7 can exist within the cavity. We shall consider in detail the fundamental TE_{01} mode. The boundary conditions used in the development of the solutions to the field equations within the waveguide also apply in the cavity, which puts an additional constraint on the field pattern. The conducting walls at $z = 0$ and $z = d$ (the end walls) impose the condition that at these walls the tangential electric field, E_y, must be zero. Thus d must be an integral number of half guide-wavelengths long. The lowest order cavity mode is when $d = \lambda_g/2$. Then the E field is zero at the end walls and has one maximum, in the centre of the cavity.

If the width of the guide is a and the free-space wavelength is λ_0, the first resonant length can be found from Eqn (9.2) to be

$$d = \frac{\lambda_g}{2} = \frac{\lambda_0}{2[1 - (\lambda_0/2a)^2]^{1/2}} \qquad (9.124)$$

The resonant frequency corresponding to a cavity of width a and length d in this fundamental (TE$_{101}$) mode is found directly from

$$f_{r_{101}} = c/\lambda_{0_{101}}$$

where

$$\lambda_{0_{101}} = 2ad/(a^2 + d^2)^{1/2} \tag{9.125}$$

Clearly, equations similar to Eqns (9.124) and (9.125) can be derived easily for TE$_{10p}$ modes where p can, in principle, have any integer value in the range $1 \leqslant p \leqslant \infty$.

In the TE$_{10p}$ modes there is no variation of E field with y and hence the height of the cavity b is not involved in the last two equations.

The E field inside the cavity in the TE$_{101}$ mode can be considered as the sum of the incident and reflected waves, i.e.

$$E_y = [E_1 \exp(-j\beta_g z) + E_2 \exp(j\beta_g z)] \sin\left(\frac{\pi x}{a}\right) \tag{9.126}$$

and, by analogy, the H_x and H_z components are

$$H_x = -\frac{\sqrt{[1-(\lambda_0/2\alpha)^2]}}{\eta}[E_1 \exp(-j\beta_g z) - E_2 \exp(j\beta_g z)] \sin(\pi x/a) \tag{9.127}$$

$$H_z = \frac{\lambda_0}{2a\eta}[E_1 \exp(-j\beta_g z) - E_2 \exp(j\beta_g z)] \cos(\pi x/a) \tag{9.128}$$

These equations can be simplified if we apply some of our earlier statements about the properties of the cavity. Assuming that the wall losses are negligible, the magnitude of the incident and reflected waves at the end wall will be equal, although of opposite sign, i.e. $E_1 = -E_2$. Also, because we are using the TE$_{101}$ mode, in which $E_y = 0$ at $z = 0$ and $z = d$, and there is only one E field maximum between the end walls, $\beta_g = \pi/d$. Then Eqn (9.126) can be written

$$E_y = E_1\left[\exp\left(-j\frac{\pi}{d}z\right) - \exp\left(j\frac{\pi}{d}z\right)\right]\sin\frac{\pi x}{a}$$

$$= -2jE_1 \sin\frac{\pi}{d}z \sin\frac{\pi}{a}x \tag{9.129}$$

and using this expression in Eqns (9.127) and (9.128) we have

$$H_x = \frac{2E_1}{\eta}\frac{\lambda}{2d}\sin\frac{\pi x}{a}\cos\frac{\pi z}{d} \tag{9.130}$$

$$H_z = -\frac{\lambda}{2a}\frac{2E_1}{\eta}\cos\frac{\pi x}{a}\sin\frac{\pi z}{d} \tag{9.131}$$

where

$$\eta = \sqrt{(\mu/\varepsilon)}$$

From these three expressions the field patterns inside the cavity can be plotted for the fundamental mode. Figure 9.13(b) shows a representation of these fields.

Higher-order modes

We can extend our discussion of the TE_{101} mode to the more general case of the TE_{mnp} mode where m, n and p represent the number of half-sinusoidal variations of field in the x, y and z directions, respectively. In theory, each parameter can have an infinite number of values.

The rectangular waveguide TE_{mn} mode has an H_z field of the form.

$$H_z = [A \exp(-j\beta_g z) + B \exp(j\beta_g z)] \cos\frac{m\pi}{a}x \cos\frac{n\pi}{b}y \qquad (9.132)$$

As in our discussion on the TE_{101} mode, we can assume that $A = -B$ and that $H_z = 0$ at $z = 0, d$ making $\beta_g = p\pi/d$.

Then Eqn (9.132) can be written

$$H_z = -2jA \sin\frac{p\pi}{d}z \cos\frac{m\pi}{a}x \cos\frac{n\pi}{b}y \qquad (9.133)$$

For the general TE_{mnp} mode the relationship between the transverse field components E_x, E_y, H_x and H_y, are the same as those given by Eqns (9.58) to (9.62) in Section 9.9.

Substituting Eqn (9.133) into those equations gives

$$H_x = \frac{2jA}{k_c^2}\left(\frac{p\pi}{d}\right)\left(\frac{m\pi}{a}\right)\sin\frac{m\pi}{a}x \cos\frac{n\pi}{b}y \cos\frac{p\pi}{d}z \qquad (9.134)$$

$$H_y = \frac{2jA}{k_c^2}\left(\frac{p\pi}{d}\right)\left(\frac{n\pi}{b}\right)\cos\frac{m\pi}{a}x \sin\frac{n\pi}{b}y \cos\frac{p\pi}{d}z \qquad (9.135)$$

$$E_x = \frac{2A\omega\mu}{k_c^2}\left(\frac{n\pi}{b}\right)\cos\frac{m\pi}{a}x \sin\frac{n\pi}{b}y \sin\frac{p\pi}{d}z \qquad (9.136)$$

$$E_y = -\frac{2A\omega\mu}{k_c^2}\left(\frac{m\pi}{a}\right)\sin\frac{m\pi}{a}x \cos\frac{n\pi}{b}y \sin\frac{p\pi}{d}z \qquad (9.137)$$

where as before

$$k_c^2 = \left(\frac{m\pi}{a}\right)^2 + \left(\frac{n\pi}{b}\right)^2 \qquad (9.138)$$

For the TE_{mnp} cavity mode

$$\beta_g = \frac{p\pi}{a} = \left[\left(\frac{2\pi}{\lambda_0}\right)^2 - k_c^2\right]^{1/2} \qquad (9.139)$$

Rearranging Eqns (9.138) and (9.139) we can derive the resonant free-space wavelength for a cavity of dimensions a, b, d operating in the TE_{mnp} mode, viz.

$$\lambda_0 = 2\pi \bigg/ \left[\left(\frac{m\pi}{a}\right)^2 + \left(\frac{n\pi}{b}\right)^2 + \left(\frac{p\pi}{d}\right)^2\right]^{1/2} \qquad (9.140)$$

It is worth noting that, for a particular resonant frequency, the resonant length of the cavity, d, depends also on the cross-sectional dimensions of the waveguide appropriate to the mode of interest [Eqns (9.125) and (9.140)].

One of the main applications of multimode cavities is in microwave heating;[4] for example, in microwave ovens, where the object is to heat the sample uniformly, there is a need for a uniform field over the central space to be occupied by the sample, and this uniformity can be achieved by a combination of high-order modes and the use of a mode-stirrer, which is a metal plate, made to change its position and so vary the resonant length, and therefore the position of field maxima in the oven.

9.20 CAVITY Q – TE$_{101}$ MODE

As in all resonant systems, there are some losses, and eventually the oscillations in the cavity will decay. It is shown in other texts[5,6] that the cavity Q, which is a measure of the sharpness of the resonance curve, can be expressed as a ratio of the energy stored in the cavity to the mean loss at the walls, or

$$Q = \omega_r \frac{W_s}{W_L} = \omega_r \times \frac{\text{energy stored in the cavity}}{\text{average power loss}} \qquad (9.141)$$

where ω_r is the angular resonant frequency of the cavity.

W_s is the energy stored in the fields within the cavity. It is shared between the electric and magnetic fields, and it can therefore be found by calculating the maximum of one, when the other is zero.

The energy in an electric field is $\frac{1}{2}\varepsilon E^2$ integrated over the volume of the cavity:

$$W_s = \frac{\varepsilon}{2} \int_0^d \int_0^b \int_0^a |E_y|^2 \, dx \, dy \, dz \qquad (9.142)$$

since there is only one field component, E_y in the TE$_{101}$ mode.

Substituting for E_y from Eqn (9.129),

$$W_s = \frac{\varepsilon}{2} \int_0^d \int_0^b \int_0^a \frac{\varepsilon}{2} E_0^2 \sin^2 \frac{\pi x}{a} \sin^2 \frac{\pi z}{d} \, dx \, dy \, dz$$

where E_0 depends on the initial field in the cavity.

Remembering to replace $\sin^2 \theta$ by $\frac{1}{2}[1 - \cos 2\theta]$, this expression is integrated and reduces to

$$W_s = \frac{\varepsilon abd E_0^2}{8} \qquad (9.143)$$

The average power loss at the walls is $\frac{1}{2} I^2 R_s$ where I is the peak wall current induced by the fields in the cavity, and R_s is the resistivity of the walls. The current on any wall is directly related to the tangential magnetic field at the wall surface. This field will vary with position in the cavity and therefore integration is required over each of the wall surfaces. The symmetry of the system means that opposite walls will have identical losses, hence

$$W_L = \frac{R_s}{2} \left\{ 2 \int_0^b \int_0^a |H_x|^2_{z=0} \, dx \, dy + 2 \int_0^d \int_0^b |H_z|^2_{x=0} \, dy \, dz \right.$$

$$\left. + 2 \int_0^d \int_0^a (|H_x|^2 + |H_z|^2) \, dx \, dz \right\} \qquad (9.144)$$

i.e.

W_L = loss in (end walls + side walls + top and bottom walls)
If we now develop Eqn (9.144) using

$$H_x = -j \frac{E_0}{\eta} \frac{\lambda_g}{2d} \sin \frac{\pi x}{a} \cos \frac{\pi z}{d}$$

and

$$H_z = j \frac{E_0}{\eta} \frac{\lambda_g}{2a} \cos \frac{\pi x}{a} \sin \frac{\pi z}{d}$$

We have

$$W_L = \frac{R_s \lambda_g^2 E_0^2}{8\eta^2} \left[\frac{ab}{d^2} + \frac{bd}{a^2} + \frac{1}{2}\left(\frac{a}{d} + \frac{d}{a}\right) \right] \tag{9.145}$$

Then, from Eqns (9.143) and (9.145) and after some rearranging, the Q factor becomes

$$Q = \frac{\pi \eta}{R_s} \left[\frac{2b(a^2 + d^2)^{3/2}}{ad(a^2 + d^2) + 2b(a^3 + d^3)} \right] \tag{9.146}$$

9.21 CIRCULAR CAVITY

By analogy with the previous section we can see that a length of circular waveguide with short-circuited ends will also form a cavity. The separation of the ends, d in Fig. 9.14(a), determines its resonant frequency; resonance occurs when d is an integral number of half-guide-wavelengths long. For each of the circular waveguide modes (Section 9.17) there is a corresponding family of cavity resonating modes, e.g. the fundamental TE_{11} circular waveguide mode will give rise to the TE_{11l} family of cylindrical cavity modes, where l indicates the number of half-guide-wavelengths in the length of the cavity. The fundamental cavity mode corresponds to $d = \lambda_g/2$, and it is designated the TE_{11} mode. In terms of the dimensions of the cavity, the resonant frequency of the TE_{111} mode is given by

$$f_{111} = \frac{c}{2\pi} \left[\left(\frac{\pi}{d}\right)^2 + \left(\frac{p'_{11}}{a}\right)^2 \right]^{1/2}$$

where p'_{11} is the Bessel function coefficient of Section 9.17.

Another mode which has several applications is the TE_{011}. Its field pattern, shown in Fig. 9.14(b), produces no current flow between the cylindrical wall and the cavity ends. This means that the cavity can be constructed to allow the distance d to be varied (in applications where the cavity is to be tuned, as in a wavemeter), without imposing very rigorous constraints on its fabrication; good contact between the wall and the end faces is not essential.

The Q factor for TE modes has a form which is a function of the dimensions of the cavity, the mode number, and the skin depth of the wall material, δ_s.

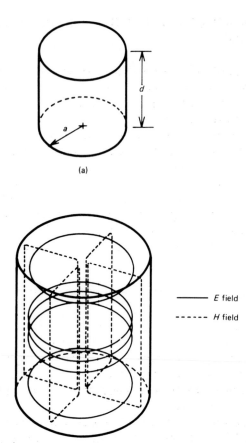

(a)

(b)

Fig. 9.14 (a) Circular waveguide cavity; (b) field pattern for TE_{011} mode.

From Collin,[5] for the TE_{mnl} mode

$$Q = \frac{\lambda_0}{2\pi\delta_s}\left[1 - \left(\frac{n}{p'_{nm}}\right)^2\right]\left[(p'_{nm})^2 + \left(\frac{l\pi a}{d}\right)^2\right]^{3/2} \Bigg/ \left[(p'_{nm})^2 + \frac{2a}{d}\left(\frac{l\pi a}{d}\right)^2 + \left(1 - \frac{2a}{d}\right)\left(\frac{nl\pi a}{p'_{nm}d}\right)^2\right]$$

where δ_s is related to the conductivity of the walls, σ, by

$$\delta_s = \sqrt{(2/\omega\mu\sigma)}$$

Coupling into the cavity depends on the mode required in the particular application. The aperture between the cavity and the waveguide feed has to be so placed that, if possible, only the mode of interest is generated. Some modes are degenerate and special techniques may be necessary to absorb or avoid those which are unwanted.

A full discussion on circular waveguide cavities can be found in Harvey[7] and Marcuvitz.[8]

9.22 RECTANGULAR WAVEGUIDE COMPONENTS – TE$_{10}$ MODE

In any practical application, several components are essential to provide a useable microwave system. A typical layout is shown in Fig. 9.15. In this section we will discuss briefly the passive devices shown; generators have been considered separately in Chapter 8.

Nearly all components are designed to operate in a particular mode. In our discussion we will assume that the fundamental TE$_{10}$ mode is being transmitted. The purpose of most passive microwave devices is self-evident from their names, but their operation is not always quite so obvious. In most cases there has been considerable analytical work done to evaluate how these devices work, and their operating performance characteristics. Here we shall not consider any analysis, and if the reader would like to explore this area, reference should be made to the specialist texts available.[5,7]

9.23 WAVEGUIDE – COAXIAL TRANSFORMER

To launch a wave into the guide some mechanism is required that will transform the RF energy on the feed from the generator (often coaxial cable) into microwave energy. Alternatively, and usually at higher frequencies, the generator is mounted directly on to the waveguide and indeed may be integral with it. Examples of a coaxial–waveguide transformer and a directly coupled feed are shown in Fig. 9.16.

In Fig. 9.16(a), energy is coupled from the generator via a simple probe, the centre of a coaxial feed. The probe penetrates into the guide through a broad face, and acts like an aerial. The electric fields induced between the probe and the waveguide propagate in all directions. By placing a short-circuit behind

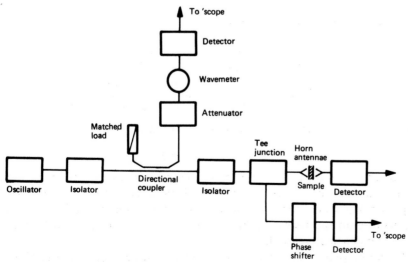

Fig. 9.15 Typical microwave circuit.

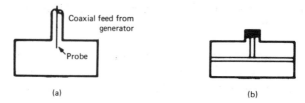

Fig. 9.16 Coaxial–waveguide coupler.

the probe, energy reflected from the short is propagated in the forward direction, in phase with that leaving the probe. The degree of coupling between the probe and the guide depends on the depth of penetration of the probe into the guide. The guide itself acts as a filter, which causes the many spurious (often evanescent) modes generated at the probe to be attenuated within a short distance.

Sometimes more rigidity or a greater bandwidth is required at the transformer than can be achieved by a simple probe. A horizontal bar [Fig. 9.16(b)] attached to the probe makes the feed mechanically more robust, and at the same time increases the bandwidth of the device.

In general, these feeds are reciprocal devices that can be used as either transmitters or receivers.

In the above paragraphs we have looked only at electric field probes, but the TE_{10} mode can also be generated by inducing a magnetic field in the guide. This is done by placing a loop through an end wall. The current variation produced in the loop by a signal from the generator induces a magnetic field, and again the filtering action of the guide creates the conditions for propagation of the TE_{10} mode.

9.24 ATTENUATOR

If the amplitude of the wave in the guide is too high, it must be reduced by using an attenuator, which consists of a vane made of a lossy material. The degree of attenuation depends on the strength of field at the vane; by varying the position of the vane, different levels of attenuation are possible. Two common forms of attenuator are shown in Fig. 9.17. In Fig. 9.17(a) the vane is attached to a micrometer through the sidewall of the guide. Attenuation is a minimum when the micrometer is fully extended, and the vane is against the sidewalls of the waveguide, where the electric field is at a minimum. Maximum attenuation occurs when the micrometer is screwed in and the vane lies along the axis of the guide in a position of maximum electric field. Good quality attenuators are tested in the manufacturer's laboratory and a calibration chart is issued to relate the reading on the micrometer screw to the attenuation introduced by the vane. For some applications, an attenuator is used to reduce the level of the signal in the guide to a workable value, and in that case a simple ungraduated screw will suffice. In order to avoid reflections the ends of the attenuator vane may be either stepped or tapered. The

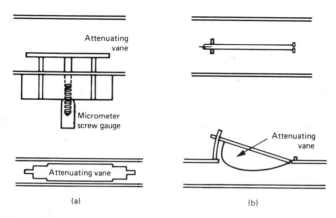

Fig. 9.17 Waveguide attenuators.

attenuator in Fig. 9.17(b) shows a different type of mechanism – often used in the high-frequency microwave, or millimetre, range. The vane has a cycloid profile and it is introduced into a slot along the centre of a broad face of the guide. The degree of attenuation increases with the depth of penetration into the waveguide, achieving a maximum when the vane is fully inserted.

9.25   DIRECTIONAL COUPLER

In many systems a proportion of the field is taken from the main waveguide to be measured or sampled in some way. A directional coupler allows a known fraction of the power in the primary guide to be filtered off. It consists of a second waveguide, contiguous with the first, and coupled to it by one or more holes, which may be circular, rectangular or cruciform in shape. The degree of coupling depends on the size of the hole and its position relative to the axis of the waveguide, if it is in the broad face. A double (four-port) coupler is often used in which energy is passing from port 1 to port 2 in the primary guide. Port 3 couples out a fraction of the forward wave and port 4 couples out a similar proportion of the backward wave. In this type of coupler (Fig. 9.18) the coupling holes are separated by $\lambda_g/4$. Hence, part of the forward wave from port 1 to port 2 is coupled through hole A and an equal part through hole B. The coupled fields add and are detected at port 3. Power from the forward wave coupled through hole B and travelling towards port 4 is

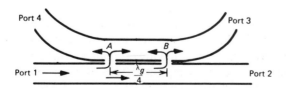

Fig. 9.18 Directional coupler.

cancelled by the power from hole A travelling in that direction. The effect of the coupling holes is given by two parameters:

(i) Coupling factor. With matched loads at ports 2 and 3, any power at port 4 is due to that coupled from the primary guide. The coupling factor is given by $10 \log (P_1/P_4)$ dB. The typical values are 3, 10, 12, 20 dB.

(ii) Directivity. The coupling holes will themselves produce reflections, and there will not be total isolation between ports 3 and 4. The directivity measures the ratio of the power in these ports for a matched load at port 2, i.e. directivity $= 10 \log (P_4/P_3)$ dB and this value would not be expected to be less than 30 dB.

9.26   TEE JUNCTIONS

Rectangular waveguides can be joined by using simple tee devices. Three commonly used, fundamental mode, junctions are shown in Fig. 9.19. To distinguish between junctions into the broad or narrow side of the guide, the first [Fig. 9.19(a)] is called an E plane tee, and the second [Fig. 9.19(b)] is called an H plane tee. Apart from the difference in geometry, these tees have different characteristics. In the E plane tee a wave entering port 1 divides equally into ports 2 and 3, but in phase opposition, so that at a given distance from the junction the signal out of port 2 is 180° out of phase with that from port 3. The H plane junction does not exhibit such phase reversal. The E field change across the E and H plane junctions is shown in the diagram, and explains the difference between them.

The third tee is a four-port, known as a hybrid, since it combines an E and an H plane junction. It has the property that a wave entering one port divides into the two adjacent ports, but does not appear at the opposite port. This characteristic allows the junction to be used in any systems in which two signals are to be processed, but must be isolated. For example, as we see in Fig. 9.19(d), this device (sometimes called a magic-tee) can be used to isolate the receiver from the transmitter, in a radar system that has a common transmit–receive aerial. The hybrid-tee has the disadvantage that half of the power is lost – in this example half of the received power goes to the receiver, but the other half is lost in the transmitter, and in a radar system that would degrade the sensitivity of the system considerably.

9.27   WAVEMETER

Traditionally, frequency has been measured by first measuring the wavelength of the signal in the waveguide. This can be done easily by using a wavemeter that is a waveguide cavity having one end wall attached to a piston. The position of the piston can be determined from the micrometer gauge to which it is attached. Energy from the waveguide is coupled into the cavity via small coupling holes that, in the case of a cylindrical cavity, would be via the fixed

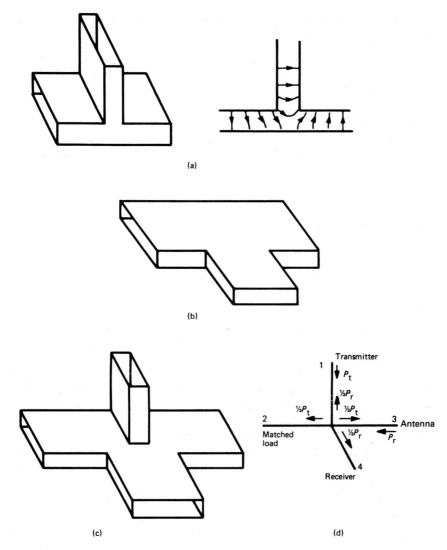

Fig. 9.19 Rectangular waveguide Tee junctions: (a) *E* plane; (b) *H* plane; (c) hybrid.

end wall, or for a rectangular cavity via a broad face of the cavity guide [Figs. 9.20(a) and (b)]. Connection into the cavity can be one-port (absorption), or two-port (transmission). The single port wavemeter absorbs energy from the waveguide when it is adjusted to a resonant frequency, and consequently the output from a detector at D in Fig. 9.20(a) will dip sharply at resonance. Conversely, in a transmission-type cavity, energy passes through the cavity most effectively at resonance; hence the detector in Fig. 9.20(b) will detect a surge of current at that frequency. In either case, the Q of the wavemeter will normally exceed 3000, and for high-quality wavemeters it may be greater than 10 000.

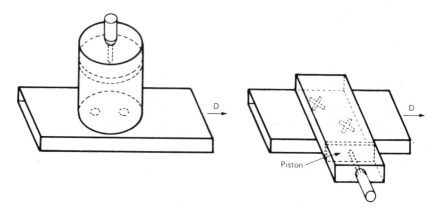

Fig. 9.20 Wavemeters: (a) cylindrical cavity; (b) rectangular cavity.

9.28 MATCHED LOAD

There are several positions in a microwave circuit, e.g. at one side of a directional coupler, beyond a wavemeter, or at the opposite end of a sample to be measured, when a matched termination is required. A termination provides a match if it does not induce a reflected wave. Hence most matched loads operate on the principle of completely absorbing incident energy so that no reflections can take place. Figure 9.21 shows a typical matched load. The absorbing material is wood, or some other lossy dielectric, and it is shaped to offer a gradual taper to the incident wave, and therefore no reflecting surface. To further reduce the possibility of reflection, the ends of the legs of the double wedge are displaced by $\lambda_g/4$, so that any reflections from these ends cancel. The dielectric insert is several wavelengths long to ensure that any energy reflected from the distant end is absorbed on the return path.

9.29 SHORT CIRCUIT

In Section 6.10 the use of stubs in transmission lines was discussed and there it was shown that the input impedance of a short-circuited stub can be altered by varying the stub length. A short circuit can be used in the same way in microwave systems. By attaching a short circuit to a plunger, the length of the

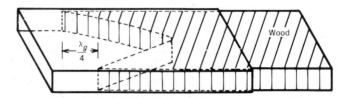

Fig. 9.21 Matched load.

short-circuited waveguide can be adjusted to provide the required input impedance. A typical short circuit is shown in Fig. 9.22.

9.30 ISOLATOR

Here we are not able to develop the theoretical basis for ferrites; we will acknowledge that a ferrite is a material that, when subjected to a magnetic field, will cause a polarized electromagnetic wave passing through it to rotate. The direction of rotation is not dependent on the direction of propagation of the wave. Hence a device can be constructed such that in one direction the rotation assists in the transmission of the wave, whereas in the opposite direction the rotation blocks propagation. Reference to Fig. 9.23 will explain how this can happen. Let us assume that the ferrite and magnetic field are so chosen that the electric field vector is rotated clockwise by 45°. By using a 45° anticlockwise twist in the waveguide input before the ferrite, the rotation by the ferrite restores the wave to its original polarization and it can leave the output guide in the normal way. For a wave travelling in the opposite direction, however, the rotation through the ferrite is still clockwise, but the rotation through the 45° twist is clockwise also, resulting in the wave approaching the input guide horizontally polarized, and thus unable to pass

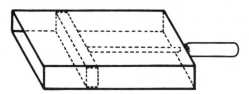

Fig. 9.22 Variable position short circuit.

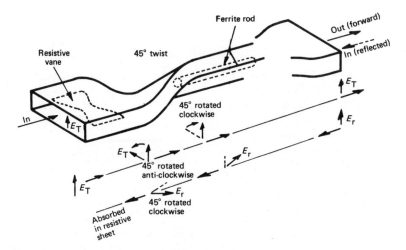

Fig. 9.23 Principles of operation of isolator.

through the system. There would be a reflection in such a case, but the resistive vane absorbs the energy in this backward wave and prevents reflections occurring. The loss in the forward direction is very small, whereas that in the backward direction might be as high as 30 dB. This device is used chiefly to isolate a generator from a system, to prevent mismatches in the system from causing reflections which might interfere with the performance of the generator.

Another type of isolator is shown in Fig. 9.24. Here the magnetic field is across the guide, produced by a substantial permanent magnet.

9.31 ⌐ MICROSTRIP

The development of microwave integrated circuits, involving substrate-based sources, detectors and passive components, has necessitated the development

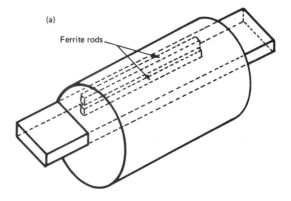

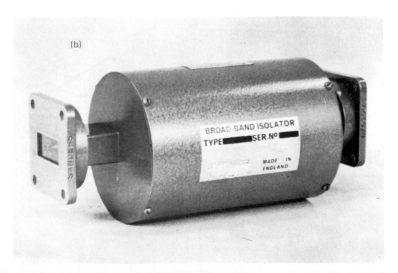

Fig. 9.24 Ferrite isolator. Reproduced from Marconi catalogue by permission of Sanders Division Marconi Instruments Ltd.

of waveguiding techniques suitable for integrated circuit applications. Some of the structures investigated for guiding waves on substrates are shown in Fig. 9.25. The most commonly used is the microstrip line and that is the structure that we will consider here. The dimensions of interest are indicated in Fig. 9.26.

The ground plane and the strip are of good conducting material such as copper, while the substrate is of a dielectric material with low loss and high permittivity, typically $\varepsilon_r \approx 10$.

We noted earlier, in Section 9.7, that in pipe waveguides, the operating modes are TE and TM. In microstrip, the modes are extremely complex, but when the dielectric has a high permittivity they are almost TEM. The complication arises from the fact that the propagation is not entirely through the dielectric material. Figure 9.27 shows the E field pattern about the line and it is clear that while most of the field is in the substrate some of it is not. The analysis is therefore difficult.

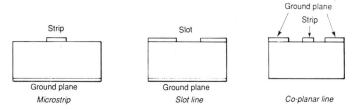

Fig. 9.25 Microwave integrated circuit structures.

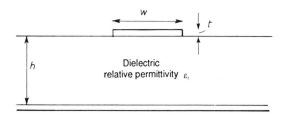

Fig. 9.26 Microstrip dimensions.

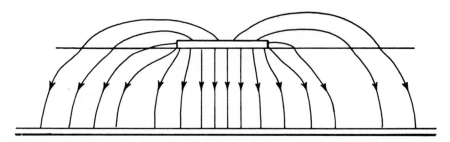

Fig. 9.27 E-field distribution in microstrip.

Here we will not develop the underlying theory of propagation on a microstrip system; if the reader wishes to do so, excellent treatments are found in references 9 and 10.

If the energy propagated along the microstrip were confined entirely to the dielectric, the analysis would be relatively straightforward, consisting of calculating the velocity along the microstrip from

$$v = c/\sqrt{\varepsilon_r}$$

and from that would follow good approximations for the other parameters of interest, in particular the capacitance and characteristic impedance.

Given that only a small amount of the energy propagates through air, the line can be modelled by a system in which all the energy travels in the substrate, but for which the dielectric permittivity is modified from its real value to an effective value ε_e. The difficulty is then to devise an expression ε_e in terms of w, h, and ε_r which can be verified by experiment. Many workers have tackled the problem of providing a suitable expression for ε_e and in the literature there is a variety of different equations available. One such equation is the following for the characteristic impedance, Z_0:

$$Z_0 = \frac{42.4}{\sqrt{(\varepsilon_r + 1)}} \ln \left\{ 1 + \left(\frac{4h}{w}\right) \left[\left(\frac{14 + 8/\varepsilon_r}{11}\right)\left(\frac{4h}{w}\right) \right. \right.$$
$$\left. \left. + \sqrt{\left(\left(\frac{14 + 8/\varepsilon_r}{11}\right)^2 \left(\frac{4h}{w}\right)^2 + \left(\frac{1 + 1/\varepsilon_r}{2}\right)\pi^2\right)} \right] \right\} \qquad (9.147)$$

Clearly Eqn (9.147) is an empirical relationship. From an analytical point of view it gives an expression for Z_0 which can be used to derive other parameters. However, a designer is usually required to determine the dimensions of the strip line given specified values for ε_r and Z_0. To do that Eqn (9.147) would be rearranged with w/h as a subject. The equations used for ε_r can vary with the value of w/h and therefore care must be taken in their application. Reference 11 gives one of the clearest expositions of the basic features of microstrip design.

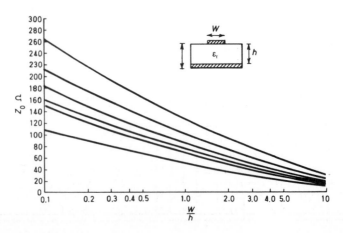

Fig. 9.28 Characteristic impedance of microstrip line.

In order to design a microstrip line it is necessary usually to use design curves of the type shown in Fig. 9.28. There it can be seen that the characteristic impedance of the microstrip is portrayed as a function of the dimensions w and h for a variety of values of dielectric permittivity. Hence the curves can be used either to find Z_0 for a particular strip line or to determine the ratio of w/h once the required characteristic impedance is known.

REFERENCES

1. Bleaney, B. I. and Bleaney, B., *Electricity and Magnetism*, Oxford University Press, Oxford, 1978.
2. Paul, C. R. and Nasar, S. A., *Introduction to Electromagnetic Fields*, McGraw-Hill, Maidenhead, 1982.
3. Karbowiak, A. E., *Trunk Waveguide Communication*, Chapman & Hall, London, 1965.
4. Okress, E. C., *Microwave Power Engineering*, Vol. 2, Academic Press, London, 1968.
5. Collin, R. E., *Foundation of Microwave Engineering*, McGraw-Hill, Maidenhead, 1966.
6. Atwater, H. A., *Introduction to Microwave Theory*, McGraw-Hill, Maidenhead, 1962.
7. Harvey, A. F., *Microwave Engineering*, Academic Press, London, 1963.
8. Marcuvitz, M., *Waveguide Handbook*, McGraw-Hill, Maidenhead, 1951.
9. Gupta, H. C. and Sing, A., *Microwave Integrated Circuits*, Chapter 3, Wiley Eastern, 1974.
10. Edwards, T. C., *Foundations for Microstrip Circuit Design*, Wiley, Chichester, 1981.
11. Combes, P. F., Graffeuil, J. and Santerean, J. F., *Microwave Components, Devices and Active Circuits*, Wiley, 1987.

PROBLEMS

9.1   Find the values of $\lambda_0, \lambda_g, \lambda_c, v_{ph}$ and v_g for a rectangular waveguide of inner cross-section $2.3\,\text{cm} \times 1.0\,\text{cm}$ when it carries a signal of $10\,\text{GHz}$ in the TE_{10} mode.

Answer: $3\,\text{cm}$, $3.96\,\text{cm}$, $4.6\,\text{cm}$, $3.96 \times 10^8\,\text{m/s}$, $2.27 \times 10^8\,\text{m/s}$.

9.2   A rectangular waveguide of cross-section $a \times b$ is filled with a dielectric material of permittivity $\varepsilon = k\varepsilon_0$. Show that for the TE_{10} mode the guide wavelength is given by

$$\lambda_g = \frac{\lambda_0}{[k - (\lambda_0/\lambda_c)^2]^{1/2}}$$

9.3   (a) Show that the superposition of two rectangular TE_{10} modes of equal amplitude and opposite direction of travel (as in a short-circuited waveguide) produces standing wave maxima and minima of field along the guide.
(b) Show also that the E field and H field maxima are displaced along the guide by $\lambda_g/4$.

9.4 Calculate the minimum value of a, in a rectangular waveguide of width a, operating in the TE_{10} mode over the frequency range from 10 to 11.5 GHz if the variation in v_g is not to exceed 20% of its value at 10 GHz.

Answer: 1.87 cm.

9.5 A TE_{10} signal at 10 GHz is propagating down a waveguide of width $a = 2$ cm. What is the change in phase velocity if the width is increased to 2.4 cm?

Answer: 0.70×10^8 m/s.

9.6 Find the distance between two adjacent wave minima in a 2.3 cm $\times$ 1.0 cm rectangular air-filled waveguide which is propagating in the TE_{10} mode at the frequency at which the TE_{21} mode could just begin to propagate.

Answer: 0.799 cm.

9.7 Sketch the electric and magnetic field patterns in both the cross-sectional and longitudinal planes of a rectangular waveguide operating in the TE_{20} mode. From the model of two interfering plane waves, find the cut-off wavelength for that mode in terms of the cross-sectional dimensions $a \times b (\lambda_c = a)$.

9.8 An air-filled rectangular waveguide has a TE_{10} mode cut-off frequency of 8 GHz. The signal frequency is 15 GHz, and the ratio $a/b = 2$. If the guide is now filled with perspex of relative permittivity 2.5, find the guide wavelength and phase velocity of the signal in the guide.

Would you expect higher order modes to be propagated in the dielectric-filled guide?

Answer: 1.34 cm, 2.02×10^8 m/s; yes.

9.9 A wave of frequency 9.95 GHz is travelling in the TE_{10} mode down a rectangular air-filled waveguide of width 1.5 cm. How far along the guide will the wave travel before it is attenuated to $1/e$ of its initial values?

Answer: 5.15 cm.

Telephony $\boxed{10}$

The telephone system is the largest integrated system in the world. Individual subscribers, separated by half the globe, can, in many cases automatically, make contact using this system. There is very rapid progress towards a complete direct-dialling international network, and it now seems very likely that the development of microcircuits, telecommunications satellites and new transmission media will allow a telephone link to be established to an individual in the remotest part of the earth via his own radio transceiver, identified by his unique world telephone number and accessed via satellite.

As a subject, telephony is usually restricted to considerations of the way in which subscribers are inter-connected, and an analysis of the traffic they generate. Traditionally, voice traffic, limited to a 3.4 kHz bandwidth, is assumed to be the nature of the signal sent over the telephone system, and we shall take that view in this chapter. However, the rapid growth of data traffic, generated by offices, banks, computers, airlines, etc., and the introduction of digital telephone systems in many countries, have raised the question of putting data over the telephone network and thus providing an integrated service. We will take a brief look at this area at the end of the chapter.

Telephony introduces several new concepts that appear strange at first, and tele-traffic theory (the study of telephone traffic and its relation to the design of switching and network systems) uses probabilistic models that need to be studied carefully in order to be understood, although the level of difficulty is well within the scope of an undergraduate statistics course, which most engineers follow.

We start with a section on traffic theory to establish a mathematical foundation. Systems were developed without the use of such theory, and indeed a mathematical analysis for many aspects of modern computer-controlled exchanges has not yet been developed, but if analysis can be applied it allows designers to make full use of their system, or an administration to predict how a manufacturer's equipment will perform.

10.1 TELEPHONE TRAFFIC

The use made of a telephone exchange, or a group of trunk circuits, is determined by both the rate at which calls arrive and the length of time for which they are held. The term 'traffic' takes these two quantities into account. Two units of traffic are employed: the CCS (hundred-call seconds) which is used principally in the USA; and the erlang which is used in Europe and other parts of the world.

If a circuit carries one call continuously for one hour it is said to carry one erlang (= 36 CCS) of traffic. The erlang, named after the Danish pioneer of tele-traffic theory, A.K. Erlang,[1] is expressed in mathematical terms as

$$A = \lambda s \qquad (10.1)$$

where A = traffic in erlangs, λ = mean call arrival rate (= calls per unit time) and s = mean call holding time measured in the same time units as λ.

10.2 INTER-ARRIVAL AND CALL HOLDING TIMES

Usually the quantities λ and s in Eqn (10.1) are mean values because in practice calls arrive at random and they last for a random length of time. The associated inter-arrival and holding time distributions could be determined by taking measurements over a long period, and building up a pattern for these parameters. However, apart from being a lengthy process, this approach has the disadvantage that the results would be accurate only for the time and place of the measurements; it would not necessarily follow that such distributions were applicable in general. To have some idea of the shape of the inter-arrival time and holding time distributions is, however, most important from the traffic theorist's point of view, and if they can be approximated by well-understood mathematical models, the development of a theoretical foundation to the subject can follow.

Over the years there have been many series of measurements on traffic from subscribers, and there is widespread agreement that the negative exponential distribution, with an appropriate choice of mean value, is satisfactory for both parameters. Figure 10.1 compares the negative exponential holding time curve with the results of a typical measurement. We notice that the ends of the distribution depart from reality, but, taking into account some of its attractive features, it is sufficiently accurate to stand as a valuable theoretical model.

The negative exponential distribution can be handled with comparative ease, but the property that is most appealing is that it is 'without memory', i.e. the probability that a call arrives in an interval dt is not dependent on the time

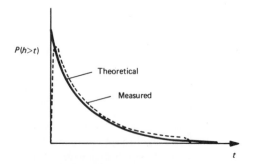

Fig. 10.1 Call holding time distribution.

that has elapsed since the last call or, for the holding time distribution, the probability that a call releases in a given interval d*t* is not related to the time for which the call has been in progress. Some thought will satisfy you that analysis would be significantly complicated if the history immediately before d*t* had to be taken into account.

The numerical values used for the mean inter-arrival time and the mean holding time depend on the application. The mean inter-arrival time is the inverse of the mean calling intensity, and therefore depends on the level of offered traffic. For the mean holding time, telephone administrations assume some reasonable value, typically 120 or 180 seconds. Calls arriving with a negative exponential inter-arrival time distribution are said to be Poissonian. A formal discussion of the Poissonian distribution is given in Section 13.3.

10.3   TRAFFIC VARIATION

The amount of telephone traffic passing through a particular exchange depends on several factors: the nature of the subscribers on the exchange (business, residential or mixed), the time of day, the month of the year, the occurrence of holidays and catastrophies, and the tariff in operation.

Figure 10.2 gives a typical distribution of traffic on a weekday in an exchange with a preponderance of business subscribers.

We see from the diagram that the night-time traffic is low, that there is a morning peak between 9.30 am and noon and that there is an afternoon peak between 2.00 pm and 4.30 pm. There is a slight increase in traffic between 6.00 pm and 10.00 pm owing to cheap-rate trunk calls from the residential subscribers. The heavy afternoon peak is typical of a system that offers the business subscriber a financial incentive for leaving trunk calls to the afternoon – by charging more for morning calls. The difference between the morning and afternoon peaks could be reduced by lowering the tariff differential. From the point of view of efficient use of the system, the distribution of traffic should be as uniform as possible throughout the day, and one of the functions of having a variable tariff scheme is to induce subscribers to use parts of the day that would not be their first choice.

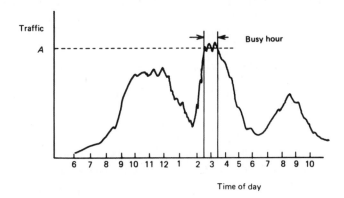

Fig. 10.2 Typical daily traffic distribution

10.4 BUSY HOUR

In the last two sections we have discussed the variation of traffic and its dependence on several factors. However, the designer must be able to use some traffic value against which he can design his exchange or determine the number of trunk circuits required. The traffic level is an average, taken over several days, and over the busiest period. The period is usually one hour, and the average traffic over that hour is called the 'busy hour' traffic. In Fig. 10.2 the busy hour is 2.30 pm to 3.30 pm, and the mean traffic over that period, A, would be used as the value of the busy hour traffic. In general, when traffic levels are quoted for routes or switching systems, it is implied that they are busy hour traffics. For example, if a circuit is said to carry 0.6 erlangs, it will be busy, on average, for 0.6 h (36 min) during the busy hour (not necessarily consecutive periods of time).

10.5 LOST-CALLS-CLEARED SYSTEMS

There are two major types of system used in telephony – lost-calls-cleared, and delay; although in a delay system a waiting subscriber may become impatient and leave, or the system may run out of holding positions. In a lost-calls-cleared system, if a call arrives to find there is no free path available to its destination it is cleared down. Once the system has tried to set up the call and has been unable to do so, there is no mechanism for putting the call in a queue to await a free path. Most of the theory developed in this chapter will concern lost-calls-cleared systems.

10.6 EQUATIONS OF STATE

Calls arrive at the network and depart from it at random, and the number of calls in progress, that is the number of devices busy, will vary in a random fashion. In the following we use N to be the maximum number of devices in the system, and i to indicate the number of calls in progress. Clearly, i is equal to the number of busy devices, and it is often called the state of the network. When $i = N$ the network is full and we use the term 'congested' to describe it; any further calls arriving when the network is in state N will be unable to find a free device.

In developing a mathematical model to allow us to analyse the properties of the system (the most important being the probability that the network is congested), we make the following assumptions:

(i) the calls are independent;
(ii) the rate of call arrivals when the network is in state i is λ_i;
(iii) the rate of call departure when the network is in state i is μ_i;
(iv) only one event (departure or arrival) can occur at a particular instant of time.

One common method of representing the states of a system, and the transitions between them, is by the state transition diagram of Fig. 10.3. There are

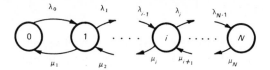

Fig. 10.3 State rate transition diagram.

$N + 1$ states, going in unit steps from no calls in progress (state 0) to a congested system (state N), i.e.

$$0 \leqslant i \leqslant N \qquad i = 0, 1, 2 \ldots$$

The parameters λ_i and μ_i are the rates of call arrival and departure, respectively, when the network is in state i. As we saw earlier, λ_i is related directly to the traffic offered to the system, and μ_i is determined by the nature of that traffic. Our objective is to develop an expression for the probability that the system is in state i, i.e. we want to find the state probabilities.

The probability that the system is in state i at some time $t + \mathrm{d}t$ is equal to the sum of the following:

(i) the probability that the system was in state i at time t and no call arrived or departed in time $\mathrm{d}t$;
(ii) the probability that the system was in state $i - 1$ at time t and a call arrived in time $\mathrm{d}t$;
(iii) the probability that the system was in state $i + 1$ at time t and a call released during time $\mathrm{d}t$.

As before, we are assuming here that no more than one transition can take place in the interval $\mathrm{d}t$.

In symbols we can write this as

$$[i]_{t + \mathrm{d}t} = [i]_t (1 - \lambda_i \mathrm{d}t - \mu_i \mathrm{d}t) + [i + 1]_t (\mu_{i + 1} \mathrm{d}t) + [i - 1]_t (\lambda_{i - 1} \mathrm{d}t) \quad (10.2)$$

where $[x] \equiv$ probability that the network is in state x.

Rearranging this expression

$$\frac{[i]_{t + \mathrm{d}t} - [i]_t}{\mathrm{d}t} = -(\lambda_i + \mu_i)[i]_t + \mu_{i + 1}[i + 1]_t + \lambda_{i - 1}[i - 1]_t \quad i = 1, 2, 3, \ldots$$

$$(10.3)$$

$i = 0$ is a special case, and

$$\mu_0 = 0 \text{ (no departure from state 0)}$$
$$\lambda_{-1} = 0 \text{ (state 1 does not exist)}$$

Then Eqn (10.3) becomes

$$\frac{[0]_{t + \mathrm{d}t} - [0]_t}{\mathrm{d}t} = -\lambda_0 [0]_t + \mu_1 [1]_t \quad (10.4)$$

As $\mathrm{d}t \rightarrow 0$ Eqns (10.4) and (10.3) can be written in differential form:

$$\frac{\mathrm{d}[0]_t}{\mathrm{d}t} = -\lambda_0 [0]_t + \mu_1 [1]_t \quad (10.5)$$

and from Eqn 10.3

$$\frac{d[i]_t}{dt} = -(\lambda_i + \mu_i)[i]_t + \mu_{i+1}[i+1]_t + \lambda_{i-1}[i-1]_t \qquad (10.6)$$

10.7 STATISTICAL EQUILIBRIUM

In section 10.4 we discussed busy hour traffic. If the mean traffic over that period is used for design purposes we can assume that there will be no long-term change in the state probabilities with time. Therefore we can assume that during the busy hour the time dependence of the state probabilities is zero, and $d[i]/dt = 0$.

Applying this condition to Eqns (10.5) and (10.6) gives

$$\lambda_0[0] = \mu_1[1] \qquad (10.7)$$

$$(\lambda_1 + \mu_1)[1] = \lambda_0[0] + \mu_2[2] \quad \text{etc.}$$

In general

$$(\lambda_i + \mu_1)[i] = \lambda_{i-1}[i-1] + \mu_{i+1}[i+1] \qquad (10.8)$$

Substituting successive equations gives

$$\lambda_1[1] = \mu_2[2]$$

$$\lambda_2[2] = \mu_3[3] \quad \text{etc.}$$

The general expression becomes

$$\lambda_{i-1}[i-1] = \mu_i[i] \qquad (10.9)$$

or, in words,

(probability of being in state $i-1$) × (rate of call arrivals in state $i-1$)
   = (probability of being in state i) × (rate of call departure in state i)

In the long run, then, we can say that the number of calls leaving the system is equal to the number of calls arriving. If this were not so, as time passed the state of the system would gradually creep towards 0 or N and stay there.

Statistical equilibrium is a concept that lies at the heart of the models we are going to develop in the next few pages.

10.8 STATE PROBABILITY

We have made considerable use of the notion of state probability $[i]$ in the last section, and it is worthwhile remembering, before developing the theory further, that this probability is equivalent to the proportion of the busy hour that the system is in state i.

In Fig. 10.4 the call arrivals and departures are plotted over the busy hour for a system having $N = 5$. This diagram shows the way in which the system changes state; note that only one transition can occur at any instant.

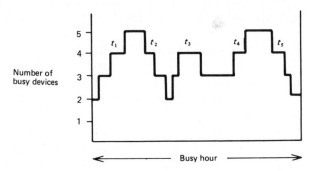

Fig. 10.4 Changes of state for a 5-device system.

Taking state 4 as an example:

$$[4] = \frac{t_1 + t_2 + t_3 + t_4 + t_5}{3600} = \frac{\sum \text{Times spent in state 4}}{\text{Busy hour}}$$

which is the fraction of the busy hour during which exactly four devices are busy.

We can also see from that diagram that

$$[0] + [1] + [2] + [3] + [4] + [5] = 1$$

In general, if there are N possible states

$$\sum_{i=0}^{N} [i] = 1 \tag{10.10}$$

Although this may appear to be rather obvious, it is a most important equation, as we shall see.

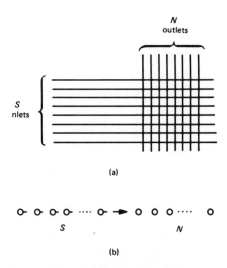

Fig. 10.5 Full availability system: (a) switch representation; (b) theoretical model.

10.9 FULL-AVAILABILITY MODELS

A full-availability system is one in which there are S sources of traffic and N outlets and any free source can seize any free outlet, regardless of the state of the system. The simplest example is a matrix switch with $S \times N$ cross-points as shown in Fig. 10.5(a). A straightforward representation, which we shall use, is shown in Fig. 10.5(b).

In order to evaluate $[i]$ for a full-availability group we need to have information about the offered traffic and how it depends on the state of the system; that in turn depends on the size of S in comparison with N.

10.10 ERLANG DISTRIBUTION: $S \gg N$

Constant traffic

A　　　　　N

Fig. 10.6 Erlang full-availability model.

If the number of sources is very much greater than the number of devices, it can be assumed that, no matter how many devices are busy, the rate of call arrivals will be constant. We can therefore ignore the number of sources, and our model becomes that shown in Fig. 10.6. A erlangs are offered to N devices.

Assuming the system is in statistical equilibrium, Eqn (10.9) holds. In this case the rate of call arrivals is constant, and, using Eqn (10.1),

$$\lambda_i = \lambda = \frac{A}{s} = \frac{\text{Mean offered traffic}}{\text{Mean holding time}}$$

$$\text{and the rate of call departures} = \frac{\text{Number of calls in progress}}{\text{Mean holding time}}$$

i.e.

$$\mu_i = \frac{i}{s}$$

Then from Eqn (10.9)

$$[i]\frac{i}{s} = [i-1]\frac{A}{s}$$

Putting in values for i:

$$[1] = A[0]$$

$$[2] = \frac{A^2}{2}[0]$$

$$\vdots$$

$$[i] = \frac{A^i}{i!}[0] \qquad\qquad (10.11)$$

$$\vdots$$

$$[N] = \frac{A^N}{N!}[0]$$

Using Eqn (10.10)

$$[0] + A[0] + \cdots + \frac{A^N}{N!}[0] = 1$$

Hence

$$[0] = \frac{1}{\displaystyle\sum_{v=0}^{N} \frac{A^v}{v!}}$$

and using this in Eqn (10.11), the probability of being in state i is

$$[i] = \frac{A^i}{i!} \bigg/ \sum_{v=0}^{N} \frac{A^v}{v!} \tag{10.12}$$

Thus we can find $[i]$ for all values of i if we know A and N.

    Equation (10.12) is called the probability distribution of the states of the system, and it is a truncated form of the Poisson distribution (Section 13.3).

10.11   TIME CONGESTION

Time congestion E is defined as the proportion of the busy hour for which the network is in state N, i.e.

$$\text{Time congestion } E = [N]$$

It follows directly from equation (10.12) that

$$E = [N] = \frac{A^N}{N!} \bigg/ \sum_{v=0}^{N} \frac{A^v}{v!} = E_N(A) \tag{10.13}$$

This is known as Erlang's loss formula. It has been tabulated extensively and it can be evaluated simply on a computer by using the iteration

$$E_i(A) = \frac{A E_{i-1}(A)}{i + A E_{i-1}(A)} \tag{10.14}$$

and the initial condition that $E_0(A) = 1$.

10.12   CALL CONGESTION

Another type of congestion is call congestion, B. It is the probability that a call arrives to find the system fully occupied, i.e.

$$B = \frac{\text{Number of calls arriving when the system is in state } N}{\text{Total number of calls arriving in the busy hour}}$$

In the Erlang case,

$$B = \lambda \cdot [N] \bigg/ \sum_{i=0}^{N} \lambda \cdot [i]$$

and since λ is constant

$$B = [N] \bigg/ \sum_{i=0}^{N} [i]$$

Again applying Eqn (10.10), we have

$$B = [N]$$

Thus for the Erlang case

call congestion = time congestion

This model is used extensively in practice. It is applied whenever the assumption can be made that the calling rate is independent of the number of calls in progress.

10.13   BERNOULLI DISTRIBUTION: $S \leqslant N$

If the number of sources is less than the number of devices, a different model has to be used becase we cannot assume that the traffic intensity is indepenent of the number of calls in progress. The model must now take account of the number of sources (Fig. 10.7). The simplest way of taking account of the state of the system is to relate the traffic offered to the traffic offered per free source and the number of free sources available, i.e.

$$\lambda_i = (S - i)\alpha/s \qquad (10.15)$$

where α is the traffic per free source, $S - i$ is the number of free sources and, as before, the mean holding time is s.

Applying the statistical equilibrium Eqn (10.9), once more

$$\lambda_{i-1}[i-1] = \mu_i[i]$$

$\mu_i = i/s$ as in the Erlang case, and if this is used, along with λ_{i-1} from Eqn (10.15), we have

$$(S - i + 1)[i - 1]\frac{\alpha}{s} = \frac{i}{s}[i]$$

or

$$[i] = \frac{S - i + 1}{i}\alpha[i - 1]$$

giving

$$[1] = S \cdot \alpha \cdot [0]$$

$$[2] = \frac{S-1}{2} \cdot \alpha \cdot [1] = \frac{S(S-1)\alpha^2}{2}[0] \quad \text{etc.}$$

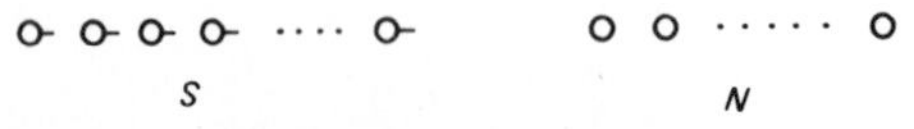

Fig. 10.7 Limited source full-availability model.

and generally

$$[i] = \frac{S!}{(S-i)!i!} \alpha^i[0] = \binom{S}{i}\alpha^i[0]$$

Summing over all i and applying Eqn (10.10),

$$\sum_{i=0}^{S} [i] = 1 = \left[1 + S\alpha + \frac{S(S-1)\alpha^2}{2} + \frac{S(S-1)(S-2)\alpha^3}{3!} + \cdots + \alpha^S\right][0]$$

(Note that the summation is from $0 \leqslant i \leqslant S$ since $S \leqslant N$.)

$$[0] = \frac{1}{1 + S\alpha + S(S-1)\frac{\alpha^2}{2!} + \cdots + \alpha^S} = \frac{1}{(1+\alpha)^S}$$

and the general term can now be written

$$[i] = \binom{S}{i}\alpha^i\frac{1}{(1+\alpha)^S}$$

$$= \binom{S}{i}\left(\frac{\alpha}{1+\alpha}\right)^i\left(1 - \frac{\alpha}{1+\alpha}\right)^{S-i}$$

or

$$[i] = \binom{S}{i}a^i(1-a)^{S-i} \qquad (10.16)$$

where

$$a = \frac{\alpha}{1+\alpha}$$

If $S < N$ there will be no time congestion because the state N is never reached; there will always be at least $N - S$ free devices.

If $S = N$ then the time congestion is

$$[N] = a^S = a^N \qquad (10.17)$$

But the call congestion will be zero because there are no calls arriving when the network is in state N, all the sources being busy.

The quantity a in Eqn (10.16) is the carried traffic per device. Note that Eqn (10.16) is the classical form of the Bernoulli distribution.

10.14   ENGSET DISTRIBUTION: $S > N$

This case models the situation in which the number of sources is greater than the number of devices, but not so large that the traffic offered is constant. As in the Bernoulli case we shall assume here that the traffic offered is related directly to the number of free sources, allowing us to start with the same state equation, viz.

$$(S-i+1)[i-1]\frac{\alpha}{S} = \frac{i}{S}[i] \qquad (10.18)$$

from which

$$[1] = S\alpha[0]$$

$$[2] = \frac{S(S-1)\alpha^2}{2}[0] \quad \text{etc.}$$

and generally

$$[i] = \binom{S}{i}\alpha^i[0] \tag{10.19}$$

We can now sum $[i]$ over all states $0 \leqslant i \leqslant N$ and apply Eqn (10.10), i.e.

$$\sum_{i=0}^{N} \binom{S}{i}\alpha^i[0] = 1$$

or, changing the variable to avoid confusion later,

$$[0] = 1 \bigg/ \sum_{v=0}^{N} \binom{S}{v}\alpha^v$$

Then, inserting this expression in Eqn (10.19) gives

$$[i] = \binom{S}{i}\alpha^i \bigg/ \sum_{v=0}^{N} \binom{S}{v}\alpha^v \tag{10.20}$$

10.14.1 Time congestion

From Eqn (10.20)

$$[N] = \binom{S}{N}\alpha^N \bigg/ \sum_{v=0}^{N} \binom{S}{v}\alpha^v \tag{10.21}$$

which is denoted by $E(N, S, \alpha)$. Values of E for given N, S and α can either be calculated from Eqn (10.21) or be obtained from tabulated results.

10.14.2 Call congestion

From the previous defintion

$$B = (S - N)\frac{\alpha}{S}[N] \bigg/ \sum_{i=0}^{N} [i](S - i)\frac{\alpha}{S}$$

Substituting from Eqn (10.21) for $[N]$ and rearranging gives

$$B = \binom{S-1}{N}\alpha^N \bigg/ \sum_{v=0}^{N} \binom{S-1}{v}\alpha^v$$

$$= E(N, S - 1, \alpha) \tag{10.22}$$

10.15 PROBABILITY OF OCCUPANCY OF PARTICULAR DEVICES

We have just developed the three most important full-availability models, and in each case the general term has been $[i]$, the probability that any i devices are busy.

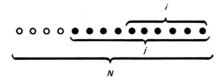

Fig. 10.8 Model to find $H(j)$.

There are some applications where a distinction must be made between the two probabilities of i busy devices:

(i) that any i devices are busy, denoted by $G(i)$;
(ii) that a particular i devices are busy, denoted by $H(i)$.

The first is what we have calculated so far. To find the second, $H(i)$, we consider the situation shown in Fig. 10.8.

The problem is, given that j out of N devices are busy, what is the probability, $H(i)$, that a particular i are busy?

We can argue as follows:

Given that j devices in total are busy, there are $j - i$ busy devices not included in our i, and these $j - i$ can be distributed over the $N - i$ positions available for them in $\binom{N-i}{j-i}$ ways. Then

$$\frac{\text{Number of ways the particular } i \text{ are busy}}{\text{Total number of ways of distributing } j \text{ over } N} = \frac{\binom{N-i}{j-i}}{\binom{N}{j}}$$

Hence

$$H(i) = \sum_{j=i}^{N} \left[\binom{N-i}{j-i} \Big/ \binom{N}{j} \right] P(j) \tag{10.23}$$

noting that the summation can only be over the range for which $j \geqslant i$. The value of $H(i)$ depends of course on the quantity $P(j)$ which, in the terms introduced at the beginning of this section, is equivalent to $G(j)$. Applying the models of Section 10.2, there are three cases.

10.15.1 Erlang

$$H(i) = \sum_{j=i}^{N} \left[\binom{N-i}{j-i} \Big/ \binom{N}{j} \right] P(j)$$

$$= \sum_{j=i}^{N} \left[\binom{N-i}{j-i} \Big/ \binom{N}{j} \right] \cdot \left[\frac{A^j}{j!} \Big/ \sum_{v=0}^{N} \frac{A^v}{v!} \right] \tag{10.24}$$

from Eqn (10.13)

$$= \sum_{j=i}^{N} \frac{(N-i)!}{(j-i)!(N-j)!} \cdot \frac{j!(N-j)!}{N!} \cdot \left[\frac{A^j}{j!} \Big/ \sum_{v=0}^{N} \frac{A^v}{v!} \right]$$

$$= \left[A^N \Big/ N! \sum_{v=0}^{N} \frac{A^v}{v!} \right] \cdot \frac{(N-i)!}{A^N} \sum_{j=i}^{N} \frac{A^j}{(j-i)!}$$

$$= \left[\frac{A^N}{N!} \Big/ \sum_{v=0}^{N} \frac{A^v}{v!} \right] \cdot \frac{(N-i)!}{A^N} \left[\frac{A^i}{1} + \frac{A^{i+1}}{1} + \cdots + \frac{A^N}{(N-i)!} \right]$$

$$= E_N(A) \cdot \frac{A^i(N-i)!}{A^N} \left[1 + A + \cdots + \frac{A^{N-i}}{(N-i)!} \right]$$

$$= \frac{E_N(A)}{E_{N-i}(A)} \tag{10.25}$$

Showing that $H(i)$ in the Erlang case can be found from two Erlang functions.

10.15.2 Bernoulli

In this case

$$P(j) = \binom{N}{j} a^j (1-a)^{N-j}$$

and

$$H(i) = \sum_{j=i}^{S} \left[\binom{N-i}{j-i} \Big/ \binom{N}{j} \right] \binom{N}{j} a^j (1-a)^{N-j}$$

Expanding and rearranging terms, this reduces to

$$H(i) = \frac{\binom{S}{i}}{\binom{N}{i}} a^i \quad \text{for} \quad S < N \tag{10.26}$$

In the special case when $S = N$, which is common in practice,

$$H(i) = a^i \tag{10.27}$$

10.15.3 Engset

From the earlier analysis

$$P(j) = \binom{S}{j} \alpha^j \Big/ \sum_{v=0}^{N} \binom{S}{v} \alpha^v$$

so

$$H(i) = \sum_{j=i}^{N} \left[\binom{N-i}{j-i} \Big/ \binom{N}{j} \right] \cdot \left[\binom{S}{j} \alpha^j \Big/ \sum_{v=0}^{N} \binom{S}{v} \alpha^v \right] \tag{10.28}$$

which becomes

$$H(i) = \frac{E(N, S, \alpha)}{E(N-i, S-i, \alpha)} \tag{10.29}$$

Summarizing the results obtained for the full availability models considered above, we can tabulate the formulae we have derived as shown in Table 10.1.

Table 10.1

	$G(i)$	$H(i)$
Erlang	$E_i(A) = (A^i/i!)\Big/ \sum_{v=0}^{N} (A^v/v!)$	$\dfrac{E_N(A)}{E_{N-i}(A)}$
Bernoulli	$\dbinom{N}{i} a^i (1-a)^{N-i}$	$\dfrac{\dbinom{S}{i}}{\dbinom{N}{i}} a^i \quad (a^i \text{ for } S=N)$
Engset	$E(N,S,\alpha) = \dbinom{S}{i} a^i \Big/ \sum_{v=0}^{N} \dbinom{S}{v} \alpha^v$	$\dfrac{E(N,S,\alpha)}{E(N-i,S-i,\alpha)}$

10.16 LINK SYSTEMS

A link system is a multi-stage switching network that consists of a set of first-stage switches, connected to a set of second-stage switches, which in turn are connected to a set of third-stage switches, and so on. The number of stages of switching used depends on the purpose of the system. We will consider first a two-stage system as shown in Fig. 10.9(a). There we see that there are four inlets and five outlets on each of the six first-stage switches, and six inlets and six outlets on each of the five second-stage switches.

The connections, or links between the two switching stages are shown in Fig. 10.9(b). We see that the diagram is very messy, and even if a link is drawn only when there is a call in progress along it, the lines indicating a particular call will be difficult to trace when the number of calls is high. A different type of representation suits our purposes better, i.e. one that was first introduced by workers at the Swedish company L.M. Ericsson, and is called a chicken diagram. The chicken diagram representation of the network in Fig. 10.9(a) is shown in Fig. 10.9(c). Now the link connections are not shown, but are implied by the diagram. A column on the A matrix represents the inlets of a first-stage switch; the corresponding column on the B matrix represents the outlets of the first-stage switch. A row on the B matrix corresponds to the inlets on a second-stage switch, and a row on the C matrix corresponds to the outlets on the same second-stage switch. Calls can be traced through the network by placing a number on the appropriate place in each matrix. An example of the network representation, and a chicken diagram representation corresponding to it are shown in Fig. 10.10. In addition, several calls have been placed in the system, and the method of representing them is shown clearly in each diagram. The reader should study these diagrams carefully until the relationship between them becomes clear; the network is discussed further in the next section.

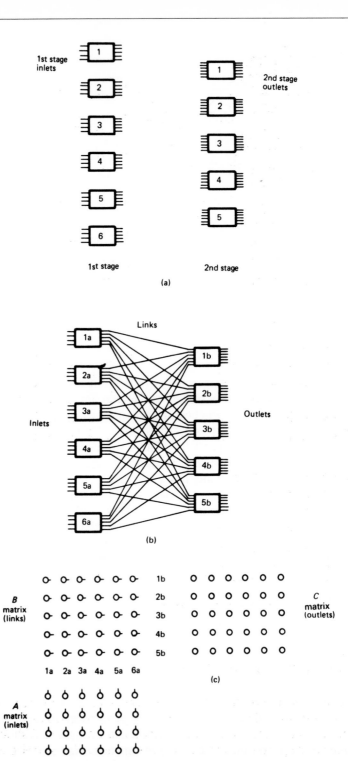

Fig. 10.9 Two-stage links system; (a) switches; (b) complete with links; (c) chicken diagram equivalent.

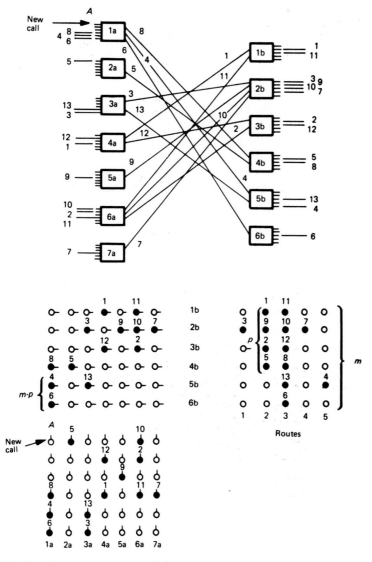

Fig. 10.10 Blocking in a two-stage link system.

10.17 PROBABILITY OF BLOCKING

In order to determine the equipment requirement of the system or, putting that in another way, to find out if the system will perform as it should, some measure of performance is required. That most often used is the probability of blocking, which is the proportion of calls, in the long term, that are rejected. If we consider the general case of a two-stage link system in which the corresponding outlets on the second-stage switches form a route (which is a group of circuits going to the same destination), as indicated by the chicken diagram in Fig. 10.10(b), we can see that there are two ways in which a call may be blocked:

(i) there may be no free route circuit;
(ii) there may be no path available between the inlet carrying the incoming call and the free route circuits.

The diagram shows both of these conditions; route 3 has no free circuits available, whereas if inlet A on first-stage switch 1 is wanting to be connected to route 2, on which there are free route circuits, it cannot be, because there is no available link.

If we now look at this particular connection between inlet A and route 2 further, we can develop from it an analytical approach to the probability of blocking. The part of the network involved in the connection from A to route 2 is shown in Fig. 10.11.

There are m links to a route from any first-stage switch, since any route has one circuit on each of the second-stage switches. Thus a route consists of m outlet circuits, one on each second-stage switch.

Let us assume that there are p busy circuits on the route, then a call will not be able to be connected if, at the same time, the particular $m - p$ links, having access to the free $m - p$ route circuits, are busy, as they are in the diagram.

The first probability, that there are any p route circuits busy, we denote by $G(p)$, and the second probability, that $m - p$ particular devices are busy is denoted by $H(m - p)$. The probability of blocking, that is the probability that a call cannot be connected, is the joint probability of these two events, summed over all possible values of the variable p. Denoting the probability of blocking by E, we have then that

$$E = \sum_{p=0}^{m} G(p)\, H(m - p) \tag{10.30}$$

This apparently simple expression, known as the **Jacobaeus equation**, after its development by Christian Jacobaeus of L.M. Ericsson[2] in 1950, has been the cornerstone of blocking calculations on link systems for many years. There are some assumptions implicit in its form which become important, and indeed limiting, under certain circumstances: the most significant is that the

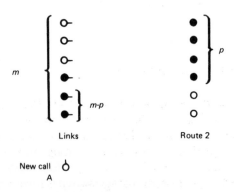

Fig. 10.11 Paths from a particular inlet to a route.

two probabilities $H(m-p)$ and $G(p)$ are independent. Provided that the traffic level in the network is low, this independence assumption can be made without concern, but at higher traffic levels the analysis must be modified to take the conditional dependence of $G(p)$ on $H(m-p)$ into account.

The evaluation of Eqn (10.30) depends on the probability distributions appropriate to $G(p)$ and $H(m-p)$. In most cases one of those summarized in Table 10.1 is chosen. Often $G(p)$ will be Erlang because the number of route circuits (m) will be much smaller than the number of potential traffic sources (all the inlets on the first stage). The distribution used for $H(m-p)$ will depend entirely on the relationship between the inlets and outlets of the first-stage switches. An example will demonstrate how the Jacobaeus equation can be applied to a two-stage system.

EXAMPLE: In a two-stage link system the outlet circuits are combined into routes – each route has one circuit from each outlet switch. Using a chicken diagram representation, find the internal link congestion if the total traffic offered to the network is 42 erlangs, and if the outlets are formed into routes, as explained in the first sentence. Assume that there is equal demand for each route.

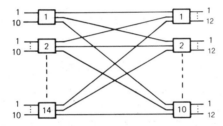

The chicken diagram representation is

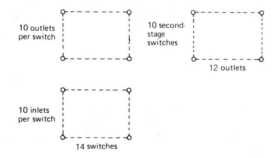

There are 12 routes, and 42 erlangs of total offered traffic; therefore

each route is offered $42/12 = 3.5$ erlangs

The Jacobaeus equation gives the probability of blocking as

$$\sum_{p=0}^{m} G(p)\,H(m-p)$$

where $G(p)$ = probability the p devices are busy on the route and $H(m-p)$ = probability that $m-p$ particular links are busy. Since there are 140 inlets to the network, and each route consists of 10 devices, we can assume an Erlang distribution on the route, i.e.

$$G(p) = \frac{A^p}{p!} \Big/ \sum_{v=0}^{m} \frac{A^v}{v!}$$

There is the same number of inlets as outlets on a first-stage switch, and therefore the Bernoulli distribution is appropriate for $H(m-p)$, giving, from Table 10.1,

$$H(m-p) = a^{m-p}$$

where a = traffic carried per link.

There are 140 links in total; therefore, assuming the congestion to be small,

$$a = 42/140 = 0.3 \text{ erlangs}$$

Hence the total blocking, E, is

$$E = \sum_{p=0}^{10} \frac{(3.5)^p/p!}{\sum\limits_{v=0}^{10} 3.5^v/v!} (0.3)^{10-p}$$

Evaluating this expression for each value of p, and summing, gives the total

$$E = 0.0082$$

The internal link congestion = total congestion − route congestion.

In any route there are 10 circuits, and we calculated earlier that the traffic offered per route is 3.5 erlangs. Therefore the route congestion = $E_{10}(3.5)$ = 0.0023. Hence internal congestion = $0.0082 - 0.0023 = 0.0059$.

Although we shall not do more here, the Jacobeaus approach can be used for a wide variety of link system structures, but they do not involve any change in the principles we have developed.

10.18   DELAY SYSTEM

In the foregoing sections we dealt exclusively with lost-calls-cleared systems in which a call is lost if it arrives when all the devices are occupied. In such systems there is no mechanism for allowing calls to wait until a free device becomes available.

If queueing is allowed, however, the call can be delayed, or held within the system until it has access to a free device. In this section we shall briefly examine the simplest delay model. It has the following attributes:

(i) calls arrive at random with constant mean traffic and negative-exponential distribution of inter-arrival times (Poisson offered traffic);
(ii) the holding times follow a negative-exponential distribution;
(iii) there are n devices (servers);
(iv) the queue can accept an infinite number of waiting calls;

(v) the service discipline is first in–first out, i.e. calls are passed to devices in order of arrival.

Queues satisfying (i), (ii) and (iii) are usually classified as $M/M/n$ in books on queueing theory, after the nomenclature of Kendall[3] who proposed a systematic classification of queues.

The queue we will now study is idealized through the assumption (iv), which could not be achieved in practice, but there are many real situations which are sufficiently close to it to make the analysis applicable.

If, as before, the mean calling rate and the mean departure rate at any instant are given by λ_i and μ_i, where i is the number of calls is the system, we can draw a transition diagram as shown in Fig. 10.12.

If we compare this diagram with that of Fig. 10.3 we can see two differences. In the delay system there is an infinite number of states, and the departure coefficient μ_i is dependent on the number of calls in progress when $i \leqslant n$, and thereafter it is equal to μ_n for all i.

We can use the same approach as we did in the analysis of the Erlang loss formula, since here too we have assumed that the mean traffic offered is constant.

Starting with the relevant birth and death equation, assuming that we have statistical equilibrium (which in a queueing system requires that the mean traffic offered per device is less than unity, i.e. $A/n < 1$), we have

$$\mu_i[i] = \frac{A}{s}[i-1]$$

and

$$\mu_i = \frac{i}{s}$$

Therefore

$$[i] = \frac{A}{i}[i-1]$$

Putting in values

$$[1] = A[0]$$

$$[2] = \frac{A}{2}[1] = \frac{A^2}{2}[0]$$

$$[3] = \frac{A}{3}[2] = \frac{A^3}{3!}[0]$$

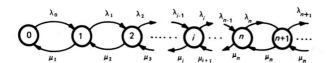

Fig. 10.12 State transition diagram for a delay system.

etc. to

$$[n-1] = \frac{A^{n-1}}{(n-1)!}[0]$$

$$[n] = \frac{A^n}{n!}[0] = \frac{A}{n}[n-1]$$

For $i > n$ the factor A/n will remain constant, so we have

$$[n+1] = \frac{A}{n}[n]$$

$$[n+2] = \left(\frac{A}{n}\right)^2 [n]$$

and in general

$$[i] = \left(\frac{A}{n}\right)^{i-n} [n] \qquad (10.31)$$

Again we use the normalizing condition that the sum of all state probabilities is unity:

$$\sum_{i=0}^{\infty} [i] = 1 = [0]\left\{ \sum_{i=0}^{n-1} \frac{A^i}{i!} + \sum_{k=0}^{\infty} \left(\frac{A}{n}\right)^k \frac{A^n}{n!} \right\}$$

where the variable k is used instead of i in the range $i \geqslant n$.

$\sum_{k=0}^{\infty}(A/n)^k$ is the infinite sum of a geometric series of ratio A/n, which has the value $1/[1-(A/n)]$.

Therefore

$$1 = [0]\left\{ \sum_{i=0}^{n-1} \frac{A^i}{i!} + \frac{A^n}{n!} \cdot \frac{1}{1-(A/n)} \right\}$$

or

$$[0] = 1 \left/ \left[\sum_{i=0}^{n-1} \frac{A^i}{i!} + \frac{A^n}{n!} \frac{n}{n-A} \right] \right. \qquad (10.32)$$

The probability that a call is delayed is the same as the probability that all devices are busy, which is the sum of the state probabilities $[i]$ in the range $i \geqslant n$, i.e.

$$\text{Prob (call is delayed)} = \sum_{i=n}^{\infty} [i]$$

$$= \sum_{i=n}^{\infty} \left(\frac{A}{n}\right)^{i-n} \frac{A^n}{n!}[0]$$

$$= \frac{A^n}{n!} \frac{n}{n-A}[0]$$

$$= \frac{(A^n/n!)\, n/(n-A)}{\sum_{i=0}^{n-1} A^i/j! + (A^n/n!)\cdot n/(n-A)}$$

This expression, symbolized by $E_{2,n}(A)$, is known as the Erlang delay formula. For purposes of computation it can be expressed in terms of the Erlang loss formula as

$$E_{2,n}(A) = \frac{nE_n(A)}{n - A + AE_n(A)} \tag{10.33}$$

where $E_n(A)$ is given by Eqn. (10.13).

In queueing systems, several parameters other than the probability that a call is delayed may be of interest, e.g. the mean queue length, the mean delay of all calls, the mean delay of delayed calls, maximum delay, and so on. Further treatment of queues related to telephony can be found in Cooper[3] and Fry.[4]

10.18.1 M/M/1 queue

Modern systems depend on software and processes for their control and management. In many of these, and in packet switched systems, buffers are used to queue tasks awaiting processing. All buffered systems can be analysed by using queueing theory, which is a set of appropriate mathematical models from which such metrics as delay time can be determined. The following analysis, expressed in terms of telephone calls to make it consistent with our discussion, is applicable to any analogous queueing system.

Earlier we determined the probability that a call is delayed. Here we will use a simple model to find the mean delay of a call in a queue with an infinite buffer space. With no end to the size of the buffer, it is not possible for calls to be lost due to blocking; although they may disappear through impatience.

In addition to having an infinite length queue we also reduce the number of servers to one, hence we are analysing the M/M/1 queue.

We noted earlier that the probability that the system is in state i is given by

$$\mu_i[i] = \frac{A}{s}[i-1] \tag{10.34}$$

Since there is only one server, μ is fixed and independent of i. In addition, we can use Eqn (10.1) to express A in terms of λ. Thus,

$$\mu_i = \mu \quad \text{and} \quad A = \lambda.s$$

therefore

$$\mu[i] = \frac{\lambda.s}{s}[i-1] \tag{10.35}$$

Putting in values

$$\left.\begin{aligned}[1] &= \lambda/\mu\,[0] \\ [2] &= (\lambda/\mu)^2\,[0]\end{aligned}\right\} \tag{10.36}$$

and so on to

$$[i] = (\lambda/\mu)^i\,[0] \tag{10.37}$$

We assume here that the queue can grow indefinitely, since $0 < i < \infty$. The normalization equation therefore becomes:

$$\sum_{i=0}^{\infty} [i] = 1$$

i.e.

$$\sum_{i=0}^{\infty} (\lambda/\mu)^i [0] = 1$$

giving

$$[0] = \frac{1}{\sum_{i=0}^{\infty} (\lambda/\mu)^i}$$

hence

$$[i] = \frac{(\lambda/\mu)^i}{\sum_{i=0}^{\infty} (\lambda/\mu)^i} \tag{10.38}$$

We know that λ must be less than μ to avoid the queue growing without limit. The quantity λ/μ is known as the utilization ρ. Using this form,

$$[i] = \frac{\rho^i}{\sum_{i=0}^{\infty} \rho^i}$$

The denominator is the infinite series representation of $(1 - \rho)^{-1}$; hence

$$[i] = \rho^i (1 - \rho)$$

where $[i]$ is the probability that there are i customers in the system at any time.

The mean number of customers in the system, N, is given by the weighted sum of the state probabilities:

$$N = \sum_{i=0}^{\infty} i[i]$$

$$= \sum_{i=0}^{\infty} i.\rho^i (1 - \rho)$$

$$= \frac{\rho(1 - \rho)}{(1 - \rho)^2}$$

$$= \frac{\rho}{1 - \rho}$$

or

$$N = \frac{\lambda}{\mu - \lambda} \tag{10.39}$$

The mean time a customer spends in the system is the average time for delayed and undelayed customers combined, and is the sum of the queueing and service times. This sojourn time we will denote by T. It can be related to N

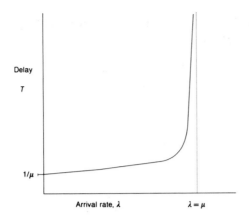

Fig. 10.13 Delay in an M/M/1 queue.

and λ by Little's Law:

$$\lambda T = N \qquad (10.40)$$

Therefore

$$T = N/\lambda = (\mu - \lambda)^{-1}$$

Using this expression, we can see from Fig. 10.13 that if the mean service time μ is fixed, then T is a non-linear function of λ. We can also see that the minimum mean sojourn time is the mean service time $1/\mu$ and that as $\lambda \to \mu$ the delay becomes very large indeed. This last condition is sometimes called the point of instability of the queue.

Although the M/M/1 is the simplest of all queues to analyse, and although the analysis depends on the assumption of infinite buffer space, it gives a very good insight into the behaviour of queues in general.

10.19 SIMULATION

Modern systems involve a considerable amount of control by microprocessors dedicated to a particular small task and operating on an individual bit or word basis. The sequence of tasks involved in passing a call through the system can be represented as a series of queues, and at each queue there is the possibility of delay. If the traffic is very high, the total delay may become so great that the processors cannot cope and the throughput falls dramatically. Tele-traffic theory has to provide the tools to predict this degradation in performance. However, sufficient analytical power is not available in queueing theory. It provides the facility to study fairly simple problems, but when there are several queues interacting the models can become too difficult to analyse. Resort must then be made to simulation. A computer program is written to represent the behaviour of the system to be studied and it is used in conjunction with some form of simulation package which generates calls in a prescribed way (e.g. with negative exponential inter-arrival and call holding time distributions), provides random number streams and organizes the simulation of the system operation. Simulation is a most powerful tool and it can be used for any system. It relies on:

(i) sufficient computer power;
(ii) a satisfactory software model of the system being studied;
(iii) the correct choice of the behaviour of the traffic offered;
(iv) a proper evaluation of which results to take;
(v) sufficient length and frequency of runs to account for the effects of correlation.

The last condition is one that must be in mind when a simulation is being organized; correlation can produce errors in results if it is not taken into account. In particular, the urge to reduce cost by making runs short should not inhibit the need for long runs to make the results reliable.

Simulation also has disadvantages. It is costly because it requires long runs and therefore considerable computer time, it is time consuming for the analyst because the programs to model the system being studied are often very complex, and the results are not of general applicability since they are related to a particular set of initial conditions such as system dimensions and traffic level.

Until recently, simulation programs were written on the assumption that the traffic offered to the network would be telephony, the representation of which is well understood, as describe earlier. Thus the creation of traffic models was straightforward. Now the interest is in developing systems that can handle many types of traffic, some of which will not be well understood at all. In addition, it used to be assumed that the traffic would be more or less balanced across the network. If the network was a switch then the traffic on each inlet would be the same, and if the network was a representation of the interconnection of exchanges, then the offered traffic at the nodes was assumed to be uniform. This again can no longer be justified. There are two questions here. First, what is an appropriate distribution for each of the traffic types that might be carried, and second, how can simulations be compared when it may not be clear what variation in traffic load is assumed across the network?

There is a general move towards using proprietary software rather than custom-built simulation packages. The advantage is that the proprietary packages are usually well written and well supported, giving a high degree of reliability. The mechanisms for obtaining results are also well developed and simple to use. The disadvantages are that in general such packages are slow, and to some degree inflexible; they do not give the user complete control. Often, experienced users will combine some element of their own programming into the system.

One further difficulty with simulation as a methodology is that it is easy to simulate the model rather than the system. This is of particular concern when using simulation to validate analytical results. If the simulation merely parallels the analysis, creating a representation of the same model, it will not be surprising if the results agree very closely; indeed it would be a cause for some concern if they did not.

10.20   SWITCHING AND SIGNALLING

When a subscriber makes a telephone call a series of events takes place which have the following sequence, illustrated in Fig. 10.14.

1. Calling subscriber lifts handset.
2. Exchange detects demand for call.
3. Check is made to find free equipment in the exchange.
4. When equipment becomes free, dial tone is returned to the caller. If

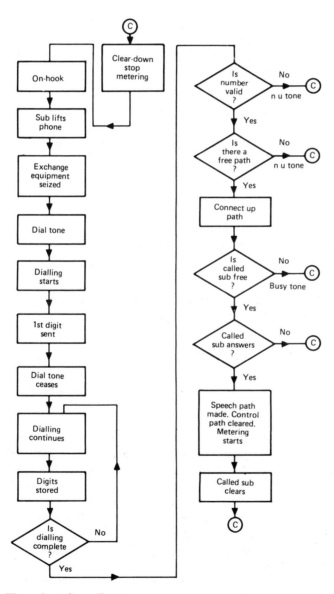

Fig. 10.14 Flow-chart for call set-up.

equipment is not immediately available at the exchange there may be a long dial-tone delay.

5. Subscriber dials digits.
6. Exchange interprets the digits and routes call to its destination.
7. Check is made to determine whether or not the called subscriber is free.
8. If called subscriber is busy, busy tone is returned to caller.
9. If called subscriber is free, ringing tone is returned to caller and ring current is sent forward to ring called subscriber's bell.
10. If called subscriber is unavailable, caller hangs up and equipment clears down.
11. If called subscriber answers, ring tone and ring current cease.
12. Speech path is set up and metering commences.
13. When conversation ceases called subscriber hangs and metering ceases.
14. Calling subscriber hangs up and system clears down.

Note that if an unallowed number is dialled, or service to the number dialled has been discontinued, number-unobtainable tone is returned to the caller.

From the above description we can see that the process of signalling has at least three functions:

(i) To indicate the state of the call to each subscriber by

   (a) dial tone,
   (b) ring tone and ring current,
   (c) busy tone or number-unobtainable tone.

(ii) To tell system what to do next, by indicating the path for the call, for example.

(iii) To initiate a billing procedure – usually by tripping the calling subscriber's meter at the correct charging rate – to enable the administration to gather the revenue needed to provide the service.

In the early days of telephony when calls were handled by an operator, these signalling functions were done manually. Indeed in small communities having up to 200 lines, or so, the operator provided an automatic call-transfer facility of some sophistication: the more interest the operator had in local affairs the better that part of the service could be!

The flow-chart of Fig. 10.14 applies particularly to the setting up of calls that share the same exchange. Although most calls are still of that type, in many large cities an increasing number of calls involve inter-exchange working, and the use of direct-dialling systems has created a greater demand for trunk calls. Inter-exchange signalling is necessary to allow the calling subscriber's local exchange to keep control of the call, and later we shall look in some detail at the signalling systems used for the purpose. Evidently, inter-exchange systems are not restricted in speed as are the subscriber's signalling systems by the dialling process, and they are not required to produce the tones which the local exchanges must produce to keep the subscriber informed of the progress of the call.

10.21   TYPES OF SIGNALLING SYSTEM

We shall not look at any of the historical systems, but we shall look at those which are in use. They fall into four main categories:

(i)   loop disconnect – dc signalling;
(ii) multi-frequency – ac signalling;
(iii) voice frequency – ac signalling;
(iv) common channel signalling.

10.21.1   Loop disconnect signalling

Until recently, the universal means at the subscriber's disposal for indicating the number he wished to call was the telephone dial, and it is still present on many handsets. It was developed to its present state many years ago, but although robust, it has a major disadvantage: it operates very slowly by the standards of modern electronics, and its slow motion places a definite limit on the speed at which signals can be sent to the exchange.

When the dial is rotated the finger hole corresponding to the required digit is pulled round to the finger guard and then the dial is released. A governor inside the dial causes it to rotate back automatically at a fixed speed, causing a series of pulses to be sent down the subscriber's line. After the dial has returned to its rest position the next digit is dialled. The time between the last of the pulses for one digit, and the first pulse of the next is called the inter-digit pause. It is this pause that allows the exchange to recognize the end of a digit. A typical sequence, with average operating times (nominally the pulses are sent at a rate of 10 per second) is shown in Fig. 10.15. Although the dials are designed to operate at a nominal 10 pulses per second, the exchange equipment will still function reliably if the pulse speed is between 7 and 12 pulses per second and the break is between 63 and 72% of the pulse period.

10.21.2   Multi-frequency (mf) signalling

Modern handsets are fitted with key-pads instead of dials to facilitate a much more rapid transfer of signals between handset and exchange. Generally the keypads send out frequencies instead of pulses to represent a digit. Most systems use two frequencies to represent a particular digit. Whether or not the much increased speed of mf signalling can be realized depends on the local exchange. An mf key-phone working into a Strowger-type exchange gives no overall improvement, it just changes the position in the process at which the

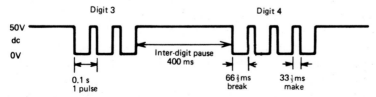

Fig. 10.15 Ideal output from telephone dial.

caller must wait: when a dial is used there is a long inter-digit pause and short post-dial delay, whereas in an mf system there is little inter-digit pause because the numbers can be keyed in very quickly, but if the exchange equipment operates at 10 pulses/s the mf signal must be converted into normal digits producing a lengthy post-dial delay.

However, in modern exchanges the equipment will respond directly to the mf signals and the call can be set up very quickly.

10.21.3 Voice-frequency (vf) signalling

The use of dc signalling is limited by several factors. In particular, over long lines the variation of equipment performance, and the effect of the characteristics of the lines used to transmit between exchanges, degrade the pulse shape of the dialled digits to such an extent that errors can be caused. In addition, the receiving equipment requires a higher voltage than would be necessary if the system used alternating current signals. Apart from these considerations multi-channel working makes dc signalling inappropriate.

The normal telephone channel occupies a bandwidth of 300–3400 Hz, from the range of 0–4000 Hz which it is allocated. If ac signalling is to be used it must operate at frequencies within this range, and it is therefore known as voice-frequency signalling. The frequencies used can be either inside the normal speech band (in-band signalling) or outside that band and within the 0–4000 Hz range (out-of-band signalling). As the frequencies used for vf signalling are the same as those used in speech, special care must be taken to ensure that the two functions do not interfere. By understanding something of the characteristics of speech it is possible to specify signalling systems that will not be operated erroneously by speech frequencies. Since signalling is done by tones within the baseband of the telephone channel it is possible to use vf signalling when the channels are multiplexed on to a common carrier – either line or radio.

10.21.4 Common channel signalling

An alternative to a signalling system being associated individually with a speech path is to couple the signals for a large number of calls together and send them on a separate common signalling channel. This method of signalling is particularly useful for inter-exchange working, when the signals will be sent over a data-type high-speed link, and in PCM systems. In the latter, one channel is dedicated to signalling in a 32 channel system. The signals must be coded so that they can be related to a specific speech channel. There is an inherent security problem in common channel signalling, so provision should be made for a second signalling channel to be used in the event of failure, and an error-detection facility must be provided.

10.22 VOICE-FREQUENCY SIGNALLING

As noted in Section 10.21, vf signalling can be either in-band or out-of-band (although some writers refer to in-band as vf signalling, and treat out-of-band separately).

10.22.1 In-band signalling

Since the signalling frequency is within the 300–3400 Hz speech bandwidth there are obvious problems associated with this signalling method. It cannot be operated during speech and the equipment must be able to distinguish between a speech pattern and a signal. There are two parameters available for variation: the signal frequency and the signal recognition time. Other considerations that will assist in distinguishing between speech and signal are:

(i) speech at the signal frequency is accompanied by other frequencies;
(ii) more than one signal frequency could be used;
(iii) the signals could be coded bursts of the signal frequency.

10.22.2 Choice of frequency

This is related to the frequency characteristics of speech. As shown in Fig. 10.16, the energy in English is predominant at lower frequencies (maximum at 500 Hz or thereabouts) and it falls gradually over the rest of the band. This suggests that a high signal frequency should be used to reduce the possibility of imitation by speech frequencies. However, there are other considerations that suggest that frequencies at the upper end of the band should not be used. At the higher frequencies there is an increase in crosstalk, so low-level transmission is necessary. If the amplitude is low the receiver must be very sensitive, raising the prospect of imitation signalling by low-level speech. The channel characteristics may vary at the upper end of the band on different links so that on some older links the cut-off may be below 3400 Hz. In addition, the variation of amplitude with frequency is quite marked at the high frequencies so any change in signal frequency will result in a significant change in signal amplitude. Between these considerations some compromise is necessary and in practice the frequency chosen lies in the range 2040–3000 Hz.

10.22.3 Signal duration

By delaying the recognition of the signal until it has persisted for some time the chance of signal imitation is reduced significantly. Using a guard circuit (see below) and a 40 ms recognition delay the probability of the signal receiver responding to a speech frequency is reduced to a low level. Figure 10.17(a) shows the degree of improvement that may be possible as the recognition time is increased.

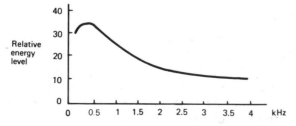

Fig. 10.16 Variation of energy with frequency in English speech.

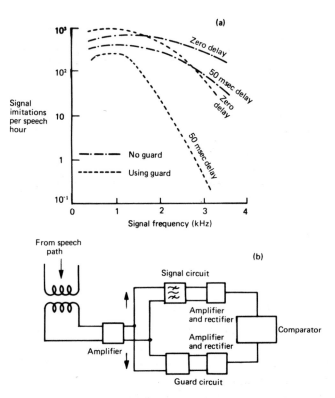

Fig. 10.17 (a) Effect of guard circuits, with and without delay; (b) block diagram of system using guard circuit.

10.22.4 Guard circuit

As can be seen in Fig. 10.17(a) the effect of signal recognition delay is greatly enhanced if a guard circuit is used in the receiver to increase the rejection of imitation signals. Figure 10.17(b) shows the basic features of such a system. The circuit shown there forms part of the vf receiver shown in Fig. 10.18.

The energy coupled from the speech channel is amplified and then passes along two paths:

(i) through a bandpass filter tuned to the signal frequency, forming part of the signal circuit;
(ii) through the guard circuit which includes a bandstop filter allowing all but the signal frequency to pass.

The outputs from (i) and (ii) are compared. When a signal arrives the signal circuit responds strongly and the guard circuit weakly, leading to a strong positive signal being detected by the signal-detector circuit. When speech is present, any imitation signals passing through the signal circuit are attenuated by the strong signal passing through the guard circuit, thus reducing the chance of spurious imitation.

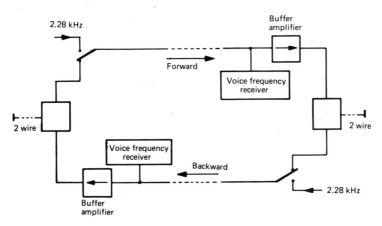

Fig. 10.18 Voice-frequency signalling system.

10.22.5 Basic layout of the system

Two-way working is achieved by combining two forward wires and two backward wires into a four-wire system (Fig. 10.18). The tone generators are switched in as required and the line is monitored constantly by the vf receivers for the presence of a signal.

Echo suppression can be used if required.

When employed in a multi-link connection, signalling is generally done on a link-by-link basis and it is important that signals from one link do not leak over to the next. To prevent such leakage, line splitting is used. There are various methods of performing the split; a common one is the buffer amplifier which, on the signal being detected by the vf receiver, is biased below cut-off, which effectively breaks the line after the receiver while the signal is in progress. The buffer also protects the receiver from local-end interference. Because the line splitting is initiated by the receiver, it is necessary to reduce the chance of signal imitation, so recognition delay is also used at this point. The receiver on detecting a signal frequency, waits for 2 ms before initiating the line split. During that 2 ms there will be leakage into the next link, but it will not be a problem as it is much less than the normal recognition delay of 40 ms used in the receiver.

Line splitting may also be inserted at the transmit end of each direction to prevent switching transients passing down the line at the same time as the signal.

10.22.6 Types of vf signalling

There are two basic types, pulse and continuous, and the recognition of a particular signal can be based on:

(i) the signal direction;
(ii) the position of the signal in a sequence;

(iii) the frequency content;

(iv) its length.

Pulse signalling

The recognition of a signal is determined from its length and its sequence. The following points can be made about pulse signalling:

(i) It has a higher signal repertoire than continuous signalling.

(ii) It can be transmitted at a higher voltage level and therefore provides a better SNR.

(iii) It is less influenced by interference.

(iv) It complicates the dc/ac and ac/dc conversions because the pulses have to be carefully timed.

(v) It requires a memory facility at the receiver for pulse recognition.

EXAMPLE: The UK AC9 signalling system is typical of that used in many countries. The signal frequency is 2280 Hz and the pulses used are shown in Table 10.2.

The calling subscriber's clear signal should initiate the backward release-guard signal. If the release guard is not received, the outgoing end sends seizure followed by forward clear until release guard is received.

A continuous tone busies the sending end relay set (busy back signal) and raises an alarm if it continues for several minutes.

The release-guard facility ensures that calls are not passed down to the receiving end if there is a fault. When a fault occurs and the forward-clear signal is sent, the outgoing equipment is guarded from accepting another call until the release-guard signal is received. The repeated use of the seizure and calling-subscriber-clear signal will initiate the release-guard signal once the fault has cleared.

Note that the backward answer signal and the called-subscriber-clear signal have the same pulse length, but they are distinguishable by their place in the sequence.

Metering is stopped by either the forward-clear or backward-clear signals.

Continuous signalling

There are two types of continuous vf signalling: (i) compelled and (ii) two-state non-compelled. Both are used, (i) in CCITT system 5 and (ii) in Bell SF. In (i) a

Table 10.2

Signal	Duration (ms)
Seizure (forward)	65
Dial pulsing (forward) (loop disconnect)	57
Called subscriber answers (backward)	250
Called subscriber clears (backward)	250
Calling subscriber clears (forward)	900
Release-guard (backward)	1000
Busy (backward)	Continuous

signal ceases when an acknowledgement is received, whereas in (ii) the signal information is carried by the change in state, and acknowledgement is not used.

In terms of reliability (i) is preferable to pulse, which is preferable to (ii), but (i) is much slower than (ii) because it depends on acknowledgement signals, and hence on propagation time. This slowness of (i) means that it is used only in special applications. In the second continuous mode, (ii), the tone is on for much of the non-speech time, which can lead to overload on the transmission system if the signal level is not limited.

10.23 OUT-BAND SIGNALLING

This term is generally applied to a signalling system in which the signal frequency is in the range 3400–4000 Hz. The CCITT recommended frequency is 3825 Hz, although 3700 and 3850 Hz are also used. This method is applicable only to carrier systems because the equipment used for baseband transmission may attenuate the signal frequency.

Compared with in-band signalling there are two advantages:

(i) There is no need to take steps to avoid signal imitation from speech-guard circuits and line splits are not required.
(ii) Signals and speech can be transmitted simultaneously.

The general layout of an out-band signalling system is shown in Fig. 10.19.

Signalling across the switching stage is usually dc, hence out-band signalling is link by link with ac/dc, dc/ac converters at each end. Filtering is used as

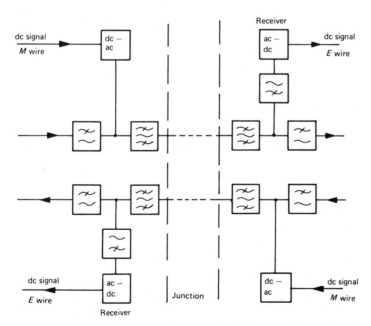

Fig. 10.19 Transmitter and receiver for out-band signalling.

Table 10.3

Signal	Tone	
	Forward	*Backward*
Idle	Off	Off
Seizure	On	Off
Decadic address	Off (pulse breaks)	Off
Release	Off	On when clear forward first
Answer	On	On
Clear Back	On	Off
Release guard	Off	On
Blocking	Off	On

shown to ensure that there are no spurious frequencies and to isolate the signal before the receiver.

This type of signalling has the advantage of simplicity and consequent cheapness. It can be either two-stage continuous or pulse. There are inherent difficulties in each mode; continuous cannot be at too high a level because of the risk of overload, so the receiver needs to be sensitive, whereas pulse requires more sophisticated circuitry to perform the memory function. On balance, the continuous mode is preferred because of its simplicity (compared with in-band signalling where the pulse mode has advantages).

Out-band signalling is more attractive than in-band, but cannot always be applied without considerable expense to existing transmission plant; consequently it is commonly used in new fdm systems, but vf is the most common on existing systems.

There are two modes of continuous signalling, tone-off idle and tone-on idle, and there is little to choose between them for many applications, although tone-on idle would be preferred if the system were in general use because tone-off idle would tend to overload the transmission system.

A typical signal code for a tone-off idle system is shown in Table 10.3.

10.24 SWITCHING NETWORKS

In switching systems the number of cross-points is a good measure of cost, and part of the effort involved in switch design is concerned with reducing the number.

An idea of the way in which the number of cross-points can be reduced, and the effect of that reduction on network performance, can be obtained from the following simple approach.

If there are N lines into and N lines from a single switch the number of cross-points required is N^2, e.g. the matrix shown in Fig. 10.20 requires 10 000 cross-points.

This large number of cross-points allows any free inlet to be connected to any free outlet regardless of the connections made between other inlets and outlets – a full availability lossless system.

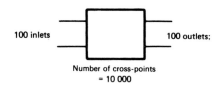

Fig. 10.20 Full-availability switch.

However, each inlet, if it is a subscriber's line, will carry a very small load, probably less than 0.1 erlangs, so it will be busy for only 6 min on average during the busy hour. This means that there will be fewer than 100 calls in progress simultaneously and therefore the system could operate with many fewer cross-points.

As a first step towards reducing the size of the network the matrix can be split into two parts; the first part having 100 inlets and 25 outlets and the second part having 25 inlets and 100 outlets [Fig. 10.21(a)]. This reduces the number of cross-points by 50%.

A further saving can be made if the inlet and outlets switches are further subdivided, as in Fig. 10.21(b). However, this last division raises serious problems. Each inlet switch has access to only one outlet switch so, for example, inlet switch 3 cannot route a call to outlet switch 5, etc. One way round this diffficulty would-be to interconnect the links between the inlet and outlet switches in the manner shown in Fig. 10.22, but that itself has the drawback that only one of the calls from an inlet switch can be connected with a particular outlet switch.

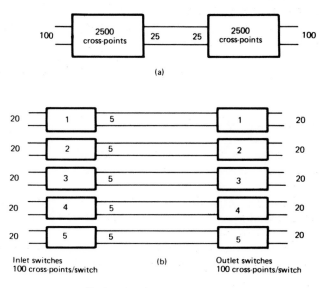

Fig. 10.21 (a) Cross-point reduction by vertical partitioning, (b) Cross-point reduction by vertical and horizontal partitioning.

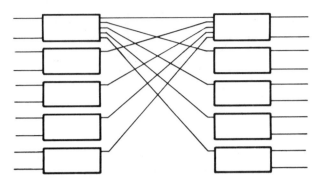

Fig. 10.22 Diagonal link connections to increase availability.

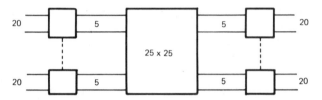

Fig. 10.23 Three-stage link system.

It seems that the reduction in availability has been too great and that more cross-points need to be used to make the system less restrictive. A middle stage can be included to make the distribution of paths between the inlet and outlet stages more satisfactory. This distributor stage will, in this example, have 25 × 25 cross-points as shown in Fig. 10.23.

In the same way as was done with the inlet and outlet stages, the distributor stage can be subdivided into five 5 × 5 switches and the link pattern between first and second stages, and second and third stages arranged to provide five paths through the network from any first stage (inlet) switch to any third stage (outlet) switch (Fig. 10.24).

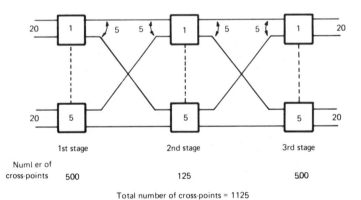

Fig. 10.24 Cross-point reduction on middle stage. Total number of cross-points = 1125.

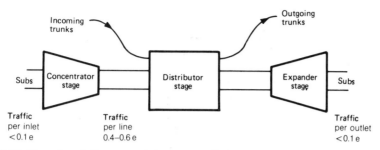

Fig. 10.25 Functional diagram of three-stage link system.

Clearly there has been a very large reduction in the number of cross-points. The chance that a call is lost because of the reduced availability is a question that can be answered by calculating the internal blocking of the network at given traffic levels, using the Jacobaeus equation, or some other suitable method.

Considering once again the three stages of the network in Fig. 10.24, we can see that the first stage has many more inlets than outlets (100 and 25, respectively), the third stage has many more outlets than inlets, and the second stage has an equal number of inlets and outlets. Figure 10.25 represents this aspect of the three stages diagrammatically. For obvious reasons the stages are named:

First stage   many more inlets than outlets – concentrator
Second stage   same number of inlets and outlets – distributor
Third stage   many more outlets than inlets – expander

This division into concentration, distribution and expansion stages typifies many switching systems. Subscribers' lines, with very low traffic per line, are concentrated to high-usage circuits operating at about 0.4–0.6 erlangs, and the distribution stage carries that traffic. Subscribers' lines are attached to the outlets of the system so there must be an expansion stage between the distribution stage and the subscribers.

Trunk circuits, which are designed for high traffic, come directly to, and go directly from, the distribution stage, as shown.

10.25   BASIC ANALOGUE SWITCHING SYSTEMS

Switching systems can be classified in several different ways, but here we shall restrict our interest to separating them on the basis of their method of control. Two methods are in common use: (i) step-by-step; or (ii) centralized control.

10.26   STEP-BY-STEP

In step-by-step systems, the control path and speech path are the same. As each digit is dialled, the connection is made to one further stage, until the final selection when two digits are required.

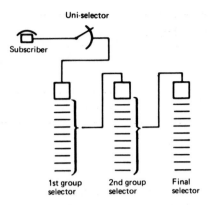

Fig. 10.26 Diagrammatic representation of a step-by-step system.

Strowger-type systems operate in this way. As an example consider a local exchange call on a four-digit number system (Fig. 10.26).

When the caller goes off-hook, the line circuit connects his line to a free first numeral selector (group selector). This connection may take some time, and not until the first selector is seized is dial tone returned to the caller. The dial-tone delay is a measure of system capacity from the subscriber's point of view, and most administrations specify a maximum value for this time. If it is too long, the subscriber feels that the system is not responding to his demand for service and he may hang-up.

Having seized the first selector the control waits for the first digit to be dialled; on its receipt the first selector wipers are racked up to the corresponding level and hunt round for a free outlet to a second selector. The second digit causes the second selector to go through a similar process and seize a final selector. The final selector has subscribers' line circuits attached to it, and so each of its hundred contacts represents a different subscriber. Consequently it needs two digits for its operation.

If at some stage during the setting up of the call, after the dial tone has been received a free selector is not available, a busy tone is returned to the caller; the wipers, having hunted over all the outlets to find a free selector at the next stage, without success, automatically return to their home position. In this method of control, part of the path is set up and then the system waits for more information before extending the path a little further and waiting again, thus setting up the call in a step-by-step manner.

10.27 COMMON CONTROL

The step-by-step system described earlier operates on the principle that the control of the call follows the same path as the speech circuit. This means that all the control equipment is provided on a per-call basis. However, some of it need not be, and savings can be made by separating some of the control from the speech path. It can then be used only as required by a call, and can then

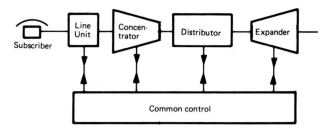

Fig. 10.27 Use of common control.

become available for other calls. For example, the register–translator, which is only required during the routing, or path selection process, could be placed in a common-control area, used to set up a path between caller and called, and then released for use in setting up another call. Figure 10.27 shows a diagram of a common-control system.

The switching block will not be of the Strowger-type, but either a cross-bar switch or a reed-relay system. A cross-bar switch consists of a matrix of horizontal and vertical conductors that can be made to interconnect at any required cross-point by the vertical bar trapping a metal finger attached to the horizontal bar when the latter is tilted by a relay operated by the marker-circuit in the common-control. The reed-relay switch is an 'electronic' cross-point consisting of two contacts inside an evacuated envelope and surrounded by an inductive coil. When the coil is energized it induces a magnetic field which forces the contacts together. The reed-relay switch is also of the matrix type with horizontal and vertical wires, the interconnection again being controlled by the marker.

Although systems differ in their arrangements and in the names used to identify particular units, Fig. 10.28 shows the basic form of a common-control exchange.

Subscribers are attached directly to a subscriber's line unit which recognizes a request-for-call condition. When the calling subscriber goes off-hook, the line unit indicates the request to the common control that seeks a free outlet from the first switching stage, which is a concentrator, and returns dial

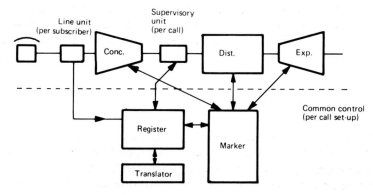

Fig. 10.28 Centralized control exchange.

tone to the caller. Each outlet from this first stage of switching is connected to a call-monitoring (supervisory) unit that maintains an awareness of the call for metering purposes, and reacts to a clear-down signal. The calling subscriber dials and the digits are stored in a register in the common control. If translation is required, the register passes the digits to a translator when one becomes available, and the translation is sent to the marker. The marker examines the switching units for a free path to the called outlet and tests the outlet to find out if it is free. If it is not, busy tone is returned to the caller. If it is free, a path through the switches is set up, ring-current is sent to the called subscriber and ring-tone is sent back to the caller. The common control is released and the progress of the call is noted by the call-monitoring unit.

There will usually be fewer translators provided than registers and many fewer registers than call monitoring units, the provision being proportional to the relative holding times of these elements except that in an exchange which requires only one translator, two may be provided for security reasons.

There is considerable saving in expense if the most provided control equipment, the subscriber's line unit, is made as simple as possible, and more complex features of the control are placed in those units that are required on a per set-up basis only, like the registers and markers.

10.28 MULTI-FREQUENCY SIGNALLING

The rather tedious procedure of dialling long numbers can be alleviated by using a key-pad, which allows the caller to press the required number instead of operating the dial (Fig. 10.29). This method reduces the dialling time considerably, and since the signalling is no longer at 10 pulses/s it too can be speeded up. A digit is signalled by a unique combination of two frequencies, one from each of the rows in Fig. 10.29. Detection at the exchange is carried

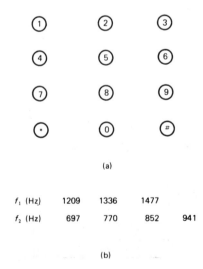

(a)

f_1 (Hz)	1209	1336	1477	
f_2 (Hz)	697	770	852	941

(b)

Fig. 10.29 (a) Key-pad arrangement; (b) frequencies used in key-pad.

out by using frequency filters. The frequencies are within the voiceband and care must be taken to reduce the risk of speech imitation. As we saw in our discussion on in-band vf signalling, there are two approaches; one involves a guard circuit and the other, signal recognition delay. Both of these methods are used.

Another benefit of the key-pad, which will be exploited in digital systems, is that there are more combinations of frequencies than the ten required for digits. If there are 16 possible combinations – one frequency from four, twice – there are five potential combinations that could used to signal other information to the exchange. Some systems have only two additional unused combinations because the pair of frequencies are chosen from a group of four and a group of three, giving 12 in all.

The development of common control based on computer operation removed some of the constraints on inter-exchange signalling. Very rapid reception, detection and processing were possible, and for signalling between processors care need not be taken to counter speech imitation. CCITT No. 4 was developed as a fast-inter-exchange mf signalling method based on transmitting signals as a combination of two from five or two from six, depending on the repertoire required.

10.29   THE TELEPHONE NETWORK

The telephone network has developed dramatically in recent years so that it is now possible to make calls automatically between subscribers separated by thousands of miles. Long-distance calls pass through several stages of switching and several possible transmission links before reaching their destination, and to make such calls possible many facets of telecommunications must be integrated and reasonable compromises made by systems designers. In practice, there are very few occasions when a telecommunications administration has the opportunity to design a network from the beginning. Usually some sort of network exists already and there is a demand for increased service: more lines, better performance, more facilities, more comprehensive direct dialling, etc. This demand has to be satisfied by grafting on to the existing system new switching and transmission equipment that may be totally different in type and in its principles of operation. Recent years have seen two revolutionary developments: the integrated circuit leading to cheap, reliable, small and very powerful digital systems; and the optical fibre, which is much better in many respects than conventional transmission media. These two have led to an entirely new approach, and enormously enhanced facilities for future systems. One of the major planning problems is how to graft such new systems on to existing equipment. We shall return to this point later.

Although the structure of telephone networks has developed piecemeal as demand has increased, it still has some identifiable form. The passage of a call through a national network can be represented by the multi-level diagram shown in Fig. 10.30. Before going further, some comment is necessary on the terminology to be used. The words used in North America and the UK differ

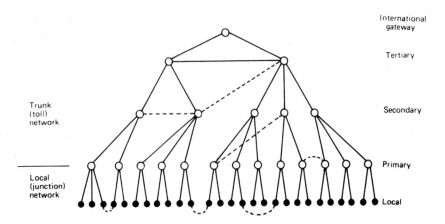

Fig. 10.30 Hierarchical routing system.

Table 10.4 Terminology

CCITT	USA	UK
Primary centre	Toll centre	Group switching centre
Secondary centre	Primary centre	District switching centre
Tertiary centre	Sectional centre	Main switching centre
Trunk exchange	Toll office	Trunk exchange
Trunk network	Toll network	Trunk network
Trunk circuit	Trunk	Trunk (circuit)
Local exchange	End (central) office	Local exchange
Junction circuit	Inter-office trunk	Junction

significantly in their meaning. Here we will refer to the various levels of switching by using the CCITT recommended terms. Table 10.4 relates them to those commonly used in North America and the UK.

From Fig. 10.30 we can see that there are several levels of switching that combine to form the complete network. It is usual to think of the systems in two parts. The first is the junction network, serving the subscriber and consisting of the link from subscriber to local exchange, from local exchange to primary switching centre, and back to the called subscriber via another local exchange. The second part of the system is the trunk network, which is concerned only with calls passing at primary centre level and above. Thus the primary centre is associated with both parts of the network.

The structure shown by the solid lines represents the basic hierarchical network of the system, and telephone calls routed along these links are said to be travelling on a backbone or final route. When there is much traffic between exchanges, either belonging to the same cluster, or at different levels, direct routes may be installed to save a stage of switching. Such routes are shown as broken lines in Fig. 10.30. These direct routes are dimensioned to a different grade of service from the final routes. Because a call finding all the direct-route circuits busy can be placed on the backbone route, the direct route can be dimensioned to a higher grade of service than the final route, and thus work more efficiently.

The number of exchanges at the various levels depends on several factors: the physical extent of the network, the number of subscribers, the amount of traffic, the forecast growth and the transmission methods used. Beyond the top level of the national system is a layer that gives access to the international network. This layer may consist of one or more international (usually called gateway) exchanges.

10.30 NUMBERING SCHEMES

In modern systems, the numbering scheme used by a telephone administration to allocate subscribers' numbers has an underlying plan, and there are a few constraints on the development of the plan that must be taken into account:

(i) It must provide each subscriber with a unique number within the national network.
(ii) The allocation to areas must be able to meet forecast growth for several decades.
(iii) The number of digits should not exceed that recommended by CCITT.

In principle, (i) is easy to satisfy if (ii) and (iii) have been met. The length of the number recommended by CCITT is $11 - n$ where n is the country code (see below). If, for example, n is 2, then the national number should not exceed nine digits in length. These digits are used to denote the subscriber's number on the local exchange, the exchange within a given area, and the area within the national numbering scheme. In many local exchanges there is a maximum capacity of 10 000 lines, thus the last four digits of the national number are allocated to the subscriber's number in the exchange. Of the remaining five digits in our example, the first two would denote the area and the remaining three the exchange within the area. Thus the number has the form shown in Fig. 10.31.

The division of digits between area exchange and subscriber's code may not be the same as that shown – but all three components exist in every number.

For automatic long-distance dialling a prefix is necessary to indicate to the exchange equipment that a trunk call is being made. In many countries a '0' is used, but any other digit would do. In calculating the length of the national number the prefix is not included.

Fig. 10.31 National telephone number.

10.30.1 International number

In the introduction to this chapter we imagined a future when person-to-person calls could be made very easily via individual instruments and satellite links. For such availability of connections to be possible each subscriber in the world must have a unique number. To achieve that, some agreement between countries is essential; each country must be identified by a number different from that of all other countries. That is achieved by agreement through CCITT on the way in which these codes, called country codes, are allocated.

The first digit of the country code is the zone code; the world is divided up into nine zones and each country belongs to a zone. The relationship between zone number and geographical area is shown in Table 10.5.

In all but two of the zones, one or two digits are added to the zone number to produce the country code. For example, Brazil has a zone number 5 and an additional country code digit 5 to give its country code 55. Brunei is in zone 6 and two further digits are added to give a country code 673.

The two exceptions are zones 1 (North America and the Caribbean) and 7 (USSR). Throughout each of these zones there is a linked numbering scheme that means, for example, that no subscriber in Canada has the same national number as a subscriber in the USA. Consequently, to connect to anyone in zone 1 the digit 1 is followed by the national number. A similar situation exists in the USSR. Europe is at the other extreme; there are many countries with large national networks that have nine digits in their national numbers. For these, a two-digit country code is required, and that can only be achieved by having two zone numbers allocated to Europe.

The division of the world into the zones shown in Table 10.5 is intended to be satisfactory until early in the next century, but clearly, as some large countries develop their telephone networks, some adjustment will be necessary at some future time.

The national and international numbering schemes we have discussed above are the simplest. However, in several parts of the world there are small exceptions, particularly in regard to local calls. In the scheme where the national number is used for all calls within a country, it can lead to irritation on the part of the subscriber and long set-up times for the exchange equipment. Consequently, in many countries local calls use a shorter code. For calls within the same area, the area code is omitted, and for small single-exchange areas no exchange code is used for own-exchange calls.

Table 10.5 Zone numbers

1 North America	6 Australasia
2 Africa	7 USSR
3 Europe	8 Eastern Asia
4 Europe	9 Far East and Middle East
5 South America	0 Spare

Coupled with this last arrangement will be a very short code for calls to adjacent exchanges; these arrangements are particularly well suited to rural areas. The disadvantage of short codes is that they change with the location of the calling subscriber, and therefore a short code directory must be available in each exchange area.

The above description relates to the current practice of allocating numbers to premises. In countries where the telephone network is a public utility, the creation of a numbering plan and the allocation of numbers are the responsibilities of the network operator, the PTT. However, in a privatized, non-monopolistic environment the responsibility for numbering will fall to some other organization. It could be the regulator, or a government ministry, but it is not feasible to allow one operator, among several, to have control of numbering since that would create an intolerable degree of market advantage.

As the number of operators, services and service providers increases there is a substantial demand for the allocation of more numbers. For example, the widespread use of facsimile, dialled up in the same way as a telephone call, has placed a considerable pressure on numbering plans.

There is likely to be an even more significant change in the near future. In anticipation of a general use of mobile systems, plans have been developed to relate numbers to people, rather than to premises. A personal numbering scheme will probably be introduced so that access can be made via any handset, identification of the user being established by using a smart card or similar electronic code that can be entered by the user; this mechanism would also ensure correct billing.

10.31 ROUTING CALLS

The early type of switching equipment, called step-by-step or Strowger, operated by using the dialled pulses to move the selectors to the position corresponding to the digit dialled. In many ways this was an excellent system, but one major disadvantage was that it allowed no flexibility in the way calls were routed – the route was predetermined by the dialled digits. Although some systems were modified to overcome this problem it was not until common-control equipment became widely used that the path a call took between calling and called subscribers could be chosen to allow the most efficient use of the available capacity in the system. The function of the common control in the routing process was to store the dialled digits in a register and then translate them into routing digits which would indicate to the switching system the path to take through the network. This register–translator combination is essential to automatic trunk and international dialling schemes; it allows the telephone administration to manage the system efficiently by changing routes as circumstances alter without having to change subscribers' numbers. This therefore separates the subscriber from the system. The subscriber dials the national number from any location and the register – translator automatically selects an appropriate route.

10.32 DIGITAL SYSTEMS

The telephone network worldwide is essentially digital, having developed from being predominantly analogue in the 1970s. The change was brought about by a combination of technological and theoretical improvements, and it gave rise to the notion that by using a digital representation of voice, the same system could be used to carry data.

The first mainstream digital exchanges were installed in the late 1970s, and they will be superceded by the new systems, based on packet rather than circuit switching, that are being introduced on an experimental basis in the mid-1990s. They are the initial steps towards a high-speed, multi-media network that will carry various forms of voice, data and video traffic.

In this chapter we will concentrate on the switching and transmission used in the public switched telephone network (PSTN), including the integrated services digital network for voice and data, known as ISDN. However, the concluding sections will introduce the intentions and principles underlying the new broadband network, B-ISDN that will depend on the synchronous digital hierarchy, SDH, and use a high-speed packet format, called asynchronous transfer mode, ATM.

At the outset it is worth emphasizing the distinction between ISDN and B-ISDN. There are several differences, but in the context of this chapter the most imporant is that ISDN is a natural extension of the fixed channel, circuit-switched, 64 kb/s telephone network, whereas B-ISDN uses the totally different technology of high-speed packet switching.

10.33 PCM TELEPHONY

In Chapter 3 the principles of PCM were discussed so the details will not be repeated here. However, it is worth reiterating why the building block for the telephone network is a 64 kb/s channel. The basic system is shown in Fig. 10.32. The analogue output from the microphone in the handset is filtered to ensure that it is band-limited to the range 300–3400 Hz. The filter output is sampled at 8 kHz/s, giving 125 µs between samples. Later, we will refer to this sample separation time (or frame time as it is usually called) as one

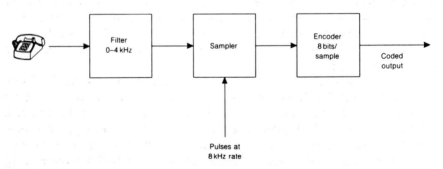

Fig. 10.32

of the parameters that limits the flexibility of the network; it is essential that samples from a particular telephone arrive at precisely 125 μs intervals.

Returning to the PCM system, a sample is changed from an amplitude into a series of 8 bits in an encoder. Thus the bit rate used to carry the signal representing the speech sample is 8000 × 8, or 64 kb/s. For the standard PSTN this channel bit rate is fixed.

There are two transmission systems used world-wide, one with a 24 channel format and the other based on 32 channels. The former is used in North America and Japan, and the latter in the rest of the world. The larger format operates with a frame of 32 timeslots, each one equivalent to a speech channel, although only 30 of the slots carry speech. The other two are used for sychronisation and signalling as described below. All 32 timeslots must fit into the 125 μs frame. A timeslot is therefore approximately 3.9 μs long. Sixteen frames are combined to form a multiframe which takes 2 ms to transmit.

The network uses the signalling information to ensure that the speech channels are sent to the correct destination and that release, cleardown and metering are carried out at the appropriate times.

Two modes of signalling are used in digital systems: channel associated and common channel. The second, which relates to digital links between exchanges, will be described later. Channel associated signalling is used on the PCM transmission path when traffic is arriving from different sources such as analogue connections, or a multiplexor. In this case, signals are associated with each channel by allocating in each multiframe one signalling half-word in timeslot 16 to each channel; thus the 8 bit signalling slot contains signalling information of 4 bits for two channels in each frame.

Figure 10.33 shows that in all but the first frame (frame 0), timeslot 16 is used to carry signalling. Hence in frame 1, timeslot 16 carries the signals related to speech channels 1 and 16, in frame 2 it carries those for channels 2 and 17, and so on until frame 15 when it has signals for channels 15 and 30. The next frame is the first of the following multiframe and the sequence is repeated.

Timeslot 0 in each frame, and timeslot 16 in frame 0 carry synchronization and alignment words to ensure that the transmission and reception of the system are in step.

When common channel signalling is used the above arrangement is not relevant. Instead, the capacity of timeslot 16 is made available to the signalling packets as required, and it is likely that some of the signalling information is not related to the traffic being carried on the voice channels.

The detailed relationship between the multiframe and the timeslots is shown in Fig. 10.33.

Digital exchanges consist of the basic components shown in Fig. 10.39. Most of the control of the system is handled by microprocessor devices, singly or in clusters, which are driven by software. Here we are not able to discuss the huge new field of telecommunications software engineering, but it provides the most important challenge in modern system design. The software must be efficient, reliable, secure, understandable and well documented. In theory it affords a degree of flexibility in the operation of the system which is

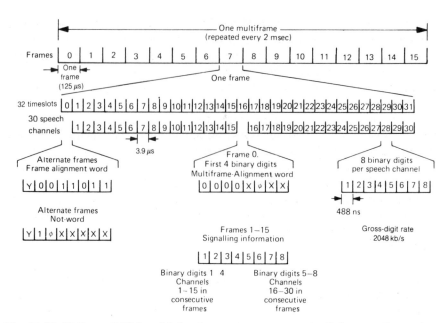

Fig. 10.33 32-frame PCM multiplex frame arrangement: X, digits not allocated to any particular function and set to state *one*; Y, reserved for international use (normally set to state *one*); ϕ, digits normally *zero* but changed to *one* when loss of frame alignment occurs and/or system-fail alarm occurs (timeslot 0 only) or when loss of multiframe alignment occurs (timeslot 16 only).

much higher than is possible in hard-wire control. However, such is the complexity of large telephone networks that software development presents the most difficult problems to designers, for it must last for tens of years and although made up of very long programs, it must be able to cope with dramatic changes in hardware technology.

In analogue exchanges the usual figure of merit used is the grade of service, or probability of blocking, but in digital switches the blocking is virtually zero. In these systems the major problems concern delay in the processing of calls caused by the processor units becoming overloaded. The analysis of such problems is very difficult and if simple queueing theory models do not apply resort must be made to computer simulation. One of the difficult tasks of the software engineer is to produce a satisfactory compromise between short efficient programs and those that are longer, more complex and more reliable and secure.

10.34 DIGITAL TRANSMISSION HIERARCHY

In the last few paragraphs we have discussed PCM transmission at baseband. In practice, in order to provide the required capacity, and to exploit the bandwidth available on cable or fibre, channels must be put together in an orderly manner. The basic multiplexed element is the 2 Mb/s multiframe as

	24 channel	32 channel
Primary rate	1544 kb/s	2048 kb/s
Secondary rate	6312 kb/s	8448 kb/s
Tertiary rate	44736 kb/s	34368 kb/s
Quaternary rate	139264 kb/s	139264 kb/s

Fig. 10.34 Digital transmission hierarchies.

discussed earlier. Multiframes can be multiplexed into superframes, and superframes into hyperframes, and this sequence can continue until there is sufficient capacity, or the bandwidth of the transmission medium is exceeded. Figure 10.34 shows the hierarchies used in telephone networks based on 24 and 32 channel frames.

10.35 COMMON-CHANNEL SIGNALLING (CCS)

End-to-end digitization, processor control of exchanges, the demand for a larger range of facilities and services, and the need for automatic, high-speed network management, all place heavy demands on the signalling arrangements, and they cannot be met by the channel-associated signalling mechanisms discussed above. A flexible, digital approach is required that has the capacity for a large repertoire of signals and that can handle a variety of traffic types. By general agreement the system used is that recommended by CCITT as Signalling System No. 7. We shall refer to it by the shorter title SS7 for convenience.

The flexibility of SS7 derives from three factors; the dis-association of the signalling from the speech channels, the consequent opportunity to have a logically separate signalling network, overlaying that for transmission, and a packet-based message transfer protocol, allowing the channel to be used for information, management and control, as well as for signalling for individual calls.

SS7 is used to send information between exchanges. In principle it is comparatively simple, but its implementation consists of a large amount of detail, and there are small but significant differences between the various systems in use, making it necessary to introduce interface circuitry between national implementations in order to facilitate international connections.

10.35.1 Overlay network

Although in practice the physical implementation of the signalling network is usually co-incident with the user network, it does not have to be so. It is quite feasible for the signalling network to be entirely detached from the traffic network. We can represent this detachment by considering the logical position. The logical representation of the two networks is as shown in Fig. 10.35. Note that the signalling network is connected to a digital exchange via a signalling point (SP). Because the signalling network is logically separate, there can be signalling transit units. Such units are known as signalling transfer points (STPs) and they act as intermediate nodes.

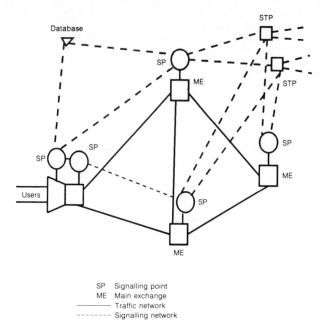

SP Signalling point
ME Main exchange
———— Traffic network
------- Signalling network

Fig. 10.35 SS7 non-associated overlay network.

10.35.2 Reference model

Knowing that signalling information is sent in the form of packets, and that SPs and STPs are the nodes of a signalling network, it is not surprising that the protocols used in SS7 can be expressed as a multi-layer architecture. The reference model used to describe SS7 pre-dates the OSI seven-layer model, but the two can be related. Traditionally SS7 is described by a four-layer model, as shown in Fig. 10.36.

Layer 1 is the physical layer. It specifies the interface conditions and functions across the boundary to layer 2, making SS7 independent of the specific physical link medium. In digital systems operating in standard PCM format the physical layer is provided by the signalling channel, TS 16, which may be transmitted over a wide variety of physical media including coaxial cable and optical fibre.

Layer 2 is the signalling link layer. The packet to be sent over the physical layer is given some form, and incorporates codes to ensure that it can be transmitted to the correct destination with a high probability that it will be error free. Source and destination codes, and error detection and correction codes are incorporated, along with codes that allow flow control mechanisms to be used to restrict the transmission of signalling packets to a rate that can be handled by the receiver.

At this level there are three types of packet known as

MSU Message signalling unit
LSSU Link status signalling unit
FISU Fill-in signalling unit

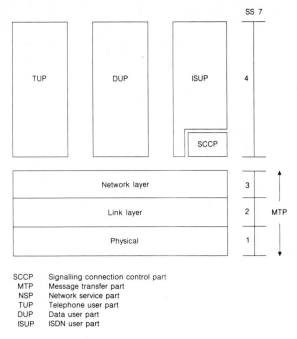

SCCP   Signalling connection control part
MTP   Message transfer part
NSP   Network service part
TUP   Telephone user part
DUP   Data user part
ISUP   ISDN user part

Fig. 10.36 SS7 reference model.

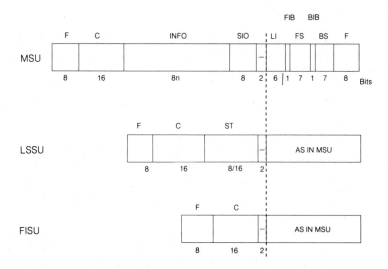

Fig. 10.37 SS7 layer 2 data units.

The frame format for each of these units is shown in Fig. 10.37, where it can be seen that much of the overhead information is common to all three types.

The functions of these frames are as follows:

MSU   Transports higher-level information
LSSU   Sets up the link and is used in flow control

FISU   Is transmitted when no other information is being sent in order to maintain the link; i.e. it ensures that the receiver is correctly set up, and that it does not clear down the connection too early.

The framing structure of the MSU is as follows:

Flag	indicates the start and finish of the frame and it has the fixed form 01111110
FS, FIB	
BS, BIB	used in the error control mechanism
LI	indicates the length of the frame from C to LI, and the type of SU
C	uses a standard CCITT check sequence
SIO, INFO	carry information from higher levels
ST	carries information on the condition of the link and its alignment status.

Layer 2 functionality is sufficient for passing control information across a single link between exchanges. The information field is not interpreted at this level, so it is passed unprocessed from level 1 to level 2.

Layer 3 is the signalling network layer. As in data networks, layer 3 is concerned with network functions, in particular with providing routing information for the STPs to ensure that the packets are transmitted efficiently through the network. When there is more than one link between source and destination signalling point the layer 3 functions are required.

Layers 1–3 constitute the message transfer part (MTP).

Layer 4 in the SS7 model depends on the application. It is known as the user part, in general, but there are several such parts: the telephone user part (TUP), data user part (DUP), and ISDN user part (ISUP) are the three of present interest, but as new services are introduced, additional user parts may be defined.

The user parts were designed with a functionality that cannot be mapped easily onto the OSI reference model. This raises problems for future developments in an open system environment, particularly for user applications that are not based on circuit switching, such as network management.

The telephone, data and ISDN user parts are unlikely to be redefined. They operate directly into the message transfer part of the SS7 model and their overall structure is established.

It is perhaps worth emphasizing that from the layered model perspective the upper layer users, normally thought of in data networks as the users of the system, are the control processors in the exchanges at the ends of the signalling link path. To benefit fully from the scope of SS7 these processors use all the upper layers and can therefore be regarded as the users of the system.

The early specification of SS7 not only treated the upper layers as one set of functions, but chose a functionality for layer 3 of the MTP that is slightly different from layer 3 in the OSI model. To map the SS7 model onto the 3/4 interface of OSI requires an additional sublayer, called the signalling connection control part (SCCP). By adding this sublayer to the top of MTP, new

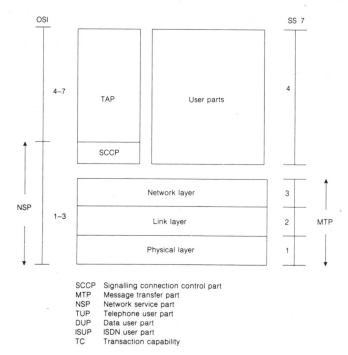

SCCP   Signalling connection control part
MTP   Message transfer part
NSP   Network service part
TUP   Telephone user part
DUP   Data user part
ISUP   ISDN user part
TC   Transaction capability

Fig. 10.38 SS7/OSI reference model.

user parts can be developed in an OSI framework. The upper layer, 4 in SS7, is given more structure too by subdividing it into transaction capabilities sublayers.

In Fig. 10.38 the complete layered structure of SS7 is given, with the associated OSI layers indicated for comparison.

The universal acceptance of SS7 as the digital signalling system for the telephone network has enhanced considerably the global operation of the network. It is essential that there is a common understanding and interpretation of the recommendations. In practice there are three ways in which this requirement proves to be difficult. The recommendations may be interpreted ambiguously. This may be because they are ambiguous in the obvious sense; not sufficiently explicit and therefore leave scope for different, equally valid, interpretations. The ambiguity may also arise, however, because different developers interpret the same terminology in different ways; the writers of the recommendations having been wrong in assuming that their semantics had only a single meaning.

Secondly, there may be some implementation decisions that in practice do not produce the precise functionality at an interface as set out in the recommendations, leading to an inability of pieces of equipment to interwork.

Finally, there is a need, for various reasons, to include options within the recommendations, for example when a specific feature is required by one operator but not by others. The provision of options must be made rarely since, unless the effects on interworking can be comprehensively evaluated, unforeseen difficulties will occur.

The mutual self-interest of having a universally recognized signalling system has lead the interested parties to become committed to establishing an efficient mechanism, through CCITT, for iterating towards a well-defined system by reporting problems, and suggesting solutions.

10.36   DIGITAL SWITCHING

The digital switch can have many structural forms depending on the application, the number of connections required, and the technology used. The system shown schematically in Fig. 10.39 is a local exchange and it shows that there are two types of switch involved. A subscriber switch to act as a concentrator, and a central switch which has a distribution function.

The architecture of the subscriber switch will depend to a large extent on the number of subscribers to be attached to the exchange. If a small rural exchange is being connected the switch may act as no more than a multiplexor, being the mechanism whereby up to 30 speech channels are time division multiplexed on to a single PCM carrier. In such a case, since a group of 30 channels is the smallest PCM link available, it may be sensible to allow all subscriber lines full availability access. However if the number of subscribers is somewhat larger a concentrator will be necessary. As in analogue systems, a concentrator provides significantly fewer lines on the outlet side than are attached at the input. The fact that the inlets are subscribers indicates that generally the mean traffic per inlet line will be less than 0.1 erlang thus a very low loss probability can be achieved even if the number of outlets is no more than a fraction of the number of inlets.

For digital systems the interworking between the digital switch (exchange) and the digital transmission link (PCM line) is straightforward, using open system signalling SS7. However, if the switch has to carry traffic originating on an analogue line then the signalling will be one of many possible analogue signalling protocols, such as those described earlier. In that case interworking is difficult, and special interface units are required to ensure smooth oper-

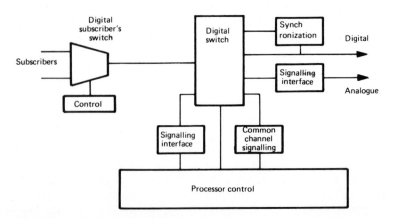

Fig. 10.39 Digital telephone exchange.

ation. These interface circuits can be exteremely expensive, adding considerably to the cost of the switch. In practice, since there are many types of analogue signalling, there may be a need for more than one interface in a particular installation. Because of the variations in the detail of signalling system protocols it is highly likely that interfaces have to be developed on a one-off basis; often, a general design to interface SS7 to each analogue signalling system will not be a feasible solution.

10.36.1   Time switching and space switching

The last section dealt with the notion that subscribers lines could be connected to PCM links via a concentrator or a multiplexor. However, that is not the only function of digital switching. It is often used to interconnect a calling subscriber on an incoming line with a called subscriber on an outgoing line, or as a distribution stage as discussed earlier.

We saw in Section 10.33 that the channels on a PCM link are structured into a frame and multiframe arrangement so that any attempt to interconnect calling and called subscribers will require a timeslot adjustment; see Fig. 10.40 for an example. Five subscribers on the inlet side of a multiplexor are able to call subscribers on the output side; if, for example, subscriber C on inlet timeslot 3 wants to be connected to subscriber V on outlet timeslot 4, the inlet sample must be delayed in the switch for one timslot before being sent out on timeslot 4 for the connection. The type of switch designed to provide the appropriate delay is called a time switch. The sample from the inlet channel is stored in a buffer and then read out when the appropriate output timeslot arrives. Figure 10.41 illustrates the way in which this is done. Synchronization is obtained from the frame and multiframe alignment signals in the PCM format (Section 10.33). Each input timeslot word is stored in a buffer. A control store holds information on the time at which each sample has to be read out. At the appropriate time this control store connects the data buffer to the output line. The control store instructions are derived from a central processing unit which responds to the call request from subscribers.

Time switching is clearly not enough if there is more than one incoming and outgoing PCM line. Any channel on an inlet line must be able to obtain access to any channel on any outgoing line and to do this, space switching is necessary to provide the linking between the particular inlet and outlet lines.

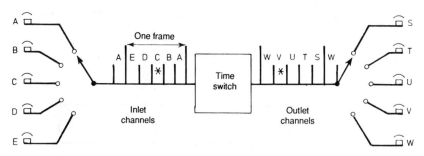

Fig. 10.40 Time switching.

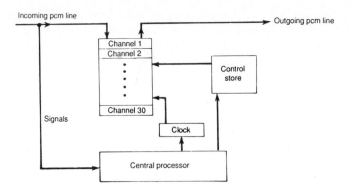

Fig. 10.41 Buffer delay for time switching.

For complete flexibility in the operation of the digital switch some combination of time and space switching is required. Although it is possible to achieve full switching with just one time switch (T) in combination with one space switch (S), such a switch would probably suffer internal blocking. More commonly a three-stage arrangement is used, either STS or TST. As an example we will consider the TST configuration, shown in Fig. 10.42. Each inlet time switch has one four-channel inlet line. A frame has four timeslots, and on the outlet side of the system there are three outlet time switches each connected to a PCM line of four channels. A, B and C are the inlet lines and P, Q and R are the outlet lines. Assume the connections required are as follows:

Inlet Channel	A1	A2	A3	A4	Bl	B2	B3	B4	C1	C2	C3	C4
Outlet Channel	R4	R3	Q4	R1	P3	P4	P2	Q2	R2	Q1	P1	Q3
Delay (timeslots)	3	1	1	1	2	2	3	2	1	3	2	3

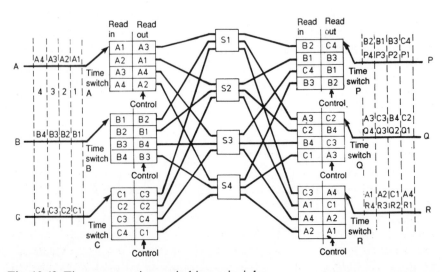

Fig. 10.42 Time–space–time switching principles.

Also shown is the number of timeslots delay that are necessary to make the input/output connections required. Some of these connections require a delay which crosses the frame boundary, e.g. to connect B3 to P2 requires a delay of three timeslots, meaning that a sample which arrives in timeslot 3 has to be held until timeslot 4 and timeslot 1 of the next frame, have passed; the sample is then released into timeslot 2 of the new frame to make the connection.

On the inlet side of the first-stage time switches, the samples in the speech channels are stored in cyclic order – as shown by the read-in column of the inlet time switches. Similarly the channels are read out from the outlet time switches in cyclic order, from timeslot 1 to timeslot 4 as shown. Consequently, any time delay required must be produced by the combined delay between the two time switching stages.

Consider the call in channel A1. It is to be connected to channel R4, requiring a delay of three timeslots between the input and output of the system. From Fig. 10.42 we can see that A1 is read into timeslot 1 and out in timeslot 2. (The choice of which timeslot to read it out is somewhat arbitrary, being determined to some extent by the calls already set up in the system.) A1 is switched in the second space switch to time switch R and is read into that switch in timeslot 2. The additional delay of two timeslots is provided in time switch R since A1 is stored until timeslot 4 arrives, when it is read out to line R as required.

In the above description we noted that Al was switched in timeslot 2 by space switch S2, and was read into time switch R in the same timeslot. Generally there is a logical space switch for each timeslot; space switch 1 for timeslot 1, space switch 2 for timeslot 2 and so on. Thus there can be no change of timeslot across a space switch.

In reality a physically separate space switch is not required for each timeslot. Since the timeslot switching arrangements occur sequentially it is possible to use one space switch for all the switching required since the cross-connections inside the switch can be changed at the end of each timeslot. Thus, provided the switching speed is sufficiently high, a single space switch will suffice. This does not alter Fig. 10.42 in which the space switch was shown as a separate entity for each timeslot; that diagram gives a correct notional impression of the way in which the switch operates.

A more general TST switch of the type found in a distribution stage is shown in Fig. 10.43. The time switch is split into several units, each having M PCM links of L channels. Consequently, if the time switch is non-blocking it will have an outlet highway of $N = ML$ time slots. The space switch is square with R inlet highways and R outlet highways. The purpose of the TST unit is to allow a particular call, which occupies a specific channel into one of the time switches, to be connected to a particular outlet channel. Basically the switching is between highways on either side of the space switch. Each highway has N timeslots and in order for a particular call to be connected say from $H1$ to $H3$ it must find a timeslot which is free in both highways. This slot may not be the same as the required incoming and outgoing slots for the call, and so some time delay, provided by the time switches, is necessary. The method used to produce the delay again depends on the technology em-

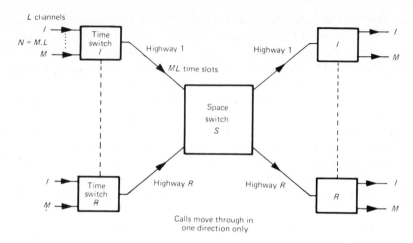

Fig. 10.43 Network representation of a digital switch.

ployed, but the delay may be from 1 to 31 timeslots depending on the relative positions of the timeslots in and out of the unit, and the chosen free timeslot in the space switch.

To understand the behaviour of the TST switch in terms of the link systems considered earlier it is important to appreciate that for each timeslot the interconnections in the space switch will be different; at each timeslot there will be a different set of calls in progress and the connections between the highways will last for only one timeslot period then new connections will be established. This can be represented by having N space switches (Fig. 10.44), one for each timeslot.

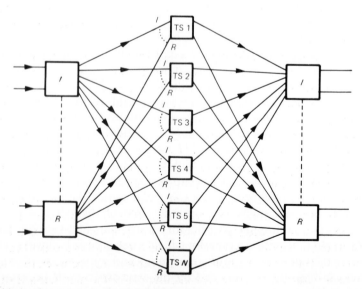

Fig. 10.44 Analogue equivalent of TST switch.

Whether or not blocking occurs in the TST unit depends entirely on the dimensions of the space switch, and since in modern systems switches are comparatively inexpensive they are usually large enough to make blocking negligible. For total non-blocking there must be at least as many outlets as inlets on the time switches, and the space switch highways must have $2N-1$ timeslots, where N is the number of timeslots in a link to a time switch.

Digital switches are uni-directional, and that implies that two paths are required to connect two channels X and Y, one for conversation from X to Y and the other for conversation from Y to X. To reduce the control process, the X to Y slot is chosen according to whatever rules are used by the designer, and the Y and X interconnection is allocated a fixed number of timeslots from it, e.g. one, or half a frame. By this method, if the X to Y connection is available, the Y to X must also be free.

The digital switch just described is situated in a main exchange, forming part of the trunk network. Interconnections between exchanges are made, for the information paths, via PCM links. However, signalling is carried over a common channel, using signalling system CCITT No. 7 as described in Section 10.35. The inter-exchange signals will be concerned not only with setting up calls between exchanges, but with accounting, administration fault diagnosis and maintenance. As described earlier, the CCITT No. 7 system transmits signals as messages that can have variable length. Each message is preceded by labels that identify its originating and destination exchanges, the type of message (call handling, fault, etc.) and includes error-detection and acknowledgement bits. If an error is detected in a message, that and all subsequent messages are retransmitted to ensure that the sequence received at the far end is in the correct order.

The error rate has an important bearing on the capacity of the signalling channel; retransmission of incorrectly received messages obviously takes up time that could be used for new signals and consequently slows down the overall process. The capacity is specified in terms of number of messages per busy hour, given that the delay from end to end is not greater than some predetermined value.

On the subscriber side of the digital switch there will be a local unit of some description. For large areas a local digital exchange would be used with mf signalling from subscriber to exchange where conversion to a PCM format would take place before concentration through a digital switch. Alternatively for very small units no exchange facility would be available, but a simple digital concentrator would be used to take in the analogue channels, convert them to PCM and multiplex them on to a single highway to the nearest local exchange. Calls between subscribers on the same concentrator would then have to pass through the local exchange. Signalling in these PCM links would be on TS16 and the control, software and firmware at the main exchange would convert it to common channel if a trunk call were required.

The introduction of digital systems is rarely a starting point for the telephone system. Usually a system exists and the digital equipment has to be grafted on to it. Whatever method is used, interworking between the old and new systems is required, and one area of difficulty is the interfacing of various analogue signalling systems with the new equipment designed to operate on

TS16 or common channel. This interfacing can be a severe problem if there are many existing signalling schemes in a particular network, and the development of satisfactory units can add considerably to system costs.

10.37   INTEGRATED SERVICES DIGITAL NETWORK (ISDN)

Alongside the digitization of the telephone network there were parallel developments concerning data communications. Coming from a main-frame computer background, and hence based on a data processing environment, the transport of data was determined by the need to transfer large quantities of information between different sites or institutions within a market sector. This lead to the establishment of large private data networks, using modems over that part of the system linked by the PSTN. Once computing developed into a more distributed, ubiquitous environment, and workstations and PCs replaced mainframes for many uses, the attraction of a common communications network instead of separate networks for data and voice was obvious.

Since the telephony network had by far the larger penetration, and because it was going digital, the obvious step was to modify it to be able to carry digital data as well as voice. In principle this would provide the means to simplify the user's system, enabling one desktop unit to handle both voice and data traffic simultaneously.

The approach used was to establish a set of standards that would be generally acceptable and; based on the 64 kb/s channel, offer the facility to carry either voice or data. Thus, by coupling together a number of these channels, both types of traffic could be exchanged simultaneously.

There are two types of ISDN connection available. The smaller is known as basic rate in which the network allocates three channels to the link; of the total bandwidth of 192 kb/s thereby allocated, two 64 kb/s channels and one 16 kb/s channel are available to the user. The 64 kb/s channels are known as B channels, whereas the 16 kb/s channel is known as a D channel.

The larger system (known as primary rate ISDN) takes up a full multiframe and offers to the user 30 B channels and one D channel. In this case the D channel has a bandwidth of 64 kb/s.

In each case the D channel carries signalling which, between exchanges, is based on SS 7. If there is spare capacity on the D channel it can be used for low-speed data such as information for process monitoring.

10.38   ISDN TERMINAL EQUIPMENT REFERENCE MODEL

It is not possible to examine ISDN in detail here. It is specified by the extensive CCITT Recommendations in the I-series. In general, these Recommendations have been agreed by most operators, manufacturers and service providers, but because of their complexity, and because at one or two points they leave scope for different interpretations on detail, there are in fact small differences between the systems being implemented in different countries. This leads to interworking problems that have to be resolved through the use of interface circuitry. With time, the differences will be reduced.

In what follows we will disregard any of these small variations, and describe the main features and functions of ISDN.

In developing the Recommendations, CCITT was concerned not to influence more than necessary the way in which ISDN systems would be implemented. Its Recommendations are therefore based on functions, interface reference points and frame structures, rather than on the circuitry and software that might be used in a particular system. This has the advantage that ISDN can be discussed without reference to the technology to be used, but it has the disadvantage that at times the description is apparently vague and difficult to relate to a physical system.

In its simplest form, the reference model is shown in Fig. 10.45.

10.38.1   User-network interface functions

The functions of the boxes in Fig. 10.45 are described in CCITT Recommendations in general terms. The terminal equipment that can be connected to ISDN is of two types, TE1 and TE2. TE1 terminals are those that satisfy the ISDN Recommendations, and can therefore be connected directly to the network, whereas TE2 terminals have interfaces at the R reference point that correspond to a non-ISDN (though CCITT) standard. These terminals are connected to the network via a terminal adapter which provides the necessary standards interface.

NT1 and NT2 are the network units that terminate the operator's network and interface with the terminals:

NT1 provides physical layer functions, isolating the user from the loop to the local exchange. Some maintenance functions may be provided in NT1 and it forms the connection interface into the user's premises; perhaps offering a multi-drop connection from several terminal types (e.g. telephone, data source) on the user's premises to one controller, or indeed combinations of connections between these internal networks.

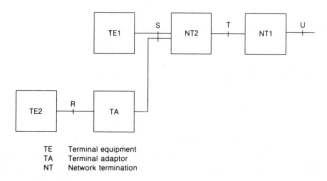

TE   Terminal equipment
TA   Terminal adaptor
NT   Network termination

Fig. 10.45 ISDN terminal reference model.

NT2 can provide a variety of functions; switching, services within the user premises, multi-point connections, and the functionality of the first three OSI layers. Hence, NT2 could represent a PABX or a LAN or a terminal controller, or indeed combinations of connections between these networks.

In a particular installation there are several possible arrangements for NT1 and NT2. NT1 may be missing, or both NT1 and NT2 functions may be integrated. The recommendations contain examples of such combinations, as shown in Fig. 10.46.

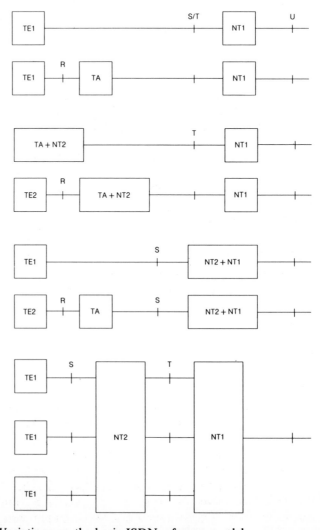

Fig. 10.46 Variations on the basic ISDN reference model.

10.38.2   Reference points

The reference points identify particular positions in the system against which the various parties involved can identify the functions of the blocks discussed above.

The S reference point, separating the terminal equipment from the network termination, provides, outside the US, the demarcation between the operator's equipment and that of the user.

The T reference point provides the interface between the two network termination functions, NT1 and NT2 described earlier.

We have assumed that the interface between the user and the network appears at point T. However, in the USA, in order to satisfy the intention of the regulatory arrangements, that attempt to minimize the part of the network that is regulated, the network access point is considered to be U. In that case, NT1 and NT2 are considered to be part of the user's equipment and the network is responsible for all functions to the right of NT1.

10.39   FRAMING

The three access channels in the basic rate service are multiplexed onto one line. In addition to the two B channels and the D channel, synchronization, framing and some management function information, are sent outside these channels and it is this overhead that uses the additional 48 kb/s capacity available in the transmission rate of 192 kb/s on the basic rate system.

Figure 10.47 shows the framing arrangement at the physical layer on the line between the exchange and the NT1 interface. It consists of 48 bits transmitted in 250 µs, corresponding to a bit rate of 192 kb/s. Examination of the frame shows that the B channels each have 16 bits, and the D channel 4 bits, equivalent to 64 kb/s and 16 kb/s, respectively. The remaining bits ensure that the AMI transmission (Section 3.8), works correctly, and that framing is synchronized at each end of the link.

In the primary rate system a complete 64 kb/s channel is available to carry overhead bits.

10.40   SIGNALLING CHANNEL PROTOCOLS

ISUP – Integrated services user part

In Section 10.36 we discussed user parts as the upper layers of the SS7 architecture, with the MTP layers below them. The user parts DUP, TUP and ISUP are related to circuit switching, being concerned with the automatic setting up, clearing down and management of calls.

The ISUP is more flexible than the other two, and will eventually supercede them if ISDN becomes the normal type of connection. It differs from DUP and TUP in that it relies more on the use of optional fields, thus providing flexibility, but at the cost of more processing at the end exchanges.

The details of the codes and protocols used to establish and monitor

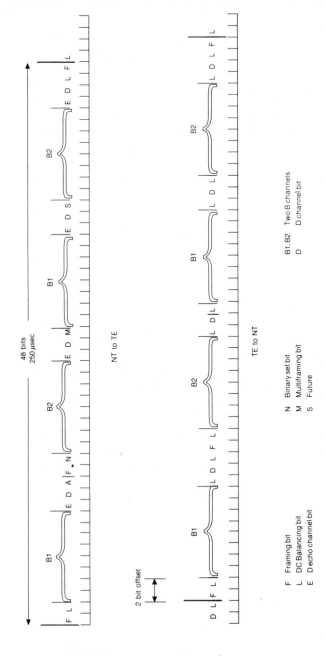

48 bits
250 μsec

NT to TE

2 bit offset

TE to NT

F   Framing bit
L   DC Balancing bit
E   D echo channel bit

N   Binary set bit
M   Multiframing bit
S   Future

B1, B2   Two B channels
D   D channel bit

Fig. 10.47 Bit pattern for basic ISDN transmission.

Fig. 10.48 ISUP, frame structure.

connections in normal and abnormal conditions is to be found in the appropriate CCITT Recommendations, but essentially the ISUP frame, that fits into the MTP frame, has the form shown in Fig. 10.48.

CIC is a circuit identification code linked to the end-to-end virtual circuit being established. The message type code is a necessary prelude to the other fields, indicating the function and structure of the message being sent. Similarly, the fixed and variable segments of the mandatory fields are essential, and related to the message type field. The optional field can be of any length up to 131 octets, accounting for more than a dozen parameters. There is a trade-off necessary between the complexity involved with the field and the flexibility that a wide repertoire presents.

10.40.1 Local access D channel

The information sent over the D channel is carried on an HDLC type protocol known as LAP-D (channel D link access protocol). Two mechanisms are available: (a) acknowledged information transfer; and (b) unacknowledged information transfer. The difference being that in the first case the protocol allows for the identification and rejection of erroneous frames, and provides error and flow control facilities. Frames have sequence numbers to identify order. In the unacknowledged mechanism there is no error or flow control facility and hence there is no need to number frames. Erroneous frames are discarded, but there is no attempt made to correct errors or initiate retransmission.

The frame format is shown in Fig. 10.49. The significant difference between LAP-D and HDLC is the inclusion in LAP-D of the terminal endpoint identifier (TEI) and the services access point identifier (SAPI) fields. The TEI

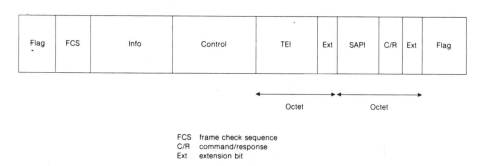

Fig. 10.49 LAP-D frame structure.

field is used to identify the terminal equipment in the connection, since one ISDN line may be accessing several terminals. The SAPI field indicates the signalling point to which the message is addressed in the particular connection.

10.41   ISDN SERVICES

Two distinct groups of services are identified in the Recommendations. These are known as:

(1)   Bearer services
(2)   Teleservices

In general bearer services are associated with, and limited to, those services that can be offered over the first three protocol layers. Hence the name bearer.

Teleservices are more complex and in principle involve the higher layer levels of the OSI architecture.

With reference to Fig. 10.46, bearer services access the network at the NT1 or NT2 points, whereas teleservices, using the non-communication layers, access the network via either TE1 or TE2. Because the bearer services are limited to the first three layers, they are described within the recommendations (Fig. 10.50) in terms of communication channel characteristics. Any service that fits into one of the categories of bearer services may be transmitted in this class.

Teleservices, also listed in Fig. 10.50, are described more clearly in terms of the application. Teleservices will be extended as more services become available, but at present they include the voice, text and image services in common use, and which will be available over the ISDN network. Interactive video will be accommodated, and voice storage and message retrieval services will be added; these are already available as separate services on some telephone networks.

Essentially the teleservice function is used to provide a high level exchange of information efficiently and reliably, incorporating the communications facilities of the lower levels with the intelligence available in a variety of

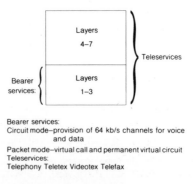

Bearer services:
Circuit mode—provision of 64 kb/s channels for voice
        and data
Packet mode—virtual call and permanent virtual circuit
Teleservices:
Telephony Teletex Videotex Telefax

Fig. 10.50 ISDN bearer services and teleservices.

computer-based terminals that add the high-level functionality. From the user's point of view, teleservices make the communications link transparent, and provide a convenient means of establishing, maintaining and clearing down a connection that transfers information in the format required by the user, between terminals that are compatible, and offers a range of input and output presentations.

10.42 BROADBAND COMMUNICATIONS

There is an expectation in the communications industry that video traffic on the network will gradually increase to form the bulk of the information carried. If that happens the 64 kb/s network will not be able to continue to provide a sufficiently high quality of service. Even if several concatenated channels could be allocated readily, the present network would be inadequate because of intrinsic limitations that would create insurmountable problems.

In the late 1980s, the need to circumvent these restrictions for future communications was addressed. The result was a proposal by AT&T of a system called SONET which, after some modification, was adopted by CCITT under the name synchronous digital hierarchy, or SDH for short. At first the title seems strange, but it refers to the main feature of the technique which is to provide a simple method of multiplexing and demultiplexing information streams of various bit rates; in contrast to the existing system which is now referred to as the plesiochronous network, or PDH.

Ultimately the large variety of information streams – voice, many forms of data, and various classes of video – will be carried over a packet-based network known as asynchronous transfer mode, or ATM. SDH is a transport mechanism which can accommodate the existing network, but which is also capable of carrying ATM packets. Thus SDH not only offers distinct advantages over the plesiochronous network, but it also avoids the effects of obsolescence something that is essential in any feasible replacement system.

Generically, the future high-speed broadband network is referred to as B-ISDN (broadband-ISDN) but it is almost certain that its implementation will be in the form of ATM, so we will restrict our attention to that.

Communications networks are pluralistic in several ways. Traditionally, in each nation state there was one public service operator, the national Ministry for Posts, Telegraph and Telecommunications (the PTT), and one type of traffic, telephony. Now there are in some countries several operators, private rather than public, and many types of traffic. The complexity of networks is increasing rapidly, for in addition to a variety of operators and sources, there are many new services and features being made available. In addition there are increasing demands for complex automatic network management facilities. These all place a considerable burden on the software that controls and supports the network, and on the signalling system that provides the means of communicating between exchanges.

In this chapter we will examine SDH and ATM in some detail, and put them in the context of current developments.

10.42.1 Limitations of the present network

The plesiochronous network has served us well since the advent of digital switching in the 1970s. However, as the demand for higher and higher bit rates occurs, the limitations of the system become apparent. The nature of the PCM multiframe structure is that the timeslots and frames have to be maintained in synchronism by constant reference between the exchange and network clocks. However, while superficially that constant comparison should provide a high degree of synchronism, in fact there can be considerable drift because of different propagation requirements and characteristics over different links. This drift has to be corrected by inserting bits in order to adjust the time in a bit stream so that it regains network synchronization.

The process of adding bits to adjust the position of the start of a word or frame is called bit-stuffing. Although this mechanism avoids slippage at the transmitter end, it causes other serious problems with regard to demultiplexing. For many applications it is desirable to be able to extract one of the component traffic streams without having to demultiplex all of the streams that make up the high-speed channel. In a plesiochronous system that is not possible because the individual bit streams cannot be identified easily, due to the effect of the bit-stuffing.

Figure 10.51 shows the stages in a PDH multiplexor. In order to isolate one 2 Mb/s stream it is necessary to completely demultiplex the hierarchy and then multiplex it up again. For POTS (plain old telephone system, a term used to describe the network prior to the development of ISDN) this clumsy structure did not cause much of a problem, but as various traffic types are linked on to the network there is a frequent need to be able to isolate a particular channel, or to insert a new one along the transmission path. The expense of having multiplexing equipment at each access point is prohibitive in terms of both complexity and cost.

SDH offers a cheaper, more convenient and more flexible approach.

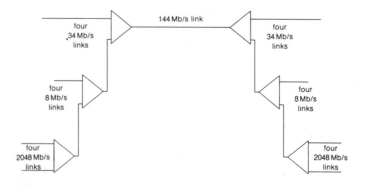

Hierarchy shown is based on 2048 Mb/s lines
i.e. on 32 channel 64 kb/s pcm multiframe

Fig. 10.51 Plesiochronous hierarchy for 2 Mb/s systems.

10.43 SYNCHRONOUS DIGITAL HIERARCHY (SDH)

Built around a basic 125 μs frame, SDH can support both 64 kb/s PCM channels and asynchronous ATM cells. Thus it is attractive as a link between POTS and broadband multi-media networks.

The 125 μs frame is known as STM-1 and consists of 2430 bytes. In discussing the operation of SDH, and its versatility, the frame is understood most easily by representing it as a rectangular matrix having 270 columns and 9 rows, Fig. 10.52. The frame payload is restricted to 261 × 9 bytes, the first nine columns being an overhead, used for control. The payload area is called a virtual container, VC. Both synchronous and asynchronous working can be used; in synchronous mode the frame in column 1 of row 1 is a start of frame byte, thus providing a reference for all other bytes.

The overhead part of the frame is used to indicate the 'address' of each byte. Given 2430 bytes in 125 μs, the information rate is 19 440 000 bytes/s, or 155.52 Mb/s.

The payload available to the user is reduced by the nine-column overhead noted earlier, thus the channel capacity of an STM-1 frame is 155.52 × 261/270 = 150.34 Mb/s. This payload area is called a virtual container level 4, or VC-4 to distinguish it from smaller container sub-sets.

The virtual container gives a high degree of versatility. It can be subdivided into tributary units, and the timing of the start of the VC can be made flexible. By using a pointer within the STM-1 overhead field the start of the VC-4 can be indicated, and provided that this pointer is updated, VC-4 can float within the STM-1 frame. Indeed, the VC-4 does not have to be completely contained within one STM-1 but can spread over a frame boundary, as shown in Fig. 10.53. By using this facility the timing between the STM-1 frame and the VC-4 can be adjusted to accommodate transmission delay at various points in the network.

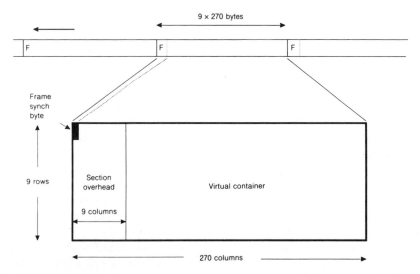

Fig. 10.52 SDH frame format.

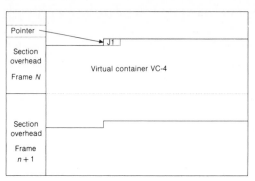

J1 is the first byte of VC-4 and its location is indicated by
the pointer in the section overhead of frame *n*

Fig. 10.53

The pointers can also be used to provide timing adjustment at a synchronizing element to modify the timing of several incoming SDH links so that they are properly aligned at the output.

Although the container VC-4 has a nominal capacity of 150.34 Mb/s, it is not all available to the user. VC-4 remains intact from transmitting node to receiving node, which may be separated by intermediate nodes. To ensure the integrity of the transmission over the whole path, a path overhead is included in VC-4, occupying the first byte of each row. Thus the payload capacity of VC-4 is reduced to 149.76 Mb/s. Evidently VC-4 can accomodate without difficulty the current multiplex rate of 139.26 Mb/s (nominal 140 Mb/s). Thus VC-4 is designed to accept input tributaries at this rate, which corresponds to that required to transmit broadcast quality television.

We discussed earlier that one of the benefits of SDH is the ease with which, compared to the plesiochronous network, tributaries of different bit rate can be multiplexed into or out of the system. This is due to the virtual circuit mechanism.

We have just discussed the 140 Mb/s tributary. Another CEPT bit rate that is well used is 2.048 Mb/s. This is included in the STM-1 frame as a set of 63 tributaries, each of 4 columns width. Such tributary units (called TU12) provide a rate of $4 \times 9 \times 8000 \times 8 = 2.304$ Mb/s.

Other lower bit rates are accommodated by using appropriate TUs to fill the basic VC-4 container. The North American system requires some different sizes of TU from the European, and these are given specific labels: e.g., TU11 occupies three columns and therefore represents 1.728 Mb/s, sufficient to support the DS1 rate of 1.544 Mb/s, and TU2 is 12 columns wide, giving a sub container of 6.912 Mb/s, sufficient for the North American DS2 rate. These TUs are represented diagrammatically in Fig. 10.54. By using the pointing system in the frame overhead, particular channels can be identified, and then abstracted or inserted into the appropriate TU.

Similarly, the 34 Mb/s nominal bit rate is accommodated in a tributary unit that occupies 36 columns; hence seven such TUs can be placed in one VC-4. For each of these bit rates there is a difference between that required by the

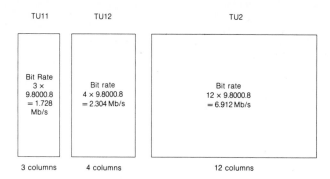

Fig. 10.54 Transmission units for an STM-1 frame.

tributary traffic and that available within the TU in the frame. The difference is made up by adding stuffing bits to the tributary traffic to fill the TU. Additional stuffing bits are needed to completely fill the VC-4 subframe when all of the TUs are in place.

Figure 10.55 shows the functional requirements of the assembler and dis-assembler. To the tributary traffic, overheads to control and manage the connection are added, and the bits are stuffed into the stream to fill the VC-4. At the dis-assembler the reverse procedure occurs. As far as the payload of the STM-1 is concerned, the VC-4, it is transmitted intact across the network, whereas the overhead bits on the STM-l frame are interpreted and adjusted at each node.

STM-1 is the basic SDH unit, but as we have seen, it has a maximum nominal capacity of 140 Mb/s. Higher rates will be required for some applications, and in principle $n \times$ STM-1 can be provided by byte interleaving n STM-1 frames. In practice only two have been defined, STM-4 and STM-16. An STM-4 frame consists of 4 byte-interleaved STM-1 frames, as shown in Fig. 10.56. The result is a frame having an overhead field of 36×9 bytes, and a payload of 1044×9 bytes. The payload is therefore equivalent to $1044 \times 9 \times 8000 \times 8 = 601.344$ Mb/s.

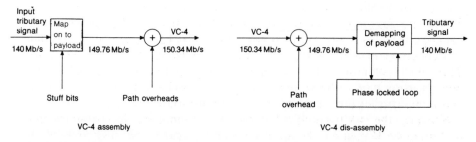

Fig. 10.55 Functional schematic for an SDH assembler/dis-assembler.

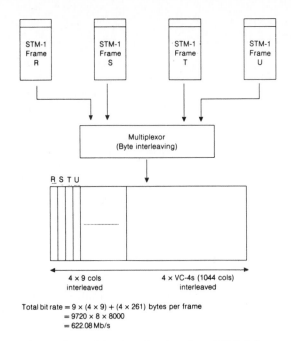

Total bit rate = 9 × (4 × 9) + (4 × 261) bytes per frame
= 9720 × 8 × 8000
= 622.08 Mb/s

Fig. 10.56 STM-4 frame formed by interleaving four STM-1 frames.

10.44   ASYNCHRONOUS TRANSFER MODE (ATM)

It is clear that the 64 kb/s timeslotted PCM transmission system is ideal for voice traffic, but severely restricted if other sources are to be attached to the network. There are two problems with it. First, the timeslot–channel mapping is totally inflexible, and when there is the possibility of a wide range of traffic speeds this is a significant limitation. Secondly, while in principle it is possible for channels to be concatenated to provide for broader bandwidth requirements, in practice the switching and control mechanisms do not make such merging a practical proposition if the service must be available more or less on demand.

To provide for multiple bandwidths an entirely different system must be developed. One that is backward compatible with POTS, but which offers the means of carrying synchronous and asynchronous traffic, from low-speed data to high-definition video.

ATM is one mechanism that will meet that objective, and it is most likely to be the one which will be adopted worldwide. There is one transport mechanism, and several potential types of traffic so it is not surprising that ATM, like any other contenders, is a compromise; offering a reasonable means of transporting most types of traffic. It will not be quite as good as PCM for voice, and perhaps not as good as X25 for data, particularly in terms of error rates and error control, but it will be good enough for both, and certainly better than PCM or X25 as a means of carrying voice, data and video on the same network.

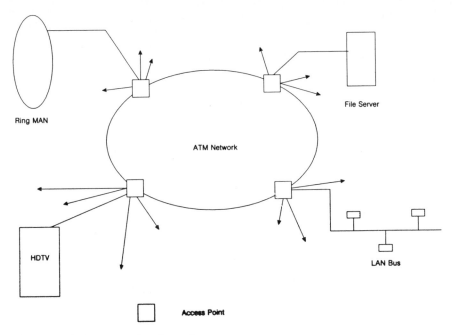

Fig. 10.57 ATM network schematic.

The layout of the network is shown in Fig. 10.57. Because the ATM network itself will have high capacity links, it is likely that the access points will be connected to some concentrating or capacity sharing system such as a LAN, MAN or PBX, or to terminals requiring direct high-capacity access such as large file servers, or high-definition television. The access point will have various functions: mulitplexing, flow control, admission control, and policing, all from the network's point of view, and suitable interfacing software to allow interaction with the access network.

Transport over the communications subnet is via short packets, known as cells. The size of the cell is chosen to give a reasonable compromise that produces a satisfactory performance for all types of traffic.

The transport mechanism is similar to virtual circuit operation in a packet switched network. At set-up the connection is established from end-to-end and that path is maintained for the duration of the exchange of information. The cell, like all packets, consists of two parts, the payload field that contains the information to be sent, and the header that carries mainly addressing and error control. The cell is shown in Fig. 10.58. The information field is 48 bytes long and the header is 5 bytes. This fits exactly the size of the packet in DQDB (see section 13.10).

The detailed structure of the header is also shown in Fig. 10.58. At present some of the fields are not yet completely defined in the CCITT Recommendations, or their purpose has yet to be fully examined. The VCI (virtual channel identifier) and the VPI (virtual path identifier) are both used to determine the route taken by the cell from end to end. The VPI is semi-permanent. It indicates the path to be taken between a paticular source–destination pair,

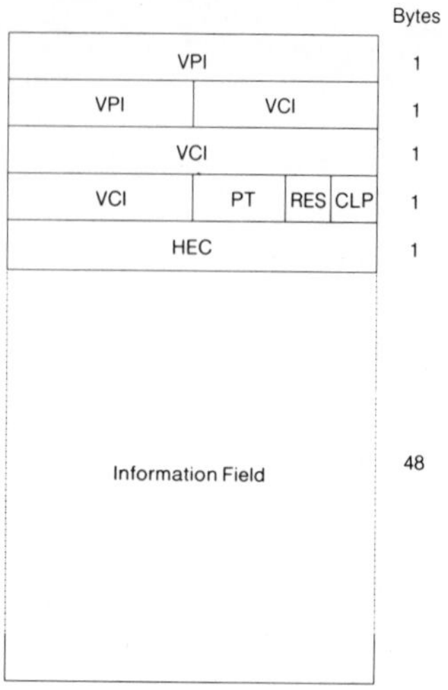

			Bytes
VPI			1
VPI	VCI		1
VCI			1
VCI	PT	RES CLP	1
HEC			1
Information Field			48

Fig. 10.58 ATM cell structure.

and this is fixed in the short term. The VCI indicates the particular channel to be used by the cell on a specific link between two nodes. At a node the incoming VCI is translated to that appropriate for the required output link. One approach is to have a look-up table at the node. The table is updated from time to time by means of a signalling channel.

The error control field is used for the header only. It ensures that the addressing is unlikely to contain errors. Unlike in X25 there is no link-by-link error control on the information field. Voice traffic is not very sensitive to cell loss and it has substantial inherent redundancy so error control is unnecessary. For data however, error control is essential since data has no inherent redundance, so any errors have to be corrected. In X25 this is done at each node traversed. That procedure is slow and not suitable for high-speed broadband applications. In ATM error control is done on an end-to-end basis. All the information is transferred to the destination address and if there are some bits in error the whole message, or that part that has been corrupted, is retransmitted. This reduces to a minimum the time used in processing at the nodes, at the cost of having to wait rather a long time for errors to be corrected. In practice, for many applications, error control is the concern and responsibility of the end user, and occasionally the requirements imposed by the user are very stringent, making it wasteful if the network itself were also to provide protection.

The cell is carried by a broadband transmission medium, probably optical fibre. The cells travel in a continuous stream. If the cell is occupied the destination address is clear. When there is no information wanting to transmit the system sends dummy cells to ensure that propagation is continuous. In theory it would be possible for the link between two nodes to be completely filled with continuous live cells, but in practice a utilization of about 80% would be the maximum that could be realistically used.

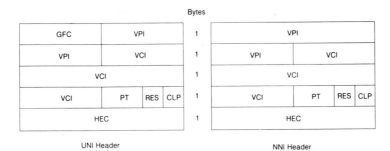

Bytes

Fig. 10.59 UNI and NNI headers for ATM cells.

The payload type field is used to describe in coded form the type of information being carried, and in particular it indicates when a cell is carrying signalling or management information.

The priority bit provides a two-level priority facility. In default value the cell is liable to loss if buffer space is exceeded, whereas in priority mode the cell is processed in preference to those with no priority status.

There is a slight difference between the headers of cells working over the user/network interface (UNI), and those over the network/network interface (NNI), as can be seen in Fig. 10.59.

10.44.1 Statistical multiplexing

Above we have considered the salient features of ATM as a transport mechanism for a broadband communications network. The reason for starting with the network is to establish in general terms how it works, and now we can look at it more from the perspective of the user and the services it offers

Remember that the justification for changing from the current time division multiplexed system to the packet-based ATM is that there is a need to be able to carry a large number of different types of traffic. That being so, how is that traffic put on to the ATM network in such a way that it can be granted sufficient bandwidth in a flexible manner without interfering with other traffic, either already on the network, or likely to use it in future?

The link between traffic sources and the network is a multiplexor. On the user side there are lines to the potential sources, and on the network side there is a very high bit rate link into the network itself, Fig. 10.60. Putting that in another way, the multiplexor can be seen as a resource allocator. It controls a large bandwidth link and has the function of allocating access to that link in as efficiently as possible. The objective is to give each traffic type the bandwidth it needs, without being wasteful and thereby reducing the number of sources that can be served.

Because the cell payload size is very small, a particular message will have to be accommodated by sending it in many cells. Clearly a high-definition television message will require many more cells than a low-speed data

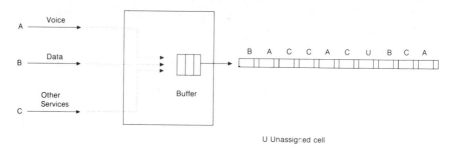

Fig. 10.60 Statistical multiplexor.

message. The statistical multiplexor works in such a way that, provided the collective mean demand of the traffic sources does not exceed the mean capacity of its outlet link, all the traffic can be handled automatically, each being given, in the long term, the bandwidth it needs.

As messages arrive at the multiplexor they are structured into cell-sized packets and queued in a first-come-first-served buffer. The outlet of the buffer goes directly to the outlet link of the multiplexor. Thus there will be many more packets from high-speed sources than from those operating at low bit rates. Consequently, in the outlet stream there will be many more cells from the high-bandwidth source.

Although the outlet provides enough capacity for all possible inlet sources, there is an implicit variation in the delay experienced by different cells. For example, if voice traffic is being transmitted, the number of cells between adjacent voice channel cells in the output stream will not be uniform because of the high probability that consecutive samples will have to be accommodated along with traffic from other sources. The statistical nature of the arriving traffic streams will produce variation in the arrival time of the voice stream cells, unlike in time division multiplex systems where the arrival of voice samples is at precise intervals of $125\,\mu s$. This variation in arrival times can be particularly troublesome for voice traffic since its efficient demodulation depends on this precision of timing. In ATM steps are taken to ensure that the delay variation does not exceed specified limits. Cells from the multiplexor are contiguous. They may be from any source, or from none. If the buffer is empty an unassigned cell is transmitted. The buffer is dimensioned so that the probability that a cell is lost due to the buffer being full is less than one in 10^{-9}

As we might expect in a data-communication type of mechanism the description of the system can be consistently described in terms of the OSI reference model. Figure 10.61 shows that which applies to ATM. The upper layers are the responsibility of the user or service provider. The network provides the physical layer, probably optical fibre, and the ATM layer which contains the mechanisms described above. On top of the ATM layer there is a higher level 2 sublayer known as the adaptation layer (AAL). This provides the interface between the ATM layer and the higher layer functions that may be user-related, signalling or management services.

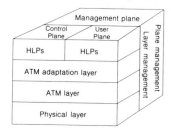

HLPs   Higher layer protocols

Fig. 10.61 Layered model for ATM.

The AAL is responsible for mapping the protocol data units (PDUs) of the higher layer functions into the ATM cell information fields, and vice versa, thus performing segmentation and reassembly of the PDUs. The AAL has a different function for each class of service that might be transported over the ATM network. Earlier we considered in very general terms the variety of sources that could generate traffic, such as low-speed data, voice, and video. The network must also carry management and control traffic and that will have particular characteristics.

The CCITT Recommendation considers four classes of traffic. In essence, traffic may require source and destination to be synchronized in order that information can be correctly reconstructed at the receiver; such traffic is sometimes called real time. Data services, in general do not require source and destination users to be aware of any time relationship. Although bit-rate requirements can vary, the actual rate is not significant in determining the class of traffic. More important is the constancy of the rate. Traffic is divided into two groups, that with constant bit rate, and that in which it varies. Finally, most traffic is transported in connection mode, but in principle there are connection-mode and connectionless mode services.

Although if each of the above variables is considered there are eight possible combinations (from three binary conditions) the CCITT Recommendation considers only four classes as shown in Fig. 10.62, known as Class A, B, C and D.

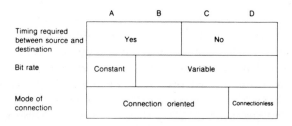

Fig. 10.62 ATM service classes.

Class A traffic is typically voice, or fixed bit rate video. In both there is a synchronization required between source and destination, requiring timing information to be sent over the virtual circuit, and the bit rate used is fixed and non-varying.

Class B is similar to Class A except that there is variation in the bit rate. Some modulation techniques used in audio or video systems result in variation of bit rate as a means of conserving bandwidth; high bit rates occur when there is a strong change in the source audio or video signal.

Class C traffic is not time sensitive, and the bit rate may vary.

In Classes A, B, and C the mode is connection oriented.

Class D traffic is connectionless and, as in Class C, the bit rate is variable.

10.44.2 Traffic management

The large variety of traffic rates and distributions that users may generate introduce entirely new problems for the management and control of broadband networks compared with POTS. There are four main functions:

acceptance
flow control
policing
congestion limitation.

Call acceptance is the mechanism whereby the characteristics of the traffic to be transmitted are relayed to the manager, and a decision taken on whether or not there is sufficient capacity to accept the traffic.

Flow control is related to call acceptance and refers to the management of the network in such a way that if traffic is accepted, it is guaranteed a high probability of successful transmission.

When a call is accepted for transmission a contract is established whereby the user agrees to supply traffic with characteristics within some stated limits, and the network agrees to transmit that traffic. Policing is the fuction of monitoring the traffic actually sent to ensure that the user does not break the contract by sending traffic that falls outside the agreed specification. If the traffic is not policed a user could send more traffic than agreed and therefore defraud the operator, or create difficulties by interfering with other users.

Congestion control concerns re-routing traffic if there is some bottleneck within the network. Normally it would consist of re-routing traffic to avoid the obstacle.

Source traffic is described in terms of its probability distribution function. Normally that function will not be known and therefore some approximation is used. The normal approach is to consider only the first two moments of the distribution, the mean and the variance. This is not ideal for some applications, and the peak bit rate is also used as an important metric.

The traffic management mechanisms raise central questions about the application and implementation of ATM. The principles discussed earlier are straightforward and understandable. However, because of the versatility of the network in being able to carry any type of traffic, there are serious questions still to be resolved about the most appropriate way of carrying out these management functions. The problems all revolve around the difficulty of defining traffic. Experience with telephony has shown that to describe its traffic as having a negative exponential holding time and a negative exponential inter-arrival time is entirely consistent with observations over a long period of time. The traffic arriving from a number of telephone handsets into an exchange can be accurately described as Poissonian, as discussed at the beginning of this chapter.

In ATM, however, things are not so simple. The complexity and range of traffic types creates a serious tension between the need to characterize them accurately in terms of probability distributions, in order that the system can be designed, and can operate efficiently, while at the same time there is a need to make the characterization as simple as possible, so that the resulting analysis and control mechanisms do not become too complicated or expensive.

It is not surprising that these issues have proved to be a rich source of research projects as a large variety of techniques are suggested and evaluated. As is usually the case in complex activities, there is no single best way. Some mechanisms are simple and robust, but relatively unsophisticated, while others can be made sensitive to changes in the traffic, but lack practical attraction. However, there will shortly come a time when either the standards bodies or the industry will agree on particular techniques and further discussion will be to a large extent academic.

There are a few parameters that are used to characterize performance that are relatively uncontroversial. How important each of them are will depend on the traffic type. Cell delay, variation in delay and cell loss probability are the most important.

10.45 CONCLUSION

There are periods of significant change and periods of consolidation in the history of many systems. In communications, significant change has been driven by a combination of technical development and market need. The relatively recent developments of optical fibre transmission and VLSI devices made possible a large number of improvements that could not have happened otherwise. The significant system changes have been from analogue to digital working, the use of cellular and satellite communications, the enormous capacity of optical links, and the power and versatility of software control.

We are rapidly moving towards a stage when the traditional constraints on telecommunications systems are removed. Very large switches are now commonplace, extremely high bandwidth is easily provided, and there are sophisticated services and facilities readily set up. The next clear step is the introduction of multi-media networks. Data communications and telephony are moving closer together, via ISDN and LANs, and soon video will be transported over the same network. The development phase of B-ISDN is giving way to practical devices, and networks will become increasingly packetized.

Eventually the communications network will be capable of carrying all types of traffic of whatever speed. The problems for systems designers will then be related to the automatic, effective and efficient management of a network that carries huge amounts of traffic, that has to provide complex features, and that is an integral part of the economies of companies and countries.

REFERENCES

1. Brockmeyer, B., Halstron, H. L. and Jensen, A., *The Life and Works of A. K. Erlang*, Copenhagen Telephone Company, 1943.
2. Jacobaeus, C., 'A study on congestion in link systems', *Ericsson Technics*, No. 48, 1950.
3. Cooper, R. B., *Introduction to Queueing Theory*, Edward Arnold, London, 1981, Chapter 5.
4. Fry, T. C., *Probability and its Engineering Uses*, Van Nostrand Reinhold, Wokingham, 1965.
5. Fishman, G.S., *Principles of Discrete Event Simulation*, Wiley, Chichester, 1978.
6. Flood, J. E., *Telecommunication Networks*, Peter Perigrinus, London, 1975.
7. Bear, D., *Telecommunication Traffic Engineering*, Peter Perigrinus, London, 1976.
8. Hills, M. T., *Telecommunication Switching Principles*, George Allen & Unwin, London, 1979.

PROBLEMS

10.1 A traffic-recording machine takes measurements of the number of busy devices in a group every three minutes during the busy hour. If the sum of the devices busy over that period is 600, what is the value of traffic carried?

Answer: 30 erlangs.

10.2 A loss-system full availability group consists of five devices. If the mean call holding time is 180 seconds, and the call intensity is 80 calls/hour, what is the mean load per device?

Answer: 0.8 erlangs.

10.3 In a particular system, it was found that during the busy hour, the average number of calls in progress simultaneously in a certain full availability group of circuits was 15. All circuits were busy for a total of 30 seconds during the busy hour. Calculate the traffic offered to the group.

Answer: 15.13 erlangs.

10.4   A group of eight circuits is offered 6 erlangs of traffic. Find the time congestion of the group, and calculate how much traffic is lost.

Answer: 0.122; 0.73 erlangs.

10.5   The overflow traffic from the eight circuits in Problem 10.4 is added to a ninth circuit. What traffic will it carry?

Answer: 0.42 erlangs.

10.6   A system of six telephones has full availability access to six devices. Find the probability that $1, 2, \ldots, 6$ devices are busy. What is (a) the call, and (b) the time congestion of the system if the carried traffic is 2.4 erlangs?

Answer: (a) 0; (b) 0.004.

10.7   (a) Two erlangs of traffic are fed to three devices. What is the congestion, and how much traffic is lost?
(b) Two erlangs of traffic are fed to one device. The overflow is fed to a second device, and the overflow from that to a third. What is the overall congestion, and is the value of the traffic lost the same as in (a)?

Answer: (a) 0.2105, 0.421 erlangs; (b) 0.165, No.

10.8   Show that, if the assumption of statistical equilibrium is valid, the probability of a system being in state i is given in terms of the probability that it is in state 0 by

$$[i] = \frac{\displaystyle\prod_{j=0}^{i-1} \lambda_j}{\displaystyle\prod_{j=0}^{i} \mu_j} [0]$$

10.9   The state transition diagram below represents a system with an infinite number of devices subjected to calls arriving at random with fixed mean arrival rate, λ.

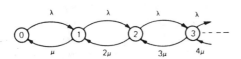

If $\lambda/\mu = A$, the mean offered traffic, use the birth and death equations, and assume statistical equilibrium, to show that

$$[i] = \frac{A^i \exp(-A)}{i!}$$

10.10 An infinite number of sources feed 5 erlangs of traffic into eight devices. Find the probability of the network being in each of its possible states, i, and check that $\sum_i [i] = 1$. Plot $[i]$ against i.

Answer: 0.0072, 0.0362, 0.0904, 0.1506, 0.1883, 0.1883, 0.1569, 0.1121, 0.0700.

10.11 Repeat question 10 for (i) 2 erlangs, and (ii) 7 erlangs of traffic, noting the variation in both the congestion and the distribution of $[i]$, with increasing traffic.

Answer: (i) 0.1354, 0.2707, 0.2707, 0.1805, 0.0902, 0.0361, 0.0120, 0.0035, 0.0009;
(ii) 0.0013, 0.0088, 0.0306, 0.0715, 0.1251, 0.1751, 0.2044, 0.2044, 0.1788.

10.12 A telephone route of n circuits has to carry a normal load of 3 erlangs. If the grade of service must not exceed 0.03, what is the smallest value that n can have? In an emergency, there is a 20% increase in offered traffic. What will be the grade of service in this overload condition?

Answer: $n = 7, 0.0438$.

10.13 Draw the chicken diagram equivalent of the two-stage link network shown below, and calculate the internal blocking.

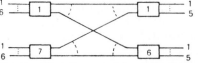

Corresponding outlets from second-stage switches form a route.

Traffic offered per route = 2.5 erlangs
Traffic carried per inlet switch = 1.8 erlangs

Answer: 0.0411.

10.14 A time–space digital switching configuration is shown. What are the functions of the time and space elements in relation to the speech channels on the PCM inlets and outlets? Draw the analogue equivalent of this switch.

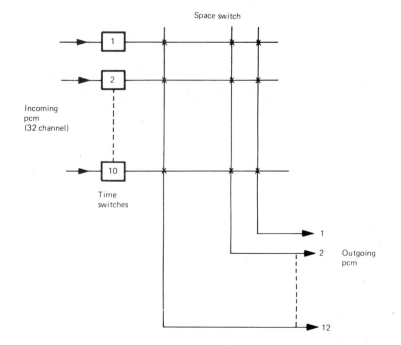

Television systems 11

11.1 INTRODUCTION

This chapter is concerned primarily with the fundamentals of domestic broadcast television systems which now includes direct broadcast satellite systems. There are, of course, many specialized closed-circuit television systems, but these are all based upon the same principles as the broadcast system and will not therefore be considered separately.

Television has progressed over a period of about 70 years from John Logie Baird's mechanical scan system to the modern highly sophisticated colour systems that can transmit both entertainment programmes and teledata information. During this period the quality of sound and visual image reproduction has increased steadily and digital techniques, in particular, have made rapid inroads in what has been basically an analogue technology. The basic principles governing the transmission of visual information by electrical means have, however, remained unchanged. These principles rely upon a property of the human eye/brain combination known as 'persistence of vision'. This means that under certain conditions, the brain is unable to differentiate between a moving image focused on the retina of the eye and a rapid sequence of still images. Television is specifically designed for the human eye/brain combination, and it is appropriate to begin by considering the subjective response of this combination to incident light energy.

11.2 MEASUREMENT OF LIGHT AND THE RESPONSE OF THE EYE

Light is a form of electromagnetic radiation and the power radiated from any source is measured objectively in watts. Light energy is visible only over a very restricted range of wavelengths, from 400×10^{-9} to 7×10^{-9} m. The subjective sense of 'intensity' varies with the wavelength of incident light, where in this context intensity means the impression of an observer of whether a light source is bright or dim. The response of the eye varies between individuals but the variation in this response has been found to be small enough to allow a 'standard observer response' to be defined. This standard response is shown in Fig. 11.1.

A black body radiator radiates equal energy at all wavelengths. The luminous intensity of such a source is obtained by weighting the uniform spectral density of the source by the response of the standard observer. The resulting subjective unit is the candela and is defined as 1/60th of the luminous intensity per cm^2 of a black body radiator maintained at a

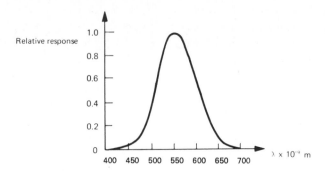

Fig. 11.1 The standard observer response.

temperature of 2042 K. The lumen is the amount of light passing through unit area at unit distance from a point source of 1 candela, i.e. the total emission from a source of 1 candela is 4π lumens. The lumen is the unit of luminous flux and, unlike the power output from a light source, depends upon the response of the standard observer.

It is clear from Fig 11.1 that two light sources of equal power but different colour do not necessarily have the same luminous flux. Two light sources of different colour that appear equally bright to the standard observer do have the same value of luminous flux measured in lumens.

The significance of the last statement will become clear when colour theory is considered in Section 11.8. As far as monochrome television is concerned the response of the camera when viewing a coloured scene should closely resemble the response of the standard observer, i.e. the electrical output should be proportional to the brightness of the scene measured in lumens. Such an output signal is termed the luminance signal.

11.3   THE MONOCHROME TELEVISION WAVEFORM

In Europe broadcast television pictures are transmitted using the 625 line standard (it should be noted however that vigorous efforts are being made to define a universal European high-definition standard). The 625 line signal waveform will be considered as the means of explaining the transmission of a three-dimensional vision signal (i.e. one that is a function of time and both horizontal and vertical position) over a one-dimensional channel (i.e. a channel in which voltage is a function of time). The three-dimensional visual image to be transmitted is focused onto a photosensitive (in the electrical sense) surface by an optical lens system. The photosensitive surface may be considered as a large number of separate photoelectric transducers. Each of these transducers produces an electrical output proportional to the intensity of the image that falls upon it. The original image is thus decomposed into a large number of picture elements known as pixels (or pels). The electrical output corresponding to each pixel is one-dimensional since it is a function only of time. The three-dimensional signal, which is the combined output of all pixels, is actually transmitted as a rapid sequence of one-dimensional outputs. The original optical image is then reconstructed by displaying each pixel in its correct spatial position.

This image is produced, in a domestic television receiver, by modulating the beam current of a cathode ray tube (CRT). The screen of such a tube is covered with a photo-emissive layer and will have the same number of pixels as the photosensitive plate in the image capture system. The image is reproduced by effectively connecting each pixel in the image capture system to the corresponding pixel in the CRT in turn. This is achieved by scanning the photo-emissive screen in the CRT with an electronic beam synchronized to the scanning beam in the image capture system. Sequential scanning of the image capture system and CRT are shown diagramatically in Fig. 11.2. The electron beam is deflected in both vertical and horizontal directions. The scan begins at point A in Fig. 11.2 and proceeds to point B. During flyback, which occurs very rapidly, no information is transmitted. The next line is scanned from C to D and the process is repeated until point F is reached. At this point flyback occurs to point A and the scanning of the complete picture is repeated.

The above explanation is based on the assumption that both image capture and display systems are based on vacuum tubes with scanning electron beams. In many personal computer systems, for example, the CRT has been replaced by a liquid crystal display in which the individual pixels are addressed by a switching waveform generated from a digital circuit and such displays are also employed in miniature TV receivers. It is also commonplace for image capture devices to be based on solid-state technology (see Section 11.18) and in such cases the scanning electron beam is replaced by a similar switching waveform. The scanning electron beams are thus replaced by synchronized switching waveforms, but the scanning process illustrated in Fig. 11.2 is still relevant to the production of television images.

It can be seen from Fig. 11.2 that each picture scan takes a finite time. If this time is longer than the duration of persistence of vision the eye will perceive flicker on the picture. If the scan time is very short the number of pixels/second and hence the signal bandwidth will be large. The optimum scan rate is therefore one that is just fast enough to avoid flicker. The picture frequency at which flicker occurs is related to the luminance of the picture being viewed. For television systems, flicker is avoided when the picture rate is as low as 25 complete scans/second. This is made possible by use of interlaced scanning, which is shown in Fig. 11.3.

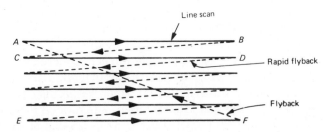

Fig. 11.2 Sequential scanning.

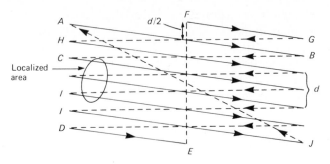

Fig. 11.3 Interlaced scanning.

The scan now occurs in two fields, the odd field starts at *A* and the line is scanned until point *B*. Flyback then occurs from *B* to *C* and so on. It will be seen from Fig. 11.3 that the last line is scanned from *D* to *E*, i.e. field flyback occurs from *E* to *F*. (In fact this flyback takes the equivalent of several line scan periods so will not be a straight line as shown in Fig. 11.3.) The even field starts half-way across the picture at *F* and the scan of the first line finishes at *G*, flyback occurring from *G* to *H*. The last line of the even field is scanned from *I* to *J*, flyback then occurs to point *A*, which is the start of the next odd field.

Interlaced scanning avoids flicker because although each complete picture is scanned 25 times/second, a localized area of the picture is scanned by both odd and even fields. Thus each such area is apparently scanned 50 times/second. This deception relies on the fact that the variation in brightness in a vertical direction usually occurs gradually. If there is a significant difference in the brightness between one line and the next, some localized flicker will be observed. It should be noted that two field frequencies are in common use i.e. 50 fields/second or 60 fields/second depending on the local power supply frequency.

It has been pointed out that it is necessary to synchronize the scans in the image capture system and the display device. To achieve this, a line-synchronizing pulse is transmitted at the end of each line and a field-synchronizing pulse is transmitted at the end of each field. The synchronizing pulses trigger the flyback circuits in the television receiver. The transmitted video signal thus has distinct components, i.e. the picture signal that represents the variation in brightness of each line and the synchronizing pulses that are transmitted below black level. The composite waveform is shown in Fig. 11.4.

The 625 line transmissions in the UK use negative amplitude modulation of the vision carrier, i.e. an increase in picture brightness produces a decrease in signal level. A portion of the envelope of the modulated signal is shown in Fig. 11.5. The advantage of negative modulation is considered to be that the black spots produced on the screen by some forms of ignition interference are less objectionable than the white spots that would be produced in the same circumstances if positive modulation were used. The ratio *WB/BS* shown in Fig. 11.5 is known as the picture/sync ratio and has a value of 7/3. This is a compromise figure that produces a reasonable picture quality and adequate

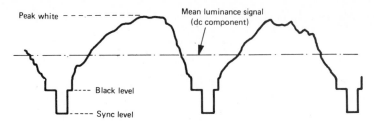

Fig. 11.4 Brightness and synchronizing information.

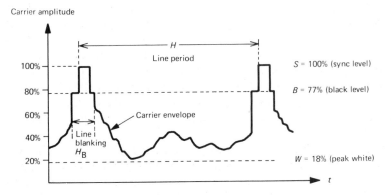

Fig. 11.5 Transmitted carrier envelope.

synchronization in poor SNR conditions. Figure 11.5 shows that a portion of each line before, and just after, each line synchronizing pulse is at black level. This allows the line flyback to be suppressed and hence no visual output is produced during this interval. A similar provision is made at the end of each field, but in this case the duration of black level extends for 25 line periods. This means that in one complete picture (two fields) there are 50 lines which are suppressed. These lines are actually used for various forms of data transmission which is considered in Section 11.8.

11.4 BANDWIDTH OF A TELEVISION WAVEFORM

The bandwidth of a television waveform is directly related to the number of pixels transmitted per second. In each 625 line picture 25 lines are blanked off at the end of each field and the number of lines seen by a viewer is thus $(L - L_B)$ where $L = 625$ and $L_B = 50$. The vertical resolution of the picture is equal to the number of lines seen. Assuming equal horizontal and vertical resolution the number of pixels per line is $A(L - L_B)$ where $A(= 4/3)$ is the picture aspect ratio (width/height). Since each line is blanked out for a period of H_B, the number of pixels transmitted per second is

$$P_s = \frac{A(L - L_B)}{H - H_B} \tag{11.1}$$

For a 625 line system the line period $H = 64\,\mu s$ and $H_B = 12.05\,\mu s$ giving a value of $P_s = 14.76 \times 10^6$ pixels/second. The maximum signal bandwidth occurs when adjacent pixels alternate between black level and peak white. The video signal is then represented by a square wave as shown in Fig. 11.6. It can be seen from this figure that it is possible to transmit adjacent black and peak white levels by a sine wave with a period equal to the period of two pixels. The bandwidth required is then $1/T$ which, in the 625 line system, is 7.38 MHz. It is clear from Fig. 11.6 that if the transmission bandwidth is restricted to this figure then the sharp edges of the square wave will be lost. This is virtually undetectable in practice because of the finite size of the scanning beam, which does not allow detail of this resolution to be displayed in any case. Statistically speaking, very few pictures require the degree of resolution allowed by a bandwidth of 7.38 MHz. Use is made of this fact to reduce the signal bandwidth even further. The actual bandwidth allowed for 625 line transmissions is $K \times 7.38$ MHz. The constant K is known as the Kell factor. The value of K depends upon several elements, e.g. the width of each line on the CRT and the low-pass filtering effect of the eye (which itself depends upon how close the viewer is to the CRT). The value of K used in Great Britain is 0.73, which gives an acceptable subjective picture quality and results in a video bandwidth of 5.5 MHz.

11.5   CHOICE OF NUMBER OF LINES

There have been various standards in operation throughout the world, but this has now been essentially reduced to two, 525 lines in North America, Greenland and Japan and 625 lines elsewhere. All the systems that have been used have one feature in common, the use of an odd number of lines. The reason for this can be explained by reference to Fig. 11.3. If each picture has $2n + 1$ lines, then each field will have $n + \frac{1}{2}$ lines. If it is assumed that the distance between adjacent lines of the same field is d the odd field scan finishes halfway across the last line. This means that field flyback (if assumed instantaneous) would cause the even field to start halfway across the first line and would therefore be a distance of $d/2$ above the first line of the previous field. Interlacing is thus achieved automatically with the same field deflection on both odd and even fields.

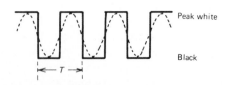

Fig. 11.6 Television picture bandwidth.

11.6   SYNCHRONIZING PULSES

Pulses are required to trigger both line and field timebase generators in the
receiver. In early receiver designs the transmitted synchronizing pulses were
used directly and this imposed a number of requirements on these pulses. The
detail of the line synchronizing pulse is shown in Fig. 11.7. The line flyback is
triggered by the leading edge of the synchronizing pulse. To avoid a ragged
picture the triggering must be fairly precise. This means that the rise time of

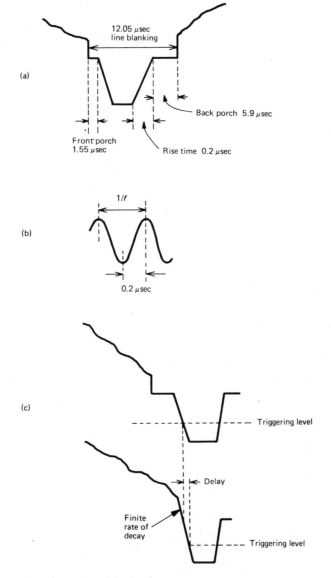

Fig. 11.7 Details of synchronizing pulses.

the pulse should be as short as possible and trigger level should not be affected by the level of the luminance signal just before the synchronizing pulse. In the 625 line system a rise time of 0.2 µs is specified and it may be seen, by reference to Fig. 11.7(b), that this requires a signal bandwidth exceeding $f = 2.5$ MHz. This is well within the allotted bandwidth of 5.5 MHz.

It will be seen from Fig. 11.7(a) that the video signal is reduced to black level for a short interval before and after the line sync pulse; these intervals are known as the front and back porch, respectively. The function of the front porch is illustrated in Fig. 11.7(c). The luminance signal can have any value between peak white and black level at the end of each line. In the absence of the front porch the time taken for the video signal to drop to the triggering level would vary with the amplitude of the luminance signal just before the leading edge of the synchronizing pulse. The timing of the flyback would then vary from line to line giving a ragged picture. The front porch ensures that each line is reduced to black level before the leading edge of the synchronizing pulse. The flyback then occurs at the same instant at the end of each line.

The back porch has two functions – it ensures that the beam is blanked off during line flyback and also provides a convenient reference for restoring the zero frequency component to the signal which has been passed through ac coupled amplifiers.

The field synchronizing pulses occur at the end of each field and their purpose is to trigger the vertical deflection system. It is important that precise triggering is affected to ensure correct interlace of the odd and even fields. The field synchronizing pulses must obviously be distinguishable from the line synchronizing pulses. The picture sync ratio is fixed as 7/3, and hence to distinguish the two types of pulses the field synchronizing pulses are made much wider than the line synchronizing pulses. These pulses have, in fact, a duration of approximately 2.5 line periods.

Modern receivers do not use the transmitted synchronizing pulses directly but use a form of line synchronization known as flywheel synchronization. Instead of using individual sync pulses to trigger each line scan the frequency and phase of locally generated line synchronizing pulses are compared with those present on the incoming video waveform in a phase detector. Any phase error is averaged and fed back to a voltage-controlled oscillator which pulls the two streams of synchronizing pulses into phase coincidence. This averaging effect minimizes the effect of noise and interference on the incoming sync pulses and produces superior picture quality. It is therefore no longer strictly necessary to provide line synchronization pulses during field flyback for modern TV receivers.

Figure 11.8 shows the detail of the field and equalizing pulses transmitted in the UK. The equalizing pulses also relate to early TV designs in which line and field pulses were separated using an integrator. (Integrators produced a higher output for wide field pulses than for narrow line pulses). The equalizing pulses were inserted to ensure that the residual output of the integrator was the same on odd and even fields, thereby ensuring accurate interlacing. In modern receivers line and field pulses are separated in a single IC which also accommodates flywheel synchronization. In such devices simple counter circuits produce the field synchronizing pulses at the correct instants.

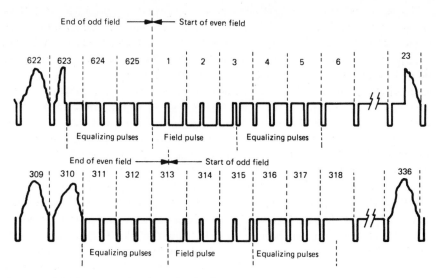

Fig. 11.8 Field sync period in UK 625 line transmissions.

11.7 THE TELEVISION RECEIVER

The domestic television receiver is required to receive signals over a wide bandwidth and is based on the superheterodyne principle. The television receiver is considerably more complicated than a broadcast radio receiver because the former is required to reproduce a video signal, synchronized scanning waveforms for the CRT, and a sound signal. The block diagram of a typical monochrome receiver is given in Fig. 11.9.

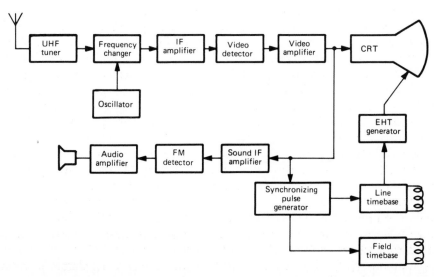

Fig. 11.9 Superheterodyne television receiver.

The British 625 line system transmits the video signal using amplitude modulation of the vision carrier and frequency modulation of a sound carrier 6 MHz above the vision carrier. (The original FM sound carrier is now accompanied by an additional carrier 6.552 MHz above the video carrier which is used for digital transmission of stereophonic sound. The coding scheme used is known as NICAM, which is covered in detail in Section 3.6.) The local oscillator in the receiver operates at 39.5 MHz above the vision carrier, the difference frequency being used as the vision IF. The relative positions of sound and vision carriers will of course be reversed in the IF signal. The spectrum of the IF signal and the frequency response of the IF amplifier are shown in Fig. 11.10.

The asymmetric response of the IF amplifier around the frequency of 39.5 MHz is required to compensate for the vestigial sideband transmission of the vision signal. The detection of VSB signals is discussed fully in Section 2.12. The IF amplifier response is 30 dB down at the FM sound carrier frequency. This is required because both sound and vision carriers are applied

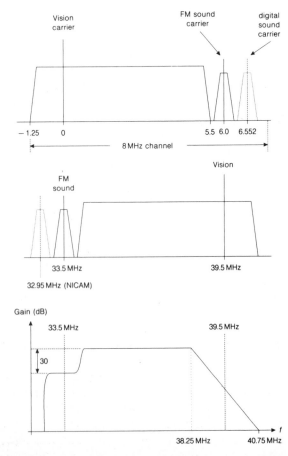

Fig. 11.10 Signal spectra: (a) transmitted signal spectrum; (b) spectrum of 625 IF signal; (c) IF amplifier response.

to the vision detector, which then produces sum and difference frequencies between these two carriers. The difference frequency of 6 MHz is used as the sound IF and this is known as the 'intercarrier sound' principle. This component will be modulated in both amplitude and frequency. The 30 dB drop in the IF amplifier response at the sound carrier frequency ensures that the amplitude of the sound carrier is much less than the vision carrier. The resulting depth of amplitude modulation of the 6 MHz component is then consequently small and easily removed by limiting. (Note that in receivers designed to accommodate NICAM the digital sound carrier is extracted from the mixer output by a filter with a centre frequency of 32.95 MHz.)

Clearly the intercarrier sound principle requires both sound and vision carriers to be present at all times. This requirement is met by ensuring (see Fig. 11.5) that the vision carrier has a minimum amplitude of not less than 18% of its peak value. The demodulated vision signal is the drive signal for the cathode ray display tube. This is a thermionic device, in which the intensity of the electron beam emitted from the cathode is a function of the cathode-to-grid voltage. The electron beam is electronically focused on to a screen coated with a photo-emissive phosphor, which produces the displayed image. The light output of a CRT is not a linear function of the cathode-to-grid voltage, but is equal to this voltage raised to power γ. This means that the output voltage E_Y produced by a television camera would not produce an acceptable image on a CRT. The non-linearity of the display device is equalized by pre-distorting the image capture system output voltage. The transmitted gamma corrected signal is $E_Y^{1/\gamma}$ and has an important but unwanted effect on colour images, as will be considered in Section 11.13.

In addition to gamma correction, the vision signal must be further modified before it is applied to the cathode and grid terminals of the CRT. The dc component of the vision signal, which represents the average brightness of a picture, is removed when the vision signal is transmitted via ac-coupled amplifiers. The dc component also affects the synchronization of line and field timebase circuits, as illustrated in Fig. 11.11 and must be restored to the vision signal. A full discussion of the operation of dc restoration circuits is given by Patchett.[1] After dc restoration the complete vision signal, including synchronizing pulses (which are actually below black level), is applied to the control grid of CRT.

The synchronizing pulses are separated from the picture information and are then used to trigger the line and field scan circuits. The scanning process in television tubes is produced using magnetic rather than electrostatic deflection, the electron beam being deflected at right angles to the magnetic field direction. There are several advantages associated with magnetic deflection systems, the main one being that the deflecting force is proportional to the electron beam velocity and thus increases as the anode accelerating voltage increases. Electrostatic systems, which are commonly used for oscilloscope applications, produce a force that is independent of beam velocity. This means that a high accelerating potential, required to give a bright image, would be accompanied by a small deflection because individual electrons would be influenced by the deflecting force for a shorter interval. The current in the deflection coils has a sawtooth waveform, and during flyback the

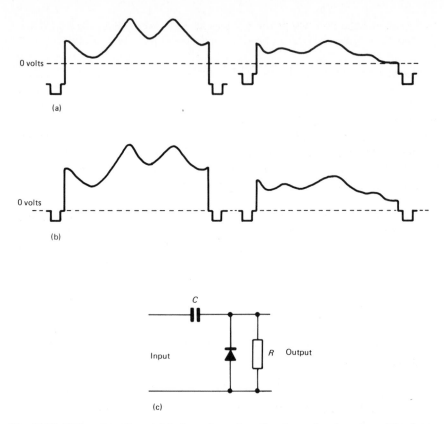

Fig. 11.11 DC restoration: (a) before dc restoration (sync level varies with picture brightness); (b) after dc restoration (sync level fixed); (c) dc restoring circuit.

current changes very rapidly. This produces a considerable induced voltage, which is stepped up by a transformer action and used to form the anode accelerating voltage. A typical accelerating voltage for a colour tube is 15 kV.

Colour television transmission uses the same standards as monochrome transmissions. There are, however, several specific requirements, which must be met to produce acceptable colour images, and in order to understand these requirements it is first necessary to consider the properties of coloured light.

11.8   COLORIMETRY

The colours produced by the display tube in a colour television receiver depend upon the principle of additive colour mixing. This is based on the postulation that a large range of the subjective sensation of colour may be produced by adding certain primary colours in different proportions. This principle is quite different from the principle of the subtractive colour mixing used by artists to create different colours by mixing pigments. A wide range of primary colours can be chosen for the additive colour process, but the primary colours used in television transmission are red, green and blue.

Examples of some colours produced by these primaries are:

$$\text{red} + \text{green} = \text{yellow}$$
$$\text{red} + \text{blue} \ = \text{magenta}$$
$$\text{blue} + \text{green} = \text{cyan}$$
$$\text{red} + \text{blue} + \text{green} = \text{white}$$

Coloured light is usually described in terms of hue (the actual colour), saturation (the dilution of the colour with white light) and luminance (the brightness). The response of the eye is additive, i.e. the luminance of a colour produced by adding three primary colours is the sum of the luminance of the individual primary colours. This algebraic relationship is referred to as Grassman's Law.

White light can be defined in several ways; one convenient definition is equal energy white, in which the energy at all wavelengths has equal brightness because, as is apparent from Fig. 11.1, the sensation of brightness varies with wavelength. Using the response of the standard observer, equal energy white can be expressed in terms of the three primaries, R, G, B as

$$1\,\text{lumen}\ W = 0.3\,\text{lumen}\ R + 0.59\,\text{lumen}\ G + 0.11\,\text{lumen}\ B \qquad (11.2)$$

In colour television trichromatic units (T units) are used to simplify Eqn (11.2). White light is then said to be composed of equal quantities of red, green and blue light when the latter are expressed in T units:

$$1\,\text{lumen}\ W = 1T(R) + 1T(G) + 1T(B) \qquad (11.3)$$

where $1T$ unit of red $= 0.3$ lumen, $1T$ unit of green $= 0.59$ lumen, $1T$ unit of blue $= 0.11$ lumen.

In order for Eqn (11.3) to balance, evidently 1 lumen of white $= 3T$ units. Equation (11.3) can be adapted to represent any colour in terms of R, G and B:

$$1T(C) = x\,T(R) + y\,T(G) + z\,T(B) \qquad (11.4)$$

The coefficients x, y and z in Eqn (11.4) are the trichromatic coefficients of the colour C and it is evident that since this equation obeys Grassman's Law then $x + y + z = 1$. This means that the colour C is actually defined in terms of two trichromatic coefficients. If x and y are known then z can be obtained from $z = 1 - (x + y)$. Colour is often represented graphically in the form of a colour triangle as shown in Fig. 11.12, in which white is defined by the point $x = 0.33$ and $y = 0.33$ (hence $z = 0.33$). The point $x = 0, y = 0$, on the other hand, represents saturated blue as in this case $z = 1$. The colour triangle therefore represents both the hue (the actual colour) and the saturation (dilution with white light), for example the point $x = y = 0.25$ represents 50% desaturated blue as $z = 0.5$ is the dominant component. The hue and saturation of any colour which can be produced using RGB primaries lies within the colour triangle of Fig. 11.12, but the range of colours produced in a colour television display system will ultimately depend upon the phosphors used in the CRT. Red emitting phosphors are based on a rare earth material yttrium/oxysulphide/europium, green emitting phosphors are based on zinc sulphide mixed with copper and aluminium, blue emitting phosphors are based on zinc sulphide mixed with silver.

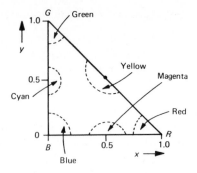

Fig. 11.12 Graphical representation of hue and saturation.

In reality no three primaries exist that can produce the full visible range, and to overcome this problem a universal colour triangle based on hypothetical super-saturated primaries has been defined. All visible colours can then be defined in terms of these hypothetical primaries. This colour triangle is shown in Fig. 11.13.

The primaries used in display tubes in the UK lie within the visible spectrum which can be produced using the hypothetical primaries. The theoretical range of colours that can be produced by a colour television display tube then lies within the dotted triangle shown in Fig. 11.13. It should be noted that although the colour receiver can produce only a limited range of colour, this range is considerably greater than that of high-quality colour film. Equal energy white has the coordinates $x = 0.33$, $y = 0.33$, but the white used for television is slightly different and has the coordinates $x = 0.313$ and

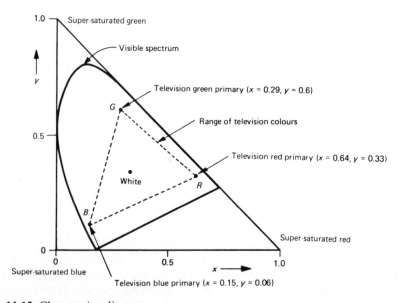

Fig. 11.13 Chromacity diagram.

$y = 0.329$, which is known as illuminant D. This in no way affects the validity of Eqn (11.2).

11.9   REQUIREMENTS OF A COLOUR TELEVISION SYSTEM

It may be concluded, from Eqn (11.4), that in order to transmit colour pictures it is necessary to split the light from the scene being viewed into its red, green and blue components, which are then recombined in the receiver. The bandwidth of the three signal components will be comparable to the figure of 5.5 MHz derived for monochrome signals in Section 11.4. Apart from the excessive bandwidth that such a transmission would require, it would not be compatible with monochrome receivers, i.e. the monochrome receiver would not be able to produce an acceptable black and white image. The requirement of compatibility is an important one and has dominated the development of terrestrial colour television system. However, compatibility is much less of an issue in satellite broadcast systems and this has led to the development of superior colour systems.

This chapter deals primarily with the PAL system which is used in Western Europe (except France and Greece), Australasia and some South American Countries. PAL is based upon the American National Television Standards Committee (NTSC) system, but there are several important differences that will be highlighted at the appropriate point in the text. There are also some minor differences in the PAL standard used in various countries, for example in Australia the sound carrier is 5.5 MHz above the vision carrier. In addition to PAL a brief description of the SECAM system which is used in France, Greece, Cyprus and Eastern Europe is given in Section 11.16, and a brief description of the MAC system, used for some satellite direct broadcast transmissions, is given in Section 11.17.

11.10   THE PAL COLOUR TELEVISION SYSTEM

The basic concept used in all three terrestrial systems mentioned in the previous section is that the colour signal is split into two distinct components. The luminance signal carries the brightness information and is identical to the monochrome signal described earlier. The chrominance signal transmits the colour information (i.e. hue and saturation). Splitting the transmitted information into these categories provides the necessary compatibility because the luminance signal produces an acceptable image in monochrome receivers.

In a monochrome television system, the camera response closely approximates the response of the standard observer. In a colour receiver the red, green and blue signals should approximate to the energy in the primary colours. This is because the eye actually sees the colour image and performs its own relative attenuation on this image, hence when displaying white light red, green and blue signals should be equal in magnitude. The luminance signal can be derived in two ways. One method is to use a standard monochrome camera tube; the other method is to combine the outputs of three separate camera tubes, each tube producing an output corresponding to a different primary colour. When viewing white light the outputs of these three cameras E_R, E_G and E_B are adjusted to be equal. The camera output voltages are therefore interpreted directly in trichromatic units. The luminance signal is derived by combining these outputs according to Eqn (11.2).

$$E_Y = xE_R + yE_G + zE_B \qquad (11.5)$$

where $x = 0.3, y = 0.59$ and $z = 0.11$.

The colour receiver requires a separate knowledge of the E_R, E_G and E_B signals. Ideally the chrominance signal should not transmit any brightness information, as this is already transmitted in the luminance signal. If the effects of gamma correction are ignored it may be shown that this condition is satisfied when the chrominance signal is transmitted in the form of colour-difference signals. The colour-difference signals are:

$$(E_R - E_Y)(E_G - E_Y)(E_B - E_Y)$$

where E_Y is the luminance signal defined by Eqn (11.5)

i.e.

$$\begin{aligned}
(E_R - E_Y) &= (1 - x)E_R - yE_G - zE_B \\
(E_G - E_Y) &= (1 - y)E_G - xE_R - zE_B \\
(E_B - E_Y) &= (1 - z)E_B - xE_R - yE_G
\end{aligned} \qquad (11.6)$$

When transmitting a black and white picture $E_R = E_G = E_B$ and since $x + y + z = 1$, then

$$(E_R - E_Y) = (1 - x - y - z)E_B = 0$$

This result is true for the other colour-difference signals also, indicating that these signals do not contain any brightness information. The original E_R, E_G and E_B can, of course, be reproduced at the receiver simply by adding the luminance signal to the colour-difference signals in turn.

The chrominance signal is in fact completely defined by any two of the colour-difference signals. This can be shown as follows, from Eqn (11.5):

$$E_G = \frac{1}{y}E_Y - \frac{x}{y}E_R - \frac{z}{y}E_B$$

Therefore

$$(E_G - E_Y) = \frac{1-y}{y}E_Y - \frac{x}{y}E_R - \frac{z}{y}E_B$$

but $(1 - y) = x + z$. Hence

$$(E_G - E_Y) = -\frac{x}{y}(E_R - E_Y) - \frac{z}{y}(E_B - E_Y) \qquad (11.7)$$

The green colour-difference signal can thus be derived from the other two colour-difference signals and need not be transmitted. In fact any one of the colour-difference signals could be reproduced from the other two. The values of $(E_R - E_Y)(E_G - E_Y)$ and $(E_B - E_Y)$ may be calculated for a range of transmitted colours and such calculations reveal that the mean square value of the green colour difference is significantly less than the mean square values of the other colour difference signals. The SNR performance will therefore be greatest when the $(E_R - E_Y)$ and $(E_B - E_Y)$ signals are transmitted.

In addition to providing compatibility, the colour television signal must be confined to the same bandwidth as a monochrome signal, otherwise adjacent channel interference would be a serious problem. The question then has to be answered as to how it is possible to fit the chrominance signal into a bandwidth which is already fully occupied by the luminance signal. Fourier analysis of the luminance signal shows that it does not completely occupy the bandwidth allocated to it. The scanning process may be regarded as a sampling operation, and the spectrum produced will be composed of the sum and the difference frequencies centred on harmonics of the sampling frequency (in this case the line frequency). Using the sampling analogy, and noting that the sampling frequency is much larger than the maximum frequency of the signal being sampled, it is clear that there will be periodic gaps in the spectrum of the sampled signal. This is the case with the luminance signal, which has a spectrum of the form shown in Fig. 11.14.

The chrominance signals will have a similar spectrum; the actual amplitude of the components will of course be different. If it assumed that the red and blue colour-difference signals can be combined, it is apparent from Fig. 11.14 that the chrominance spectrum can be slotted into the gaps in the luminance spectrum. To achieve this it necessary to shift the spectrum of the combined chrominance by modulating a sub-carrier of frequency equal to $\frac{1}{2} nf_L$ where n is an odd integer.

The effect of such a modulated sub-carrier on a monochrome receiver is shown in Fig. 11.15. This component will produce an unwanted modulation of the CRT in the receiver. As the chrominance sub-carrier is an odd multiple of the half-line frequency the unwanted modulation on any line should cancel

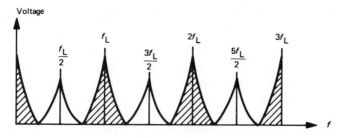

Fig. 11.14 Amplitude spectrum of a luminance signal.

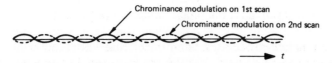

Fig. 11.15 Effect of chrominance signal on a monochrome receiver.

when that line is scanned a second time. In fact this cancellation is not complete. There are several reasons for this, one of the more important being that the light output from a CRT is not linearly related to the grid–cathode voltage (i.e. the decrease in brightness of negative half-cycles of modulation is not equal to the increase in brightness on positive half-cycles of modulation). The result is that a very fine dot pattern is produced on the CRT of a monochrome receiver. This is only visible if the picture is viewed closely and it is usually ignored.

11.11 TRANSMISSION OF THE CHROMINANCE SIGNALS

In the previous section it was assumed that the red and blue colour-difference signals were combined to form a single chrominance signal, the spectrum of which could be fitted into the gaps in the luminance spectrum. The technique used to combine the colour-difference signals uses quadrature amplitude modulation of the chrominance sub-carrier and is similar to the QAM described in Section 3.20. The resulting QAM in this case is given by Eq (11.8)

$$h_c(t) = h_R(t)\cos(2\pi f_{sub}t) + h_B(t)\sin(2\pi f_{sub}t) \tag{11.8}$$

where $h_R(t)$ and $h_B(t)$ represent the red and blue colur-difference signals, respectively. The composite signal may be conveniently slotted into the gaps in the luminance spectrum. Each of the two colour-difference signals is obtained from the chrominance signal at the receiver by coherent detection, which requires the local generation of $\cos(2\pi f_{sub}t)$ and $\sin(2\pi f_{sub}t)$. The outputs of the coherent detectors are:

$$h_c(t)\cos(2\pi f_{sub}t) = \tfrac{1}{2}h_R(t)[1 + \cos(4\pi f_{sub}t)] + \tfrac{1}{2}h_B(t)\sin(4\pi f_{sub}t) \tag{11.9}$$

and

$$h_c(t)\sin(2\pi f_{sub}t) = \tfrac{1}{2}h_B(t)[1 - \cos(4\pi f_{sub}t)] + \tfrac{1}{2}h_R(t)\sin(4\pi f_{sub}t) \tag{11.10}$$

It is clear from these equations that $h_R(t)$ and $h_B(t)$ may be readily obtained by filtering.

Coherent detectors require the locally generated components to be in phase with the suppressed chrominance sub-carriers. The maximum phase error that can be tolerated is in fact $\pm 5^\circ$. The local oscillator thus requires synchronization which is accomplished by transmitting 10 cycles of sub-carrier tone on the back porch of each line synchronizing pulse. The tone is used to synchronize the local oscillators. The detection process is illustrated in Fig. 11.16.

The QAM process is effectively the addition of two double sideband suppressed carriers in phase quadrature. The phasor diagram for the QAM signal may be derived from the phasor diagram of two DSB-AM carriers in phase quadrature as in Fig. 11.17. When the carriers are suppressed the resultant is a single modulated component that varies both in amplitude and phase. It may be shown that the phase of this component represents the transmitted colour (i.e. the hue) and the amplitude of the component

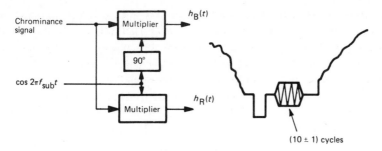

Fig. 11.16 Recovery of colour-difference signals.

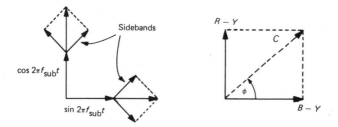

Fig. 11.17 Phasor representation of chrominance sub-carrier.

represents the saturation. The amplitude of the resultant phasor C is proportional to $\sqrt{[(E_R - E_Y)^2 + (E_B - E_Y)^2]}$ and the phase ϕ is proportional to $\tan^{-1}\{(E_R - E_Y)/(E_B - E_Y)\}$. It is convenient to represent various hue and saturation combinations in tabular form. In this context the percentage saturation of any colour is defined in terms of the RGB primaries as

$$S = \frac{\text{maximum amplitude} - \text{minimum amplitude}}{\text{maximum amplitude}} \times 100 \quad (11.11)$$

The hue and saturation for three representative colours are given in Table 11.1 together with the resulting amplitude and phase of the chrominance

Table 11.1 PAL chrominance sub-carrier amplitude and phase

Colour (Hue)	Saturation (%)	E_R	E_G	E_B	E_Y	$E_R - E_Y$	$E_B - E_Y$	C	ϕ
Red	100	1.00	0.00	0.00	0.30	0.70	-0.30	0.76	113.19
Red	50	1.00	0.50	0.50	0.65	0.35	-0.15	0.38	113.19
Red	25	1.00	0.75	0.75	0.83	0.18	-0.08	0.19	113.19
Yellow	100	1.00	1.00	0.00	0.89	0.11	-0.89	0.89	172.95
Yellow	50	1.00	1.00	0.50	0.95	0.06	-0.45	0.45	172.95
Yellow	25	1.00	1.00	0.75	0.98	0.03	-0.22	0.22	172.95
Magenta	100	1.00	0.00	1.00	0.41	0.59	0.59	0.83	45.00
Magenta	50	1.00	0.50	1.00	0.71	0.30	0.30	0.42	45.00
Magenta	25	1.00	0.75	1.00	0.83	0.15	0.15	0.21	45.00

sub-carrier. The last two columns in the table indicate that the phase of the sub-carrier represents the colour transmitted (i.e. the hue) and the amplitude of the sub-carrier represents the saturation. This being so, it is clear that any phase error in the chrominance sub-carrier which may occur during the transmission will thus produce an incorrect hue at the receiver. The PAL system is specifically designed to compensate for this type of phase error and is a development of the NTSC system, which is described in Section 11.14. The sub-carrier frequency used in the British system is 4.433 618 75 MHz, which is equal to $(283.5 + 0.25) \times$ line frequency $+25$ Hz. This differs from the odd multiple of the half-line frequency specified earlier (which is the value used in the NTSC system). The reason for the difference is related to the effect of the PAL compensation for sub-carrier phase error. If a sub-carrier frequency of $\frac{1}{2}nf_L$ is used with PAL an objectionable dot pattern is found to 'crawl' across the display. This is avoided by modifying the sub-carrier frequency as indicated.

11.12 THE TRANSMITTED PAL SIGNAL

The bandwidth of the luminance (i.e. monochrome) signal is fixed at approximately 5.5 MHz. In order to fit the modulated sub-carrier into this bandwidth it is necessary to restrict the sidebands to a bandwidth of 1 MHz. The resolution of the chrominance signal is therefore considerably less than the luminance signal. This is not usually detectable by the eye which is insensitive to high-definition colour. The combined luminance and modulated chrominance signal as shown in Fig. 11.18 is then used to modulate the main vision carrier.

This combined signal is less than ideal for several reasons, for instance in the monochrome receiver there is no provision for removing the chrominance sub-carrier and this signal will consequently be applied to the CRT along with the luminance signal. In the colour receiver it is not possible to separate the luminance and chrominance signals completely which gives rise to a form of distortion known as cross-colour distortion. Before discussing these effects in detail it is necessary to consider the significance of gamma correction in the case of colour television.

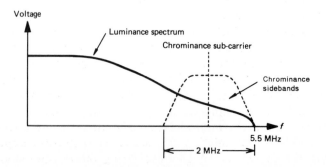

Fig. 11.18 PAL combined lumninance and chrominance signals.

11.13 GAMMA CORRECTION

Most commonly used image capture systems produce an output voltage proportional to the light input and are said to have a gamma value = 1. However, it has already been indicated in Section 11.7 that the light output of a CRT = (grid to cathode voltage)$^\gamma$, where γ has a numerical value of 2.2. To compensate for the non-linearities of the CRT the transmitted luminance signal should be $E_Y^{1/\gamma}$, i.e.

$$E_Y^{1/\gamma} = (0.3E_R + 0.59E_G + 0.11E_B)^{1/\gamma} \qquad (11.12)$$

In a colour system each of the separate colour signals is individually gamma corrected, the resulting luminance signal being

$$E_Y^{1/\gamma} = 0.3(E_R)^{1/\gamma} + 0.59(E_G)^{1/\gamma} + 0.11(E_B)^{1/\gamma} \qquad (11.13)$$

The actual light output produced at the CRT is obtained by raising Eqns (11.12) and (11.13) to the power γ. Equation (11.12) produces the correct luminance value for all values of $E_R\, E_G\, E_B$ but Eqn (11.13) produces the correct luminance value only when $E_R = E_G = E_B$. This occurs only on black and white scenes. For coloured scenes the light output produced by Eqn (11.13) is less than the light output produced by Eqn (11.12). This means that intensely coloured parts of a picture will be reproduced on a monochrome receiver with a lower luminance than the correct value. This error is reduced, to some extent, by the presence of the chrominance sub-carrier which will have a large amplitude on saturated colours, and no attempt is made to remove this component in monochrome receivers. This is regarded as an acceptable compromise from the compatibility point of view.

The situation in a colour receiver is quite different. The individual gamma-corrected colour signals are obtained and applied simultaneously to the CRT. The light output obtained by raising these individual signals to the power γ is therefore correct, i.e.

$$(E_R^{1/\gamma})^\gamma = E_R$$

The consequence of this is that if the light output is correct in the colour receiver then part of the luminance information in the gamma-corrected colour signals must be contained in the chrominance signal, which is not used in the monochrome receiver. Therefore the gamma-corrected signals do not provide the required monochrome compatibility.

The signal which is used to modulate the vision carrier is E_Y' + chrominance sub-carrier (which is equivalent to a sinusoidal component of varying amplitude and phase). In monochrome transmissions the maximum depth of modulation of the vision carrier occurs on peak white. In colour transmissions maximum modulation occurs on bright yellow, the total video signal amplitude being 1.78 times the peak white value. The next peak occurs at bright cyan, the total amplitude being 1.46 times the peak white value. It is clear from Fig. 11.5 that overmodulation of up to 78% of the vision carrier could result. This overmodulation is reduced to a practically acceptable value of 33% which only occurs rarely. Considering the transmission of bright yellow and taking $\gamma = 2.2$ the values of the luminance and chrominance components are:

$$E'_Y = 0.92 \qquad (E_R^{1/\gamma} - E'_Y) = 0.08 \qquad (E_B^{1/\gamma} - E'_Y) = -0.67$$

For bright cyan the figures are:

$$E'_Y = 0.78 \qquad (E_R^{1/\gamma} - E'_Y) = -0.53 \qquad (E_B^{1/\gamma} - E'_Y) = 0.22$$

Multiplying factors m and n are chosen for the gamma-corrected colour-difference signals to restrict total amplitude of the vision carrier to 1.33, i.e.

$$E'_Y + \{n^2(E_R^{1/\gamma} - E'_Y)^2 + m^2(E_B^{1/\gamma} - E'_Y)^2\}^{1/2} = 1.33 \qquad (11.14)$$

The numerical equations for bright yellow and cyan are

$$0.92 + \{(0.08n^2) + (0.67m)^2\}^{1/2} = 1.33$$

$$0.78 + \{(0.53n)^2 + (0.22m)^2\}^{1/2} = 1.33$$

Solving these equations gives $m = 0.877$ and $n = 0.493$. The γ-corrected colour-difference signals transmitted in the PAL system are called U and V signals and are given by

$$U = 0.493\,(E_B^{1/\gamma} - E'_Y)$$

$$V = 0.877\,(E_R^{1/\gamma} - E'_Y) \qquad (11.15)$$

The original components $E_R^{1/\gamma}$, $E_G^{1/\gamma}$ and $E_B^{1/\gamma}$ may then be reproduced in the receiver for applying separately to the CRT.

It is common practice to use a colour bar test signal known as the CVBS (chroma, video, blanking and syncs) in setting up television receivers and such a signal clearly demonstrates the concept of the allowable overmodulation

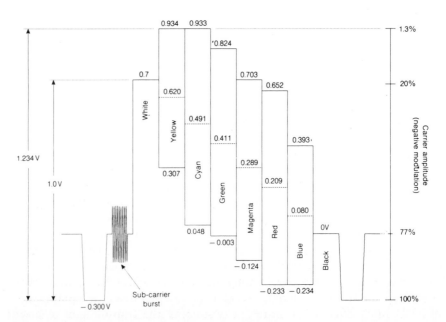

Fig. 11.19 Standard colour bar CVBS reference signal.

outlined above. This standard signal has a nominal value of 1 V peak-to-peak and is the reference signal at the video input and output sockets of cameras, video recorders, monitors and so on. The reference signal is shown in Fig. 11.19. It may be seen from this figure that peak white has a value of 0.7 V, bright yellow has a maximum value of 0.934 V (1.334 × peak white) and bright cyan has a maximum value of 0.933 V (1.332 × peak white).

11.14 THE NTSC CHROMINANCE SIGNAL

The NTSC chrominance signal differs significantly from the PAL equivalent. The NTSC system transmits I and Q signals instead of colour-difference signals. The I and Q signals and their relationship to colour difference signals are shown in Fig. 11.20. The rationale for the use of I and Q signals is based on the observation that the human eye is most sensitive to colour detail in orange and cyan hues and is least sensitive to colour detail in green and magenta hues. This means that the Q signal may be transmitted in a relatively narrow bandwidth, double sideband transmission being employed. The I signal is transmitted in a much larger bandwidth with vestigial sideband transmission.

The I and Q signals may be fully separated at the receiver over the bandwidth of the Q signal, using normal coherent detection with quadrature carriers, as two sidebands are present for both signals. Compensation is necessary for the I signal for the frequencies present in one sideband only, as is usual for vestigial sideband detection.

Restricting the bandwidth of the Q signal in this way means that the highest possible chrominance sub-carrier frequency may be used which produces the finest possible dot pattern on monchrome receivers. In colour receivers the high chrominance sub-carrier frequency means that spurious colour effects

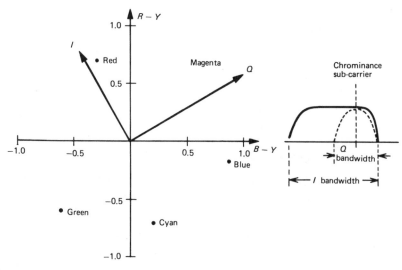

Fig. 11.20 NTSC chrominance signals.

due to adjacent luminance components (which are not fully removed by the sub-carrier detection circuits) are minimized because the luminance energy decreases at the higher video frequencies. This is also true for PAL and is illustrated in Fig. 11.18.

The I and Q signals are derived from the red and blue colour-difference signals according to the relationship specified in Eqn (11.16). These colour-difference signals, and subsequently the colour signals themselves, may be derived from the I and Q signals at the receiver.

$$I = 0.74\,(E_R - E_Y) - 0.27\,(E_B - E_Y)$$

$$Q = 0.48\,(E_R - E_Y) + 0.41\,(E_B - E_Y)$$

(11.16)

11.15 THE PAL RECEIVER

The PAL system is a development of the NTSC system, in which special provision is made to correct any phase errors that may occur in the chrominance sub-carrier. Such phase errors are mainly due to incorrect synchronization of the local oscillator and sub-carrier generator in the television receiver or to differential phase distortion in the transmission system. It is clear from Table 11.1 that the phase of the chrominance sub-carrier is related to the hue of the image and any phase error will produce incorrect colours. The PAL system compensates for this by reversing the phase of the V signal on alternate lines (PAL in fact means 'Phase Alternation, line by Line'). In the receiver the correct phase of the V signal on alternate lines is restored by averaging the phase error over a period of two lines.

A sub-carrier that has a phase error of θ is illustrated in Fig. 11.21. When the V signal is inverted on alternate lines by the receiver the phase error is also inverted (i.e. a positive phase shift of θ on the received signal becomes a negative phase shift of θ after the phase of the V signal is inverted). If the chrominance information on two adjacent lines is the same the average phase error over this period is zero. In practice the chrominance signal on adjacent lines will not be identical and complete cancellation of phase error will not result.

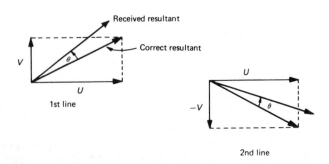

Fig. 11.21 Alternate line inversion of the V signal.

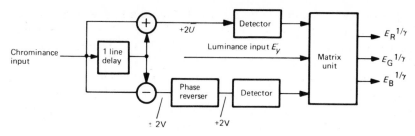

Fig. 11.22 Demodulation of the PAL chrominance sub-carrier.

The averaging is carried out in most modern receivers by adding and subtracting the chrominance signal from its value on the previous line. This requires a delay of 64 µs, which is easily produced using a charge coupled delay line similar to the analogue shift registers shown in Fig. 11.28. A block diagram of the phase error averaging procedure is shown in Fig. 11.22. The matrix unit combines the U and V signals in the required ratios to produce the three gamma-corrected colour-difference signals. These signals are added to the luminance signal to produce the gamma-corrected primary colour waveforms which are fed to the tri-colour display device.

It is clear from the discussion of the PAL system that many compromises are made at various stages of production, transmission and detection of the composite video signal. Non-ideal circuit components invariably add further distortion to the eventual colour image produced. Most of the compromises which occur within the PAL system have been dictated by the requirement of compatibility with monochrome receivers. Ironically the monochrome receiver is now something of a rarity but the limitations due to the requirements of this compatibility remain. In satellite television systems monochrome compatibility was not an issue and much higher quality colour images are possible as a result. Before discussing satellite systems a brief consideration of a third widely used terrestrial system known as SECAM will be given. As with NTSC and PAL this system has also been constrained by the requirements of monochrome compatibility.

11.16   THE SECAM SYSTEM

The *Séquential Couleur à Mémoire* (SECAM) system differs from PAL and NTSC, which both transmit two colour-difference signals simultaneously, by transmitting two colour-difference signals sequentially. The red and blue colour-difference signals are transmitted separately on alternate lines using frequency modulation of the chrominance sub-carrier. There is no local chrominance sub-carrier required in this system, and hence phase distortion is not a problem. However both colour-difference signals are required at the receiver at the same time. SECAM meets this requirement by delaying each colour-difference signal by 1 line interval and using the delayed signal on the next line. Hence each transmitted colour-difference signal is used on two adjacent lines. This obviously halves the colour definition, but this is virtually

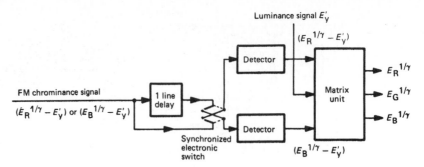

Fig. 11.23 The SECAM decoder.

undetected since the luminance signal transmits the full definition. A block diagram of the SECAM action is given in Fig. 11.23.

The inherent simplicity of the SECAM system suggested by the block diagram of Fig. 11.23 is not in fact realized in practice. The frequency modulated chrominance sub-carrier produces a visible dot pattern on both monochrome and colour receivers which varies with the frequency deviation. Several signal-processing techniques are used to reduce this effect, which consequently increases the complexity of the receiver. Discussion of these techniques is beyond the scope of this text, and the reader is referred to Sims[2] for further information.

11.17 SATELLITE TELEVISION

Satellite communications are covered in detail in Chapter 14. In this section consideration will be confined to the specific issues associated with the transmission of television signals. Direct broadcasting from satellites, in geo-synchronous orbit, to individal homes became established during 1989 with the advent of the Astra 1A satellite. In order to be geo-synchronous, satellites are required to orbit at an altitude of approximately 35 800 km above the equator with a speed of 11 000 km/h, the orbital path is known as the Clarke Belt, after the man who first suggested using satellites for communications (see Section 14.1). Since the geo-synchronous orbital path is fixed it is necessary to regulate the number of satellites permitted and individual satellites are allocated a slot in the orbital path by the World Administrative Radio Conference (WARC). The main satellites visible from Europe are shown in Fig. 11.24.

Satellites are equipped with a number of transponders, receiving programmes and signalling information from an earth station on one frequency and transmitting the programmes back to earth on a different frequency, the frequencies used for the up and down links are in the range 10.95 GHz to 14.5 GHz.

Direct broadcast satellites are usually equipped with up to 10 transponders each of which has a transmitter power of about 100 W. Satellite TV transmissions therefore use frequency modulation to capitalize on the higher

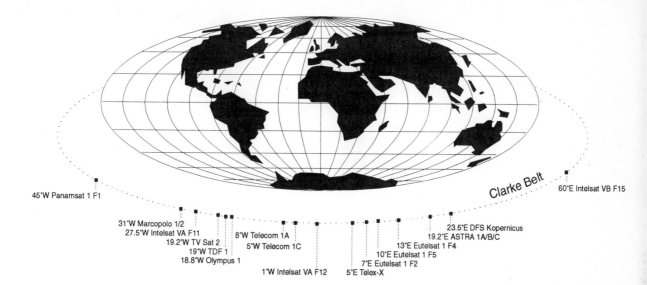

Fig. 11.24 European direct broadcast satellites.

immunity of such transmissions to noise and interference (see Section 4.9). Satellite channels are typically 27 MHz wide and often use the same transmission standards as terrestrial systems in the countries for which they are intended (i.e. PAL, NTSC etc.). These standards are far from ideal as they were designed specifically for amplitude modulation of the vision carrier. However, they do preserve compatibility which reduces the cost of receiving equipment. However, some satellites use an entirely different form of transmission known as multiplexed analogue components system or MAC.

11.17.1   The MAC standard

The terrestrial systems covered earlier in this chapter all exhibit significant compromises which are necessary to maintain compatibility with monochrome transmissions. For example in the PAL system it is not possible to completely separate the chrominance and luminance signals and this produces a form of distortion, in areas of the picture with fine definition, known as cross-colour. This appears in the form of spurious blue/yellow and red/green herringbone patterns. The MAC system is one of several new coding schemes that have been developed for broadcast systems which do not sacrifice quality for the sake of monochrome compatibility and is currently used on the Marcopolo satellite, positioned at 31 °W relative to the Greenwich meridian.

The main attribute of the MAC signal is that the chrominance and luminance information are separated completely during transmission so that much higher quality images may be reproduced at the receiver. In addition MAC systems give good quality reception at a signal-to-noise ratio of 11 dB, compared to a figure of about 40 dB for amplitude modulated systems.

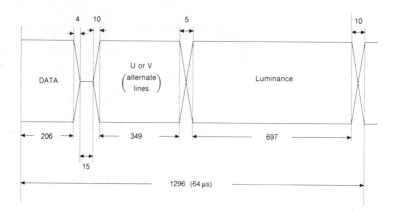

Fig. 11.25 MAC TV line format.

Current MAC transmissions are based on a 625 line standard and therefore are required to maintain the 64 μs line period. In the MAC system this interval is divided into four distinct slots (data, clamp, chrominance and luminance) and time compression techniques are used to accommodate both luminance and chrominance components. The line format for the MAC TV signal in Fig. 11.25.

In this figure the 64 μs line period is divided up into 1296 clock intervals of which 206 intervals are used for data transmission, 15 intervals are used for clamping purposes, 349 intervals are used for the analogue chrominance signal and 697 intervals are used for the analogue luminance signal. The remaining clock intervals are distributed between the necessary ramp-up and ramp-down periods between the four slots. The luminance signal is compressed in the ratio 3:2 and the chrominance signal in the ratio 3:1, a technique for achieving the required time compression is described in Section 11.18 under solid-state image capture devices. The time-compressed luminance and chrominance signals are used to frequency modulate the carrier. The uncompressed luminance signal in MAC has effectively the same bandwidth as in the PAL system. The uncompressed MAC chrominance signal has a bandwidth of 2.8 MHz compared with the 1 MHz figure for PAL.

The MAC digital data signal uses the same carrier frequency as the vision signal but the transmission mode is by PSK. The digit rate is therefore 20.25 Mb/s which fits within the allocated channel bandwidth of 27 MHz. The 206 bit data slot contains one run-in bit, a 6 bit word for line sync, 198 bits of data and one spare bit. The data bits are used to provide several high-quality voice channels, with multi-lingual options and many pages of telextext. The effective data rate is therefore 198 bits for each line period which is 3.09 Mb/s.

The satellite transmission reaches the receiver dish at microwave frequencies (12 GHz for example) and it is necessary to down-convert this frequency to a value suitable for transmission over a coaxial cable to the MAC decoder. The initial amplification and down-conversion to a first IF (usually between 950 MHz and 1.7 GHz) takes place at the point of reception (i.e. at the dish) in

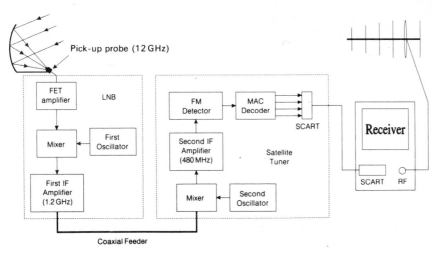

Fig. 11.26 Satellite television receiver system.

a unit known as the low-noise block (LNB). The LNB has a noise figure between 1 dB and 1.6 dB with an overall gain of in the region of 60 dB. The output of the LNB is fed by coaxial cable to a satellite tuner which performs a second down-conversion to an IF of about 480 MHz. The coaxial feeder is also used to supply power to the LNB (typically 200 mA at 15 V). The signal at 480 MHz is then fed to a frequency demodulator and the original luminance and chrominance signals are produced in a MAC decoder. In order to achieve the full potential of MAC transmissions the final output of the decoder can be made available as *RGB* signals and the majority of modern TV receivers on sale in Europe can accept such signals via a EURO-AV (SCART) socket, which means that such a receiver can accommodate both satellite and terrestrial transmissions.

The satellite decoder system actually performs many ancillary functions. For example many broadcasts are scrambled and encrypted to prevent unauthorized reception. Authorized decoders often contain a 'smart card' which may be purchased from the programme company and will facilitate descrambling and decryption enabling normal reception for a fixed period. A typical satellite receiving system is shown in Fig. 11.26. The MAC system is able to accommodate various enhancements, in particular an increase in definition provided by the 1250 line high defintion TV standard with an aspect ratio of 16:9. Using digital techniques this can provide compatibility with the 625 line standard with an aspect ratio of 4:3.

11.18 IMAGE CAPTURE AND DISPLAY SYSTEMS

Image capture systems are the electronic components of the television camera which transform incident light energy into electric signals. The television camera also consists of complex optical systems which are outside the scope of this chapter and will not therefore be covered. There are essentially two forms of image capture systems in widespread use. One is based on a photoconductive vacuum tube, known as the vidicon, and the other form is based on charged-coupled solid-state electronic devices.

11.18.1 The vidicon tube

The construction of the basic vidicon tube is shown in Fig. 11.27. Incident light is focused on to a target disc of continuous photoconductive semiconductor material. The resistance of this material is inversely proportional to the intensity of incident illumination. The photo conductive target is scanned by a low-velocity electron beam, emitted from the cathode, which results in the rear surface of the target being stabilized at approximately cathode potential. The vidicon has electrostatic focusing supplemented by a magnetic focusing coil which is coaxial with the tube. The effect of these two focusing arrangements is to cause individual electrons in the scanning beam to move in a spiral path which coincides with the paths of other electrons at regular distances from the cathode. The focusing arrangements are adjusted so that one of these points of convergence coincides with the target and this produces a scanning spot size diameter in the region of 20 micron.

The front surface, or signal plate, is held at a potential of about 20 V, which produces a current flow through the resistance R_L. When there is no illumination this current is of the order of 20 nA and is known as the dark current. When an image is focused onto the target disc the conductivity of the disc rises in proportion to the intensity of illumination. This allows an electric charge to build up on the rear surface of the target disc and, between scans, the target disc gradually acquires a positive voltage relative to the cathode. The scanning beam deposits sufficient electrons to neutralize this charge and, in so doing, generates a varying current in R_L (typically 200 nA). Since charge is conducted to the rear surface of the target disc over the whole of the interval between scans the vidicon is very sensitive.

However, the vidicon suffers from the two problems of dark current and long persistence (or lag). The dark current tends to produce shading and noise on the low intensity parts of reproduced images. Image persistence arises from the fact that the scanning electron beam does not fully discharge the target plate and this problem is particularly noticeable when levels of

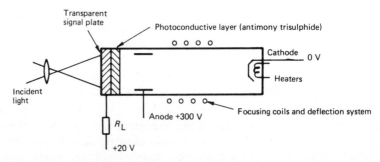

Fig. 11.27 The vidicon tube.

illumination are low. A number of techniques are available to reduce these problems, one of which is the excitation of the face plate by low-level red light generated within the camera assembly.

The vidicon tube described is essentially a monochrome device and will produce a luminance signal. For colour transmission it is necessary to generate the individual RGB signals. This is achieved by using a matrixed face plate in which the front glass surface is covered with thin vertical stripes of RGB colour filter. The vidicon target is similarly divided into vertical strips each one precisely aligned with the corresponding colour filter. All strips on the target corresponding to a primary colour are connected together and brought out to a separate load resistor. Hence by this means it is possible to derive the individual RGB signals from a single tube. These signals are fed to a matrix unit which can produce the gamma-corrected luminance and chrominance signals.

There are several variations on this theme, for example some amateur equipment uses green, cyan and clear filter strips. The target strips behind the green filter produce a G signal, the strips behind the cyan filter produce a $B + G$ signal and the strips behind the clear filter produce a $R + B + G$ signal. It is therefore possible to generate the individual RGB signals by means of a simple matrix unit.

11.18.2 Solid-state image capture devices

Solid-state devices for image capture first appeared in amateur and professional equipment around about 1985. In such devices the photosensitive surface is not continuous but arranged as many thousands of separate silicon photodiodes arranged in horizontal rows equivalent to the lines in a television picture. Each photodiode is therefore equivalent to one pixel and during the 20 ms field period builds up a charge proportional to the light falling on it. Each photodiode is connected to the input of one cell of an analogue shift register, known as a charge-coupled (or bucket brigade) device, by a MOS-FET which is normally OFF. Analogue voltages may be shifted through these devices in the form of the charge on a capacitor. The shift registers are arranged so that adjacent parallel inputs are connected to adjacent photodiodes in a vertical direction as shown in Fig. 11.28.

At the end of each field a transfer pulse is applied to the gate of each of the MOSFETs which causes the charge on each photodiode to be transferred to the appropriate input of the vertical analogue shift register (the number of vertical shift registers is equal to the number of pixels on each television line). Therefore at the end of each field the charge on each pixel is transferred to a cell of one of the vertical analogue shift registers. The complete images is thus stored in the shift registers. The last cell in each vertical analogue shift register is connected to the parallel input of a horizontal analogue shift register. Shift pulses are applied to each vertical shift register simultaneously causing the contents of each cell in the vertical registers to be shifted to the next cell above. The charge in the topmost cell of each vertical register is shifted into one of the cells of the horizontal register.

At this point the cells of the horizontal shift register contain the charge

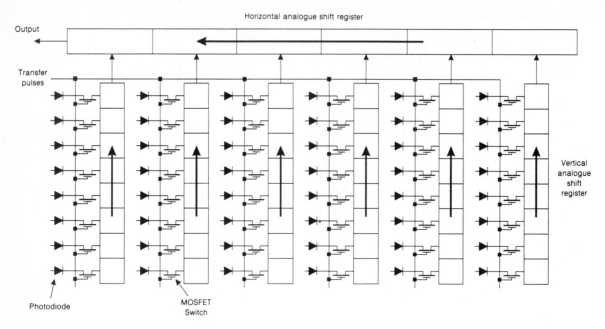

Output

Horizontal analogue shift register

Transfer
pulses

Vertical
analogue
shift
register

Photodiode

MOSFET
Switch

Fig. 11.28 Charge-coupled analogue shift registers in a solid-state image capture device.

values of the pixels of one complete horizontal line. The complete contents of the horizontal register are shifted out serially in an interval corresponding to the line scan interval and, after appropriate filtering, form the analogue video waveform corresponding to one line. When this shift operation is complete the next shift pulse is applied to the vertical registers, thereby loading the horizontal register with the charge values of the pixels of the next complete horizontal line, and so on. The horizontal shift pulse waveform has a frequency of approximately 14.76 MHz, which corresponds to the number of pixels per second given by Eqn (11.1).

It may be noted here that the horizontal shift register may be simply employed to perform time compression of the luminance signal as is required in the MAC transmission system described in Section 11.17. A time compression of 3/2 would be achieved by clocking out the contents of horizontal shift register (one line) at 1.5×14.76 MHz.

Clearly there is no scanning waveform or deflection system required with this type of device and the clock and drive pulses are produced in a timing/divider integrated circuit driven by a precision crystal oscillator. The shift pulses required for the analogue shift registers are more complex than the simple diagram of Fig. 11.28 would suggest. In practice a four-phase switching waveform is required. Techniques for generating colour signals from charge-coupled devices are based on similar principles to those of the vidicon tube.

11.18.3 Display devices

Display devices in colour television systems are largely based on cathode ray tubes, but intense research is ongoing to perfect a large screen solid-state equivalent. This section will be confined to current practice and the basic construction of the colour display tube is as shown in Fig. 11.29.

The colour display tube is required to produce three separate images in the primary colours. The formation of a single colour image is then dependent on the averaging effect of the eye. The electron gun in the colour display tube contains three separate cathodes in line abreast formation. Each cathode produces an electron beam and the three beams are deflected magnetically to form the usual scanning action, but each beam is arranged to strike only the screen phosphor corresponding to its colour. These phosphors are arranged in vertical strips on the screen of the tube each strip producing either red, green or blue light when excited by the electron beam. To ensure that the output from each cathode strikes only its own colour phosphor a mask is placed about 12 mm in front of the screen. This mask is composed of elongated slots which allow the passage of electrons in the scanning beams. The slots are positioned so that electrons passing through strike only the appropriate phosphor and a colour picture is therefore produced.

The mask is clearly a major source of inefficiency in the tube as only about 20% of the incident beams actually reach the phosphor. This means that the mask itself absorbs considerable energy and, in so doing, heats up. Special arrangments are made so that expansion of the mask does not result in inaccuracy when the electron beams strike the screen. Special mounts are incorporated within the tube so that the mask expands in an axial direction.

A variant on the display tube described above is known as the Trinitron. The main difference is that the shadow mask is composed of slots which run the complete height of the tube (rather than the elongated variety shown in

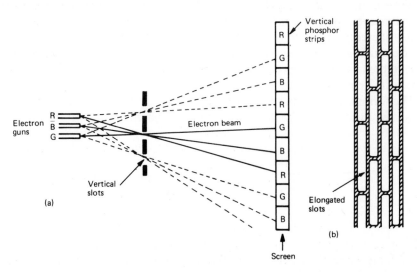

Fig. 11.29 Precision in-line display tube: (a) plan view of tube layout; (b) front view of mask.

Fig. 11.29). There are also some differences in the electron gun assembly and focusing arrangements which allow a smaller spot size than is possible with the arrangement of Fig. 11.29. However, the operating principles of the in-line display tube and the trinitron are broadly similar.

11.19 TELETEXT TRANSMISSION

There are essentially two forms of data transmission for television receivers in the UK known as teletext and viewdata. Teletext is transmitted directly by the broadcast companies and viewdata is transmitted over the switched public telephone network (and therefore requires a modem). Telext has been developed by individual broadcast companies to a common standard and it is possible to display many hundreds of pages of information on each network. It was noted in Section 11.4 that 25 lines on each field are blanked out to allow for the field flyback to return the scan to the top of the picture. Some of these lines (2.5 per field) are used to transmit field synchronizing and equalization pulses and receiver circuits are adjusted so that the remaining blank lines do not appear on the screen. These remaining lines are therefore available for the transmission of data pulses that may occur above the black level. The teletext specification allows for 16 lines on each field to be used for data transmission logic 0 being equivalent to black level and logic 1 being equivalent to 66% of peak white level. These data signals are undetected on a standard receiver but can be separated from the normal video signal in specially equipped receivers.

Each page of teletext information contains up to 24 rows of text with 40 characters per row. Each character is represented by a 7 bit international code with an odd parity check (7 information bits + 1 parity bit ⇒ 8 bit byte). The odd parity has an additional receiver synchronization function when all 7 bits of the standard code have the same value. Unlike the asynchronous data transmission, described in Chapter 4, teletext transmission is synchronous and is therefore more efficient because start and stop bits are not required. Synchronizing information is required, however, and this is transmitted during the first five bytes of each row. This means that each row contains a total of 45 eight bit bytes.

The format of each line is shown in Fig. 11.30. The bits in each byte are identified by sampling the data at the centre of each bit interval, the sampler

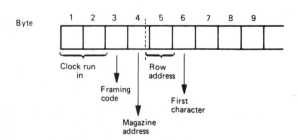

Fig. 11.30 Format of a telext row.

being driven by a locally generated clock. This means that each row contains 360 bits, these bits being transmitted during one line scan interval using NRZ pulses. The bit rate used is 444 × line frequency or 6.9375 Mb/s which means that 360 bits occupy 51.89 μs which allows 12.1 μs for line blanking purposes. The television receiver has a local clock running at this frequency which is synchronized with the incoming data stream by the first two bytes of each line. These bytes contain alternate 1s and 0s at the clock frequency and produce what is known as **clock run in**.

Once the local clock is synchronized it is necessary to detect the beginning of each individual 8 bit byte in order that correct decoding can take place. This is achieved by transmitting a special framing code 11 100100 as byte 3. The data stream is used as a serial input to an 8 bit shift register. When the register contains the framing code a flag is set indicating that the next bit will be the first bit of byte 4, i.e. byte synchronization is achieved. Rows 4 and 5 contain a row address (see later), the first character byte is byte 6 and the first step in decoding a character is to check the parity. A circuit that will produce an 8 bit parity check is shown in Fig. 5.9 and is composed of seven exclusive-OR gates. If a parity check fails (i.e. if the output of this circuit is a binary 0) the character is not decoded, but is replaced by the code for a blank.

Each page of text is composed of 24 rows, hence it is necessary for the decoder to be able to select individual pages and once a page has been selected, to assemble the rows of the page in the correct order. The rows within each page are recognized by a row address code that is transmitted in bytes 4 and 5 of each row. Since there are 24 rows in each page, a minimum of 5 bits is required to specify the number of row addresses. It is necessary at this point to consider the effect of occasional errors. An error in a character byte can be tolerated simply by omitting the character when this occurs. However, an error in the bytes containing the row address is much more serious, as this can cause a row to be wrongly placed within a page. To reduce the probability of such errors occurring, the page and row address are error-protected using a Hamming code. Hamming codes and error detection and correction are discussed in detail in Chapter 5. The code used for row addressing has a Hamming distance of 4, which means that it can correct a single error and can detect a double error.

To provide a Hamming distance of 4 each address bit is accompanied by a parity check bit, hence a total of 10 bits (address + parity) is required. Eight of these bits are transmitted in byte 5 of each row and the other two in byte 4. The remainder of byte 4 is used for a Hamming-coded magazine address which forms part of the page identification. Both row and magazine address codes are transmitted least significant bit first. The format is shown in Fig. 11.31.

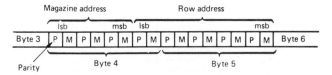

Fig. 11.31 Format of magazine and row address bits.

The function of the row address is to direct the following 40 bytes of text data to the appropriate locations in the page memory. Before this is done the required page must be selected. The top row of each page is called the header row and has the row address 00000. This row contains only 32 text characters instead of the usual 40. The first 8 bytes of text data in the header row carry a page number code, a time code and a control code. These eight bytes are Hamming coded in the same way as bytes 4 and 5 of the other lines. The text in the header row, with the exception of the page number display, is the same for each page.

The page address code (tens and units) is transmitted as two 4 bit binary coded decimal (BCD) numbers with the appropriate Hamming error coding. BCD is used because this can be compared directly with the page selection number entered by the viewer from the decimal keyboard of his/her remote control unit. The page address code occupies bytes 6 (units) and 7 (tens) of the header row. The next four bytes are used for transmitting the minutes (units and tens) and hours (units and tens) of the time code, the Hamming coded BCD format being used for these bytes also. The next two bytes are used to transmit control information.

11.19.1 Page selection

The page is selected by keying in the required three digit number on the viewer's key-pad. Each page is identified by a combination of the page code in the header row and the magazine address code that is transmitted as part of byte 4 of every row. The magazine row contains three digits that are Hamming coded and can thus have any value between 0 and 7. This code is used as the 'hundreds' of the page number identification. The page code is selected from the header row and the magazine code is selected from each transmitted row of text. When a complete match occurs, the following data is written into the page memory. The row address codes are used at this stage to select the appropriate locations within the page memory. When another header row is detected (row address 000000) the page address code are again compared. If there is no match the following data is ignored. In this way only the data from the requested page is transferred to memory. A block diagram of the page selection hardware is shown in Fig. 11.32. If an uncorrected error

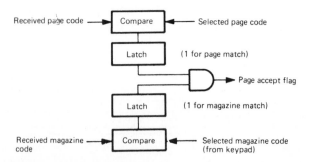

Fig. 11.32 Page selection schematic.

occurs in either the page or row address codes the following data is ignored until the requested page header is repeated.

Normally all header rows are displayed until the requested page header is received. This avoids a blank screen that could occur for up to 30 seconds in an average-sized magazine. In this case the only part of the header that changes on the display is the page number and the time. Full pages are transmitted at the rate of approximately four per second, which means that 25 seconds are required to cycle through 100 pages. This can be inefficient when several blank rows within a page exist. To increase the efficiency of transmission, the page memory is completely filled with the code for 'space' when a new page code is keyed in. Blank lines within a page can then be omitted, i.e. gaps can occur in the row address. Filling the page memory with blanks prevents the row of a previous page from being displayed when the requested page has rows omitted. This technique increases the speed of transmission by up to 25%.

If there are several pages dealing with a common subject, they are sent out in sequence using the same page number. Each new page is transmitted after a delay of about 1 minute, which is sufficient time for the viewer to read the displayed text. The whole sequence of 'self-changing' pages is then repeated continuously.

11.19.2 The page memory

In order to produce the illusion of a fixed image the data representing each page must be scanned 50 times per second. Hence there is a requirement for a complete page memory that can then be scanned sequentially at the required rate. Since each page has 24 rows of 40 characters and each character has a 7

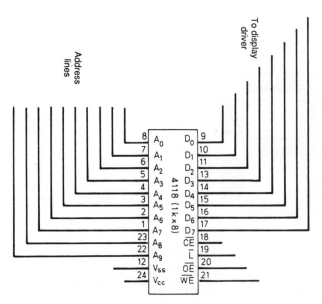

Fig. 11.33 Page memory based on a 1k × 8 bit RAM.

bit code the minimum storage requirement is 6720 bits per page. This may be readily achieved by use of a 1024 × 8 bit random access memory (RAM) as shown in Fig. 11.33.

11.19.3 Character display

The character set in a teletext display is produced using a 6 × 10 dot matrix. The format allows both upper and lower case letters and special purpose symbols to be displayed. The vertical dot resolution is made equal to the line spacing, which means that ten scanned lines are required for one row of text. A typical example is shown in Fig. 11.34.

Each character is in fact represented by a 5 × 9 dot matrix, column 6 being reserved for character separation and line 10 being used for row separation. The character patterns are stored permanently in a teletext read only memory (TROM). The scan proceeds on a line-by-line basis. During each line scan any character will have a corresponding 5 dot code. Hence the TROM must store a series of 5 bit numbers corresponding to the data code. The TROM is addressed by a combination of the character code (7 bits) and the line number (4 bits are required for one to ten lines), the appropriate 5 dot code is then stored at each 11 bit address.

As each line is scanned, the character codes in the particular row of text being displayed are placed upon the address lines of the TROM in sequence together with the 4 digit line scan code. This produces the correct sequence of 5 dot codes on the data lines of the TROM. Smoother characters than the one illustrated can be generated by producing slightly different dot codes on alternate scans. This is known as character rounding.[3]

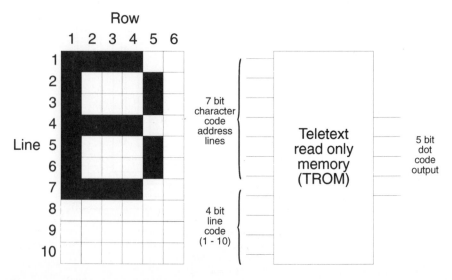

Fig. 11.34 Teletext character display.

11.19.4   Graphics

The teletext display is not restricted to text but can also operate in a graphics mode. This mode is used for the display of simple diagrams and extra large characters. The display in the graphics mode is also divided into a dot matrix, but in this case a 2×6 rather than a 6×10 matrix is used. The graphics mode is indicated by a control code and when in this mode each 7 bit character is interpreted directly as a graphics symbol. The graphics dot matrix and the corresponding bit number is shown in Fig. 11.35. The total number of graphics symbols which can be defined is $2^6 = 64$ symbols.

B1	B4
B2	B5
B3	B6

Fig. 11.35 Graphics dot matrix.

There are many other features of teletext, such as mixing of text and graphics, colour of display, super position of teletext on normal programme pictures, etc. that it is not possible to cover in this volume; the interested reader is referred to Money.[3]

11.20   VIEWDATA

This is a generic term for systems which retrieve and display computer-based information and interactive services using the public switched telephone network and a television receiver or monitor. The main difference between teletext and viewdata is that two way communication is provided between the user and database, which requires a modem to connect the television receiver with the telephone network. This makes possible such services as electronic shopping, home banking etc. The other significant difference is that viewdata services (such as Prestel, which is offered by British Telecom) operate on a menu-driven principle in which the operator selects an item in a series of branches from intermediate menus. The display format of viewdata systems uses the same standard as teletext and thus a specially adapted receiver can be used for both services. A typical set-up is shown in Fig. 11.36.

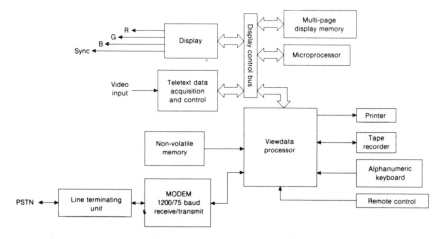

Fig. 11.36 Basic viewdata system.

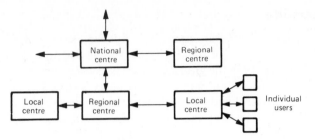

Fig. 11.37 UK national viewdata network.

In the viewdata system only one page is transmitted in response to the page number keyed in by the user. The arrangement of the British national viewdata network is shown in Fig. 11.37. Each of the local centres contains a computer with about 5×10^4 pages of information. Most of the information accessed by the average user will be stored at the local centre, hence the pattern of telephone usage will be on a local call basis. The regional centre is the next stage in the hierarchy. If the requested information is not stored at the local centre it will be automatically transferred from the regional centre over a high-speed data link. The regional centres are connected by high-speed links to a national centre. By this means any user is able to access local information from any one of the regional or local centres. The capabilities of the system in terms of the amount of information are vast, and can, of course, be extended to an international centre at a future date.

A second difference between teletext and viewdata is that the viewdata system uses a cursor instead of page and row address. The position of the cursor is, in effect, controlled by the contents of two counters. One counter represents the position of a character within a row, the other represents the position of a row within a page. The viewdata system uses a similar 7 bit code to teletext for character transmission plus a parity check bit. The viewdata transmission uses an even parity in contrast to teletext which uses odd parity. There are thus $2^7 = 128$ possible characters that can be transmitted, several of which are used for cursor control. For example the code 0001100 (form feed) causes the cursor to return to the top left-hand corner of the screen (i.e. character and row counters are set to zero) and the page memory is filled with blanks. As symbols are received the cursor is moved along the top line (i.e. the character counter is reset and the row counter is incremented by 1. The location address in the page memory RAM is formed from a combination of character and row counter. The cursor can be placed at any point on the screen by using cursor controls, this avoids transmitting blank rows and blanks within a row. For instance the code 0001101 (carriage return) causes the cursor to return to the beginning of the current line (resets character counter to zero) and the code row counter).

11.20.1　Line signals

Data is transmitted to the subscriber using frequency shift keying with asynchronous transmission as described in Section 3.8. The data rate used is 1200 baud with binary 1 being transmitted as a 1300 Hz tone and binary 0 as a

2100 Hz tone. This means that a modem is required in the television receiver to interface the FSK signals. This modem also allows transmission of the page-selection code in the reverse direction, again using FSK. The transmission speed in the reverse direction is 75 baud, a 390 Hz tone being used for binary 1 and 450 Hz for binary 0. The viewdata modem is also required to drive the line and field scan circuits of the CRT in the receiver. This means that the viewdata service, unlike the teletext service, does not require the provision of broadcast line and field synchronization. It is thus available outside normal broadcast periods.

11.21   CONCLUSION

This chapter has outlined the fundamental engineering principles of broadcast television systems and has shown that, in general, the development of transmission standards has been largely influenced by requirements of compatibility. The advent of satellite television has broken this link and the prospect now exists for the development of the full potential of colour television using digital signal processing techniques, which has recently become possible at video frequencies. The next obvious development is the introduction of a high-definition television standard and associated solid-state, ultra thin, large screen display devices. The advantages of a single global standard are clear but whether such a standard will emerge it is not yet possible to predict.

REFERENCES

1. Patchet, G. N., *Television Servicing*, Vol. 2, Norman Price, 1971.
2. Sims, H. V., *Principles of PAL Colour Television*, Newnes-Butterworth, London, 1974.
3. Money, S., *Teletext and Viewdata*, Newnes-Butterworth, London, 1979.
4. Trundle, E., *Television and Video Engineer's Pocket Book*, Newnes, London, 1992.

PROBLEMS

11.1   The transmission standards for television in the USA are as follows:

> number of lines/picture = 525
> number of fields/picture = 2
> number of picture/second = 30
> field blanking = 14 lines
> line blanking = 14µs
> displayed aspect ratio = 1.33:1

Differentiate between the transmitted and displayed aspect ratio and calculate a value for the former figure. What is the theoretical bandwidth of the transmitted video waveform?

Answer: 1.545:1, 6.85 MHz.

11.2 If the standards of the previous question were modified to accommodate 3 fields/picture what is the theoretical bandwidth of the transmitted waveform? Increasing the number of fields/picture produces a corresponding reduction in signal bandwidth. Comment on this statement and suggest why, in practice, there is a limit to the number of fields/picture.

Answer: 4.25 MHz.

11.3 The UHF television allocation allowance for channel 50 is 702 MHz to 710 MHz. A television receiver with an IF of 39.5 MHz is tuned to this channel. Calculate the frequency of the local oscillator in this receiver.

Using the channel allocation table (Table 11.2) determine which channel forms the image frequencies for channel 50.

Answer: 742.75 MHz, channel 60.

11.4 Three primary light sources *RGB* are designed to produce an output light power that is linearly proportional to a control voltage. When these three sources are used to match equal energy white light of intensity 1 lumen, the required voltages are $R = 6.9$ V, $G = 3.8$ V, $B = 8.6$ V. The same light sources are used to match an unknown colour, the corresponding voltage being $R = 5.6$ V, $G = 2.2$ V, $B = 1.5$ V. Find the trichometric units of the unknown colour and its luminance. What are the chromacity coefficients of this colour on the colour triangle?

Answer: 0.81 *T*, 0.58 *T*, 0.17 *T*, 0.6 lumen, 0.52, 0.37.

Table 11.2 UHF channels and frequencies (British Isles)

Channel	Vision (MHz)	Sound (MHz)	Bandwidth (MHz)	Channel	Vision (MHz)	Sound (MHz)	Bandwidth (MHz)
Band IV							
21	471.25	477.25	470 − 478	29	535.25	541.25	534 − 542
22	479.25	485.25	478 − 486	30	543.25	549.25	542 − 550
23	487.25	493.25	486 − 494	31	551.25	557.25	550 − 558
24	495.25	501.25	494 − 502	32	559.25	565.25	558 − 566
25	503.25	509.25	502 − 510	33	567.25	573.25	566 − 574
26	511.25	517.25	510 − 518	34	575.25	581.25	574 − 582
28	527.25	533.25	526 − 534				
Band V							
39	615.25	621.25	614 − 622	54	735.25	741.25	734 − 742
40	623.25	629.25	622 − 630	55	743.25	749.25	742 − 750
41	631.25	637.25	630 − 638	56	751.25	757.25	750 − 758
42	639.25	645.25	638 − 646	57	759.25	765.25	758 − 766
43	647.25	653.25	646 − 654	58	767.25	773.25	766 − 774
44	655.25	661.25	654 − 662	59	775.25	781.25	774 − 782
45	663.25	669.25	662 − 670	60	783.25	789.25	782 − 790
46	671.25	677.25	670 − 678	61	791.25	797.25	790 − 798
47	679.25	685.25	678 − 686	62	799.25	805.25	798 − 806
48	687.25	693.25	686 − 694	63	807.25	813.25	806 − 814
49	695.25	701.25	694 − 702	64	815.25	821.25	814 − 822
50	703.25	709.25	702 − 710	65	823.25	829.25	822 − 830
51	711.25	717.25	710 − 718	66	831.25	837.25	830 − 838
52	719.25	725.25	718 − 726	67	839.25	845.25	838 − 846
53	727.25	733.25	726 − 734	68	847.25	853.25	846 − 854

11.5   The output voltages from a three-tube colour camera when viewing equal energy white light are adjusted such that $R = G = B = 1.0$ V. When the camera is directed towards an object of uniform colour the output voltages are $R = 0.56$ V, $G = 0.21$ V and $B = 0.75$ V. Calculate the amplitude of the resulting luminance signal when:
(a) gamma correction is applied after the formation of the luminance signal;
(b) gamma correction is applied to each colour separately.

What is the percentage difference in the luminance produced when these two signals are applied to a monochrome CRT? Assume the overall gamma is 2.2.

Answer: 0.639 V; 0.617 V; 7.42%.

11.6   When the output voltages of the camera of the previous question are $R = 0.5$ V, $G = 0.9$ V and $B = 0.8$ V, find the percentage saturation for this colour.

What would be the values of the output voltages for a fully saturated colour of the same hue?

Answer: 44%, 0 V, 0.4 V, 0.3 V.

11.7   The row addressing data in a teletext transmission is coded to have a Hamming distance of 4. If the probability of a single digit error is 1×10^{-4}, find the probability that an undetected error will occur in the coded address information. State all assumptions made.

Answer: 4.33×10^{-4}.

12 Optical fibre communications

Optical fibre communications present the most exciting, and probably the most challenging, aspects of modern systems. Fibres are exciting because they seem to offer so many benefits – low cost, enormous bandwidth, very small attenuation, low weight and size, and very good security against external interference. Physically, fibres occupy very little space, and they are so flexible that they can be used in places that would not be accessible to conventional cable.

Optical fibre is still expensive in the early 1990s, and there is a strong debate about whether it can justifiably be taken to every home. The cost of replacing copper would be very large, and it would probably not be recovered by a correspondingly large increase in domestic traffic. If video services appear, and are attractive to the user, then that balance of costs could change, and the unsatisfied demand could be tapped economically by using fibre.

There was a debate in the mid-1980s about whether to use monomode or multimode. That may have been resolved. The difficulties of managing monomode fibre have been overcome, and the very significant benefits that it has over multimode make it the unreserved first choice for communications applications. As noted later, the development of fibre amplifiers has given fibre communications a significant boost, making possible transmission over any terrestrial distance, and offering scope for extensive, low-loss optical fibre distribution systems.

Eventually, all-optical systems will appear, but at present the switching elements are at the research stage, and we will not consider them further here.

The optical components of a fibre communications system are, in simple terms, a light emitter, which initiates the optical signal, a fibre which transmits it, and a detector which receives it and converts it into an electrical equivalent. If several fibres need to be joined, end to end, the couplers must ensure that the fibres are correctly aligned and butted, to reduce any joining losses to a minimum.

Each of these components has essential ancillary parts; the detector and emitter are driven by stabilized voltages, and mounted in such a way that maximum transfer of light between them and the fibre is achieved. The fibre itself must be clad in some sort of protective coating and made up into a cable that will withstand the rigours of installation over long distance. However, although these factors are essential, they are, in a sense, of secondary importance, and we shall not be considering them further here. Our interest is confined to the operating principles of the basic devices, and, for further information on system details, reference should be made to the literature listed at the end of this chapter.

Before considering some aspects of a system, we will examine the way in which the fibres, detectors and emitters work.

12.1 OPTICAL FIBRE

An optical fibre is, in essence, a dielectric waveguide. It has been known for a long time that high-frequency electromagnetic energy can be transmitted along a glass or plastic rod and, indeed, observation shows that short rods are translucent to light. However, two factors prevented that knowledge from being used to produce useful light guides:

 (i) energy leaked from the outside of the dielectric to the surrounding air, and
(ii) the attenuation was so large that worthwhile lengths could not be achieved.

The first difficulty, though virtually insurmountable at microwave frequencies, can be overcome in the optical and infrared parts of the spectrum by enclosing the guide in a cladding of similar material, but which has a slightly smaller refractive index. The boundary between the cladding and the core acts as a reflecting surface to the transmitted light. The second problem, that of high attenuation, could be reduced only by refining the methods of producing and drawing the glass so that the impurities and irregularities were reduced to a minimum. The attenuation now achievable in the laboratory is almost as low as possible, at about 0.2 dB/km.

Fibres of varying quality are used for communications, but when distances are significant, care is taken to ensure that the lowest attenuation possible is achieved. This involves choosing the best operating frequency for the particular fibre material, and ensuring that any contaminating elements are removed from the glass during manufacture. Before considering the loss mechanisms inherent in any fibre, we will look at the different fibres used, and examine, with the help of ray theory, the way in which light propagates along an optical waveguide.

12.2 STEPPED-INDEX FIBRE

As we noted earlier, light can be made to propagate down a fibre waveguide consisting of an inner dielectric, of refractive index n_1, and an outer cladding of refractive index n_2, if n_1 is slightly larger than n_2. In Fig. 12.1 we see the path taken by a beam incident on the end face of the fibre at an angle of incidence θ_0. Outside the fibre, the air is assumed to have a refractive index n_0.

By the laws of refraction,

$$n_0 \sin \theta_0 = n_1 \sin \theta_1 = n_1 \cos \theta_2 \qquad (12.1)$$

At the core–cladding boundary

$$n_1 \sin \theta_2 = n_2 \sin \theta_3$$

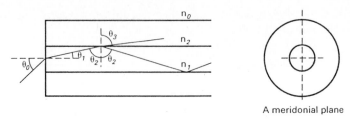

A meridionial plane

Fig. 12.1 Path of meridional ray in fibre.

Since $n_2 < n_1$, total reflection will occur at the boundary for all values of θ_2 such that

$$\sin \theta_2 \geqslant n_2/n_1 \qquad (12.2)$$

From Eqn (12.1), the maximum value of θ_2 that will produce a reflected ray at the core–cladding interface is therefore given by

$$n_0 \sin \theta_{0_{max}} = n_1 \cos \theta_2$$

where, from Eqn (12.2)

$$n_1 \cos \theta_2 = n_1 [1 - \sin^2 \theta_2]^{1/2} = n_1 \left[1 - \left(\frac{n_2}{n_1} \right)^2 \right]^{1/2}$$

Hence

$$\sin \theta_{0_{max}} = \frac{1}{n_0} [n_1^2 - n_2^2]^{1/2} \qquad (12.3)$$

The value of $\theta_{0_{max}}$ given by Eqn (12.3) is known as the maximum acceptance angle for the fibre. From that equation we can see that it is determined by the refractive indices of the core and the cladding. The relationship

$$\mathrm{NA} = [n_1^2 - n_2^2]^{1/2}$$

is known as the numerical aperture of the fibre. It is a useful indicator of the launching efficiency of the guide, and we will consider it later in relation to light sources.

When the surrounding medium is air, $n_0 = 1$, and

$$\theta_{0_{max}} = \sin^{-1}(\mathrm{NA}) \qquad (12.4)$$

If θ_0 exceeds $\theta_{0_{max}}$, θ_2 is less than the value for which total reflection takes place at the core-cladding interface, and some of the energy will be refracted out into the cladding itself, where it is absorbed.

12.3 SKEW RAYS

The ray we have just discussed, with the help of Fig. 12.1, travels in a plane through a diameter of the fibre, and it is called a meridional ray. However, rays may also be launched into the guide in other directions, and these are

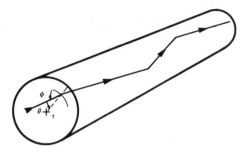

Fig. 12.2 Path of skew ray.

known as skew rays. Provided the angle of incidence of a skew ray falls within $\theta_{0_{\max}}$ it will propagate along the fibre by total internal reflection, but, rather than following a path back and forth across a diameter, it will travel in a helix, as shown in Fig. 12.2.

12.4 MODES

The ray-theory approach to the mechanism of propagation along a fibre has the attraction of being simple, and so gives a good insight into fibre transmission. It does have some disadvantages, however. It assumes that the wavelength of the signal is extremely small, and therefore that the light can travel at any angle of incidence to the cladding, provided that the maximum acceptance angle is not exceeded. In fact that is not the case. As with all guides used to transmit electromagnetic waves, the boundary imposes conditions that restrict propagation to a series of modes. The details of these modes can be found by solving Maxwell's equations for the particular guide being studied. In the case of optical fibres, the solution is complicated by both the dielectric nature of the material, and its cylindrical geometry. We will not discuss these wave-theory solutions here, and for details of how they are obtained, reference should be made to one of the many specialist texts available.[1] We can note that as a result of solving Maxwell's equations, the various propagating modes in stepped-index fibre can be shown to have different wave velocities, a property that is called dispersion. As a result of this mode dispersion, a pulse of light launched on to the guide will, because it consists of energy in several modes, gradually broaden as it travels. This dispersion places a limit on the bandwidth of the fibre, a limit that is a function of the stepped-index profile, and not of the quality of the fibre material.

12.5 GRADED-INDEX FIBRE

The effect of mode dispersion can be minimized, and in theory eliminated,[2] by using a graded profile of the shape shown in Fig. 12.3. The difference between theory and practice is that it is very difficult to produce the exact profile

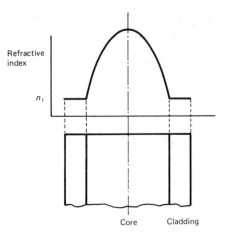

Fig. 12.3 Ideal graded profile to minimize mode dispersion.

required, and the effect of a slight deviation from it is marked. Not only is it difficult to produce the fibre to conform with theory, but there is bound to be some variation of profile with length, and this will allow dispersion to occur, with a consequent reduction in bandwidth. The ideal profile is almost a square law, i.e. the index of refraction of the core is given by

$$n^2(r) = n_1^2 \left[1 - \frac{2[n_1^2 - n^2(a)]}{2n_1^2} \left(\frac{r}{a} \right)^g \right] \qquad (12.5)$$

where $n(r)$ is the refractive index at radius r, $n_1 = n(0)$ where the exponent $g \simeq 2$.

The graded-index changes the paths followed by the waves travelling down the fibre. In terms of the ray-theory of Fig. 12.1, the path is sinusoidal in shape, with the light travelling more quickly at the core–cladding boundary than in the centre of the guide, where the refractive index is at a maximum.

Skew waves will still follow a helical path, with their velocity increasing with distance from the centre of the fibre.

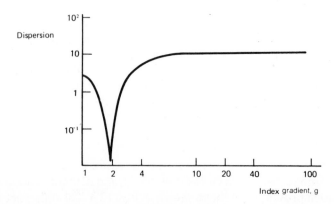

Fig. 12.4 Effect of profile index on mode dispersion.

The effectiveness of a particular profile on reducing the mode dispersion is also a function of wavelength, so it is possible to reduce the dispersion in a fibre by operating it at its most favourable wavelength.

We can see the effect of varying the profile index on the mode dispersion from Fig. 12.4. The sharp minimum shows why great importance is attached to producing fibre with the optimal value of g.

12.6 LOSS MECHANISMS

Attenuation of the signal as it travels along the guide is caused by several factors. The main ones are absorption, scatter and radiation; each is important, but as fibre production methods have improved, and the limits on the operating conditions of the fibre have been optimized, the attenuation is eventually determined by the scattering loss.

12.7 ABSORPTION

The contaminating elements in glass, particularly the transition metals (V, Cr, Mn, Fe, Co, Ni), have electron energy levels that will absorb energy from incident light. The amount of energy absorbed depends on the proportion of impurity atoms present. With care in the manufacture of the glass, and in the drawing of the fibre, these impurities can be reduced to about 1 part per billion, and at this level the loss is negligible.

The absorption produced by the transition metals is not very dependent on frequency, and therefore is not easily avoided by choosing a suitable wavelength for the light source.

The other main absorption mechanism is that due to the presence of water. The hydroxyl ion, OH^-, absorbs readily at 2.8 μm, and there are less strong,

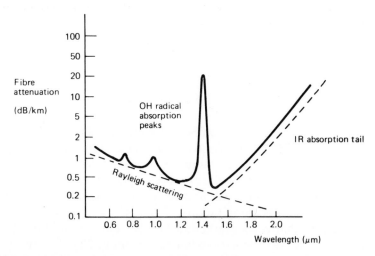

Fig. 12.5 Typical attenuation characteristic of fibres.

but still pronounced absorption peaks at wavelengths which are almost direct harmonics 1.4 μm, 0.97 μm and 0.75 μm. The position of the peaks can be seen clearly in an attenuation–frequency curve, such as that shown in Fig. 12.5. Some reduction in this absorption is obtained by pre-drying the powders used to form the glass. The strong frequency dependence of this mechanism allows an operating wavelength to be chosen that will avoid the severest levels of loss.

The two absorption mechanisms discussed are, in reality, complicated factors which depend on the composition of the glass and the method of fibre production. A deeper discussion is to be found in reference 1, and there are many specialist papers on the subject.

Optical fibres are sensitive to far-infrared radiation, and the loss peak in that region has a tail that reaches to wavelengths as low as 1.6 μm, as shown in Fig. 12.5.

12.8 SCATTER

The fine variations in atomic shape, occurring over distances very small compared with a wavelength, scatter the light by the classical Rayleigh mechanism, which in the atmosphere gives the blue colour to the sky. The loss coefficient is proportional to $1/\lambda^4$ and since it is unavoidable, provides a bound to the lowest attenuation which can be achieved. The Rayleigh loss curve is also shown in Fig. 12.5. The value of the loss is a function of the fibre and its constituent glass:

$$\alpha = \frac{8\pi^3}{3\lambda^4}[n^2 - 1]\,kT\beta \tag{12.6}$$

where T is the transition temperature at which any imperfections in the glass structure are 'frozen in', and β is the isothermal compressibility of the glass.

Other types of scatter can occur. In particular, Mie scatter, which is due to variations in the structure of the glass that occur at about one wavelength intervals, and waveguide scatter, which is caused by fluctuations in the geometry of the core along its length. These losses can be made insignificant if the fibres are produced with care.

12.9 RADIATION FROM BENDS

Energy is radiated at a bend in the fibre. The amount of radiation is usually small, but it is determined by the bend radius. As the radius is decreased there is a very rapid change, at a critical radius, R_c, from very little loss to very high loss. The value of R_c depends on the numerical aperture of the guide, and the radius of its core. Increasing NA and reducing the core size will help to decrease the minimum bending radius.

The mechanism that causes radiation results from the fact that the field pattern of the guided wave penetrates into the cladding. At a bend, the field at the outside of the bend has to travel faster than that at the inside, to maintain the phase relationship in the mode. At some distance from the fibre the velocity of the wave will reach the velocity of light; therefore the energy at a larger radius must be radiated. As the bend radius decreases, the radius at which radiation occurs also decreases, moving closer to the fibre. At the critical radius, half the light into the bend is lost. For many materials, the critical bend will be very small, and sometimes less than the limit imposed by the mechanical stress due to bending.

12.10   SINGLE-MODE FIBRE

As we have seen above, the principal reasons for attenuation and bandwidth limitations in multimode optical fibre are the various losses described, and mode dispersion. Multi-mode fibre is used because of its larger size compared with monomode fibre. The core diameter is usually $50-60\,\mu m$, with a cladding of $100-150\,\mu m$, and hence coupling into source or detector devices, and jointing fibres can be done efficiently, and easily.

Single-mode fibre has the attraction that there is no mode dispersion, and therefore the bandwidth available can be very high indeed. There is a limitation caused by dispersion, not of modes on the fibre, but of the various frequencies emanating from the source. The spread of frequencies generated by the emitter will depend on the type of device used, but there will be some, and each frequency will propagate at a slightly different velocity. This dispersion is called chromatic, and will decrease as the line-width of a source is reduced.

The size of a single-mode guide depends on the free space wavelength and the refractive indices of the core and cladding. For a particular waveguide, operating at a given wavelength λ_0, the cut-off radius for modes higher than the fundamental is a, where a is given by

$$a \leqslant \frac{2.405\,\lambda_0}{2\pi(n_1^2 - n_2^2)^{1/2}} \tag{12.7}$$

and n_1 and n_2 are the refractive indices of the core and cladding, respectively.

12.11   DETECTORS

There are many devices available for detecting optical frequency energy. Some are vacuum tube, others solid state, either semiconductor or not. There are also several photosensitive effects employed to convert incident light into a proportional electrical signal.

In communications systems the power in the incident beam may be extremely small, of the order of $10^{-14}\,W$, and this places additional requirements on suitable detectors. Although several devices have been used, and there is extensive research and development on new techniques, we will limit our attention to the PIN semiconductor, and its associated device, the avalanche photodiode (APD).

To be suitable for use in communications, the detector must be very sensitive, have high efficiency of conversion between light and electrical energy, respond very rapidly for high bandwidth, have low noise power, and good light-collecting properties. In addition, it should operate at low voltage, be easy to use, be robust and insensitive to changes in ambient conditions, have a long life, good reliability, and be inexpensive. Such an ideal specification is, of course, unattainable; in any real device compromises are necessary, and priorities must be settled between the various parameters for any application. Here we cannot look at the devices on the market; many companies produce a range of detectors, and the choice of which to use must be made by the system designer. However, we can examine some of the general features of photodiodes and the materials from which they are made.

In photodetection, light is incident on a semiconductor material that, because of the energy gap between the valence and conduction bands, will convert the light energy into electrical energy. On average, the photocurrent is related to the incident light energy by

$$I = \frac{\eta q}{h\nu} P \tag{12.8}$$

where q is the electron charge, $h\nu$ is the photon energy, P the optical power into the device and η is the quantum efficiency. η is the fraction of the incident power producing electron–hole pairs.

There is a statistical nature in the whole photodetection process that causes us to use mean values for the various parameters. These mean values, and the associated rms quantities, give us a long-term view of the device behaviour.

12.12 PIN DIODE

The diode consists of heavily doped p and n sections separated by an intrinsic layer, as shown in Fig. 12.6. The reverse bias across the device sweeps the

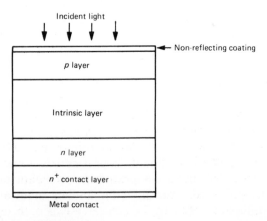

Fig. 12.6 Schematic of PIN diode.

carriers from the I region, leaving a depletion layer. Light falling on the depletion layer creates electron–hole pairs, and these carriers drift across the region to the terminals, where they leave as light-induced current. The device therefore appears to be a current source, across the depletion layer capacitance.

12.13   APD DEVICE

In an APD, the applied voltage is much higher, and there is a region where the electric field is in excess of 10^5 V/cm. At such a high level, photoinduced electrons are accelerated to a velocity at which they can ionize atoms in the intrinsic region, and produce an avalanche effect. In this way the device has an internal gain, M, which is the mean number of electrons at the output per photoinduced electron. The APD is therefore more sensitive than the PIN diode, but because of the stochastic nature of the avalanche gain, its noise level is higher.

12.14   SPEED OF RESPONSE AND QUANTUM EFFICIENCY

When discussing the PIN diode, we noted that the photoelectrons move across the depletion layer at the drift velocity produced by the applied voltage. This transit time limits the speed of response, and hence the bandwidth, of the device. If the depletion layer is narrow, a low response time can be achieved. However, the sensitivity of the device is related to the quantum efficiency, and this increases as the depletion width increases. Consequently, there is a conflict between these two parameters. In Si the depletion width is approximately 50 µm and the minimum response time is of the order of 50 ns, whereas in GaAs the depletion layer can be much narrower, because of the higher absorption coefficient of photons in GaAs compared with Si, and therefore it has a much faster response time of about 40 ps. If the depletion width is narrow, photoelectrons may be induced in the surrounding diffusion area and before being swept through the depletion layer, they will have to diffuse from the diffusion region, thus reducing the speed of response.

12.15   RELIABILITY

Silicon detectors have been developed over many years for use in the 0.8–0.9 µm range and improvements to the configuration and production techniques used have made available devices with projected lifetimes of many years, and high reliability is therefore readily available. For GaAs or other III–V devices, which are used at the longer wavelength of about 1.3 µm, experience is more limited, but long lifetimes are again anticipated. The reason for the limitation of Si to the shorter wavelength region is that its energy gap is 1.1 eV, and this cannot be achieved by photons of wavelength greater than about 1.1 μm. At the longer 1.3 μm wavelength, which is becoming increasingly attractive because of the development of very low-loss guides at that wavelength, Ge has high sensitivity but, because the ionization coefficients for electrons and holes are about the same, the noise level is high compared with that which can be achieved with GaAs, InP or some more complex compound such as InGaAs.

12.16 NOISE

The sources of noise in PIN and APD devices are well known. In PIN diodes noise is produced by:

(i) random fluctuations in the photocurrent itself,
(ii) background radiation from other than the intended light sources, and
(iii) current generated in the device when there is no light present, i.e. dark current.

In Si devices, noise produced by (ii) and (iii) can be made extremely small, leaving the shot noise of (i), and this provides a lower bound on the sensitivity of the device.

In APDs another source of noise exists – that due to the randomness of the gain process. It magnifies the shot noise by a factor $F \times M$ where M is the mean avalanche gain, as in Section 12.13, and F is a quantity called the excess noise factor. If the value of F is unity, the avalanche process is not itself noisy. In practice F is much greater than that. Smith[3] states that F is related to M, and a factor k, by the approximate relationship

$$F \simeq 2(1 - k) + kM \tag{12.9}$$

where k is the ratio of the smallest to the largest ionization coefficient in devices in which the avalanche is initiated by the carrier with the highest coefficient. In Si, for example, the avalanche is initiated by electrons that have a much higher ionization coefficient than holes, giving $k < 0.1$. By contrast, Ge is intrinsically more noisy because the electron and hole ionization coefficients are equal.

The choice between PIN and APD will depend on the priorities given to sensitivity and SNR, as well as on operating constraints. The APD can be much more sensitive than the PIN if the internal mean gain, M, is high. However, the shot noise is also a function of M, and, for some devices, this quantity will increase more rapidly than the gain as M is increased. An optimal value of M may exist at which, for a given SNR, the sensitivity is maximum. In practice, for applications in which good SNR is required, PIN devices are used, but if sensitivity is at a premium, an APD would be more appropriate. Compared with PIN diodes, avalanche devices have less attractive operating conditions. They require a much larger voltage, possibly of the order of several hundred volts, and they are sensitive to temperature variations. This means that more complicated driving circuitry is required to provide compensation if the temperature changes.

12.17   OPTICAL SOURCES

The special requirements of optical communications make many demands on optical sources, some of which may not be necessary for other applications. Apart from supplying sufficient power, at reasonable cost, over a long, reliable, lifetime, and with high efficiency, the source should have an emission wavelength that coincides with a loss minimum of the fibre. It should also show a linear output power versus drive current characteristic, emit over a narrow bandwidth, and be capable of transmitting high bit rates.

These constraints and requirements limit the types of suitable emitter to three:

  (i) light emitting diode (LED),
 (ii) semiconductor laser, and
(iii) solid-state-laser.

Here we will discuss only (i) and (ii), since they are the principal devices presently in use. The solid-state laser is likely to become an important source in the future.

Before considering the LED and the semiconductor laser separately, we will examine some general properties of these optical sources.

LED emitters are simple in their construction, and do not require complex drive circuitry. They are particularly suited to relatively short-distance links in which the bit-rate requirement is modest and the channel capacity low.

Alternatively, semiconductor lasers, which require higher drive currents and more complex circuitry, can produce higher power than LEDs, and are capable of high bit rates. They can therefore be used for high-speed, long-distance systems.

For both types of device, operating in the $0.8-0.9\,\mu\text{m}$ range, the most commonly used material is a doped GaAs, usually GaAlAs. Several III–V compounds have been used successfully. In general, to be suitable a material must have a direct energy gap in the region of $2\,\text{eV}$ so that, when electron–hole recombination takes place, a photon is released. The material must also be one in which it is possible to form a p–n junction very easily. This requirement rules out high melting point compounds. Ideally, the emission wavelength should be adjustable by changing the composition of the material. This requirement is usually satisfied by varying the mole fraction. For example, if x is varied in $\text{Ga}_{1-x}\text{AsAl}_x$, the emitted wavelength can be varied between that corresponding to GaAs and to AlAs.

The basic mechanism in both LEDs and lasers is that minority electrons in the conduction band recombine with valence band holes and release energy in the form of radiation, i.e. if the direct gap between the valence and conduction bands is Eg, the radiation produced by recombination is

$$vh = Eg \tag{12.10}$$

where h is Planck's constant and v is the radiation frequency of the photon emitted.

Although p–n junctions in III–V materials are excellent sources of light, the basic compound will absorb light easily as it passes through, and therefore the separation between the active area and the outer surface of the diode must be very small, otherwise the semiconductor will absorb some of the energy produced in the active region, and thereby reduce overall efficiency.

Here we will discuss the performance features of sources, not the physical and material aspects of their operation. Information on the choice of compounds, and the physical mechanisms involved in the devices, is given in Section 16.2 of Miller and Chynoweth,[4] and Chapters 4 and 5 of Sandbank.[5]

12.18   LIGHT EMITTING DIODES (LEDs)

In its basic, homojunction form, an LED consists of a forward-biased p–n junction. Carriers are injected, under the influence of the electric field, from the majority to the minority side where they create an excess of minority carriers. As they diffuse away from the junction, recombination occurs, and if the energy gap Eg is approximately 2 eV, optical radiation will be emitted at a frequency given by Eqn (12.10). The optical energy radiates from the device, and is launched on to a fibre guide.

For efficient performance, the recombination process must have a very high proportion of radiating events, and the optical power must not be absorbed in its path from the junction to the surface of the device, or lost between the device and the fibre.

The quantum efficiency, defined as the proportion of radiating carrier recombinations, is a measure of the internal conversion between electrical and optical power. Values of about 80% can be achieved. However, all of the generated light cannot be used. Some will be absorbed by the semiconductor material itself, as we mentioned earlier. In addition, the semiconductor–air interface will not transmit all the energy incident on it, only that which falls within the critical angle, which is about 15° for III–V materials (see Fig. 12.7). In practice this means that only about 1% of the light generated will be useful. Some improvement can be achieved in this external quantum efficiency, η_c, by using a reflecting back plane.

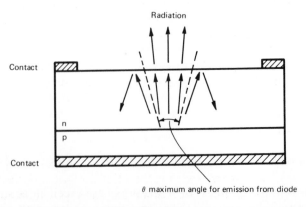

Contact

Radiation

n

p

Contact

θ maximum angle for emission from diode

Fig. 12.7 Exit of light from an LED with planar geometry.

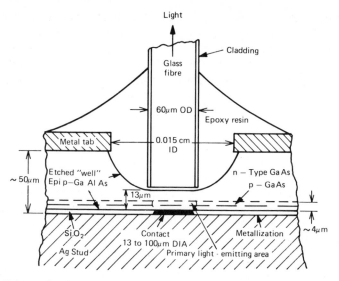

Fig. 12.8 Schematic of Burrus Light emitting diode. Reproduced from Dawson and Burrus, *Applied Optics* vol. 10, pp. 2367–2369, Oct. 1971.

The purpose of the diode is to provide a light source for fibre transmission, and one of the loss mechanisms occurs between the device and the fibre-end. The acceptance angle of the fibre is $\theta = \sin^{-1}(NA)$, where NA is defined in Section 12.2. For sources of area less than that of the fibre-end, the efficiency is approximately equal to $\sin^2\theta = (NA)^2$. This results in a fibre-launching efficiency of about 5%. The total efficiency of coupling, given as the proportion of diode input power that is launched into the fibre as light, is therefore about 1% of 5%, or 0.05%.

Clearly, considerable effort is involved in improving this figure. Some mechanical changes to the diode can be made. The simplest diode structure has a planar geometry, as shown in Fig. 12.7. By forming a well in the upper material, the fibre-end can be placed near to the active junction, thus reducing significantly the absorption of light within the semiconductor material. If, in addition, the light-generating area is smaller than that of the diode, an epoxy resin layer between the device and the fibre will increase the launching efficiency. An example of this type of device, known as the Burrus diode, is shown in Fig. 12.8. Although it has a high launching efficiency, it has the disadvantage of a relatively large spectral width. In general, the range of spectral widths produced by LEDs is from about 100 to 400 nm.

12.19 LASERS

The other important light source is the semiconductor injection laser. It employs a similar mechanism to that of the LED, but operates with laser-type stimulated emission, and this produces a different output characteristic. The relationship between the output optical power and input drive current, is

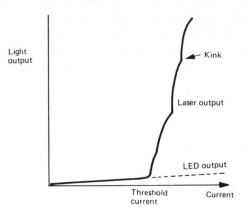

Fig. 12.9 Relationship between output optical power and input drive current for laser and LED.

shown in Fig. 12.9. We can note three features. First, the curve has two distinct sections, that below threshold and that above. The threshold level is closely related to the structure of the device. Below threshold, the action of the device is similar to that of a LED, and the output has a broad spectral width. Above threshold, the device operates under stimulated emission and the spectral width is very much reduced. The second notable feature is the steep gradient above threshold. This makes the device very fast, and this speed is exploited in high-speed digital systems. Finally, the output characteristic above threshold has non-linear features, called kinks. These kinks are thought to result from slight changes in current paths through the active region of the device, and they can be eliminated by using a stripe configuration.

Several structures have been used for injection lasers. In its simplest form, the device consists of an active p–n junction in GaAs, with planar terminal plates parallel to the junction, to which the supply voltage is applied. Two of the opposite faces, perpendicular to the junction plane, are cut to provide reflecting surfaces, between which the essential gain of the carriers across the junction can be achieved. Through one of these faces (or facets, as they are called) light is coupled to the fibre. The opposite, rear facet is sometimes used to monitor the performance of the device. Figure 12.10 shows such a simple laser. The active region has a slightly higher refractive index than the surrounding layers, and this causes a weak waveguiding mechanism, allowing the generated light to be constrained within a narrow region.

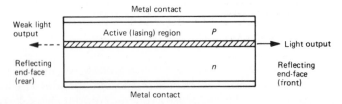

Fig. 12.10 Schematic of laser.

This waveguiding is a very useful feature because it allows the device to produce an intense light source that is more suitable for launching on to fibres than that from an LED, which produces a less directional output. To improve the containment of the light into the active region, a double-heterojunction (DH) structure is used. Figure 12.11 shows the main features. The heterojunction is an interface between GaAs and GaAlAs, and in the DH device GaAlAs is placed at each side of the GaAs active region. Apart from providing a stronger waveguiding action, the DH structure causes the threshold current to be considerably lower than that of the homojunction arrangement shown in Fig. 12.12. This reduction in threshold current is sufficient to allow the device to operate in a continuous mode, whereas in the homojunction structure the current density required to sustain the lasing action can only be achieved in a pulse mode.

In Fig. 12.12 we can see the stripe referred to earlier. Several experiments on the best width of stripe to use have been performed, and that most commonly available is about 20 μm across. The effect of the stripe is to confine the active region of the device to an area below the stripe, thus producing a two-dimensional waveguiding mechanism, and improving the stability and directivity of the generated light.

The DH GaAlAs semiconductor injection laser is only suitable for the lower, 0.8–0.9 μm, wavelength range. For the longer wavelength of about

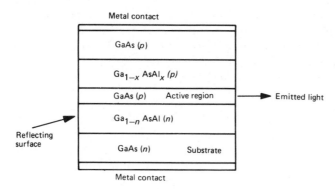

Fig. 12.11 Double heterojunction laser structure.

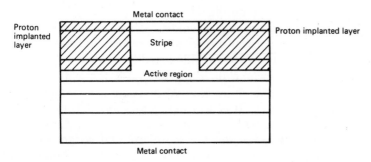

Fig. 12.12 Stripe laser.

1.3 μm, a different base material has to be developed. Present research suggests that the most promising is InP, with doped layers of GaInPAs forming the heterojunctions.

From a systems point of view, the injection laser offers the advantages of higher intensity, more directive output, with a faster response time and narrower spectral width than those of the LED. These advantages are gained at the expense of higher operating voltages and more complex drive circuitry, which is necessary to counteract the variation in threshold current produced by temperature changes and ageing in the injection laser.

12.20   COMMUNICATIONS SYSTEMS

Of the many applications of optical fibre systems, the one that will have, and indeed has had, the most significant effect on equipment development is long-distance telephony. Telephone operating companies around the world, stimulated by the paper of Kao and Hockham,[6] have investigated optical fibre systems in the hope that they could be used as a future trunk transmission system, with significant economic advantages over any alternatives. In the mid-1960s, it was clear that coaxial cable would not be the best medium to cater for the anticipated growth in telephone and television traffic, and considerable effort was directed to developing what seemed to be the most attractive alternative, overmoded circular waveguide. The special feature of an inherent reduction in attenuation with increase in frequency of the fundamental TE_{01} mode, made the circular waveguide approach very appealing, in spite of the essential expense of producing high-precision, helix-lined guides. At that time the major telephone administration research laboratories around the world were performing test trials on waveguide, before bringing it into service. Optical fibres were also interesting, but their high loss made them unable to challenge coaxial cable. Gradually, over the next five years, the considerable amount of research into the material and equipment aspects of fibre production began to be successful and the attenuation fell rapidly from 20 dB/km, or more, to 0.02 dB/km. A comprehensive understanding of the mechanism of transmission over fibre guide was being established, and it was becoming clear that there was no inherent loss mechanism to prevent transmission over very long distances. Gradually, the relative importance given to research into circular waveguide and optical fibre systems changed, and the big decision to commit the future to optical transmission was taken. The reasons why are, from our viewpoint, clear. The demand was for a large-capacity system that was cost effective, secure, comparatively easy to instal, and low-loss. Circular waveguide could be made low-loss, but otherwise it could not compete with fibre, and once the potential very low attenuation of fibre began to be realized, its other outstanding attributes made it the obvious choice. It is now highly possible that the transmission length between repeaters will be at least twenty times that which could be achieved with coaxial cable, and the small size and weight of fibre cable makes the installation a very straightforward operation. Rather than provoke additional difficulties, whose cost could be offset against its advantages, optical fibre is so light weight, flexible and slim that it can be threaded into ducts that had previously been thought to be full, and thus the lifetime of the trunk system can be extended. The cost of the fibre was, in the early days, very high. This was a reflection of the very high investment in development made by both the glass companies and the cable manufacturers, but cost is no longer a penalty. Once the large telephone administrations had decided to use fibre for all new trunk transmission, the per unit cost had to fall dramatically, to make it the cheapest medium on a capacity basis. The spin-off will be seen in other systems, as the technical properties of cheap fibres are exploited.

The system designer has several choices to make, once the decision to use fibre has been taken. The fibre type, refractive index profile, and operating frequency must be specified, and the types of emitter and detector must be chosen. The type of fibre to be chosen depends on the bandwidth required, the length of run, the future development of the system (called upgradability in reference 7), and the cost. There is a good case for installing as good a quality of fibre as possible, so that the system can be upgraded without having to replace the fibre. If high bandwidth is required, the choice lies between graded-index multimode or single-mode guides. As we saw earlier, the core diameter of single-mode fibre is about a tenth that of multimode, which is therefore easier to launch light on to, and easier to join. However, advanced designs of injection laser diodes, and microprocessor-controlled coupling jigs, have reduced the difficulties of single-mode working such that the comparison with multimode is much more balanced. Several administrations have plans for 500 Mb/s single-mode experimental trunk routes, and this development suggests that the attractive features of single-mode will outweigh its disadvantages. Another aspect of future systems that makes the choice of single-mode guide more appealing is the use of integrated processing devices, known as integrated optics. Several research laboratories have been working on integrated processing devices, such as switches, modulators and filters, for over a decade. Much of the effort has been in producing satisfactory materials, and making simple devices in a reproducible way. Gradually we are moving towards optical integrated circuits that will be used in monomode systems. When such devices are available, many additional communication applications will be possible.

As we have seen already, the choice of wavelength is governed by the behaviour of the fibre. A combination of low attenuation and low dispersion is desirable. Until 1982 the most commonly used wavelengths were 0.83 μm and 0.9 μm, or thereabouts. After that time the lower dispersion available at longer wavelengths began to be exploited with silica-based fibres, operating at 1.3 μm.

This change of frequency required new emitters and detectors. Silicon detectors had been clearly the best performers at the shorter wavelength, but at 1.3 μm it is necessary to use one of the III−V compounds, usually based on GaAs. The structure of the detector should offer a large area for light collection and the device should be sensitive, without being too noisy, as we discussed in Section 12.16.

The type of emitter used will depend on the application. The almost linear output versus drive current relationship of an LED makes it particularly

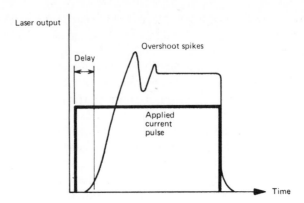

Fig. 12.13 Rise-time delay and output overshoot in laser response to pulse drive current.

attractive. It is also the obvious choice for low-speed, low-cost digital systems, particularly those using step-index fibre, which has a large value of NA. However, for launching on to graded-index fibre for high-speed systems, the injection laser might be preferred, and for single-mode systems it is essential. The small chromatic dispersion of the laser makes it necessary for long-distance communication systems. Semiconductor lasers are more temperamental than LEDs. Their performance characteristics change with temperature, and with age. To counteract these effects, some form of feedback is used, taking a monitoring signal from the rear facet of the laser, behind which is situated a simple photodiode. Two other problems with lasers are (i) rise-time delay and (ii) overshoot spikes, in response to a pulse input current (see Fig. 12.13). The dc bias used can affect the delay, and by biasing the device near its threshold level the delay is minimized. Here a compromise is required, for the spectral width increases as the bias increases, and the lifetime falls. In Section 12.19 we discussed the use of stripe conductors on the laser to stabilize its performance, and one of the benefits of this geometry is that the output oscillatory spikes are reduced.

The length of an optical fibre system will not be limited to that which can be achieved without intermediate amplification. In digital systems, regenerators will be required at intervals. The spacing will depend on the quality of the system, but it will be somewhere in the range 10–60 km. A block diagram of a regenerator is shown in Fig. 12.14. Essentially, it detects the incoming light signal, re-forms and re-times the pulses, and modulates an emitter for the next stage. Regenerators introduce additional problems for the designer. Supply voltages must be provided, and since regenerators are often placed in remote locations, the supply has to be fed along the optical cable. Monitoring, to allow remote fault finding from a control point at the end of a link, is usually included in the design. These features increase the cost of the system, so the gradual extension of the distance that can be covered without regenerators is a welcome result of fibre research.

Unlike systems that operate at lower radio frequencies, little attention has been paid to the use of free-space optical communications. This results from

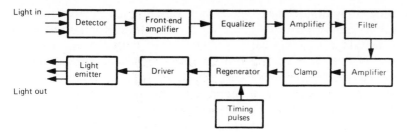

Fig. 12.14 Block diagram of optical regenerator.

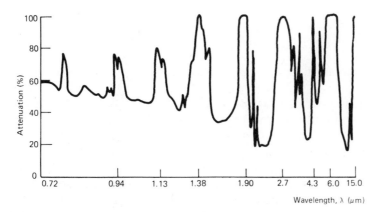

Fig. 12.15 Atmospheric attenuation plotted against frequency.

the heavy attenuation in the atmosphere due to moisture. Precipitation causes very large losses, but even under dry conditions there are attenuation peaks at certain frequencies (Fig. 12.15). In addition, the movement of air caused by localized heating produces refractive index variations that deflect the path of the light beam. Free-space links are limited to a maximum of about 2 km.

12.21   CONCLUSION

We have touched on some advantages that optical fibre systems possess. Many are attractive from the operations and management point of view, giving trouble-free performance. Some, such as no earth loops, and immunity to sparking or electrical interference, make communications links possible in environments that were previously very difficult, or even impossible. Apart from using fibres in such situations, we can expect them to form an integral part of the systems that will be required to satisfy the enormous increase in the quantity and type of traffic that will be generated as the revolution in office, computer and home communications takes place. Already there are designs for two-way, low-speed, data terminals in homes, and eventually all our domestic communications – telephony, television, automatic transactions, teletext, facsimile and systems not yet devised – will be transmitted over a single digital line – the optical fibre. Plans are being developed in the UK for national optical fibre cable television and fibre-based high capacity data systems. Similar systems are being developed in other countries. Coupled with digital representation of signals and integrated circuits of enormous density and complexity, optical fibre systems are transforming the aspirations of communications engineers, and encouraging a tremendous upsurge of innovation and enterprise. In fact, at present it seems that, for the first time in the history of telecommunications, we have more capacity and flexibility than we can use for any systems that presently come to mind.

REFERENCES

1. Midwinter, T. E., *Optical Fibres for Transmission*, Wiley, Chichester, 1979.
2. Kaminow, I. P., Marcuse, D. and Presby, H. M., "Multimode fiber bandwidth: theory and practice", *Proc. IEEE*, **68** (10), 1209–1213 (Oct 1980).
3. Smith, R. G., Photodetectors for fiber transmission system", *Proc. IEEE*, **68** (10), 1247–1253 (Oct 1980).
4. Miller, S. E. and Chynoweth, A. G., *Optical Fiber Communications*, Academic Press, New York, 1979.
5. Sandbank, C. P., *Optical Fiber Communication Systems*, Wiley, Chichester, 1980.
6. Wilson, T. and Hawkes, T. F. B., *Optoelectronics: An Introduction*, Prentice-Hall, London, 1983.
7. Kao, K. C. and Hockham, G. A., "Dielectric–fibre surface waveguides for optical frequencies", *Proc. IEE*, **113** (7), 1151–1158 (July, 1966).
8. Elion, G. R. and Elion, H. A., *Fiber Optics in Communications Systems*, Dekker, New York, 1978.
9. Wolf, H. F., *Handbook of Fibre Optics*, Granada, London, 1979.
10. Optical-fibre Communications, Special Issue, *Proc. IEEE*, **68** (10), (Oct 1980).

Packet switched networks 13

13.1 INTRODUCTION

This chapter deals with the topic of packet switched technology which is now in common use on wide area networks (WANs), local area networks (LANs) and metropolitan area networks (MANs). The most intensive use of packet transmission at the present time (1994) is on local area networks and a substantial portion of the chapter is devoted to this topic. Packet switching was, however, initially developed for connecting computers separated by long distances and some consideration of WANs will also be given. Recent years have seen the development of WANs, LANs and MANs to carry integrated digital services including data, voice and video traffic.

The main difference between wide area and local area networks is that the former cover very large distances and have limited bandwidth and substantial propagation delay. Local area networks, on the other hand, cover distances up to a few kilometres and have relatively high bandwidths and low propagation delays. This means that the packet transmission techniques for the two types of network have some important differences. It is worth pointing out here that although WANs and LANs can be categorized separately, many LANs have gateways to WANs so that global communications often make use of combinations of both type of network.

A packet is essentially a block of data which is transmitted from a source to a destination. A typical packet format for Ethernet is shown in Fig. 13.1. The data is preceded by a header which contains the destination and source address and is followed by a cyclic redundancy check. The length of the data section (for Ethernet) can be anywhere between 46 and 1500 bytes (octets). The format of the packet will be discussed in detail later in this chapter; it suffices at this point to indicate that the channel between transmitter and receiver is occupied only during the actual transmission of the packet.

The actual transmission rate is much higher than the data rate of the source, which means that packet switching may be considered as a form of

Preamble	Destination address	Source address	Type field	Data field	CRC
64 bits	48 bits	48 bits	16 bits	8 N bits	32 bits

Fig. 13.1 Ethernet packet format.

time division multiplexing on demand, i.e. transmission is only requested when sufficient data to fill a packet is available at the transmitter. At other times the transmission medium may be used to transmit packets between other sources and destinations. For this reason packetized systems are often referred to as ATDM or statistically multiplexed transmission systems; ATM (Section 10.44) is a form of packet transmission.

This contrasts with the public switched telephone network which allocates a circuit between caller and called party for the duration of the call. This is known as circuit switching and differs from packet switching in that the circuit remains allocated throughout any periods of inactivity between the communicating parties. Packet switched systems are thus generally more efficient than circuit switched systems. The analysis of packet switched networks is often very complex and, as a result, powerful software simulation tools have been developed to aid the design process. Discussion of these tools is outside the scope of a text of this nature; instead a simplified treatment will be given, where appropriate, and the results of simulations will be quoted.

13.2   WIDE AREA NETWORKS

The characteristic feature of wide area networks is that distances, and hence propagation delays, tend to be large. A typical topology is shown in Fig. 13.2 and it is usually the case that more than one path exists between a source and a destination node. Each node accepts data from computers or terminals connected to it, in a series of bursts, and assembles the data into packets with the appropriate address information.

The nodes then determine the routing, provide buffering and error control, and return acknowledgement information to the sender when the packet reaches the final destination; an early example of such a network is AR-PANET.[1] There will clearly be several hops between intermediate nodes and inter-node acknowledgements are given when a neighbouring node accepts a packet. Should an acknowledgement not be received within a specified interval (e.g. 125 ms) the packet is retransmitted. The end to end delay for short messages varies between 50 and 250 ms, depending on the packet length used. The delay can increase substantially under heavy network loads, when lengthy queues can form at intermediate nodes.

The topology of Fig. 13.2 is an example of a store and forward packet

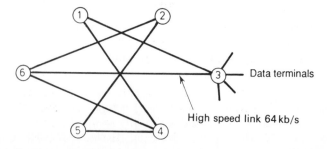

Fig. 13.2 Mesh connected wide area network topology.

switched network as each node stores data until a complete packet is received and then queues the packet for forwarding to a neighbouring node. Clearly physical links must be provided between the individual nodes and there are essentially two packet transmission techniques which can be identified. A **virtual circuit** transmission system is one in which a path is set up between soure and destination and all packets follow the same path and arrive in the sequence in which they were transmitted. A **datagram** transmission is one in which individual packets are transmitted by whatever route is available when they are presented for transmission. No fixed route is set up and each intermediate node decides on the appropriate path, according to some routing algorithm. In the case of a datagram the packets do not necessarily arrive in the order in which they are transmitted, and each packet must identify its position in the transmitted sequence.

The main feature of the datagram type of packet switched network is that the data links between nodes are used at near to their full capacity. Further, packets are scheduled for transmission so that there is no contention. Essentially each node queues packets for transmission and, since there is more than one outlet, each node can select an alternative route if the queue for the direct route becomes too long. There are many possible routing algorithms one of which is shown in Fig. 13.3.

In this example several priority ordered choices are stored in a routing table at each node. Considering Fig. 13.2, the first choice of a route between a source connected to node 1 and a destination connected to node 4 would be the direct route, i.e. 1 to 4. The second choice might then be via node 3. Once it has received an error-free packet from node 1, node 3 then selects the appropriate route for node 4. All of the nodes in a system of this type are referred to as packet switches.

In the topology of Fig. 13.2 the individual links are quite separate and routing algorithms are required. An alternative topology is one where each individual node is connected to a common broadcast channel and has to compete in some way for access, with all other nodes. This principle is used in most local area networks and in some wide area networks also. Wide area networks based on a single communications channel use either radio, to connect many geographically separate nodes, or satellite communications. This approach to wide area networks can be very cost-effective because it replaces the very expensive long distance dedicated links with a single wideband channel. A typical satellite system is shown in Fig. 13.4 in which each ground station transmits on a frequency f_{up} and receives on a frequency f_{down}. In such a system each ground station can monitor its own output for error-free transmission.

However, a single bus does introduce the problem of contention, i.e. several nodes requiring access to the channel at the same time. There is a number of techniques which have been developed to resolve such contention, and they can be categorized either as scheduled access or random access. The properties of these **access techniques** will be considered in some detail in this chapter. This will illustrate the suitability of specific access mechanisms for wide or local area application. Of particular interest will be the relationship between throughput, offered traffic and delay.

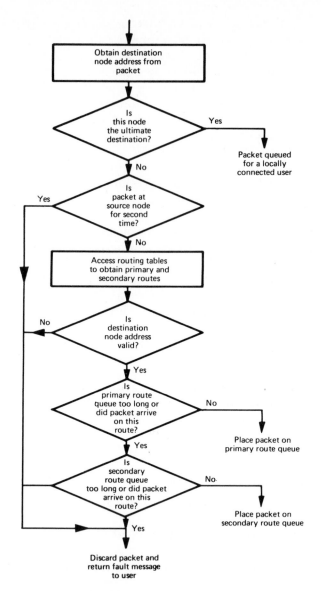

Fig. 13.3 Network routing algorithm.

13.3 THE POISSON DISTRIBUTION

Before any progress can be made in considering packet switched networks it is necessary to describe the statistics of packet arrivals on the network in any specified time interval T. To achieve this the time interval is assumed to be divided up into M intervals each of duration t seconds. As $t \to 0$ the probability of a packet actually arriving in the time interval t will be proportional to the actual value of t. Hence the probability of a packet arriving in t seconds is

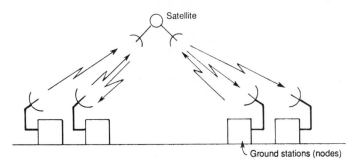

Fig. 13.4 Satellite packet broadcast system.

λt and the probability of no packet arriving in this interval is $(1 - \lambda t)$. A fundamental assumption is that the probability of a packet arriving in one interval t is completely independent of arrivals in all other similar intervals. The system is thus said to be memoryless and the probability that there will be no arrivals in the entire interval T is then

$$P(0) = (1 - \lambda t)^M = (1 - \lambda t)^{T/t} \tag{13.1}$$

As $t \to 0$ then $M \to \infty$ and $P(0) = e^{-\lambda T}$ since $\lim_{x \to 0} (l - nx)^{1/x} = e^{-n}$

The probability of k arrivals in time T is the joint probability of one arrival occurring in k of the intervals t and no arrivals occurring in the remaining $M - k$ intervals, this is

$$P(k) = \frac{M!}{k!(M - k)!} (\lambda t)^k (1 - \lambda t)^{M - k} \tag{13.2}$$

When $M \gg 1$ it may be shown that

$$M! \approx \sqrt{(2\pi)} e^{-M} M^{M + 0.5} \tag{13.3}$$

This may be verified by replacing M by any integer (greater than 2) in Eqn 13.3. Similarly if k is fixed

$$(M - k)! \approx \sqrt{(2\pi)} e^{-(M - k)} (M - k)^{M - k + 0.5}$$

Substituting into Eqn (13.2)

$$k! P(k) = e^{-k} (1 - k/M)^{-(M + 0.5)} (1 - k/M)^k (M \lambda t)^k (1 - \lambda t)^{M - k} \tag{13.4}$$

But $\lim_{M \to \infty} (1 - k/M)^{-(M + 0.5)} = e^k$ and if $M \gg k$, then $(1 - k/M)^k \approx 1$

i.e.

$$k! P(k) = (\lambda t)^k \frac{(1 - \lambda t)^{T/t}}{(1 - 1\lambda t)^k}$$

as

$$t \to 0, \quad (1 - \lambda t)^k \to 1 \text{ and } (1 - \lambda t)^{T/t} \to e^{-\lambda T}$$

Hence

$$P(k) = \frac{(\lambda t)^k e^{-\lambda T}}{k!} \tag{13.5}$$

This is known as the Poisson distribution and gives the probability of k arrivals in the interval T. In this expression λ may be interpreted as the average rate of arrivals per second.

13.4   SCHEDULED ACCESS TECHNIQUES

To illustrate the principles of scheduled access, consideration will be given to a single bus system of the type shown in Fig. 13.5 although it should be realized that the same arguments apply to ring topologies and the shared broadcast radio channel.

Scheduled access can be under centralized control or can be completely distributed. In the former case access is determined by a designated controller, and this will be considered first. The performance of such systems is usually measured in terms of access delay. This is the time which elapses between a packet being available for transmission at a node and commencement of transmission.

In order to avoid contention on the bus the controller poles each node in turn. If a node has a packet for transmission the transmission is commenced as soon as the polling signal is received. If a node has no data for transmission the controller is made aware and then poles the next node in the system. When the controller sends out a polling signal there will be a finite time which elapses before the polled node is able to reply (polling interval), followed by an interval required for data transmission (transmission interval).

The polling interval p_i will be a combination of the propagation delay on the network and the time required for the node to synchronize with the polling signal. The transmission interval t_i will depend on the amount of data a particular node has for transmission when polled. Both p_i and t_i are random variables, but it is clear that when the network extends over several hundred (or thousand) kilometres the polling delay p_i can greatly exceed the value of t_i. The efficiency of such a system is thus relatively poor when the propagation delay is very high.

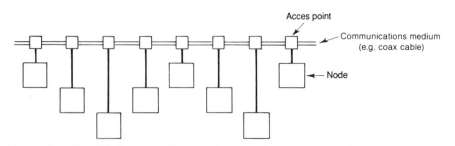

Fig. 13.5 Single bus communications channel.

A significant improvement can be achieved if central control is replaced with distributed control. When a node completes its transmission it essentially poles the next node on the network. The polling time is now, on average, significantly shorter than with centralized control, because the mean distance between a node and its immediate neighbour is much less than the mean distance between a designated controller and all other nodes. This type of distributed control is essentially the same as the control used on token passing LANs. The analysis for the token bus and token ring LAN is covered in Sections 13.8 and 13.9, respectively. Similar results are obtained for wide area token passing communications systems and the analysis will not be repeated here. It should be borne in mind, however, that propagation delay is usually insignificant in local area networks but this is not the case with long distance networks.

13.5 RANDOM ACCESS TECHNIQUES

It was stated in the previous section that when the utilization is low there can still be a considerable access delay as a result of the polling time. This problem can be overcome by allowing each node to transmit as soon as it has a complete packet ready. There are a number of variations on this theme but they all result in the possibility of contention. This occurs when two or more nodes each transmit a packet during a particular interval of time. One of the simplest random access methods is the ALOHA protocol developed by the University of Hawaii.[1]

13.5.1 The ALOHA and slotted ALOHA access protocol

The ALOHA protocol was developed for radio transmission and allows each user to transmit a fixed length packet as soon as it is formed. It relies on a positive acknowledgement to indicate that the packet was received without error. In the case of the single bus system of Fig. 13.5 the receiving node would be required to send an acknowledgement. In the satellite system of Fig. 13.4 the transmitting node can monitor the packet retransmitted by the satellite and this produces an automatic acknowledgement. Assuming that each packet requires a transmission time of P seconds then if such an acknowledgement is not received within a time $P + 2t_p$, where t_p is the maximum end to end propagation delay on a bus or the propagation delay between transmitting node and satellite, the transmitter retransmits its packet.

A packet will be successfully transmitted at a particular instant t only if no other packet is transmitted P seconds before or after t (i.e. the vulnerable period is $2P$); this is illustrated in Fig. 13.6.

The normalized channel throughput S (the average number of successful transmissions per interval P) is related to the normalized traffic offered G (the number of attempted transmissions per interval P, including new and retransmitted packets) by

$$S = Gp_0 \qquad (13.6)$$

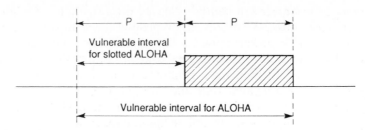

Fig. 13.6 Vulnerable period for ALOHA and slotted ALOHA.

Where p_0 is the probability that no additional packet transmissions are attempted in the vulnerable interval $2P$.

If it is further assumed that packet arrival times are independent and exponentially distributed with a mean arrival rate of λ per second, the probability of k arrivals in an interval of duration t is then a Poisson process given by

$$P(k) = (\lambda t)^k e^{-\lambda t}/k!$$

The probability of no arrivals in time t is $P(0) = e^{-\lambda t}$. In P seconds there will be $\lambda P = G$ arrivals, hence $\lambda = G/P$. Letting $t = 2P$ gives the channel throughput as

$$S = Ge^{-2G} \qquad (13.7)$$

Equation (13.7) is valid for satellite systems where all nodes are on the earth's surface and are approximately the same distance from the satellite. For earth-based packet radio the satellite will be replaced by a base station and the expression must be modified to account for the fact that some nodes will be nearer the base station than others. In such cases there will be a maximum **difference** in propagation delay of t_{dm} between near and far nodes which has a normalized value of $a = t_{dm}/P$. In such circumstances Eqn (13.7) becomes

$$S = Ge^{-2(1+a)G}$$

note that for satellite systems the normalized difference in propagation delay $a = 0$.

The maximum throughput for a pure ALOHA channel is found by differentiating Eqn (13.7) with respect to G and occurs at $G = 0.5$, i.e.

$$S_{max} = 1/2e = 0.184 \qquad (13.8)$$

The throughput characteristic for pure ALOHA is shown in Fig. 13.7.

Figure 13.7 shows that when the offered load G is low there are very few collisions and virtually all transmissions are successful and $S \approx G$. At higher values of offered traffic the number of collisions increases, which increases the number of retransmissions causing still more collisions, and so on. The characteristic is thus unstable and S actually begins to drop. The maximum throughput of the ALOHA protocol is thus limited to only 18% of the system capacity. However, this is still adequate for many purposes e.g. very bursty traffic, such as that produced by a human user at a remote terminal. If the

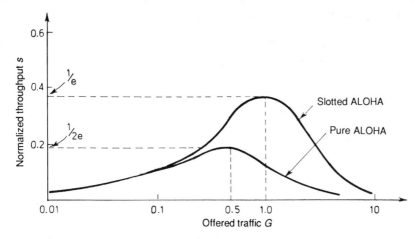

Fig. 13.7 Throughput characteristic for pure and slotted ALOHA.

system bit rate is, say, 1 Mb/s the useful capacity, including packet overheads, will be 180 kb/s.

In order to calculate the average time required to successfully transmit a packet, some knowledge of the retransmission procedure is required. A retransmission will be necessary if a node does not receive an acknowledgement within an interval $(P + 2t_p)$ after the first transmission. If a node retransmits immediately after this interval then the probability of a further collision will be unity since any other node involved in the collision will use the same retransmission procedure. To avoid a certain collision on retransmission some randomness should be introduced into the retransmission timing. One possibility is to define a retransmission interval $m \times P$ and to retransmit at time $n \times P$ where n is uniformly distributed between 1 and m. The average delay before a retransmission is attempted is then $P(m + 1)/2$. The time required to confirm a successful retransmission is thus $2t_p + P(m + 1)/2$.

If there is a total of E retransmission attempts the mean access delay is

$$t_a = P + 2t_p + E[2t_p + P(m + 1)/2] \qquad (13.9)$$

The relationship between throughput and offered traffic is

$$\frac{S}{G} = \frac{1}{1 + E}$$

If all packet arrivals are independent (which will be true only for large values of m) then eliminating S from Eqn (13.7)

$$E = e^{2G} - 1 \qquad (13.10)$$

The number of retransmissions can now be substituted into Eqn (13.9) to give the mean access delay. As an example consider the satellite system of Fig. 13.4 with a transmission rate of 1 Mb/s and a packet length of 8000 bits. The round trip delay for a synchronous satellite is approximately $2t_p = 270$ ms and the packet transmission time is 8 ms (note that the packet transmission time is much less than round trip delay, this is typical of WANs). If the normalized offered traffic is $G = 0.1$ and $m = 5$ then $E = 0.22$ and $t_a = 337.8$ ms. The major contribution to this access delay is the round trip delay of 270 ms. A typical round trip delay for a LAN would be of the order of 50 μs, and the corresponding access time would be $t_a = 8.413$ ms.

The throughput characteristic of pure ALOHA can be improved significantly if all transmissions are synchronized. The time axis is divided into intervals of time (or slots) of duration P and each user may transmit only at the start of a slot. Under these circumstances, if collisions do occur packets will overlap completely and the vulnerable period is reduced to P. The throughput of slotted ALOHA is then given by

$$S = Ge^{-G}$$

This has a maximum value when $G = 1$ given by

$$S_{max} = 1/e = 0.368.$$

The maximum throughput of slotted ALOHA is thus increased to 36.8% of the system capacity. The characteristic for slotted ALOHA is also shown in Fig. 13.7.

It is clear from Fig. 13.7 that slotted ALOHA has a similar instability when the offered load exceeds the value at which the throughput is a maximum. The access delay is developed in a similar way to that for pure ALOHA. There is, however, one important difference: a packet may be transmitted only at the beginning of a slot. This means that on average (assuming packet arrivals are independent) a packet will have to wait $P/2$ seconds until the beginning of the next slot. Hence, on average, the time required to ensure a successfull transmission is $1.5P + 2t_p$. The access delay for slotted ALOHA is thus

$$t_a = 1.5P + 2t_p + E\left[2t_p + \frac{P}{2}(m + 2) \right] \tag{13.11}$$

Hence the improved throughput of slotted ALOHA is achieved at the expense of a slight increase in access delay.

13.5.2 Carrier sense multiple access with collision detection

This access protocol is one of a group of protocols known as carrier sense multiple access, which differ from ALOHA in that a node listens to the channel and transmits a packet only if the channel is idle. Determination of whether a channel is idle or busy presents particular problems for radio systems because not all users within a radio network can hear all others, giving rise to 'hidden users'. There are methods of overcoming this problem[2], but as far as this text is concerned consideration of CSMA will be restricted to the bus system of Fig. 13.5. A coverage of several variants of CSMA is given by Kleinrock.[3] In this chapter we limit our attention to CSMA/CD which is the most widely used variant.

The CSMA/CD access protocol requires each node to continuously monitor the transmission medium, and as a result each node is able to detect a collision. Once a collision is detected all nodes stop transmitting immediately.

Retransmission is attempted at some later interval. CSMA/CD is more efficient than ALOHA since transmission is interrupted as soon as a collision is detected. The vulnerable period is also reduced to twice the end to end propagation delay on the network. This can be shown by considering the topology of Fig. 13.5. Assume a node at one end of the bus senses the bus idle and transmits a packet. A node at the other end of the bus will not be aware of this transmission until an interval t_p, equal to the end to end propagation delay, has elapsed. If it is supposed that the second node transmits a packet just prior to becoming aware of the first transmission then clearly a collision will occur. However the first node will not be aware of the collision until a further period t_p elapses; thus the minimum time required to detect a collision is $t_{sl} = 2t_p$. This argument assumes that each node will stop transmitting immediately upon detecting a collision. In practice there will be an additional period known as the jamming period. This will be ignored in the following analysis, but will be considered further in the description of Ethernet.

The characteristics of the CSMA/CD access technique are determined by assuming that a large number of nodes have a packet ready of transmission and that the length of each packet is the same (the restriction on packet length is imposed only to produce a tractable analysis and would actually be a disadvantage in practice). It is assumed that there are Q nodes with a packet ready for transmission and that the probability that any of the ready nodes will transmit within the interval t_{sl} is p. In order for this transmission to be successful one node only must transmit. The probability of a successful transmission is thus

$$P_r = Qp(1-p)^{-1} \tag{13.12}$$

The maximum value of P_r occurs when $dp_r/dQ = 0$, which gives $p = Q^{-1}$. In essence each node must determine the value of Q and it is assumed in the following analysis that this has been done (an approximate method of achieving this is considered in the Section 13.7 on Ethernet).

The probability of a successful transmission is thus

$$P_r = (1 - Q^{-1})^{Q-1} \tag{13.13}$$

As $Q \to \infty$, $P_r = e^{-1}$. Thus when the network is sensed idle each node will transmit a packet with probability Q^{-1}. The probability that a node will experience a collision will be $(1 - P_r)$. The probability that the node will be delayed by J timeslots (t_{sl}) is thus the probability of $(J - 1)$ collisions followed by a successful transmission, i.e.

$$P_J = P_r(1 - P_r)^{J-1}$$

The average number of collisions is thus

$$J_{av} = \sum_{J=0}^{\infty} J \cdot P_J = 1/P_r$$

The mean access delay is thus $t_a = 2t_p \cdot J_{av}$ and the average time required to deliver a packet is

$$t_d = P + t_p + 2t_p \cdot J_{av}$$

i.e.

$$t_d = P\{1 + a(1 + 2/P_r)\} \tag{13.14}$$

where $a = t_p/P$ is the ratio of propagation delay to packet transmission time.

If t_d is the average time to deliver a packet then the average number of packets delivered/second is $1/t_d$. The normalized throughput (the average number of packets delivered per interval P) is then

$$S = \frac{1}{\{1 + a(1 + 2/P_r)\}} \tag{13.15}$$

The limiting case occurs when the number of nodes Q tends to ∞, in which case $1/P_r = e$. This characteristic does not collapse to zero, as is the case with ALOHA. Equation 13.15 is plotted as a function of a in Fig. 13.8.

Figure 13.8 shows that for a throughput of $S = 0.6$ the required value of a is 0.1. This means that the propagation delay t_p is restricted to 10% of the packet transmission time. It may be concluded from this that CSMA/CD is not suited to long-distance communication, with corresponding large propagation delays, unless the packets are excessively long or the data rate is low. An alternative way of presenting the characteristic is shown in Fig. 13.9 where S is plotted as a function of the number of users Q.

The significance of the parameter a is once again evident. If the packet size is equal to the smallest that will allow collision detection then $P = t_{sl}$ (i.e. $a = 0.5$) and the asymptotic value of S becomes 0.24. In typical applications with much longer packets and moderate loading the throughput is of the order of 80% and this increases to approximately 98% when relatively few users generate longer packets.[4]

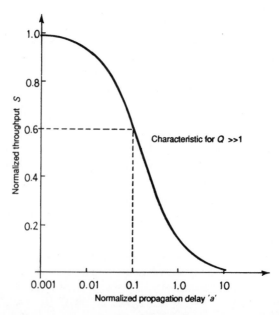

Fig. 13.8 Throughout characteristic for CMSA/CD as a formation of the normalized propagation time 'a'.

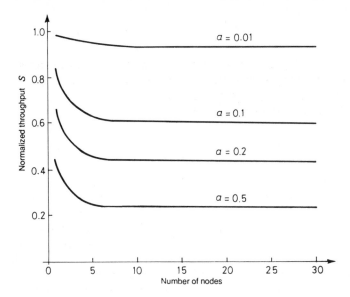

Fig. 13.9 Throughput/load for CMSA/CD.

The delay characteristic associated with CSMA/CD is very much influenced by the method of determining the transmission probability Q^{-1} and this is covered in Section 13.7 which is devoted to Ethernet.

13.6   LOCAL AREA NETWORKS

In this section we shall consider networks developed specifically for short-distance communication where the propagation delay is low and the bit rate is high. The three most common LAN topologies are shown in Fig. 13.10.

The star-connected network is a simple topology in which the central node performs the routing function for communication between nodes attached to it. For this reason the central node must have sufficient capacity to handle all simultaneous traffic and is a potential bottleneck. In particular a very high reliability of the central node is demanded, as a failure at this point completely destroys the network. In this chapter we shall be concentrating on the bus and ring structures and the access protocols which they employ.

It is useful at this point to consider some of the properties of bus and ring topologies as this will provide some insight to the reasons for their use in local area networks.

(i) A bus network is generally more reliable than a ring network as it tends to be purely passive. This means that failure of one or more nodes does not cause the failure of the network in total. In a ring network each node effectively repeats any transmission not destined for itself to the next node. This means that failure of any node in the ring can be catastrophic. One method of reducing the risk of network failure is to provide each

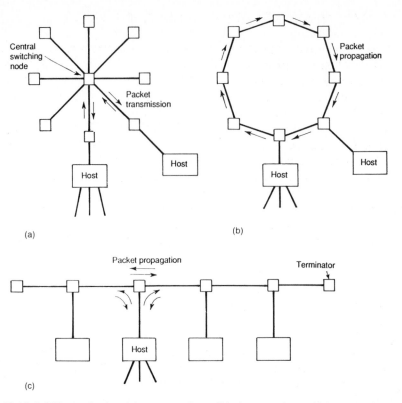

Fig. 13.10 LAN topologies, (a) star topology, (b) ring topology, (c) bus topology.

node with bypass circuitry which is brought into action automatically on a node failure.

(ii) Additional nodes can be incorporated in bus networks without disrupting network operation. Ring networks require the insertion of new cable segments which requires the network to be taken out of service.

(iii) A bus network is limited (currently) to twisted pair or coaxial cable because of the need to transmit signals in two directions, bi-directional optical couplers would be required for optical systems. Ring networks tend to transmit signals in one direction only and lend themselves to optical fibre operation with all the inherent advantages of this medium.

(iv) Bus networks are prone to reflections and care is required to avoid impedance mismatch at tapping points.

(v) Fault isolation is very difficult in bus networks but is relatively straightforward in a segmented ring structure.

Ring networks tend to support some form of token passing for network access because of their sequential nature. Bus networks, on the other hand, are suitable for both random access and sequential (or polling) access. We shall consider two common examples of bus-based LANs and compare these with the alternative ring structure.

13.7 THE ETHERNET LOCAL AREA NETWORK

Ethernet is based on a bus network operating at 10 Mb/s baseband using the CSMA/CD access protocol. This local area network was developed for computer communications, but there has been significant interest in recent times for using Ethernet for packetized voice communications[5,6] and also video. The basic CSMA/CD protocol was covered in the previous section; however, there are some further points which should be considered with reference to Ethernet and its very close derivative IEEE 802.3. The hardware configuration of a typical Ethernet node is shown in Fig. 13.11. This consists of a microprocessor, buffer memory, an Ethernet data link controller (EDLC) and its associated Ethernet serial interface (ESI) and transceiver unit. This hardware performs both transmission and reception of Ethernet packets.

Any node with a packet ready for transmission monitors the medium for ongoing transmission. If there is no ongoing transmission the ready node waits for an interval of 9.6 µs after the bus becomes idle (the interpacket gap) and then transmits immediately. This differs from the CSMA/CD discussed previously where the node transmitted with a probability Q^{-1}. Collisions can occur during the interval $t_{sl} = 2t_p$ and this 'slot time' is set equal to 51.2 µs which limits the effective maximum length of the Ethernet bus to 2.5 km. (An Ethernet system is actually composed of a number of segments, see Fig. 13.13 for details.) If a collision is detected all nodes involved abort the current transmission attempt after transmitting a 32 bit jam pattern (010101···). This jam pattern ensures that all other nodes are aware of the collision. To prevent repeated collisions with contending nodes each node involved in a collision then generates a backoff interval given by

$$T_r = \text{RND}[0, 2^r - 1]t_{sl}$$

where RND[ ] is a uniformly distributed random number between 0 and $2^r - 1$ and r is the number of successive collisions that a particular packet has

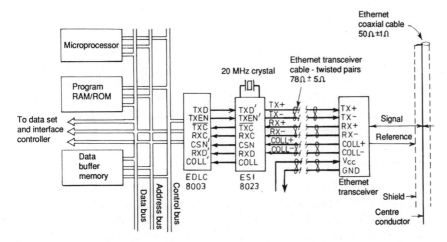

Fig. 13.11 CSMA/CD Ethernet node hardware.

experienced. The value of r is held at 10 should there be more than 10 but fewer than 16 collisions. On the 16th collision r is reset to 0 and a higher level collision warning is generated. The back-off algorithm is known as the (truncated) binary exponential back-off algorithm and is intended to facilitate the calculation of the transmission probability Q^{-1} which maximizes the chance of a collision-free despatch. In effect the binary back-off algorithm sets the probability of transmission as $1/2^r$, i.e. the number of ready users is estimated as $Q = 2^r$. For a highly loaded Ethernet with many collisions r will reach the value 10, which estimates the number of ready users as 1024. This is the maximum number of nodes specified for Ethernet. Clearly the optimum transmission probability is not implemented as when a new packet is available a transmission is attempted 9.6 µs after the network becomes idle. Further, the Ethernet packet may have a data segment between 46 and 1500 bytes long, hence the constant packet assumption is also not valid.

In reality the performance of CSMA/CD with a binary exponential back-off is virtually impossible to analyse mathematically without making gross approximations. Hence performance measures are usually obtained by computer simulation. However it should be stated that Ethernet does exhibit a stable throughput for very high loads and this is greatly influenced by the parameter 'a' considered earlier. In the case of Ethernet the packet length can vary between 576 bits and 12 208 bits (including the overhead of 208 bits). With a transmission rate of 10 Mb/s and $t_p = 25.6$ µs the parameter 'a' can vary between 0.002 and 0.44. Of particular importance, especially for speech transmission, is the access delay which increases as the traffic on the network increases. A typical access delay/throughput characteristic for speech packets derived from computer simulation is shown in Fig. 13.12.

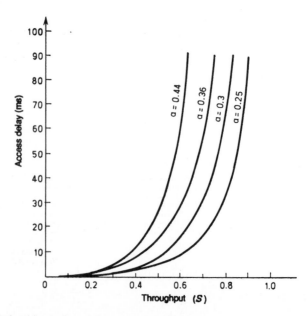

Fig. 13.12 Delay/throughput characteristic for Ethernet.

The reader's attention is drawn to the very rapid increase in delay which occurs at the higher network loads. For example when $a = 0.44$ the delay reaches 60 ms at a throughput of about 60%. The total delay in the case of speech is an important parameter. This will be composed of the access delay plus the packetization delay (the time required to collect sufficient speech samples to fill a packet).

A typical Ethernet network is shown in Fig. 13.13. This network is made up of several segments connected by repeaters or bridges. The maximum length of each segment is specified as 500 m and the maximum number of nodes per segment is 100. The maximum round trip delay for the whole system is specified as 51.2 µs.

13.8 THE TOKEN BUS

The random access nature of CSMA/CD is very attractive when the traffic levels are low because it is possible to transmit a packet almost as soon as it arrives. However, Fig. 13.12 illustrates that delays can become very large at high traffic levels owing to continued collisions between contending nodes. The contention problem can be avoided if transmission on the bus is

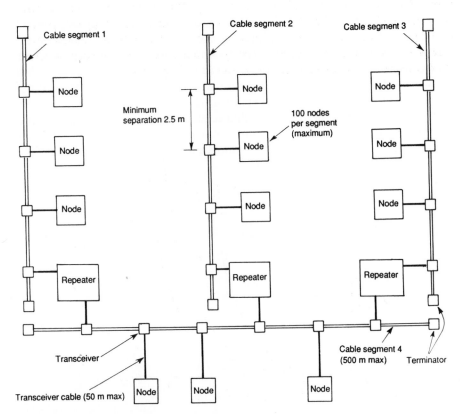

Fig. 13.13 Large-scale Ethernet installation.

scheduled and this is the basis of the token bus local area network, also known as IEEE 802.4. In effect each node waits for a token to be passed to it. When a node receives a token and it has data for transmission this data is transmitted in packetized form. The token is then passed to the next node and the process is repeated. If a node has no data for transmission the token is simply passed to the next node.

The token bus is less efficient than CSMA/CD at low traffic levels because a node must wait for the token even if the bus is free. At high traffic levels the token bus has a superior performance because contention (and the time required to resolve collisions) is avoided altogether. A simple token bus architecture is shown in Fig. 13.14. It should be noted that the token is passed from a node to the one with the next lower address, thus forming a logical ring on the bus. The token is passed only to active nodes thereby saving time. In order to receive the token an inactive node must request admission to the logical ring.

If it is assumed that the logical ring has been established the performance of the token bus LAN can be described as follows. (For information on the procedure of establishing the ring and admitting or deleting new nodes the reader is referred to Ramimi and Jelatis.[7])

A node collects packets for transmission and is able to transmit these packets when it receives the token from the previous node in the logical ring. In practice there is a maximum time allocated for transmission known as the token hold time. At the end of this period, or as soon as all waiting packets have been transmitted, the node passes the token to the next node on the logical ring. The token rotation time t_r is defined as the interval between successive arrivals of the token at a specific node. Evidently the token bus network will never be silent, as is possible with CSMA/CD, but will always be transmitting either message packets or a token.

In order to determine the delay characteristics of the token bus it will be assumed that token hold time is not limited, i.e. when a node receives the token it is able to transmit all its accumulated data. The token rotation time

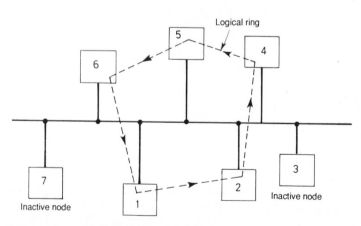

Fig. 13.14 Token passing local area bus network.

will thus be composed of the time to transmit N tokens (Nt_t) and the time required to transmit the total number of packets which accumulate during the token rotation time. If the mean packet arrival rate at each node is λ per second the total number of packet arrivals during the interval t_r is $\lambda N t_r$. If the bus can transmit packets at a rate of μ/second the token rotation time is

$$t_r = Nt_t + \lambda N/\mu \qquad (13.16)$$

Defining the bus utilization as $U = \lambda N/\mu$, gives

$$t_r = Nt_t/(1 - U) \qquad (13.17)$$

The mean access delay will thus be $t_a = t_r/2$. Note: in order to ensure stable operation $U < 1$. The minimum token rotation time will occur when $U = 0$. In this case none of the N nodes has packets for transmission, when the token arrives, and the bus traffic consists merely of token transmissions. The bus utilization U is effectively the percentage of time that the bus is transmitting data packets. The factor $(1 - U)$ is then the percentage of the time that the bus is transmitting tokens. The normalized access delay (for $Nt_t = 2$) is plotted in Fig. 13.15.

As the ultilization approaches 1 the access delay approaches ∞. It may also be noted from Eqn (13.17) that the access delay is related directly to the time required to transmit the token t_t which in turn is proportional to the end to end propagation delay on the bus and the token length, and is inversely proportional to the data rate on the bus.

If the propagation time is large (as would be the case for a wide area network) the access delay can be reduced by increasing the data rate. Hence scheduled access is far more suited to long-distance communication than is CSMA/CD which (according to Eqn (13.15)) is limited to low data rates in order to minimize the value of the parameter 'a'. In Eqn (13.15) the characteristic is given in terms of the throughput S which is defined as the average number of successful packet transmission in a specified interval. The throughput S is related to the utilization U, i.e.

$$U = \frac{\text{mean packet arrival rate}}{\text{mean packet transmission rate}} = \frac{\text{mean arrival rate}}{S}$$

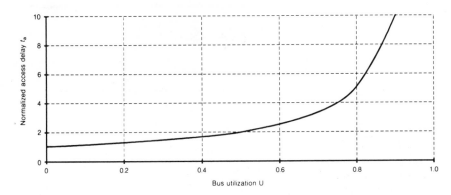

Fig. 13.15 Normalized access delay characteristic for the token bus network.

The IEEE 802.4 specification has been adopted for industrial use in the manufacturing environment. In this context a number of token bus networks operating at 1 Mb/s (known as carrier band) are connected via a bridge to a broadband cable operating at 10 Mb/s. These integrated systems operate under what is known as the Manufacturing Automation Protocol (MAP). A typical factory communications layout is shown in Fig. 13.16.

Each of the token bus subnets may be regarded as individual token bus networks in which the bridge to the broadband cable simply acts as an additional node. Tokens are not passed across the bridge so that the token passing on the subnets is independent from the token passing on the broadband cable. The subnets use phase coherent frequency shift keying with a logic low being represented by a carrier of 5 MHz and a logic high being represented as a carrier of frequency 10 MHz, the data transmission is omnidirectional. The broadband cable uses frequency division mutliplex transmission, one frequency being allocated for transmission and a second frequency allocated for reception. In this case all transmissions are unidirectional, a head end repeater being used for translating data between transmit and receive frequencies. The broadband cable is also able to handle other independent transmissions on a frequency multiplexed basis.

13.9 THE TOKEN RING

Ring networks are fundamentally different from bus networks in that the nodes on a ring network tend to act as repeaters and are therefore active

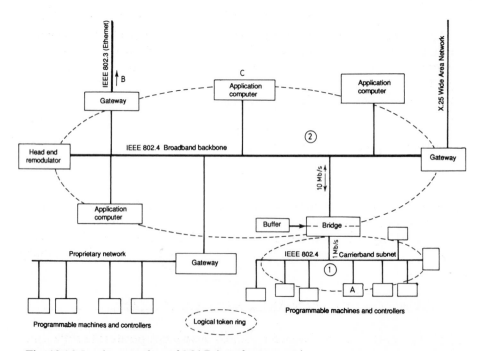

Fig. 13.16 Implementation of MAP in a factory environment.

devices. In the token ring access to the channel is controlled by passing a permission token around the ring. When the ring is initialized a designated node generates a free 8 bit token (11111111) which travels around the ring in one direction only until a node with packets ready to transmit changes the token to busy (11111110) and then transmits all its packets. The sending node is responsible for converting the token back to free, after it has travelled around the ring, and for removing its own packets. A further difference in ring structures is that each node adds delay to the circulating token and data. This is known as the node latency and is an essential requirement because each node must be able to store a bit in order to decide whether to change the last bit in a token (i.e. from free to busy or vice versa). Thus each node contributes a 1 bit delay and, if the number of nodes in a ring is large, this delay can become very considerable.

The token is usually 8 bits long which puts a lower limit on the number of nodes in a ring. If there are fewer than eight nodes on the ring the first bit of the token would return to the sending node before it had transmitted the last bit. Thus it is necessary to have sufficient delay on the ring to ensure that the complete token can be accommodated.

The mean access delay of the token ring may be obtained by assuming that there are N nodes on the ring and that each node has on average q packets ready for transmission. It is further assumed that packet arrivals have a Poisson distribution with a mean arrival rate, for all nodes, of λ packets/second. The mean arrival rate for the complete ring is thus $N\lambda$ packets/second. The time required for one bit to go all the way around the ring is termed the walk time t_w and is given by

$$t_w = N(\text{latency} + \text{propagation time between nodes}) \qquad (13.18)$$

The scan time t_s is the mean time between free token arrivals at a given node (this is equivalent to the token rotation time specified for the token bus LAN). This will consist of the walk time and the time required to service each of the packets ready for transmission. If the mean number of packets/second the ring can transmit is μ, the scan time is

$$t_s = t_w + N\lambda t_s/\mu \qquad (13.19)$$

Defining ring utilization $U = N\lambda/\mu$

$$t_s = \frac{t_w}{(1 - U)} \qquad (13.20)$$

The mean acquisition delay is then $t_s/2$ and this will approach ∞ as U approaches 1. This is similar to the characteristic for the token bus, which is not surprising as the token bus is, in effect, a logical ring. The token bus has the advantage that each individual connection to the bus is essentially passive.

13.10   METROPOLITAN AREA NETWORKS (MANs)

Local area networks are based on a shared bus which operates at a sufficiently high data rate to accommodate the needs of many users. However the LAN is restricted to a length of a few kilometres as the performance is limited by the normalized propagation delay. Metropolitan area networks are a relatively recent development which extend the operational distance of the shared bus up to distances in the range of 100 km. A number of standards have been developed for metropolitan area networks. The standard considered in this chapter was initially proposed by the University of Western Australia as the QPSX MAN.[8] This development was adopted as the basis of the IEEE 802.6 standard and is now known as the distributed queue dual bus metropolitan area network (DQDB MAN). FDDI is also considered in this category but is also often considered as a high-speed LAN.

The architecture of the DQDB MAN is based on two unidirectional buses as shown in Figure 13.17 and has been specified to operate at 155.52 Mb/s or 622.08 Mb/s. This allows full duplex communication between any pair of nodes connected to the network. The DQDB MAN is designed to support services requiring regular periodic access (isochronous services), such as digitized voice, and services requiring random access (non-isochronous services), such as packet switched data. For non-isochronous services the DQDB MAN provides a contention-free distributed queueing protocol which provides a mean packet access delay which is equal to that of a perfect scheduler. This section will concentrate on the mode of operation of the non-isochronous transport mechanism, with a brief description being given to the way in which the MAN handles isochronous services.

The principal components of the DQDB architecture are a slot (frame) generator, two unidirectional buses, a number of nodes (access units) and a slave slot (frame) generator. The slot generator creates a continuous stream of fixed size data slots on the A bus and the slave slot generator creates an identical pattern at the same rate on the Z bus. Access units are connected to both buses via read and write connections. The node reads information from the bus and writes information on to the bus by ORing its own data with that from upstream on the bus. This process is shown schematically in Fig. 13.18.

It will be noted from this figure that the data on the bus does not pass through the node, as is the case with the token ring. Hence failure of a node will not affect the operation of the remainder of the network.

The operation of the distributed queueing mechanism centres around two special bits in each slot known as the busy (B) bit and the request (R) bit. The B bit indicates that a slot is filled with user data and the R bit is used to indicate

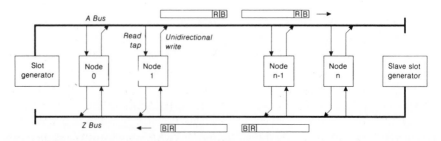

Fig. 13.17 Distributed queue dual bus MAN architecture.

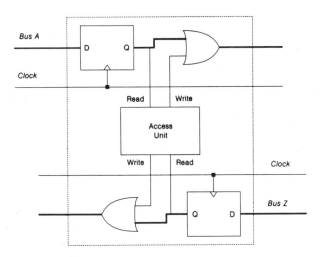

Fig. 13.18 Connection of nodes to the unidirectional buses.

that a node has a packet queued for transmission. By monitoring the B and R bits each node is able to keep a record of its position in the distributed queue. Thus when a node has a packet for transmission it will only access the bus when all other nodes that queued before it have transmitted. Access to the slots on the A bus are made using the R bit in slots on the Z bus and access to the slots on the Z bus are made using the R bit in slots on the A bus.

Consider a node n wishing to access the A bus. When the node has no packets for transmission it will be in the IDLE state and simply monitors the R bits which pass it on the Z bus. Each time an R bit which is set passes a given node on the Z bus, it indicates that a downstream node (a node with a number $> n$) on the A bus has a packet queued for transmission and is used to increment a request counter (RQ) within node n. The RQ counter is decremented each time an empty slot (i.e. one with the B bit clear) passes that node on the A bus, since that slot will be used to serve one of the downstream queued packets. Thus the value of the RQ count is at all times equal to the exact number of nodes downstream that have a packet queued for access.

When a sending node has a packet ready for transmission it will leave the IDLE state and if the value of the RQ counter is non-zero will enter a COUNTDOWN state. The contents of the RQ counter are transferred to a countdown counter CD and the RQ counter is cleared. The purpose of this state is to allow previously queued nodes to access the bus first. This is achieved by node n by decrementing the CD counter for each empty slot passing on the A bus. These empty slots will be used by nodes previously queued downstream of node n. While in the COUNTDOWN state the node continues to monitor any passing R bits which have been set on the Z bus and uses these bits to increment the RQ counter. These R bits do not affect the CD counter as they arrive after node n has queued a packet. In addition while in the COUNTDOWN state empty slots passing on the A bus decrement only the CD register and not the RQ counter as the empty slots are serving prior queued nodes.

When node *n* enters the COUNTDOWN state it will set the R bit in the first available slot on the Z bus. This R bit will be transmitted on the Z bus to all nodes which are upstream of node *n* on the A bus. This R bit indicates to the nodes upstream on the A bus that a request for an additional slot has been made on this bus. It is possible that a node could overwrite a R bit which has already been set by a downstream node. This would not affect the value of the R bit as the writing operation is achieved by a logical OR but the extra R bit generated by node *n* would not be detected by upstream nodes. To avoid this problem the read tap of each node occurs before the write tap and the node can detect whether the R bit is already set. In such a case the node would wait until a slot in which the R bit is clear passes it on the Z bus and it would then transmit a R bit.

The node remains in the COUNTDOWN state until the contents of the CD register reach zero and then enters the WAIT state. When in the WAIT state the node waits for the first empty slot on the A bus (i.e. the first slot with the B bit clear). It then sets the B bit and transmits its packet in the slot. If the node has no further packets for transmission it enters the IDLE state. If the node does have further packets for transmission it enters the COUNT-DOWN state. (Note that if a node wishes to transmit a packet on the Z bus it sends a R bit on the A bus and a similar action occurs with separate RQ and CD counters.)

The distributed queueing mechanism provides ideal access characteristics as a packet will gain immediate access if the queue size is zero or is required to wait only until previously queued packets are transmitted. Hence, unlike the other access mechanisms described in this chapter, capacity is never wasted while packets are queued and thus access delay is quaranteed to be a minimum. A typical performance characteristic of the DQDB compared with the token bus is shown in Fig. 13.19. The mean waiting time (access delay) for the DQDB is

$$t_a = \frac{P}{2(1 - U)} \tag{13.21}$$

where P is the packet transmission time.

Comparing this equation, with Eqn (13.17) it should be noted that the limitation of the token bus is the transfer of the token as nodes with packets for transmission must wait on average for an interval of half the bus latency. This is not the case with DQDB as a packet may be transmitted as soon as other packets ahead in the distributed queue have gained access. Equation (13.21) reveals that the advantage of the DQDB increases as the size of the network and the operation speed increase, as P is not dependent on any bus latency, which is in sharp contrast to the token bus performance.

An extension of the DQDB architecture is the looped bus structure of Fig. 13.20 in which the end points of the linear bus structure are combined into a single head end node. It should be noted that data does not pass through the head end node so that the looped bus does not operate as a ring. There are two specific of advantages of this structure however:

(i) The clock generator for both the A and Z bus can be derived from a common point and synchronized with the PSTN clock.

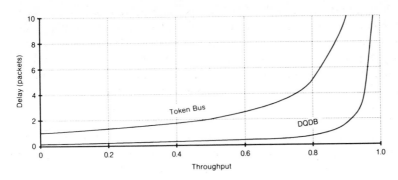

Fig. 13.19 DQDB throughput characteristic.

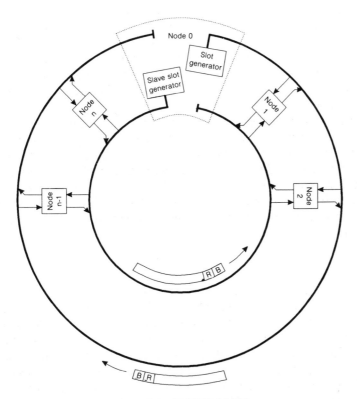

Fig. 13.20 Looped bus structure of the DQDB MAN.

(ii) In the case of a bus fault the network can isolate the fault and close the data buses through the head end of the loop. Hence the network can be reconfigured without the need for any redundant transmission hardware. This gives the DQDB looped structure a significant additional advantage in terms of reliability.

The discussions on the DQDB mechanism have concentrated on the non-isochronous transmission mode often termed the queue arbitrated (QA) mode. The DQDB also has another mode, known as the pre-arbitrated (PA) mode, which is used for isochronous services. The slots on the DQDB MAN can be made available either as QA slots or as PA slots. In the latter case the node at the head end of the bus manages the generation of PA slots which are divided into octets each of which may be allocated to particular access units for the transfer of isochronous service octets. Hence each slot can serve more than one access unit. Each octet within a PA slot has a unique offset position relative to the start of the PA slot payload. Each PA slot carries a special identifier and nodes which have requested and have been allocated an isochronous service will obtain from that identifer the offset of the slot which they been allocated. When slots are used in this way nodes are guaranteed periodic access and do not participate in the alternative distributed queueing mechanism.

13.11 OPEN SYSTEMS INTERCONNECTION REFERENCE MODEL

Packet switched networks were designed initially for computer communications. In such circumstances the computers themselves became an integral part of the communications system, performing tasks such as channel demand, addressing, routing, error checking, buffering, etc. Clearly the proliferation of various computer architectures could result in the creation of barriers to efficient communication. The early realization of this problem resulted in the adoption of an international standard for communication, in 1983, known as the Open Systems Interconnection reference model[9]. This model divides up the various aspects of communication into seven layers as shown in Fig. 13.21

The OSI model was designed to facilitate global communications, primarily for computers, irrespective of the characteristics of the particular networks of which they are part. The model does not specify how systems are implemented but rather how they communicate. This means that many different networks, using products of different manufacturers, may be coupled together by mapping the OSI model onto the complete communications path. A brief description of each layer will now be given.

(i) The physical layer (layer 1)
This layer deals with transmitting individual bits over a communications system. It deals with the way in which the bits are represented, e.g. voltage levels, bit duration, etc. on the transmission medium and whether the transmission may proceed simultaneously in two directions.

(ii) The data link layer (layer 2)
The data link layer makes use of the network layer to ensure error-free communication between the network layer at each end of the communications link. In Ethernet, for example, the data link layer is responsible for transforming data into the packet format of Fig. 13.1. Layer 2 also

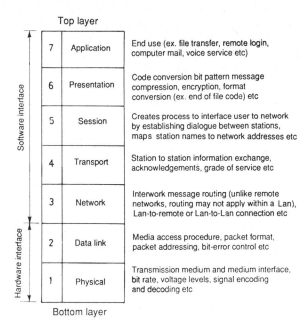

Fig. 13.21 Layered architecture of the OSI reference model.

handles acknowledgements from the receiver and responds to errors.

(iii) The network layer (layer 3)

This layer essentially deals with routing of packets over the communications link and ensures that the data is assembled in the correct order for presentation at the receiver. This layer would be concerned also with providing some form of error control.

(iv) The transport layer (layer 4)

This layer provides an end to end communication between host machines. It is not concerned with the hardware constraints of providing the actual connection, as are the lower layers, but is involved with such functions as how to create the most efficient use of the communications media for a particular throughput. The transport layer handles flow control, which is essential if a high speed transmitting host is communicating with a low speed receiving host.

(v) The session layer (layer 5)

This is the layer which deals with setting up the connection between users. This layer would handle, for example, the procedure for logging in on a remote machine and initiating a file transfer.

(vi) The presentation layer (layer 6)

This layer deals with the way in which the data is presented for transmission and is usually performed by subroutines called by the user. The presentation layer could, for example, handle data encription or the transmission of hexadecimal data in ASCII format.

(vii) The application layer (layer 7)

This layer deals with the way in which individual users' programs communicate with each other, i.e. the meaning of the transferred data.

The OSI model specifies seven layers, but not all communications connections will involve all seven layers. A special protocol known as X25 has been developed (actually before the OSI reference model) which implements layers 1, 2 and 3 of the OSI reference model. This protocol will be briefly described in the next section, as an example of the partial implementation of the OSI reference model.

13.12 THE X25 COMMUNICATIONS PROTOCOL

The X25 standard defines the interface between a host computer, known as the Data Terminal Equipment (DTE), and the carrier's equipment, known as the Data Circuit-terminating Equipment (DCE). When a host communicates, using X25, with a second host the DCEs at each end will be linked over an unspecified network by means of various switching centres. The connection between the two DCEs is given the general label of Data Switching Exchange (DSE) and is transparent to the X25 protocol. The relationship between X25 and the OSI reference model is shown in Fig. 13.22.

X25 defines three layers of communication, the physical layer (equivalent to the physical layer of OSI), the frame layer (equivalent to the data link layer of OSI) and the packet layer (equivalent to the network layer of OSI). The physical layer of X25 deals with the electrical representation of binary 1 and 0, timing, etc. and is described by a standard known as X21. This standard specifies how the DTE exchanges signals with the DCE to set up and clear calls, etc. and is illustrated in Fig. 13.23.

In many installations the DTE will be connected to the DCE over a standard analogue telephone line which may not support the full X21 specification.

The frame layer ensures reliable communication between the DTE and DCE. In effect the frame layer adds additional bits to the data stream produced by the packet layer to make the communication reliable. The additional bits and the data produced by the packet layer are combined into a frame to produce what is known as the link access procedure (LAP) protocol, a typical LAP format is shown in Fig. 13.24.

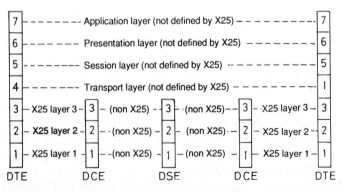

Fig. 13.22 Relationship between OSI reference model and X25.

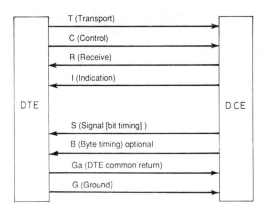

Fig. 13.23 Signal lines specified in X21.

Start byte 01111110	Address 8 bits	Control 8 bits	Data ≥ 0 bits	Checksum 16 bits	End byte 01111110

Fig. 13.24 LAP frame format.

The LAP frame structure will not be discussed in detail in this text but it should be noted that there are three kinds of frame known as information, supervisory and unnumbered. The reader is referred to Tanenbaum[10] for further information.

The third layer in X25 deals with the meaning of the various bits in the data field of the frame. The data within the frame is itself known as a packet, hence X25 packets are embedded in a LAP frame. There are actually 15 different types of X25 packet dealing with various aspects of call set-up and monitoring. As an example, when a DTE wishes to communicate with a second DTE a virtual circuit must be set up between them. The initiating DTE sends a 'call request packet' to its DCE which delivers this packet to the destination DCE which then passes the packet to the destination DTE. The calling DTE defines the logical channel on which communication is to take place by means of a channel field in the call request packet. If the distant DTE wishes to accept the call it returns a call accepted packet to the originating DTE and communication commences until a 'clear request packet' is sent to terminate the call.

There are many facilities available to X25, for example if the logical channel requested by the initiating DTE is in use at the far end the called DCE will substitute an unused channel. X25 is able to cope with collisions which could occur when a logical channel request from an initiating DTE coincides with an allocation of that channel (for a separate purpose) by the called DTE. The actual X25 protocol is quite complex and reflects the number of facilities available. A full description of X25 is given by Schwartz[11]. For the purpose of this text it is sufficient to illustrate the functional separation of the three defined layers of X25.

It is common practice to use X25 links for devices, such as simple terminals, which do not themselves have X25 capability. This has resulted in the widespread use of devices known as packet assemblers/disassemblers or PADs. A PAD is essentially a mulitplexer or concentrator which collects data from a number of asynchronous terminals and periodically outputs a correctly formated X25 packet made up of this data. The X25 packets may then be passed to a DCE in the normal fashion. The inverse operation is also required to convert X25 packets, from a host, into the asynchronous format for the individual terminals. The PAD contains significant intelligence and may itself be considered as an X25 DTE host.

13.13   CONCLUSIONS

This chapter has concentrated on the description of packet switched networks with a detailed consideration of local and metropolitan area networks. Some reference has been made to the integration of speech into local area networks. Packetization of speech and other real-time signals is a topic of intense interest both for wide area and local area communications. It is envisaged that all analogue telephone communications will eventually be replaced by integrated services digital networks. Major research projects currently being funded in Europe are directed towards replacing the fixed telephone by a completely mobile cellular digital radio telephone system. Initially this system is likely to be a PAN European system where any mobile user (which will eventually be all citizens) may communicate with any other mobile user without a prior knowledge of that user's physical location.

REFERENCES

1. Abrahamson, N., "The ALOHA System - Another Alternative for Computer Communications", *AFIPS Conference Proceedings*, 1970 Fall Joint Computer Conference, Vol. 37, pp. 281–285.
2. Dunlop, J., "Packet Access Mechanisms for Cellular Radio" *Electron. & Commun. Eng. J.*, 1993, Vol. 5, No. 3, pp. 173–179.
3. Kleinrock L *Queuing Systems Vol. II Computer Applications*, Ch. 5, John Wiley, 1976.
4. Shoch, J. and Hupp, J. A., "Measured Performance of an Ethernet Local Network" *Communications of the ACM*, Vol. 23, No. 12, 1980, pp. 711–721.
5. Dunlop, J. and Rashid, M. A., "Speech Transmission Capacity of Standard Ethernet Systems" *Journal of the IERE (UK)*, Vol. 55, No. 3, 1985, pp. 119–122.
6. Dunlop, J. and Rashid, M. A., "Improving the Delay Characteristics of the Standard Ethernet for Speech Transmission", *Journal of the IERE (UK)*, Vol. 56, No. 5, 1986, pp. 184–186.

7. Rahimi, S. K. and Jelatis, G. D., "LAN Protocol Validation and Evaluation", *IEEE Journal on Selected Areas in Communications*, Vol. SAC-1, No. 5, 1983, pp. 790–802.
8. Newman, R. M., Budrikis, Z. L. and Hullett, J. L., "The QPSX MAN", *IEEE Communications Magazine*, Vol. 26, No. 4, 1988, pp. 20–28.
9. Zimmerman, H., "OSI Reference Model-The ISO Model of Architecture for Open Systems Interconnection", *Trans. IEEE on Communications*, Vol. COM-28, 1980, pp. 425–432.
10. Tanenbaum, A. S., *Computer Networks* Prentice-Hall, 1981, Ch 4.
11. Schwartz, M., *Telecommunication Networks: Protocols, Modelling and Analysis*, Addison-Wesely, 1987, Ch 5.

PROBLEMS

13.1   A group of N stations share a 64 kb/s pure ALOHA channel. Each station, on average, outputs a 1024 bit packet every 50 seconds. What is the maximum value of N which may be accommodated?

Answer: 575.

13.2   A large number of nodes using an ALOHA system generate 50 requests/second; this figure includes new and re-scheduled packets. The time axis is slotted in units of 40 ms.

(i)   What is the probability of a successful packet transmission at the first attempt?
(ii) What is the probability of exactly four collisions followed by a successful transmission?
(iii) What is the average number of transmission attempts per packet?

Answer: 0.135, 0.076, 6.39.

13.3   An Ethernet cable operating at 10 Mb/s has a length of 1.5 km and a propagation speed of 2.2×10^8 m/s. Packets are 848 bits long which includes an overhead of 208 bits. The first bit slot after a successful transmission is reserved for the destination node to access the channel to send an acknowledgement packet (data field empty). What is the effective (i.e. useful) data rate assuming no collisions? Assume an inter-packet gap of 9.6 μs.

Answer: 5.33 Mb/s.

13.4   One hundred nodes are distributed over a CSMA/CD bus of length 4 km. The transmission rate is 5 Mb/s and each packet is 1000 bits long including the standard overhead of 208 bits. Calculate the maximum number of packets/second that each node can generate. Assuming an inter-packet gap of 9.6 μs and a propagation delay of 5 μs/km, calculate also the useful data rate of the system.

Answer: 45.45, 3.61 Mb/s.

13.5   The bit rate on the system of the previous question is increased to 10 Mb/s. Determine the effect of this increase on the number of packets/second that each node can generate and the useful data rate of the system. Repeat the calculation if the packet length is increased to 10 000 bits. Comment on the results.

Answer: 83.61, 6.62 Mb/s, 980.8, 9.6 Mb/s.

13.6 A coaxial cable of length 15 km supports the token bus protocol, the propagation speed being 2×10^8 m/s and the data rate being 10 Mb/s. There are 512 nodes connected to the bus each producing packets 12 kb long, with an overhead of 96 bits. The token, which is a standard packet with an empty data segment, is passed sequentially to each node. Assuming a latency of 8 bits at each node and that each node generates 6 packets per 10 seconds on average calculate the average token rotation time. What is the maximum rate of transmission of packets at each node?

Answer: 17.9 ms, 1.55 packets/s.

Satellite communications $\boxed{14}$

14.1 INTRODUCTION

For many years, the notion of satellite communications was a fantasy produced by the fertile mind of Arthur C. Clarke;[1] a brilliant idea but rather impractical. As with so many creative ideas, technology eventually caught up and now satellites are commonplace. For the communication engineer, however, they represent a challenging and stimulating field of work. Within satellite communication systems we find the whole gamut of technologies operating in a strange and demanding environment. Coupled with the essential requirement of demanding as little energy to run as possible, the system must be capable of withstanding the arduous journey from earth to orbit. Consequently a careful balance must be struck between the mechanical, structural, electronic, electrical and electromagnetic engineering requirements of the system[2]. However, in this chapter we will limit our attention to system considerations.

Satellite communications provide opportunities, and pose problems, in communication methods. Their large area of access (footprint) allows a single transmission to cover an enormous number of receivers, thus allowing broadcast signals to be transmitted simultaneously to large numbers of people. However, this feature itself creates difficulties; partly political and partly economic. National boundaries are no barrier whatsoever, and the charging mechanism required to allow the satellite operator to recover the cost of development and provide continuous support requires a novel solution.

Satellites can travel in a variety of orbits, basically eliptical in shape, but we will limit our attention to geostationary systems, in which the orbit is fixed, and essentially circular. A geostationary satellite maintains a fixed position relative to points on earth, and this allows a cheap receiving antenna to be set up once and then fixed in position.

The enormous attraction of satellites for a whole range of communication objectives has meant that there is considerable danger of overcrowding. More frequency bands will have to be allocated, and the satellites themselves will have to be pushed closer together so that more can be placed in the geostationary orbit. Recently there has been a compression from 3° to 2° separation.

In a sense, there is no new communications work in this chapter. What we have is an unusual application which produces unusual operating conditions and therefore requires novel developments of existing knowledge. In many of the following sections we will refer to earlier chapters in this book for explanations of principles, and we will limit our attention here to those features of communication systems which are peculiar to satellite applications.

14.2 GEOSTATIONARY ORBITS

The analysis of the satellite orbit is complex and rather beyond our scope. Excellent treatments are given elsewhere[2,3]. For our purpose, however, we only need to know that it is possible to place a satellite in a geostationary orbit at a height of approximately 36 800 km above the earth, and that the station can be maintained, provided it has the provision for slight adjustments of orientation and position by way of small rockets. Although we will concentrate on geostationary satellite communications, it should be realized that there are areas near the poles which can be reached only by satellites in other orbits. For example, satellites in tundra orbit are not geostationary, but they provide regular coverage for the extreme latitudes (Fig. 14.1).

For a geostationary orbit the angular velocity of the satellite is constant and therefore, since the angle swept per unit time (Kepler's third law) is also constant, the orbit must be circular. There is a simple relationship between the period and the radius of the orbit

$$a \propto (T^{2/3}) \tag{14.1}$$

and the equality is provided by Kepler's constant μ: i.e.

$$a = \left(\frac{T^2 \mu}{4\pi^2}\right)^{1/3} \tag{14.2}$$

Assuming that the period T is the same as a normal (sidereal) day $= 24.60.60$ s and taking $\mu = 3.986\,1352 \times 10^5\,\text{km}^3/\text{s}^2$, the above equation gives the orbital radius to be

$$a = 42\,241.558\,\text{km} \tag{14.3}$$

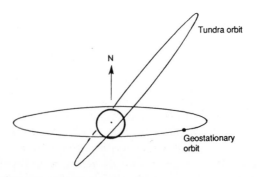

Fig. 14.1 Communication satellite orbits.

14.3   ANGLE FROM SATELLITE TO EARTH STATION

Again, the case of a geostationary satellite is a much simplified version of the general case. The point at which a line between the satellite and the centre of the earth intersects the earth's surface is called the sub-satellite point. There are two angles of intersect between the earth station and the satellite: the elevation and the azimuth. The elevation is the angle of the satellite above the horizon, from the earth station, and the azimuth is the angle between the line of longitude through the earth and the direction of the sub-satellite point. These are shown diagrammatically in Figs. 14.2 and 14.3.

From Fig. 14.2 the angle γ at the centre of the earth is given by

$$\cos \gamma = \cos L_e \cos (l_s - l_e) \qquad (14.4)$$

where L_e = latitude of the earth station and $l_s - l_e$ = difference in longitude between the earth station and the sub-satellite point.

Applying the cosine rule

$$d_s^2 = r_e^2 + r_s^2 - 2r_e r_s \cos \gamma \qquad (14.5)$$

From the sine rule

$$\frac{r_s}{\sin \psi} = \frac{d_s}{\sin \gamma} \qquad (14.6)$$

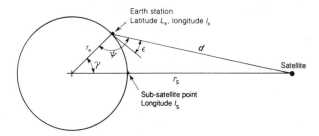

Fig. 14.2 Satellite position relative to Earth.

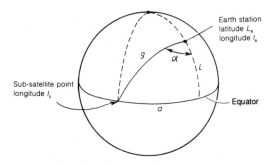

Fig. 14.3 Angle of azimuth.

giving

$$\sin\psi = \frac{r_s \sin\gamma}{d_s}$$

$$= \frac{r_s \sin\gamma}{[r_e^2 + r_s^2 - 2r_e r_s \cos\gamma]^{1/2}}$$

$$= \frac{\sin\gamma}{\left[\left(\dfrac{r_e}{r_s}\right)^2 + 1 - 2\left(\dfrac{r_e}{r_s}\right)\cos\gamma\right]^{1/2}} \qquad (14.7)$$

and since, from Fig. 14.2

$$\varepsilon = \psi - 90$$

$$\cos\varepsilon = \frac{\sin\gamma}{\left[1 + \left(\dfrac{r_e}{r_s}\right)^2 - 2\left(\dfrac{r_e}{r_s}\right)\cos\gamma\right]^{1/2}} \qquad (14.8)$$

Using $r_e = 6370\,\text{km}$ and $r_s = 42\,242\,\text{km}$ in Eqn (14.8) gives

$$\cos\varepsilon = \frac{\sin\gamma}{[1.02274 - 0.30159\cos\gamma]^{1/2}} \qquad (14.9)$$

Clearly the angle of elevation will depend on the latitude of the earth station via angle γ; ε will be large towards the equator and small towards the poles. A minimum value of about $5°$ limits the highest latitude. The azimuth angle $180 + \alpha$ in Fig. 14.3, is related to the latitude and longitude of the earth station, L_e and l_e and the longitude of the sub-satellite point l_s by

$$\tan^2\frac{\alpha}{2} = \frac{\sin(S - \gamma)\sin(S - L_e)}{\sin S \, \sin(S + L_e)} \qquad (14.10)$$

where

$$S = \tfrac{1}{2}(a + L_e + g) \qquad \text{and} \qquad a = (l_s - l_e)$$

The elevation and azimuth angles define the 'boresight' of the receiving antenna for a given receiver location. The azimuth angle uses the line from earth station to north pole as a reference.

It was noted earlier that towards the poles there is a restriction on the minimum angle of elevation considered suitable because the noise increases as $\varepsilon \to 0$. Another consideration is the additional attenuation produced by the increased distance to the satellite (slant height) with latitude.

From Eqn (14.5) the slant height is

$$d_s = [r_s^2 + r_e^2 - 2r_s r_e \cos\gamma]^{1/2}$$

14.4   SATELLITE LINKS

In satellite communications the direction of transmission is indicated by the term used: the downlink is transmission from satellite to earth station and the uplink is transmission in the opposite direction.

As we have mentioned before, communications by satellite impose enormous problems on the systems designer. In the uplink we have the availability of a high intensity transmission beam, because the transmitter is ground based and therefore power is not a problem, coupled with a sensitive receiver on the satellite, while in the downlink we have a low power source in the satellite and a highly sensitive receiver in the earth station. However, this presupposes that the earth station cost is not a constraint. For public utility applications such as telephony, cost will not be a limiting factor; the high cost of an expensive receiving station can be spread over a very large number of users. In direct broadcast television however, the story is different. A single user has to pay for the receiver and therefore its cost must be low. This means that the earth station will be 'basic'. The dish will be as small as possible and the amplifier as cheap as possible, consistent with obtaining an adequate signal for most of the year. Outage will be high, compared with that for a better quality receiver, but must still be at a reasonable level.

The preoccupation in satellite link design is the power budget; how much power can be obtained from the transmitter, how much of that power can be directed towards the receiver, how much power is lost over the link, and at the transmitter and receiver, and how much is left at the detector? What margin is necessary above the minimum detectable signal for the detector being used? Essentially the problem is to ensure that the signal level at the detector is large enough to produce a satisfactory output for a large part of the year.

Since the satellite distance is fixed, the attenuation (except for the effect of rain) is also fixed, and it is high. It cannot be avoided and so any improvement to be made must occur at the transmitter or receiver, and will be marginal, but very important. Fig. 14.4 shows the basic features of a satellite link.

We will discuss first the link transmission loss and the effect of antenna design.

Let P_t be the power from the transmitter

G_t the transmitter antenna gain

R the range from transmitter to receiver

P_r the received power

G_r the receiver antenna gain

We know from Chapter 7 that the power density, P, at a distance R from

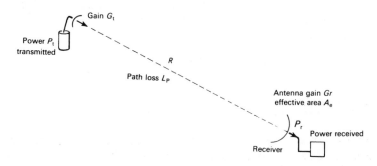

Fig. 14.4 Transmission link.

an isotropic source radiating total power P_t is

$$P = \frac{P_t}{4\pi R^2} \quad \text{W/m}^2 \qquad (14.11)$$

The transmitting antenna of gain G_t will concentrate the power in the direction of its maximum lobe to produce a flux density

$$P = \frac{P_t G_t}{4\pi R^2} \quad \text{W/m}^2 \qquad (14.12)$$

The receiver antenna, when properly lined up with the transmitter, intercepts power of density P. The effective area of the receiving antenna is $A_e = \eta A_r$ where A_r is the antenna aperture area and η is known as the antenna efficiency.

The power received = power density at the receiver × effective antenna area

$$= \frac{P_t G_t}{4\pi R^2} \cdot A_e$$

$$= \eta \frac{P_t G_t A_r}{4\pi R^2} \qquad (14.13)$$

There is a simple relationship between the effective area and the antenna gain which is

$$G_r = \frac{4\pi A_e}{\lambda^2}$$

Hence

$$P_r = \frac{P_t G_t G_r \lambda^2}{(4\pi R)^2} \qquad (14.14)$$

For a geostationary satellite, R is fixed, and for a particular application λ is also fixed. The expression naturally splits into three sections

$$P_r = P_t G_t G_r \left(\frac{\lambda}{4\pi R}\right)^2 \qquad (14.15)$$

$P_t G_t$ is called the EIRP (effective isotropic radiated power), G_r is the receiver gain and $(\lambda/4\pi R)^{-2}$ is a measure of the power dispersion between transmitter and receiver, known as the path loss, L_p. Thus

$$P_r = \frac{\text{EIRP} \times G_r}{L_p} \qquad (14.16)$$

More simply, this can be expressed in dB as

$$P_r = (\text{EIRP} + G_r - L_p) \quad \text{dBW} \qquad (14.17)$$

where

$$\text{EIRP} = 10\log_{10} P_t G_t \quad \text{dBW}$$

$$G_r = 10 \log_{10} \frac{4\pi A_e}{\lambda^2} \quad \text{dB}$$

$$L_p = 20 \log_{10} \frac{4\pi R}{\lambda} \quad \text{dB}$$

The resultant power received must be sufficient to overcome losses not included in the above equation. As mentioned earlier, there will be precipitation losses, and there will be small losses at the transmitter and receiver which will need to be taken into account. For example, it has been assumed that all the power produced at the transmitter will be radiated from the transmitting antenna; that will not be so since there will be small losses in the transmitter circuits.

In Eqn(14.15) the received power was related to other system parameters by

$$P_r = P_t G_t G_r \left(\frac{\lambda}{4\pi R} \right)^2$$

This is known as the carrier power C. Of considerable interest is the ratio of this power to the noise power N at the detector. From the discussion in Chapter 4 the noise power at a receiver is

$$P_n = k T_s B G$$

where T_s is system noise temperature
   B is receiver bandwith
   G is receiver gain
   k is Boltzman's constant

and the power received due to the carrier is $P_r G$; thus the ratio C/N is given by

$$\frac{C}{N} = \frac{P_t G_t G_r}{k T_s B} \left(\frac{\lambda}{4\pi R} \right)^2 \tag{14.18}$$

assuming that the receiver has been replaced by an equivalent system characterized by T_s, and B

or

$$\frac{C}{N} = \frac{P_t G_t}{k B} \left(\frac{\lambda}{4\pi R} \right)^2 \frac{G_r}{T_s}$$

or, assuming the subscripts to be implicit, the carrier/noise ratio at a particular earth station is

$$\frac{C}{N} = \frac{P_t G_t}{k B} \left(\frac{\lambda}{4\pi R} \right)^2 \frac{G}{T} \tag{14.19}$$

where G/T is a measure of the goodness of the earth station. This quantity is widely used as a figure of merit to provide a means of comparison between different systems.

14.5 ANTENNAS

For satellite communications the ideal antenna performance is to have high gain, high efficiency (low loss), low side lobe levels, and narrow beamwidth. The beamwidth and side lobe levels are important in determining the interference from adjacent satellites and hence have a strong impact on the proximity which adjacent satellites can have in orbit.

The efficiency of the antennas used in satellite communications is extremely important and much care is taken to restrict any losses. Factors which reduce efficiency are wide ranging. The design of the antenna and its feed are obvious elements, but equally significant is the design of the waveguide and other components from, in the case of the receiver, the antenna feed to the first stage of the amplifier.

The degree to which clever antenna and feed design can be used to improve antenna efficiency will depend on the cost budget for a system. If the antenna forms part of an earth station carrying heavy traffic then a high cost can be tolerated, and sophisticated design techniques will be justified. However, for a simple broadcast television receiving station low cost and ease of manufacture are paramount, consequently antenna loss will be comparatively high.

The simplest antenna system, Fig. 14.5, is a straightforward dish paraboloid with a waveguide feed at its focus. The sources of loss are:

(i) poor illumination profile over the dish;
(ii) blocking by the feed;
(iii) long waveguide run in the feed;
(iv) high side lobe level.

As far as the antenna itself is concerned the paraboloid, in principle, allows a parallel beam of radiation from a distant source to be reflected to a focal point (section 7.18). However the ray theory approach on which that is based assumes that the wavelength is very small compared to the diameter of the antenna, but the wavelengths used in practice in satellite communications do not satisfy that condition. As a consequence there are diffraction effects which have a significant bearing on the performance of the system. Ideally, a uniform distribution of E-field amplitude is desirable at the antenna aperture, but this can only be achieved for an antenna of infinite diameter. Usually

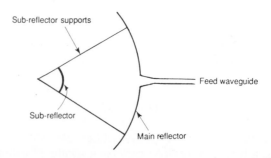

Fig. 14.5 Cassegrain dish antenna with direct feed.

some compromise is required between cost and performance. To reduce loss and limit cost, the direct feed arrangement is replaced by an indirect feed system, and in practice nearly all 'A' standard antennas for use in the INTELSAT system use this approach. Two methods are favoured, both based on the same type of principle and known after the sixteenth century inventors of the equivalent optical telescope. They are Cassegrain and Gregorian antennas. In both, the primary feed is at the centre of the paraboloid. This feed illuminates a secondary reflector; in the case of the Cassegrain of hyperbolic cross-section and in the case of the Gregorian of eliptical cross-section. These sub-reflectors in turn illuminate the main parabolic reflector. The sub-reflector and its mounting block some of the energy from the paraboloid. To reduce this effect the sub-reflector can be offset from the main axis as shown in Fig. 14.6.

The compromise between low side lobe level and high directivity (narrow beamwidth) is achieved by adjusting the illumination used across the aperture. To obtain the required distribution the primary feed, sub-reflector shape and dish shape can all be modified.[4]

Apart from the losses associated with diffraction effects and limited antenna aperture, there are other sources of antenna loss. In receiver mode the signal is transmitted from the primary feed into the receiver. For practical reasons the receiver is preferably placed away from the antenna, with the consequent waveguide run adding to the system loss. This loss can be reduced to a small value by using a beam waveguide, which is almost a free-space system enclosed in a large envelope, as shown in Fig. 14.7. This system is employed in a number of practical systems and in the example shown in Fig. 14.8 the beam waveguide can be seen clearly.

One type of primary feed which is of considerable interest is a corrugated pyramidal horn. The corrugations can be broadbanded and it has the advantage that it provides excellent cross-polarization rejection. In many systems the channel utilization can be increased significantly by using two directions of polarization to separate channels.

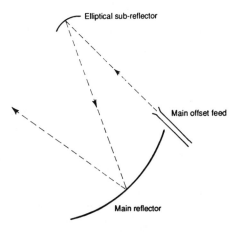

Fig. 14.6 Schematic of offset Gregorian reflector.

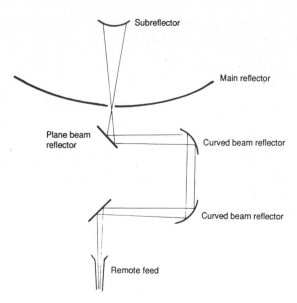

Fig. 14.7 Schematic of beam waveguide feed.

The function of spacecraft antennas is to direct the beam to those areas of the earth which have appropriate receivers. Although this may appear to be an obvious statement, it does emphasize the difficulties for the designer. Wasting power by sending a signal into areas with no receivers is undesirable, but to achieve coverage over specified areas only is also difficult.

With a standard paraboloidal dish the coverage on the earth's surface is circular. Since the angle subtended by the earth at the satellite is 17°, such a beamwidth will provide coverage over almost one hemisphere. To reduce the coverage area significantly requires a much larger antenna, and to provide a contoured (non-circular) coverage pattern involves the use of complex feed arrangements.

Gradually, linear array feeds are being introduced. They give the designer considerable flexibility in the range of patterns which can be obtained, at the cost of increased complexity. However, there are other benefits which may arise from these feeds. By making them electronically fed, the beam can be reconfigured to suit new users of the satellite who may require different coverage patterns.

Even further variability, and therefore options, can be achieved with antennas using linear arrays to illuminate a sub-reflector. The reflecting surfaces of both the sub-reflector and the main reflector can be distorted to modify the shape of the ground coverage profile.

There is a limit to the size of antenna array a satellite can accommodate. Large antennas can be deployed by using collapsible lightweight structures which can be opened out once the satellite is in orbit.

Offset feeds are often used, both to eliminate any blocking of the beam and to provide good mechanical support for the feeds. Several feeds, for different frequencies, may be mounted on the same support.

Fig. 14.8 Photograph of antenna and beam waveguide feed.

Earlier we discussed polarization of the beam to extend the utilization of the channel. This can be achieved by relying on the polarization from the feed horn being maintained, or by using a polarizer in the form of a grid of appropriately orientated wires in front of the reflector.

A future development in spacecraft antenna design, planned for business satellites in particular, is to provide space switching. Earlier we noted that additional capacity can be obtained by using polarization. Space switching will also provide more capacity over a given area by allowing frequency re-use on a spatial basis. Beams will be very narrow and once the area of

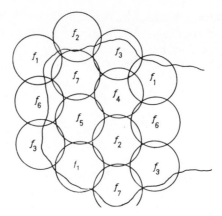

Fig. 14.9 Frequency re-use radiation 'foot-prints'.

potential interference has been exceeded the same frequencies can be used again (Fig. 14.9). This is the same basic idea as that used in cellular radio (Section 13.12).

14.6 ERROR CODING

It is in the nature of transmission channels that from time to time interference with the transmitted signal will produce errors. In digital systems these errors will result in bits being incorrectly received. Coding is a methodology that provides some protection against such errors.

Often the provision of error detection and correction is the responsibility of the satellite user; the operator merely supplies the link to meet some specification, but does not provide the means of ensuring accurate transmission. One reason for this division of functions between the operator and the user is that the operator is not concerned with the type of information being transmitted, and therefore cannot predict when designing the system the degree of correction the user will require.

Normally, speech channels are not error corrected; there is sufficient redundancy in speech to ensure that it is made intelligible in the presence of such errors (Section 1.13). In transmitting data, however, error detection is essential. There may be no redundancy in the information stream and consequently any error will result in wrong information being received and there will be no mechanism for making a correction.

Error detection is achieved by adding bits to the information packet in such a way that the receiver can, by using binary algebraic manipulation, determine whether errors are present (Section 5.6). The number of errors which can be detected in this way is related directly to the length of the information field, and the number of additional code bits provided by the encoder. The combination of information field and coding digits is called the code word. There is a trade-off, as mentioned in Chapter 5, between error detection power and transmitted capacity. Adding code bits effectively reduces the rate at which real information is transmitted, and therefore reduces the capacity of the link; the code bits represent redundancy.

Error correction is a different matter from error detection. It is possible to provide codes which will allow the receiver to detect and then correct a number of errors in a bit stream, but correction is more difficult to achieve than detection and therefore more additional bits are required. In practice this means that codes can correct a much smaller number of errors than they can detect. However, this type of correction, called forward error correction (FEC) has a significant role in satellite links. The very long transmission time between earth station and satellite discourages the use of retransmission for error correction, and therefore forward error correction in which the error is detected and corrected by the receiver is highly desirable for this application.

A simple method for correcting errors on short terrestrial links is to use a stop and wait protocol with automatic request for transmission (ARQ). The transmitter sends one packet at a time; when it is received correctly an acknowledgement signal (Ack) is returned. If the receiver detects an error no Ack is returned and the packet is retransmitted after a suitable time-out. The time-out is slightly longer than the expected time for the return of Ack, Fig. 14.10. Clearly the main drawback with this protocol is that the time between transmission of the packet and reception of the acknowledgement is wasted. For satellite applications, where that delay is about a quarter of a second, such wasted time cannot be tolerated. Instead a duplex link is used and packets are transmitted continuously. When an error occurs a retransmission mechanism causes those packets, and probably a few following, to be retransmitted.[5]

Clearly a number of mechanisms exist for handling the problem of error correction and detection. One is to provide such a strong link that the excess power over the minimum required for normal circumstances is sufficient to cope with nearly all the occurrences of poor transmission conditions. This is a very expensive and crude approach, and does not attack the problem of error correction directly. The second method is to arrange for all the incorrect packets to be retransmitted automatically (using ARQ). The long round-trip delay discourages complete reliance on this method so a third scheme is used in which a measure of forward error correction is provided at the receiver and if the number of errors exceeds the capacity of the FEC algorithm then ARQ

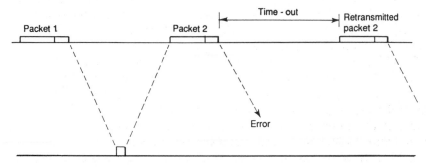

Fig. 14.10 Sample stop and wait protocol.

is used. A fourth approach would be to provide no error detection or correction facility; a satisfactory philosophy on channels of high reliability carrying information which has built-in redundancies such as voice and television.

Coding is designed to cater for channels with a low bit-error rate, consequently the improvement is best displayed in circumstances where errors are very occasional, and the error correction facility is rarely used. If the channel is very noisy the FEC system will break down, or the delay in providing the necessary retransmission will be intolerable. Indeed the receiver may not be able to decide whether or not it has received a packet. In such circumstances effective transmission breaks down. Practical systems are designed to ensure that breakdown occurs only within some specified limit, measured in terms of minutes per month.

14.6.1 Types of error

The incidence of errors is classified as random or bursty. Random errors occur as isolated erroneous bits in a stream of good bits; the probability of an error is the same for all bits, regardless of whether the previous bits were correct or not. Bursty errors are somewhat different, and in a sense more typical of some types of interference. As the name suggests, bursty errors turn up in blocks; a sequence of correct bits is followed by a sequence in which the incidence of errors is very high, followed by a stream of good bits. During an error burst the probability of bit error is high; during a sequence of good bits the probability of error is low.

Codes can also be classified into several types, but the principal division is between block codes and convolutional codes.

In a block coder the information is presented in a block of k digits, and this block is coded into a longer word of n digits; known as an (n, k) block code. The additional $k-n$ digits provide the redundancy required to enable error detection and correction to take place. Successive coded blocks are independent of each other.

Convolutional codes consist, similarly, of (n, k) blocks, but in addition there is a dependency between the codes used for successive incoming frames. The memory order of codes is denoted by m, the number of previous frames having an influence on the coder used for the present frame.

A feature of the decision making process in the detector is to determine if an error has occurred, and that could be based on a threshold-related decision (hard decision) or on a probability-related decision (soft decision). In the hard decision system there is no allowance for any doubt; either a bit has or has not been received. However, in soft decision decoders the doubt can be included as a probability and taken into account in the algorithm for decoding the received code word, although this approach can only be used in convolutional decoding.

14.7 INTERFERENCE AND NOISE

Background noise on the downlink has a significant effect on receiver performance since it reduces the C/N factor. This particular form of noise is represented as a noise temperature; and for an antenna that temperature is variable and depends strongly on the elevation of the boresight and the frequency of the channel. The quiescent noise temperature will be less than 70 K.

However, the direction of the antenna boresight is determined by the relative position of the earth station and the satellite. Consequently the antenna may pick up atmospheric noise from time to time well in excess of this value. Most noticeably, excess temperature occurs when the sun is behind the satellite, in which case the noise temperature will rise to over 10^4 K at frequencies up to about 10 GHz.

Interference increases the effective noise of the satellite channel as spurious signals from nearby links, or from terrestrial sources, are detected. However, there is also a reciprocal effect in which our channel interferes with other communication systems, both satellite and terrestrial.

14.7.1 Atmospheric attenuation

The link path between satellite and earth is not homogeneous to radio wave propagation, nor is it immune from interference from nearby radio systems. There is always some attenuation due to the excitation of electrons in the path of the radio wave, but at satellite frequencies it is a minor factor, accounting for about 0.5 dB loss. Much more important are those effects which are transitory in nature, and consequently unpredictable. On a 'clear day' there are none of these loss mechanisms, but at other times some or all of them may occur. We can consider such mechanisms in several categories: here we choose three; attenuation, depolarization and interference. The first two are properties of the radio link itself. The third results from other sources of radiation in the proximity of the link.

Attenuation

At satellite frequencies the most important variable attenuation mechanisms are those assciated with water droplets, hydrometeorites, in the form of rain, ice, snow, sleet and combinations of these. Rain is the most important type of precipitation from this point of view. Snow produces relatively little attenuation, but wet ice can impair the signal significantly.

There are several parameters associated with rainfall, but in this context the most important is rainfall rate; measured in millimetres per hour, it has a direct effect on attenuation.

The attenuation A can be related to the rainfall rate R and the effective path length, L by

$$A = kR^aL \quad \text{dB} \qquad (14.20)$$

This simple relationship assumes that the rainfall rate R is constant over the whole of the path length L. Several very difficult problems arise in utilizing this relationship. The constants k and a are highly dependent on frequency and on rain drop size, as well as depending to a small degree on temperature. The relationships between k, a, frequency and drop size that are used in practice are empirical and have been tabulated by a number of organizations such as CCIR. Resort must therefore be made to such tables[6] to find values of A. However, knowing k and a is only part of the solution. R is impossible to determine precisely. The rainfall rate will vary along the length of the path in a statistical way, and even estimating the value of R at a particular point by using a ground-based measure of quantity of water versus time may be a poor approximation to the rate at which, over the same spot, raindrops are falling through the link path. The turbulence in the atmosphere precludes a straight fall of rain and hence complicates the assessment of fall rate.

To provide designers with a procedure to calculate A, CCIR have produced an algorithm based on tabulated values for k and a, and cumulative distributions for R. This gives, for example, a figure $A_{0.01}$ based on $R_{0.01}$, where the suffix indicates the percentage of the average year for which the attenuation will exceed the value $A_{0.01}$ given that the mean rainfall rate exceeded for 0.01% of the average year is $R_{0.01}$. In this way the designer can dimension the system to meet satisfactory performance specification for all but 0.01% of the year.

The ionosphere can produce absorption and other effects. At satellite frequencies the absorption is negligible unless there is electron activity in the ionosphere in the form of clouds of free electrons which interact with the earth's magnetic field, producing not only absorption but also depolarization due to the rotational effect on the wave, and a slight misalignment caused by variations in refractive index. If the variations are rapid they are called scintillations and these can produce variations of up to 10 dB in the received signal amplitude.

Depolarization

Previously, we noted that better use of the links can be achieved by sending out two orthogonally polarized signals. The success of this approach depends on the receiver being aligned to the correct polarization. Unfortunately there are several mechanisms which make this difficult to achieve consistently. The effect of changing the direction of polarization from that expected is called depolarization and that has two results. First, the intensity of the intended signal is reduced because the polarization of the incoming signal and that of the receiver are not matched. Second, the receiver can pick up some of the signal from the other polarization, hence effectively increasing the channel noise.

The main cause of depolarization is Faraday rotation, caused by action of the magnetic field and the ionospheric free electrons on the signal. The degree of rotation depends on both the electron density, and the length of path over which the electron cloud or clouds exist. The ionosphere does not exhibit the same electrical properties in all directions; it is said to be anisotropic. This lack of homogeneity itself results in depolarization.

As estimate of the rotation, φ, produced by the Faraday effect can be obtained from the formula

$$\varphi = \int k_1 \frac{NL}{f^2} B \cos \theta \, dl \tag{14.21}$$

where k_1 is a constant

N is the ionospheric plasma electron density

B is the magnetic flux density

L is the path length

f is the channel frequency

and θ is the angle between the magnetic field vector and the link

The integration is over the length of the path in the ionosphere. In practice N will not be constant and some suitable model will have to be chosen to determine its value as a function of L.

14.8   MULTI-ACCESS

As the number of earth stations increases, and satellite communications are more widely used, the multi-access techniques used become more significant. In principle the function of multi-access is to provide the many earth stations wishing to use a particular transponder with channel access, at the required bandwidth.

In many respects the problems of multi-access in satellite systems are similar to those in local area networks, and the same mechanisms can be used in each. In particular, the traffic to be transmitted is a combination of voice and data and the same problems, and solutions, as discussed in Chapter 13 occur. There are, however, some additional issues in satellite systems, and they will be mentioned here.

14.8.1   FDMA

In FDMA the bandwidth can be divided on a multi-carrier or single carrier basis. In multi-carrier the transponder bandwidth is divided into sub-bands, each sub-band being assigned to an earth station. The sub-bands need not be all of the same width; the number of channels in each will depend on the earth station's requirements and the allocation they have been given. This arrangement gives the earth station some flexibility in the allocation of bandwidth within its own sub-band. The main disadvantage of this type of access is that the power output from the transponder must be reduced compared with the output that could be achieved with a single carrier arrangement. Except for a few satellites using solid-state amplifiers, the output device at the transponder is normally a travelling wave tube. It has a non-linear input–output characteristic, and this leads to serious inter-modulation products in the output. This effect becomes more serious as the number of carriers increases. To avoid the non-linearities from producing serious degradation on the output signal, it is necessary to impose an input back-off. This back-off has to be large enough to ensure that the output back-off brings the operating point to a position where the inter-modulating products are insignificant. The non-linear characteristic means, of course, that the output back-off will be smaller than the input back-off, Fig. 14.11. The amount of back-off required increases with the number of carriers in the system.

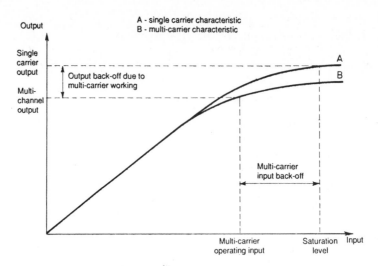

Fig. 14.11 Effect of multi-carrier working.

A single carrier FDMA arrangement, in which the available 36 MHz bandwidth is divided into uniform channels, allows more efficient use of the travelling wave tube output. If QPSK is used, and each channel is spaced by 45 kHz, there is room for 800 channels to avoid cross-channel interference, and some back-off is necessary because of the inter-modulation products between channels. Most satellites currently use FDMA access methods of one sort or another.

14.8.2 TDMA

Throughout communications systems design, the trend for many years has been to use digital representations of signals, and digital transmission. Its ease of transmission, and robustness in noise, makes it very appealing, and the same arguments in its favour in terrestrial systems also apply to satellite networks. As far as the TWT operating conditions are concerned, in digital transmission there is no reason why the full output should not be used: there is only one carrier, and since the transmission is digital, inter-modulation products are not a problem.

The discussion on TDMA in local area networks, in Chapter 13, is highly relevant here. The different characteristics of voice and data traffic, their different delay and error-correction requirements and the methods of accommodating them are all explained there.

Two issues are, however, of especial interest in satellite operations. The frame size, which has to be greater than 125 µs if we assume that normal PCM is used for speech, can have any value up to about 25 ms; the limit imposed by the maximum tolerable delay to speech traffic. In the *INTELSAT* system, for example, a frame size of 2 ms is used, with the frame structure shown in Fig. 14.12. The reference bursts are for synchronization and each traffic burst may be from a different earth station. The traffic bursts contain time multiplexed speech channels.

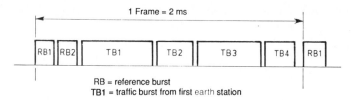

Fig. 14.12 Typical Intelsat frame.

The other issue is synchronization. The uplink packets arriving at the transponder from different earth stations must not overlap in time. Consequently the whole network related to one satellite has to be in synchronism. This can be achieved in different ways in detail, but it depends on the transponder sending out synchronizing bits to indicate the start of each frame, and to provide reference times to which earth stations relate their own transmission. The variable nature of the satellite position will produce some changes in this synchronization timing, and guard intervals are used to avoid overlap due to this cause. The guard interval adds to the transmission overhead, leaving less time available for revenue earning traffic. Other overheads which are included concern addresses, acknowledgement for data signals and network management.

14.8.3 Random access

The bandwidth allocated to each channel depends on the method of access employed. Fixed assignment means that at the outset the designer determines the bandwidth allocation, and there is no flexibility for the operator to adjust the system in use. Demand assignment does give the operator flexibility. Channel bandwidths can be adjusted to cater for the traffic mix, which may vary from time to time. For such demand assignment to be feasible, an overhead in terms of network management information is imposed.

An alternative approach is to use a random access mechanism. Instead of having a network management protocol within the system to ensure that both the channel allocation is satisfactory and there is no possibility of collision, random access is a free-for-all. Each station transmits at will, and runs the risk of collision. If collision occurs, the transmitting stations will back off, wait for some random time and then try again. The simplest random access system is the ALOHA, in which the utilization is low (about 18%); some considerable improvement is achieved with the slotted ALOHA (Section 13.5). It is worth remembering that the more sophisticated the access mechanism, the higher is the overhead and the more complex is the management protocol, so that there is a trade-off to be made.

14.8.4 Speech interpolation

One of the early developments in transatlantic submarine communications, when the demand immediately saturated the circuits provided, was to utilize the silent intervals in normal speech to allow channels to be shared. In normal speech about 45% of the time is not used, in the sense that it is occupied by silence intervals. Speech interpolation techniques allow these intervals to be utilized, and this same method can be applied to satellite links. In packet-based transmission systems, since there is no packet to be transmitted during silence intervals, speech interpolation is automatic.

REFERENCES

1. Clarke, A. C., 'Extra-terrestrial Relay', *Wireless World*, Oct. 1945.
2. Pratt, T. and Bostian, C. W., *Satellite Communications*, John Wiley, 1986.
3. Spilker, J. J., *Digital Communication by Satellite*, Prentice–Hall, 1977.
4. Evans, B. G. (ed.), *Satellite Communication Systems*, IEE London, 1987, Chapter 12.
5. Tanenbaum, A. S., *Computer Networks*, Prentice–Hall, 1981, Section 4.2.2.
6. Propagation in non-ionised Media. *CCIR, Recommendations and Reports*, Vol. V, 1982, ITU, Geneva.

Mobile communication systems 15

15.1 INTRODUCTION

In the fixed telephone network the final link with individual subscribers has traditionally been by means of a twisted pair cable, in other words the telephone handset is generally regarded as being immobile. Some minor changes to this situation became possible with the introduction of the cordless handset which allowed the subscriber limited mobility over a distance of a few metres. The concept of mobility has been extended from this limited movement to pan national and pan continental transit. Future forecasts suggest that the fixed telephone handset will eventually disappear completely and be replaced by mobile units allowing individual subscribers the facility of global travel with continuous personal communications.

This chapter considers the fundamentals of these new public mobile systems (there are numerous private mobile systems also) which are broadly divided into cellular mobile systems and cordless mobile systems. Although different in concept both cellular and cordless systems rely on radio transmission for the final link with the subscriber. The management of this radio resource is a fundamental requirement of mobile systems which therefore makes mobile systems substantially different from the traditional public switched telephone network. This chapter will consider elements of both cellular and cordless systems and will focus on their similarities and differences.

15.2 THE CELLULAR CONCEPT

Cellular systems are now well developed in many countries throughout the world. The systems in use in individual countries all have many similarities and also some significant differences. It is not possible to cover all the systems which are in use or are currently under development. To avoid this problem the coverage in this chapter will concentrate on European systems and will begin by considering what are known as 'first generation' analogue systems, and will then move on to 'second generation' digital systems.

The essential feature of all cellular networks is that the final link between the subscriber and fixed network is by radio. This has a number of consequences:

(i) radio spectrum is a finite resource and the amount of spectrum available for mobile communications is strictly limited;

(ii) the radio environment is subject to multipath propagation, fading and interference and is not therefore an ideal transmission medium;

(iii) the subscriber is able to move and this movement must be accommodated by the communications system.

The basic elements of a cellular system are shown in Fig. 15.1.

The mobile units may be in a vehicle or carried as a portable and are assigned a duplex channel and communicate with an assigned base station. The base stations communicate simultaneously with all mobiles within their area of coverage (or cell) and are connected to mobile switching centres (MSCs). A mobile switching centre controls a number of cells, arranges base stations and channels for the mobiles and handles connections with the fixed public switched telephone network (PSTN).

Figure 15.1 indicates that each base station (in a cluster) is allocated a different carrier frequency and each cell has a usable bandwidth associated with this carrier. Because only a finite part of the radio spectrum is allocated to cellular radio the number of carrier frequencies available is limited. This means that it is necessary to re-use the available frequencies many times in order to provide sufficient channels for the required demand. This introduces the concept of frequency re-use and with it the possibility of interference between cells using the same carrier frequencies.

Clearly with a fixed number of carrier frequencies available the capacity of the system can be increased only by re-using the carrier frequencies more often. This means making the cell sizes smaller. This has two basic consequences:

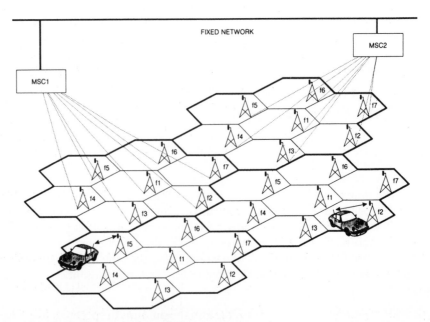

Fig. 15.1 Basic elements of a cellular system.

1. It increases the likelihood of interference (known as co-channel inter-ference) between cells using the same frequency.
2. If a mobile is moving it will cross cell boundaries more frequently when the cells are small. Whenever a mobile crosses a cell boundary it must change from the carrier of the cell which it is leaving to the carrier of the cell which it is entering. This process is known as handover. It cannot be performed instantaneously and hence there will be a loss of communication while the handover is being processed. If the cell sizes become very small (microcells) handovers may occur at a very rapid rate.

It becomes clear therefore that frequency planning is a major issue in the design of a cellular system which must achieve an acceptable compromise between the efficient utilization of the available radio spectrum and the problems associated with frequency re-use.

15.3 TYPICAL CELL OPERATION

Each cell has allocated to it a number of channels which can be used for voice or signalling traffic. When a mobile is active it 'registers' with an appropriate base station. The information regarding the validity of the mobile (for charging etc.) and its cellular location is stored in the responsible MSC. When a call is set up either from or to the mobile the control and signalling system assigns a channel (from those available to the base station with which the mobile is registered) and instructs the mobile to use the corresponding channel. This channel may be provided on a frequency division basis (typical of analogue systems), on a time division basis (typical of digital systems) or on a code division basis (also typical of digital systems).

A connection is thus established via the base station to the fixed network. The quality of the channel (i.e. radio link) will be monitored by the base station for the duration of the call and reported to the MSC. The MSC will make decisions concerning the quality and will instruct the mobile and base station accordingly. As the mobile moves around the carrier-to-interference ratio on its allocated channel will vary. This is monitored and if it falls below some threshold (either as the mobile is about to leave the cell or as a result of co-channel interference), the mobile can be instructed to handover to the strongest base station. The handover algorithm is actually significantly more complicated than this simple treatment suggests. For example it must be able to cope with:

  (i) Whether the current loss in channel quality is due to short-term fading;
 (ii) Whether a simple increase in power would be sufficient to restore the channel quality (this could however produce an unacceptable co-channel interference in other cells using the same frequency);
(iii) Whether the measurements from adjacent cells are valid (averaging is necessary to remove spurious fluctuations);
(iv) Whether the cell chosen for handover has spare channels available.

In analogue systems the fixed network (i.e. the base stations) performs all radio channel measurements and the mobile terminal is completely passive (this is known as network-controlled handover). In digital systems both mobiles and base stations make measurements and report these to the fixed network for handover decisions (this is known as mobile-assisted handover). When handover does occur communication is interrupted and the voice channels are muted. This interruption is usually short and largely unnoticed during voice communications. However, it does present serious problems if the mobile is transmitting or receiving data, especially when handover is frequent. When the call is completed the mobile releases the voice channel which can then be re-allocated to other users.

15.4   SYSTEM CAPACITY

The capacity of a system may be described in terms of the number of available channels, or alternatively in terms of the number of subscribers that the system will support. The latter measure takes account of the fact that each call has a mean duration and that not all of the subscribers will be trying to make a call at the same time.

The system capacity depends on:

  (i) the total number of radio channels;
 (ii) the size of each cell;
(iii) the frequency re-use factor (or frequency re-use distance).

The total number of voice channels that can be made available to any system depends on the radio spectrum allocated and the bandwidth of each channel. Once this number is defined a frequency re-use pattern must be developed which will allow optimum use of the channels. This, in turn, is closely linked with cell size.

The minimum distance which allows the same frequencies to be re-used will depend on many factors, for example:

  (i) the number of co-channel cells in the vicinity of the centre cell;
 (ii) the geography of the terrain;
(iii) the antenna height;
(iv) the transmitted power within each cell.

Assuming an omni-directional base station antenna it is appropriate to

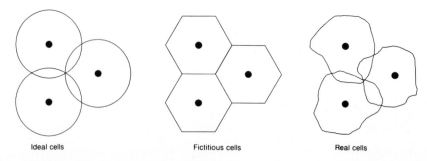

Ideal cells   Fictitious cells   Real cells

Fig. 15.2 Schematic representation of cells.

consider the cells as circles with the base station at the centre. (Such a model is actually only appropriate for a flat terrain with no obstacles). This means that the circular cells would overlap and this would make diagrams somewhat confusing. It is therefore common practice to represent the cells as non-overlapping hexagons which would fit into the corresponding circles. This fictitious model with the ideal and real models are shown in Fig. 15.2.

15.5 FREQUENCY RE-USE DISTANCE

When calculating the frequency re-use distance this is based on the cluster size K. The cluster size is specified in terms of the offset of the centre of a cluster from the centre of the adjacent cluster. This is made clearer by reference to Fig. 15.3. In this figure the cell cluster size is 7 and the centre cell is the cell marked 1. The next cell 1 is offset by $i = 2$ cell diameters to an intermediate cell and a further $j = 1$ cell diameter from that intermediate cell.

The cluster size is calculated from

$$K = i^2 + ij + j^2 \qquad (15.1)$$

Common cluster sizes are 4 ($i = 2$, $j = 0$) and 7($i = 2$, $j = 1$) for city centres, and 12($i = 2$, $j = 2$) for rural areas.

The actual frequency re-use distance is D as shown in Fig. 15.3. Assuming circular cells of radius R (based on the hexagon shape), the frequency re-use distance may be determined from

$$D = R\sqrt{3K} \qquad (15.2)$$

The corresponding re-use distances are given in Table 15.1.

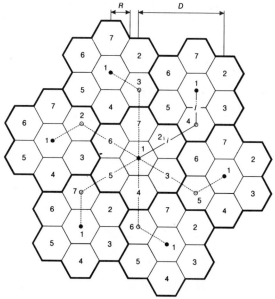

Fig. 15.3 Cell cluster with $K = 7$.

Table 15.1 Re-use distances for cellular systems

K	D
4	3.46 R
7	4.58 R
12	6.00 R

If all cells transmit the same power then as K increases the frequency re-use distance increases, thus increasing K reduces the probability of co-channel interference. However, in order to maximize frequency re-use it is necessary to minimize the frequency re-use distance. Hence the design goal is to choose the smallest value of K which will meet the performance requirements in terms of capacity and interference.

15.6 DETERMINATION OF CELL RADIUS

Figure 15.4 shows two cells using the same frequencies at a re-use distance D. Assuming that the power by each base station is fixed the received power at a distance r from the base station is proportional to $r^{-\gamma}$. For free space $\gamma = 2$, however, it is found that in the cellular environment a more appopriate value is $\gamma = 4$.

A mobile in one of the cells will receive a carrier-to-interference ratio (CIR) which, on average, is a function of $q = D/R$. (On average the CIR at a mobile receiver will be the same as at the base station receiver.) Note that the co-channel interference reduction factor (q) is independent of the actual power level P_0 which is assumed the same for all cells. The carrier to interference ratio within a cell depicted by Fig. 15.4 is thus

$$\frac{C}{I} = \left(\frac{R}{D}\right)^{-\gamma} = \left(\frac{R}{D}\right)^{-4} \tag{15.3}$$

For a fully developed cellular system based on the hexagonal model there will be six interfering cells in the first tier of surrounding clusters. (If $\gamma = 4$ it can be assumed that the interference due to cells in the second tier can be ignored.) The carrier-to-interference ratio in one of the cells will thus be

$$\frac{C}{I} = \frac{C}{\sum\limits_{k=1}^{6} I_k} \tag{15.4}$$

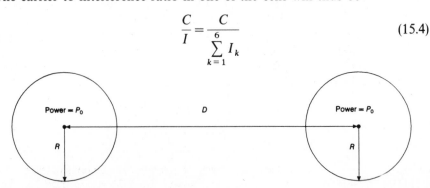

Fig. 15.4 Carrier-to-interference ratio.

Assuming local noise is much less than the interference level, this can be written

$$\frac{C}{I} = \frac{R^{-\gamma}}{\sum\limits_{k=1}^{6} D_k^{-\gamma}} \tag{15.5}$$

Assuming that all values of D_k are equal $(=D)$, this becomes

$$\frac{C}{I} = \frac{R^{-4}}{6D^{-4}} \tag{15.6}$$

In the cellular environment it is normal practice to specify that the CIR should be greater than 18 dB for acceptable performance $(= 63.1)$; i.e.

$$q^4 = 6 \times 63.1 \quad \text{hence} \quad q = 4.41$$

For the hexagonal structure $q = \sqrt{3K}$, i.e.

$$K = 6.48$$

hence the cluster size for this CIR is 7. Note that this approximate analysis closely reflects the practical case based on statistical measurements.

Having established the cluster size it is then necessary to determine the cell radius. This is based on the number of available channels and the expected density of mobile subscribers (i.e. the average number of mobile subscribers/m^2).

To illustrate this point assume that the total number of channels available is 210. This means that the number of channels per cell is $210/7 = 30$. It is necessary to find the total offered traffic in the 'busy hour'. This is related to the average number of calls/hour and the mean duration of each call. Assume that there are W subscribers per cell and that during the busy hour a fraction η_c of these subscribers make or receive a call of duration T minutes; i.e. the total number of calls in the busy hour is $Q = \eta_c W$. The offered load is then

$$A = QT/60 \quad \text{erlangs}$$

To obtain the number of channels for this traffic it is necessary to attach a 'blocking probability' for each call. A typical value for this is 2%. The relationship between offered traffic, blocking probability and number of channels is given by the Erlang B formula which is usually represented in tabular form and is given in Appendix D. For example, 30 channels can support an offered traffic of 21.9 erlangs with a blocking probability of 2% (from the table). It is possible to relate this figure to the number of subscribers which the cell can support; i.e.

$$21.9 = Q \, T/60 = (\eta_c W T)/60$$

from which

$$W = 60 \times 21.9/(\eta_c T)$$

In this expression T is the mean call duration in minutes. Extensive measurements have indicated that for cellular systems $T = 1.76$ minutes.

If it assumed that 60% of the total subscribers make a call during the busy hour then

$$W = 60 \times 21.9/(0.6 \times 1.76) = 1244.3$$

Hence with 30 channels available the cell can support 1244 subscribers. If the cell radius is R metres then user density $= 1244/\pi R^2$.

It is now possible to calculate the cell radius required for an average user density. For example assume that the number of users/km^2 is 1600. This represents a user density of 1.6×10^{-3}/m^2; i.e.

$$1244.3/\pi R^2 = 1.6 \times 10^{-3}$$

which gives $R = 497.5$ m, i.e. the approximate cell diameter is 1 km.

15.7 SECTORING

In the previous section an approximate analysis indicated that in order to achieve a CIR of 18 dB it is necessary to plan frequency re-use on a cluster size of 7. This gives a value of $q = \sqrt{21} = 4.58$. However, it is found that in areas of high traffic density this value can be inadequate. The worst-case situation is illustrated in Fig. 15.5 (a) in which a mobile is on the boundary of its serving cell and interference is produced by all six interfering cells. In this case the distances from the mobile to interfering cells varies from $D - R$ to $D + R$. The carrier-to-interference ratio in this case is given by

$$\frac{C}{I} = \frac{R^{-4}}{2(D - R)^{-4} + 2D^{-4} + 2(D + R)^{-4}} = \frac{1}{2\{(q - 1)^{-4} + q^{-4} + (q + 1)^{-4}\}}$$

Substituting for q in this expression gives a CIR $= 17$ dB, which is less than the desired value. The situation can actually be worse than this and a more conservative CIR estimate would be about 14 dB which is 4 dB less than the specified value.

Clearly, increasing the value of K to improve the CIR would reduce the efficiency of frequency re-use and this is not an attractive option. An alternative is to reduce the co-channel interference in a cell by using directional antennas at the base station and dividing the cell into a number of sectors. A three-sector arrangement is shown in Fig. 15.5(b). The original frequencies allocated to the cell are then divided between the sectors. Reduction in interference is achieved by choosing a frequency re-use pattern such that the front lobe of any base station transmitter illuminates only the back lobe of its co-channel counterpart. What this means, in effect, is that the number of interfering base stations is reduced from six to two with a corresponding increase in CIR.

This is made clearer from Fig. 15.5(b) in which it may be seen that only base stations **A** and **B** actually cause interference to mobile **M**. This produces a reduction in interference of approximately 5 dB over the omni-directional case. Hence by employing sectored antennas in areas of high traffic density it is possible to achieve the required CIR values. It should be noted that the sectored approach effectively increases the cluster size to 21 without any increase in the frequency re-use distance, however, handovers may be required between sectors of the same cell.

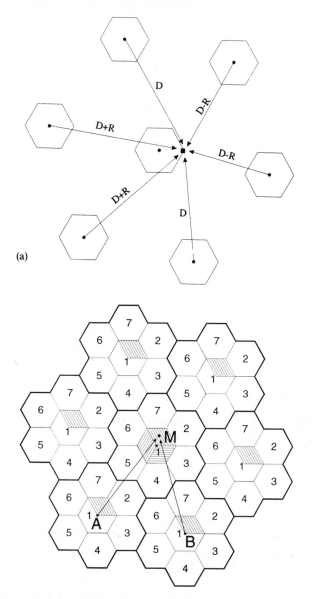

Fig. 15.5 Cellular system (a) with six interfering cells, and (b) with sectored base station antennas.

15.8 PROPERTIES OF THE RADIO CHANNEL

The radio channel in a cellular system has a major influence on the overall system design. This has already been evident in the way in which frequency re-use is implemented based on a radio attenuation proportional to D^4. Cellular radio systems are categorized by the fact that the height of antennas at both base station and mobile are usually low compared to the distance of separation. The model is shown in Fig. 15.6.

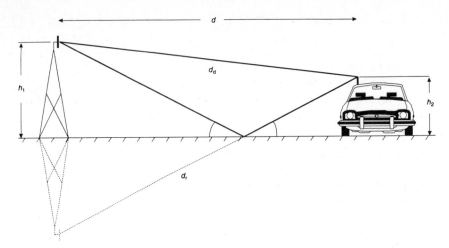

Fig. 15.6 Plane earth propagation model.

The analysis of Section 14.4 is the basis of the model in Fig. 15.6 and is repeated here for completeness. If it is assumed that an antenna radiates energy equally in all directions (isotropic antenna) it is possible to calculate the power density at a distance r from the antenna. If the antenna radiates a total power P_t the power at any distance r from the antenna is the power passing through the surface of a sphere of radius r. The surface area of the sphere is $4\pi r^2$ and the power received per unit area is thus

$$P_a = \frac{P_t}{4\pi r^2} \quad \text{watts/m}^2 \tag{15.7}$$

At sufficiently large value of r the wave becomes a plane wave. The power received by an antenna placed in this field is

$$P_r = P_a A_e \tag{15.8}$$

A_e is known as the 'effective aperture' of the antenna and is the equivalent power absorbing area of the antenna. The effective aperture of an isotropic antenna when used as a receiver can be shown to be $A_e = \lambda^2/4\pi$. Hence the power received by such an antenna is

$$P_r = P_a \times \frac{\lambda^2}{4\pi}$$

But $P_a = P_t/4\pi r^2$, i.e.

$$P_r = \frac{P_t \lambda^2}{(4\pi r)^2} \tag{15.9}$$

The isotropic antenna has unity gain in both the transmit and receive modes. A non-isotropic transmit antenna will have a gain of G_t and the product $P_t G_t$ is known as the effective radiated power (ERP). In mobile radio ERP is used as the standard method of quoting transmitted power. In effect, if the ERP is quoted as 100 W (50 dBm) and the antenna gain is 10 dB, the actual transmitted power would be 10 W (40 dBm). A non-isotropic receive antenna will have a gain of G_r and, in such cases, the received power would be given by

$$P_r = \frac{G_t G_r P_t \lambda^2}{(4\pi r)^2} \qquad (15.10)$$

This expression indicates that the attenuation is proportional to $(distance)^2$. In the case of mobile radio it is necessary to consider the height of both transmit and receive antennas above the earth's surface.

15.9   SPACE WAVE PROPAGATION

If the height of the base station antenna is h_1 and the height of the mobile antenna is h_2 the system may be represented as shown in Fig. 15.6, where the separation between transmitter and receiver is d. It is assumed that d is small enough to neglect the earth's curvature. Figure 15.6 shows that there will be both a direct and ground reflected wave. The direct path length is d_d and the reflected path length is d_r. It can be seen from the geometry of the system that

$$d_d = \sqrt{d^2 + (h_1 - h_2)^2}$$

Using the binomial expansion and noting that $d \gg h_1$ or h_2, the length of the direct path approximates to

$$d_d \cong d \left\{ 1 + 0.5 \left(\frac{h_1 - h_2}{d} \right)^2 \right\}$$

similarly

$$d_r \cong d \left\{ 1 + 0.5 \left(\frac{h_1 + h_2}{d} \right)^2 \right\}$$

The path difference is thus $\Delta d = d_r - d_d$, i.e.

$$\Delta d = \frac{2h_1 h_2}{d} \qquad (15.11)$$

The corresponding phase difference between direct and reflected path is

$$\Delta\phi = \frac{2\pi}{\lambda} \times \frac{2h_1 h_2}{d} = \frac{4\pi h_1 h_2}{\lambda d} \qquad (15.12)$$

The total received power is thus

$$P_r = P_t \left(\frac{\lambda}{4\pi d} \right)^2 \times |1 + \rho e^{j\Delta\phi}|^2 \qquad (15.13)$$

ρ is the reflection coefficient and for low angles of incidence the earth approximates to an ideal reflector with $\rho = -1$; i.e.

$$P_r = P_t \left(\frac{\lambda}{4\pi d} \right)^2 \times |1 - e^{j\Delta\phi}|^2 \qquad (15.14)$$

but $1 - e^{j\Delta\phi} = 1 - \cos \Delta\phi - j \sin \Delta\phi$, hence $|1 - e^{j\Delta\phi}|^2 = (1 - \cos \Delta\phi)^2 + \sin^2 \Delta\phi$.

If $\Delta\phi \ll 1$ then $\cos \Delta\phi = 1$ and $\sin \Delta\phi = \Delta\phi$, i.e.

$$P_r = P_t \left(\frac{\lambda}{4\pi d}\right)^2 \left(\frac{4\pi h_1 h_2}{\lambda d}\right)^2$$

hence

$$P_r = P_t \left(\frac{h_1 h_2}{d^2}\right)^2 \tag{15.15}$$

This is the 4th power law used in the frequency re-use calculation, and is known as the plane earth propagation equation. The loss is given by

$$\text{loss (dB)} = 40 \log_{10} d - 20 \log_{10} h_1 - 20 \log_{10} h_2$$

This means that the loss increases by 12 dB each time the distance is doubled. It should be noted that this equation is not dependent on frequency, which is a surprising result. This is a consequence of assuming that h_1 and h_2 are much smaller than d and that the earth is flat and perfectly reflecting. If the surface is undulating a correction factor, which is frequency dependent, must be included.

In the cellular environment it is quite likely that no direct path will exist between base station and mobile. The communication then depends on single or multiple reflections from buildings and surrounding objects. Under these circumstances field strength variations may only be derived from measurement and approximate computer modelling. In practice it is found that in the majority of cases the loss is close to that given by the plane earth propagation equation. However, it is important to realize that under these conditions the radio channel is subject to **fading**. The principle of fading can be demonstrated by reference to Fig. 15.7 in which it is assumed that there is no direct path between base station and mobile antennas.

If the phase difference between diffracted and reflected waves is a whole number of wavelengths the two waveforms will reinforce and the amplitude at the receiving antenna will (approximately) double. As the mobile moves the phase difference between the two paths changes. If the phase difference becomes an odd number of half wavelengths the two waves cancel producing a null. Thus as the mobile moves there are substantial amplitude fluctuations in the received signal known as **fast fading**. (There will also be a Doppler shift associated with the movement.) A typical variation of signal strength with distance is shown in Fig. 15.8.

Fast fading (due to local multipath) is also accompanied by a slower

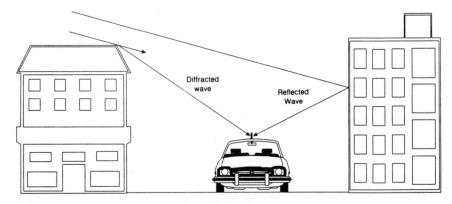

Fig. 15.7 Multipath propagation.

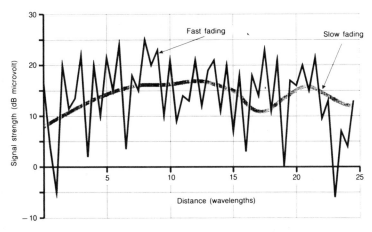

Fig. 15.8 Fading due to multipath propagation.

variation in mean signal strength known, as **slow fading**. Fast fading is observed over distances of about half a wavelength and can produce signal strength variations in excess of 30 dB. Slow fading is produced by movement over much longer distances, sufficient to produce gross variations in the overall path between base station and mobile. It should be noted that at the frequencies used in cellular radio a mobile moving at 50 km/h will experience several fast fades/second which will clearly effect the system performance. It should also be noted that fading is a spatially varying phenomenon which becomes a time-varying phenomenon only when the mobile moves.

It is clear that an exact representation of fading characteristics is not possible because of the effectively infinite number of situations which would have to be considered. Reliance therefore has to be placed on statistical methods which produce general guidelines for system design.

15.10   SHORT-TERM FADING (FAST FADING)

When a mobile unit is stationary the received signal strength will be formed by the vector sum of the various signals reaching the antenna and will have constant amplitude. When the mobile is moving it is assumed that the signal received will be the vector sum of N reflected signals of equal amplitude which arrive at the receiving antenna at a random phase angle ϕ_N. This is accepted as a reasonable model for the cellular environment where there is not usually a direct line of sight path between transmitter and receiver. (If there is a direct line of sight component this will alter the nature of the fading envelope and its statistics.) The addition of these components gives rise to a resultant with an amplitude (i.e. envelope) which varies in a random manner.

Applying the central limit theorem it can be shown that the received electric and magnetic field components have independent Gaussian distributions. This in turn leads to the conclusion that the envelope of the resultant received carrier has an amplitude which has a Rayleigh distribution given by

$$p_a(a) = \frac{a}{\sigma^2} \exp\left(\frac{-a^2}{2\sigma^2}\right) \tag{15.16}$$

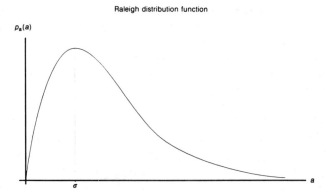

Fig. 15.9 Resultant carrier envelope distribution function.

In this expression σ^2 is the mean square value (i.e. mean power) of the carrier envelope and a is the instantaneous amplitude of the envelope. The distribution function is shown in Fig. 15.9. Note that the probability density function has a peak value of $0.6/\sigma$ at $a = \sigma$, where σ is the rms value of the received signal.

The corresponding cumulative distribution function (CDF) is

$$\text{prob}[a < A] = P_a(A) = 1 - \exp\left(\frac{A}{2\sigma^2}\right) \tag{15.17}$$

When the CDF is known it is possible to determine the average number of times per second that the signal envelope crosses a particular level in the positive direction. This is known as the level crossing rate. The level crossing rate is related to the velocity of the mobile v and the wavelength of the received carrier λ and, for a vertical monopole antenna, can be shown to be:

$$N(A_0) = \sqrt{2\pi}\frac{v}{\lambda}\rho\exp(-\rho^2) \tag{15.18}$$

where $\rho = A_0/\rho$ and A_0 is the specified level.

The situation is shown in Fig. 15.10. It should be noted that $N(A_0)$ is a

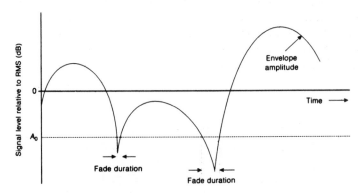

Fig. 15.10 Fading experienced by a mobile.

maximum when A_0 is 3 dB below the rms carrier level. This can be explained by the observation that if A_0 is low, the envelope is above this level for a large proportion of the time and hence the number of crossings per second decreases. A similar situation is observed when A_0 is set at a high level, the envelope being below this level for a large percentage of the time which again reduces the average number of crossings per second. At a carrier frequency of 900 MHz and a mobile speed of 48 km/h the level crossing rate at $A_0 = -3$ dB is $N(A_0) = 39$ per second. (In effect the number of fades per second is 39.) A further parameter of importance is the average fade duration. The duration of a fade is the interval of time that the envelope remains below the level A_0 and this is also shown in Fig. 15.10. The average duration of fades below the level A_0 is

$$\tau(A_0) = \frac{\mathrm{prob}[a < A_0]}{N(A_0)}$$

However, $\mathrm{prob}[a < A_0] = 1 - \exp(-\rho^2)$, hence the average fade duration for a vertical monopole is

$$\tau(A_0) = \frac{\lambda\{\exp(\rho^2) - 1\}}{\rho v \sqrt{2\pi}} \tag{15.19}$$

It is clear that fading is a frequency selective phenomenon. This effect is also apparent in the time domain and is measured in terms of delay spread. It has already been stated that the signal arriving at the antenna of a mobile (or base station) is the sum of a number of waves of different path lengths. This means that the time of arrival of each of the waves is different. If an impulse is transmitted from the base station, by the time it is received at the mobile it will no longer be an impulse but rather a pulse of width given by the **delay spread** Δ. The delay spread is, of course, different for different environments but typical values are given in Table 15.2.

The delay spread is an important parameter for digital systems as it limits the maximum data rate which can be sent. In general the time delay dispersion should be much less than the bit rate in a digital cellular system (without equalization). **Coherence bandwidth** is an additional parameter closely related to delay spread. In a wideband signal two closely spaced frequency components will suffer similar multipath effects. However, as the frequency separation increases the differential phase shifts over the various paths become decorrelated and the spectral components in the received signal will not have the same relative amplitudes and phases as in the transmitted signal. This is essentially frequency selective fading and the bandwidth over

Table 15.2 Delay spreads

Environment	Delay spread $\Delta(\mu s)$
Rural area	< 0.2
Suburban area	0.5
Urban area	3

which the spectral components are affected in a similar way (i.e. are correlated) is known as the coherence bandwidth. It is common practice to define the coherence bandwidth for a correlation coefficient of 0.5, in which case the approximate relationship between coherence bandwidth and delay spread is given by Eqn (15.20)

$$B_c = \frac{1}{2\pi\Delta} \tag{15.20}$$

The radio channel provides a hostile environment for cellular radio and much of the system design deals with overcoming these difficulties. This will become clear when examples of both analogue and digital systems are considered.

15.11 FIRST GENERATION ANALOGUE SYSTEMS (TACS)

Analogue cellular systems are operated in many countries worldwide. It has already been pointed out that the systems in use all have some basic similarities, but also have some important differences. This section will concentrate on the total access communication system (TACS), employed in the UK and a number of other countries, which is a derivative of the advanced mobile phone system (AMPS) developed by AT&T for the USA.

TACS is an analogue FM system operating in the 900 MHz waveband providing 1000 duplex channels occupying the frequencies 890–915 MHz (25 MHz) and 935–960 MHz (25 MHz). The nominal channel bandwidth is thus 25 kHz. At the present time, in the UK, 600 channels have been divided between two operators (Cellnet and Racal) and the remaining 400 channels have been reserved for the second generation digital system (GSM). The lower frequency band is assigned for transmissions from mobile to base, the upper band is assigned for transmission from base to mobile. The channels are numbered consecutively in ascending order along the frequency spectrum of each band. The duplex spacing is the difference between the transmit channel frequency and receive channel frequency and maintained at 45 MHz in the TACS system. This is illustrated in Fig. 15.11 where it should be noted that mobile transmit channel 1 has a nominal carrier frequency of

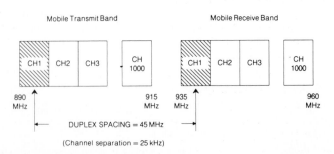

Fig. 15.11 Channel allocation in TACS.

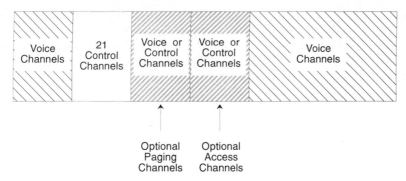

Fig. 15.12 Control and traffic channels.

890.0125 MHz, the corresponding base station transmit channel carrier being 45 MHz above this at 935.0125 MHz.

Each operator in the UK has been allocated 300 channels, 21 of which are a contiguous block of dedicated control channels. The first control channels of each block are channel 23 and 323 respectively. This structure is shown in Fig. 15.12. The channels are categorized as 'traffic channels' which carry voice or data, and 'control channels'.

The control channels carry the signalling information which is essential to the operation of the cellular system. The mobile unit has a single transmitter and a single receiver which means that once a voice channel has been allocated (at call set-up) any subsequent exchange of signalling information between mobile and base station must make use of the voice channel. When a mobile is not actively engaged in a call it is still necessary to transmit information periodically to the fixed network, for updating of the mobile's location, for example. Under these circumstances a control channel is used.

Although each mobile has only a single transmitter and receiver the channels which are used can be changed by instructions from the base station. There are actually four **signalling** paths between mobile and the base station. These are:

(i) the forward control channel (FOCC), from base to mobile;
(ii) the reverse control channel (RECC), from mobile to base;
(iii) the forward voice channel (FVC), from base to mobile:
(iv) the reverse voice channel (RVC), from mobile to base.

The FOCC and RECC are used to maintain contact between mobiles and base stations outwith a normal call and also for call set-up. The forward and reverse voice channels are used when calls are in progress. The FVC and RVC signalling messages are formatted in short bursts and are inserted from time to time into the voice path. Being of short duration the users are unaware of their existence. It will be apparent from Fig. 15.12 that, although there are 21 dedicated FOCCs and RECCs, provision is also made to use some of the voice channels for signalling information.

In a busy system using many of the available voice channels per cell the signalling data exceeds the capacity of the dedicated control channels. Hence it would not be possible for new users to access unused voice channels. Therefore provision is made to re-allocate some of the voice channels for signalling under high-load conditions. As far as system operation is concerned it is necessary for the FOCC channels to operate continuously in order that whenever a mobile enters a cell it can determine which channels are allocated for voice traffic and whether the RECC is in use. The RECC on the other hand is only activated occasionally when it is necessary to transmit information about mobiles. Such information is required when, for example, mobiles move from one cell to another or when a mobile wishes to initiate a call.

15.12   TACS RADIO PARAMETERS

15.12.1   Power levels

When considering frequency re-use it was assumed that all base stations operate at the same power levels. In the TACS system cell sizes range from a radius of 1 km (urban areas) to 15 km (rural areas) and the base station power level is chosen to ensure adequate coverage for the size of cell which it serves. The maximum ERP, for the larger cells, is restricted to 100 W.

The situation with the mobile is somewhat different as battery drain is an important consideration. In effect, the mobile operates at the lowest power level which will ensure an adequate link quality and can be instructed by the base station to alter power output as appropriate. It is also important to minimize the total ERP in the cellular environment to keep interference as low as possible, as a high power mobile can produce both adjacent channel and co-channel interference. Four classes of mobile are recognized, as in Table 15.3 (class 1 is a vehicle-mounted transceiver, class 4 is a hand-held portable). Note in this table that dBW is the power output relative to 1 watt, i.e.

$$ERP(dBW) = 10 \log_{10}\left(\frac{\text{power}}{1 \text{ watt}}\right)$$

Base stations instruct mobiles to adjust power output levels to maintain an acceptable signal level at the base station by use of a 3 bit mobile attenuation code (MAC). The codes and the corresponding outputs are shown in Table 15.4.

Table 15.3   Mobile power levels

Class	Nominal ERP		Mobile output 1.5 dB antenna gain
1	10.0 dBW	(10 W)	8.5 dBW (7.0 W)
2	6.0 dbW	(4 W)	4.5 dBW (2.8 W)
3	1.6 dbW	(1.6 W)	0.5 dBW (1.1 W)
4	−2.0 dBW	(0.6 W)	−3.5 dBW (0.45 W)

Table 15.4 Mobile attenuation codes

Mobile power level	Mobile attenuation code MAC	Nominal ERP (dBW)			
		Class 1	Class 2	Class 3	Class 4
0	000	10	6	2	−2
1	001	2	2	2	−2
2	010	−2	−2	−2	−2
3	011	−6	−6	−6	−6
4	100	−10	−10	−10	−10
5	101	−14	−14	−14	−14
6	110	−18	−18	−18	−18
7	111	−22	−22	−22	−22

15.12.2 Modulation

The modulation used for voice signals is frequency modulation. This has superior SNR properties to AM, and it is also possible to take advantage of the FM capture effect to minimize the co-channel interference. The FM capture effect is fully described in Section 4.11 and Fig. 4.12.

In this figure, X may be interpreted as the carrier amplitude received from the serving base station and Y as the resultant co-channel interference from interfering cells using the same frequency. For values of $X/Y > 10(10\,\text{dB})$ X captures the system. Hence it can be seen that if the CIR is maintained at around 18 dB (which is the figure used to decide the cluster size) the capture effect of the FM transmission has significant advantage.

In the TACS system the bandwidth of the channel is restricted to 25 kHz. Assuming the audio signal has a bandwidth of 3 kHz this means that four significant sidebands are alowed giving a modulation index of approximately

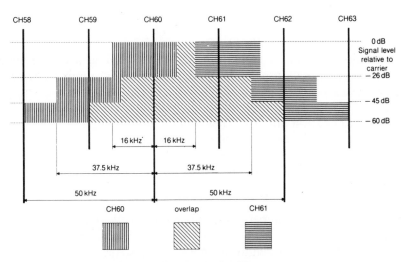

Fig. 15.13 Adjacent channel interference in TACS.

$\beta = 1.6$. The corresponding carrier deviation is

$$\Delta f_c = \beta f_m = 5 \, \text{kHz}$$

However this value is increased to 9.5 kHz ($\beta = 3.2$) in TACS to improve CIR performance, in the presence of co-channel interference. There is a penalty to be paid for this, however, as the signal bandwidth is no longer restricted to 25 kHz and **adjacent** channel interference will thus occur. The amount of interference allowed is shown in Fig. 15.13. The 26 dB bandwidth is specified as 32 kHz, the 45 dB bandwidth is specified as 75 kHz and the 60 dB bandwidth is specified as 100 kHz. Hence it is necessary to ensure, as far as possible, that allocation of adjacent channels in adjacent cells is minimized. This means that once the cluster size is fixed the channel allocation within a cluster should minimize the allocation of adjacent channels to adjacent cells. (A pre-emphasis characteristic of 6 dB/octave is used between 300 Hz and 3 kHz).

15.13   TACS MOBILE NETWORK CONTROL

It was shown in Fig. 15.11 that 21 channels from the available set are dedicated to the system management and call set-up functions. There are 21 forward control channels for base to mobile communication and spaced at 45 MHz above these there are a corresponding 21 reverse control channels for mobile to base communication. These channels are paired, i.e. if a mobile receives control information on a particular FOCC it responds on the corresponding REVC which is 45 MHz above.

Functionally there are three different types of control channel known as

1. Dedicated control channels (DCC)
2. Paging channels
3. Access channels

The DCC is the basic co-ordinating forward control channel for the network and is transmitted continuously. All mobiles are permanently programmed with the channel numbers of the DCCs and scan these channels at switch-on. This is necessary to obtain:

(i) basic information about the network;
(ii) the channel numbers of the paging channels.

The paging channels are used to:

(i) alert particular mobiles to an incoming call;
(ii) transmit channel numbers of the access channels in use in the mobile's locality;
(iii) transmit traffic area identities, etc.

Mobiles use the access channels to:

(i) obtain parameters about the required access procedure and the status of the access channels on the RECC (busy or idle);
(ii) acknowledge paging messages;
(iii) update the network with their locations by registering with the base stations offering the best radio path;
(iv) initiate outgoing calls.

All three types of functional control channels carry status information in sequences of data blocks known as **overhead messages** which contain a multiplicity of status information fields and instructions. When the control channels are multiplexed the overhead messages contain the relevant information in one continuous bit stream.

15.14. CALL MANAGEMENT IN TACS

The primary tasks of the mobile network are:

1. To have a record of the location of all active mobiles within the system at any given time. This is so that incoming calls may be directed to the appropriate cell.
2. To manage the handover process during calls as the mobile crosses cell boundaries.

In order to describe the call management procedure used in TACS it is first necessary to discuss the supervisory tones which are involved in the sequence for establishing a voice channel. Supervisory tones are necessary to monitor the progress of a call as the mobile moves in the Rayleigh fading environment. Fading will influence the level of the signal from a mobile's base station and also the level of co-channel interference from base stations operating on the same frequency in adjacent clusters. Note that if a mobile experiences a deep fade it is possible for a transmission from a different cluster to capture the receiver (FM capture effect) and an unwanted conversation would intrude. Such a possibility is prevented in TACS which uses two supervisory tones known as:

1. The supervisory audio tone (SAT)
2. The signalling tone (ST)

While a call is **in progress** a supervisory audio tone (SAT) is transmitted by the base station and re-transmitted by the mobile. Both the base station and the mobile require the presence of the SAT on the received signal to enable the audio path. The SAT is one of three frequencies (5.970 kHz, 6.000 kHz or 6.030 kHz).

In setting up the call the base station informs the mobile which SAT is appropriate to the *cluster* of cells which is handling the call. The mobile then re-transmits the defined SAT to confirm that the correct connection is made

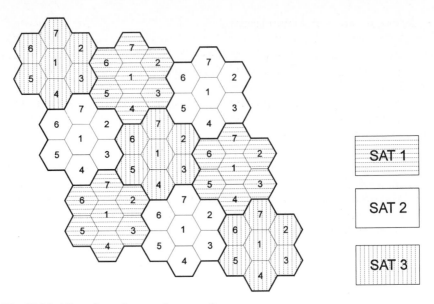

Fig. 15.14 Allocation of supervisory audio tones.

and continues to transmit this SAT throughout the call. The two other SAT frequencies are assigned to adjacent cell clusters as shown in Fig. 15.14.

If a co-channel interferer, originating in an adjacent cluster, captures the mobile receiver it will carry the wrong SAT and the audio output will be muted. The only time the SAT is itself muted is when data is being transmited from the mobile on the RECC. In such cases a digital colour code D'C'C' is transmitted instead. This is the digital equivalent of the SAT and is compatible with the signalling format used. The D'C'C' is also used on the FOCC during data transmissions.

15.14.1 Mobile scanning

When a mobile is powered up it scans the 21 control channels of its primary network. It then tunes to the strongest channel and attempts to read the overhead messages being transmitted. If successful the mobile receives information indicating the channel numbers of the paging channels in its location. If this information is not received correctly the mobile tunes to the second strongest channel and tries again.

If the second attempt is unsuccessful it is likely that the mobile is not within range of its primary network and so it repeats the scanning process on the secondary network (provided by a different operator). When the mobile receives the channel numbers of the paging channels it scans these looking for the two strongest signals. It tunes to the strongest signal and attempts to receive the overhead messages being transmitted on that channel. If sucessful it receives information about the traffic area in which it is operating and a number of parameters about the network configuration.

If unsuccessful the mobile tunes to the second strongest paging channel and again attempts to read the overhead messages. If this fails it restarts the complete scanning operation. From switch-on to locking on to a paging channel usually takes between 5 and 10 seconds. However this can be increased to 17 seconds if a second strongest control or paging channel is required. When the mobile finds a suitable paging channel it reads information concerning the access channels in use. The mobile then signals a registration on the RECC.

15.14.2   Registration

Registration is used by mobiles to inform the cellular network of their current location in order that incoming calls can be routed to the correct traffic area (i.e. group of cells). The routing within the mobile network is handled by special exchanges known as mobile switching centres (MSCs). Each MSC controls several cell clusters known as a traffic area and is connected to other MSCs and the fixed network, as shown in Fig. 15.1. The record-keeping activity of the MSCs requires that each mobile continually updates its location by a process known as **registration**.

The TACS system has two forms of registration known as **forced registration** and **periodic registration**. Either or both types of registration can be enabled by messages on the paging channels. With forced registration the mobiles are required to register every time they cross a new traffic area boundary. Whenever a mobile enters a scanning sequence to find a new paging channel it compares the received traffic area information with that of the previous paging channel. If there is a difference the mobile has crossed a traffic area boundary and so registers its location with the network. (This will also occur at power up since there is no previous traffic area.)

When it has registered the mobile updates its internal memory with the new traffic area. With periodic re-registration the mobile maintains a list of the last four traffic areas visited, together with a numeric indicator for each which tells the mobile when to re-register. The base station transmits a registration identity number on a regular basis on the paging channel. This number is incremented each time it is transmitted. The mobile compares the received number with the number in its memory appropriate to the current traffic area. If the two correspond the mobile intiates a registration with the network. When it has registered, the mobile increments the number stored in its memory and awaits the next registration period.

The network is able to control the rate at which mobiles re-register by varying the time between registration identity messages, or by changing the number by which mobiles increment their stored number. If a mobile finds itself in a traffic area for which it has no entry in its memory, it creates a new entry (deleting the oldest) and immediately registers with the network as for forced registration. Registration is achieved by the mobile performing a system access and sending a registration messsage. (The system access procedure is covered in the sub-section on call origination.)

When a mobile registers in a new traffic area its location is passed by the MSC, controlling the area, to the mobile's **home area location** MSC. This

MSC holds data on the location of all its active mobiles (all mobiles are allocated a home traffic area). Hence when a call is made to a mobile the system is able to decide, by interrogating the mobile's home traffic area MSC, which part of the network will page the mobile.

15.14.3 Call origination

When a mobile subscriber wishes to make an outgoing call the user enters the number manually from the keyboard (or automatically from an on-board memory) and initiates the call by pressing a SEND key. This causes the mobile to perform a system access in order to transmit its message to the system. The mobile first scans the access channels (whose numbers are indicated by overhead messages of the paging channels) in the same way as other scanning operations are carried out and chooses the two strongest. It attempts to read the overhead messages on the strongest access channel (of the FOCC) which contains parameters about the required access procedure. (If unsuccessful the procedure is repeated on the second channel.)

Once these parameters are read the mobile monitors the BUSY/IDLE bit stream being sent by the base station on the access channel. If this indicates an idle condition the mobile waits a random time and transmits its message on the mobile to base access channel which is 45 MHz above the base to mobile access channel to which it is tuned. The mobile continues to monitor the BUSY/IDLE bit stream transmitted by the base station. This is changed from IDLE to BUSY by the base station as soon as it receives the start of the message from the mobile.

The mobile checks the interval between its start of transmission and the transition from IDLE to BUSY. If this is too long or too short the mobile assumes that the IDLE/BUSY transition was caused by a message from another mobile and aborts its transmission. It then waits for a random time and attempts to transmit its message again. This is a form of collision resolution which is necessary on all random access systems.

When the mobile has completed its message it turns off its transmitter and continues to monitor the base-to-mobile access channel. For call originations the message from the base station is normally a speech channel allocation which contains a channel number and the SAT code (this is the digital colour code D'C'C'). On receipt of the message the mobile tunes to the required voice channel and transmits the SAT. If the SAT is correct the audio paths are enabled and the user can hear the call being set up. If the access was as a result of a registration the message received on the base-to-mobile access channel is a registration confirmation. On receipt of this message the mobile returns to the idle condition.

15.14.4 Call receipt

When an incoming call is received for a mobile its home area MSC is checked for the mobile's current registered location. A paging call is then transmitted on the paging channel of all base stations in the mobile's current traffic area. When a mobile receives a paging call it accesses the network in the same way as for a call origination, but the message sent to the base station informs it that this access is a result of receiving a page call. The mobile receives a voice channel allocation from the base station and checks the SAT received and loops it back to the base station. The base station then transmits an alert message to the mobile causing the mobile to alert the user to the incoming call and to transmit the 8 kHz signalling tone (ST). When the user answers, the 8 kHz tone is disabled and this enables the audio paths and the call proceeds.

15.14.5   Handover

Whenever the mobile is operating on a voice channel the base station monitors the received signal level. When this level falls below a threshold value the base station informs its MSC that handover to a nearby cell may be necessary. (Note that the nearby cell may be in the same traffic area or an adjacent traffic area.) The MSC then requests surrounding base stations to measure the signal strength of the mobile by using their special purpose measuring receivers. When the MSC receives the results of these measurements it decides whether any of the reporting base stations is receiving a stronger signal than the current base station.

If there is a better cell the relevant base station is requested to allocate a voice channel and the MSC requests the original base station to inform the mobile to tune to the new channel. (The new base station will be operating on a different set of carriers to the current one.) A short signalling message is sent to the mobile on the FVC giving the new channel number and the mobile tunes to the new channel. The duration of the signalling message is about 400 ms hence the user notices only a brief silence during the handover process.

15.14.6   Power control

If a mobile moves close to the base station a high signal level could result in inter-modulation in the base station receivers causing interference to other users. To avoid this happening the signal level is monitored and if found to be above a given threshold a message is sent to the mobile on the FVC to reduce its transmitter power level. The mobile acknowledges on the RVC and selects the appropriate power level. The opposite procedure is possible as the mobile moves away from the base station.

15.14.7   Additional services

The facility exists within the system to provide additional services such as three party calls, call diversion, etc. To request such facilities the user enters the appropriate code from the keyboard and presses the SEND key. The mobile sends a 0.4 s burst of 8 kHz signalling tone to the base station which responds with a digital signalling message on the FVC requesting the mobile to send its information. The mobile transmits the information to the base station in digital form on the RVC and then returns to the conversational mode while the network processes the request for the facility.

15.14.8 Call termination

When a mobile user finishes a call and replaces the handset the mobile transmits a 1.8 s burst of 8 kHz tone to the base station and then re-enters the control channel scanning procedure. If the other party on the PSTN clears down, a **release** message is sent to the mobile on the FVC. The mobile then responds by sending the 8 kHz tone after which it re-enters the control channel scanning procedure.

15.14.9 Protection of signalling messages

The radio channel is subject to interference (co-channel and adjacent channel) and fading and it is necessary to protect the signalling sequences against these imperfections. Two methods of data protection are employed in TACS and these are:

1. Repeat each transmitted data word several times and take a majority vote. This is designed to obtain a 95% accuracy in data word transmission.
2. Use an error-correcting code to improve the 95% accuracy to 99.9% accuracy.

Details of the coding scheme can be found in reference 1.

15.15 ALTERNATIVE ANALOGUE SYSTEMS

Although TACS has been considered in detail it has already been pointed out that it is not the only system currently in use. There are a number of others, e.g. AMPS used in the USA, NAMTS used in Japan and NMT used in Scandinavia.

These systems differ from TACS in relatively minor ways but they are not compatible. In NMT, for example, the frequency modulated carrier is restricted to a bandwidth of 25 kHz (compared to a value of 32 kHz in TACS). This means that in NMT adjacent channels can be used in the same cell, if required. However, because the resulting modulation index is smaller, the co-channel interference is significantly poorer. This means that NMTS requires a larger re-use distance than TACS.

The systems also differ in the number of channels dedicated to control functions. TACS has the most (21) and NAMTS has the least (1). The carrier spacing in AMPS is 30 kHz rather than 25 kHz, and so on. Hence there is no single standard in the analogue world. In the digital world, especially in Europe, considerable effort has been made to derive a common standard. The European standard is known as GSM and will be described in the next section.

15.16 DIGITAL CELLULAR RADIO

Since the introduction of first generation analogue systems an unprecedented demand for installations has occurred. This means that the systems are rapidly reaching capacity and ways of meeting the demand with second generation systems have been investigated. The capacity of the first generation systems is determined by the bandwidth occupied by the individual voice channels and the minimum CIR at which mobiles and base station receivers can operate.

Since the radio bandwidth allocated to mobile communications will always be finite, large-scale increases in capacity are only possible by utilizing the allocated bandwidth more efficiently. One example of such a technique is to use SSB instead of FM. However, such systems would not have the capture effect advantage of FM and the overall improvements likely to be achieved with SSB are not judged to be significant enough to warrant its introduction.

The alternative is to make the cells smaller which means that problems with frequent handovers would then have to be addressed. It is stressed that there is no solution which is optimum in all respects. However, there are a number of significant advantages to be achieved with digital systems which makes their adoption as second generation systems quite attractive. In assessing the available digital techniques the system(s) must satisfy the following criteria:

  (i) high subjective voice quality;
 (ii) low infrastructure cost;
(iii) low mobile equipment cost;
(iv) high radio spectrum efficiency;
 (v) capable of supporting hand held portables;
(vi) ability to support new services.

Clearly a significant advantage of digital systems is that digital signal processing can be used both to reduce voice bandwidth requirements and to extract signals in poor CIR conditions. There are other advantages, for example, digital transmissions can be encrypted to avoid unauthorized eavesdropping.

15.16.1 Advantages of broadband transmission

Digital transmission can be either narrowband or broadband and the actual choice is a compromise between many factors. It has been pointed out that the effect of multipath propagation is frequency dependent. If a null is produced at one frequency due to destructive addition a peak may be produced at a different frequency due to constructive addition. Hence if a signal has a bandwidth which is wider than the coherence bandwidth it follows that only some frequencies in the signal may experience a fade while others will not. Thus the use of wideband transmission can have some advantage in a Rayleigh fading environment (additional processing can sometimes recover the signal).

Four possible strategies which could be employed are:

1. **Narrowband**: in this case a single carrier is allocated to each channel (this is the simple frequency division multiple access (FDMA) used in the TACS system).
2. **Wideband**: in this case each channel uses all of the available bandwidth for

a fraction of the time (this is known as wideband time division multiple access (TDMA)).

3. **Intermediate**: in this case each carrier has a bandwidth which is made greater than the coherence bandwidth by time multiplexing a number of channels. The complete bandwidth is occupied by a number of such channels. This is usually referred to as narrowband TDMA or FDMA-TDMA.

4. **Spread spectrum**: in this case perfect theoretical signal extraction is possible in virtually zero CIR conditions.

The choice between these alternatives is not straightforward and each one has certain advantages and disadvantages. The advantages of FDMA (one channel per carrier) are:

1. The bandwidth of each carrier is considerably less than the coherence bandwidth. Hence equalization is not required as this will not produce any improvement in performance. Capacity increase is obtained by reducing the bit rate per channel and using efficient channel codes.

2. The technological advances required for such a system, over existing analogue systems, is modest. It is possible to configure a system so that subsequent advances in reduced rate speech coders could be easily accommodated.

3. The system is flexible and can be easily adapted to handle both large rural cells and small urban cells.

The disadvantages associated with FDMA (one channel per carrier) are:

1. The architecture is similiar to that used in analogue systems. Hence any capacity improvements would rely on operation at lower CIR values. Narrowband digital systems have only limited advantages in this respect which severely restricts any capacity improvements which would be achieved for a given radio spectrum allocation.

2. The maximum bit rate per channel is fixed at a low value. This is a considerable disadvantage for data communications, which is seen as an essential element in future systems.

The advantages of TDMA (narrowband and broadband) are:

1. It offers the capability of overcoming Rayleigh fading by appropriate channel equalization.

2. Flexible bit rates are possible, i.e. both multiples and sub-multiples of the standard bit rate per channel can be made available to users.

3. It offers the opportunity of frame-by-frame monitoring of signal strength and bit error rate to enable either base stations or mobiles to initiate handover.

The disadvantages of TDMA are:

1. On the UP link TDMA requires high peak power in the transmit mode. This is a particular problem for hand held portables with limited battery life.

2. To realize the full potential of digital transmission requires a significant

amount of signal processing. This increases power consumption and also introduces delay into the speech path.

Considering all these issues broadband digital systems appear to have the greatest potential but they also have the greatest demands on the associated signal processing requirements. Whether such systems will be developed for third generation systems is still the subject of intense research effort. Second generation systems have been developed, however, and the system adopted for Europe was chosen and specified by a specially formed group of the *Conference Europeene des Administrations des Postes et des Telecommunications* (CEPT) known as the Groupe Special Mobile (GSM). GSM is composed of about 40 members representing 17 European countries and its function is to co-ordinate and produce specifications for a Pan-European cellular mobile radio system.

The GSM carried out field trials of a number of competing systems in 1986 and chose as the standard a digital system based on a narrowband TDMA approach. The standard which has been developed is known simply as GSM, and a variation of this with low-power terminals is known as DCS 1800. The GSM standard has been adopted in countries worldwide and GSM has been renamed as Global System for Mobile communication. The choice of the GSM standard was based on the following ranked criteria:

1. Spectral efficiency (the number of simultaneous conversations/MHz/km^2).
2. Subjective voice quality.
3. Cost of mobile unit.
4. Feasibility of a hand portable mobile unit.
5. Cost of base station.
6. Ability to support new services.
7. Ability to coexist with existing systems.

In GSM the voice waveform is digitally encoded before transmission. As the system is based on TDMA individual users are given access to the radio channel for a limited period and transmit a burst of binary information. The GSM specification is very detailed and, in a text of this nature, it will be possible only to give an outline of the system. The natural place to begin is with a description of the radio interface.

15.17 THE GSM RADIO INTERFACE

The radio subsystem constitutes the physical layer of the link between mobile and base stations. The main attributes of the GSM interface are:

1. time division multiple access (TDMA) with 8 channels/carrier
2. 124 radio carriers in a paired band (890 to 915 MHz mobile to base station, 935 to 960 MHz base to mobile, inter-carrier spacing 200 kHz)
3. 270.833 kb/s per carrier
4. Gaussian minimum shift keying with a time bandwidth product $BT = 0.3$
5. slow frequency hopping (217 hops/second)
6. synchronization compensation for up to 233 μs absolute delay

7. equalization of up to 16 μs time dispersion
8. downlink power control
9. discontinuous transmission and reception
10. block and convolutional channel coding coupled with interleaving to combat channel perturbations
11. 13 kb/s speech coder rate using regular pulse excitation/linear predictive coding (RPE/LPC)
12. overall channel bit rate of 22.8 kb/s

The first pair of carriers in the GSM system are 890.2 MHz and 935.2 MHz, i.e. the spacing is 45 MHz which is the same as for TACS. GSM recommends that carriers 1 and 124 are not used due to energy of the modulated carrier lying outside the nominal 200 kHz bandwidth. The radio subsystem is the physical layer of the link between mobiles and base stations. Each cell can have from 1 to 15 pairs of carriers and each carrier is time multiplexed into 8 slots. The carriers and their associated time multiplexed slots form the **physical channels** of the GSM system. The operation of the radio subsystem is divided into a number of **logical channels** each of which has a specific function in terms of handling the transmission of information over the radio subsystem. Each of the logical channels must be mapped in some way onto the available physical channels. This is illustrated in Fig. 15.15.

The two categories are

1. Traffic Channels (TCH)
2. Signalling channels (Broadcast Control Channel (BCCH), Common Control Channel (CCCH) and Dedicated Control Channel (DCCH))

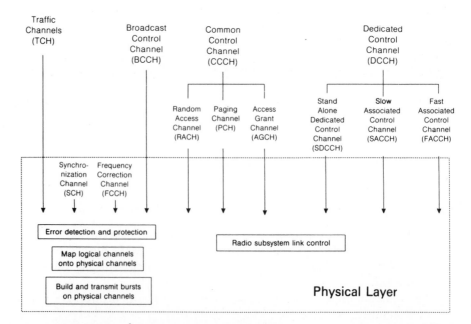

Fig. 15.15 Mapping of physical channels in GSM.

Some of these channels are divided into sub-channels. The sub-channels exist conceptually in parallel but the existence of one signalling channel may exclude the presence of another one. The Paging Channel (PCH) and Access Grant Channel (AGCH) are never used in parallel. The radio subsystem requires two channels for its own purposes. These are the **Synchronization Channel** (SCH) and the **Frequency Correction Channel** (FCCH).

The logical channels are mapped onto the basic TDMA frame structure which is shown in Fig. 15.16. The purpose of the radio subsystem is to provide to the **data link layer** a 'bit pipe' with a defined throughput, acceptable transmission delay, and a reasonable quality for each of the logical channels. To achieve this the physical layer performs a variety of tasks which can be grouped into four categories.

1. Create physical channels by building data bursts and transmitting them over the radio path.
2. Map the logical channels onto the created physical channels, taking into account the throughput needs of particular logical channels.
3. Apply error protection to each logical channel according to its particular needs.
4. Monitor and control the radio environment to assign dedicated resources and to combat changes in propagation characteristics by functions such as **handover** and **power control**.

Bursts of transmission from base station and mobile occur in the slots of the up and down carriers as shown in Fig. 15.16. The bit rate on the radio channel is 270.833 kb/s which gives a bit duration of 3.692 µs. A single timeslot consists of 156.25 bits and therefore has a duration of 0.577 ms. The recurrence of one particular timeslot on each frame makes up one physical channel. This structure is applied to both uplink and downlink. The numbering scheme is staggered by three timeslots to remove the necessity for the mobile station to transmit and receive at the same time. This is illustrated in Fig. 15.17. Data is transmitted in bursts which are placed in these timeslots. It is clear from Figure 15.16 that the length of the bursts are slightly shorter than the

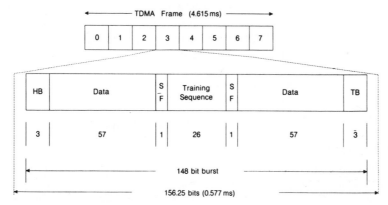

Fig. 15.16 Normal burst in GSM.

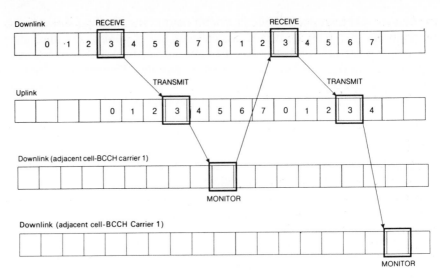

Fig. 15.17 GSM slots and scanning structure.

duration of the timeslots. This is to allow for burst alignment errors, time dispersion on the propagation path, and the time required for smooth switch on/off of the transmitter.

There are four types of burst which can occupy a timeslot, these are:

1. Normal burst (148 bits + 8.25 guard bits).
2. Frequency correction burst (148 bits + 8.25 guard bits).
3. Synchronizing burst (148 bits + 8.25 guard bits).
4. Access burst (88 bits + 68.25 bits), used to access a cell for the first time in case of call set-up or handover.

Figure 15.16 shows the data structure within a normal burst. It consists of 148 bits transmitted at a rate of 270.833 kb/s. Of these bits 114 bits are available for data transmission, the remaining bits are used to assist reception and detection. A training sequence (26 bits) in the middle of the burst is used by the receiver to synchronize and estimate the propagation characteristics. This allows the setting up of an equalizer to compensate for time dispersion produced by multipath propagation. Tail bits (3 bits) transmitted at either end of the burst enable the data bits near the edges of each burst to be equalized as well as those in the middle. Two stealing flags (one at each end of the training sequence) are used to indicate that a burst which had initially been assigned to a traffic channel has been 'stolen' for signalling purposes.

The burst modulates one carrier of those assigned to a particular cell using Gaussian minimum shift keying (GMSK). If frequency hopping is not employed, each burst belonging to one particular **physical channel** is transmitted using the same carrier frequency. A network operator can implement **slow frequency hopping**. It will be recalled that fading is frequency dependent, hence slow frequency hopping is one technique which can be used to overcome the problems of fading. If slow frequency hopping is implemented the frequency changes (among the set of carrier frequencies available within a cell) between bursts, this is discussed in more detail in Section 15.22.

15.18   MAPPING OF LOGICAL CHANNELS IN GSM

There are five different cases of mapping logical channels on to physical channels.

1. Mapping of a full-rate traffic channel (TCH) and its slow associated control channel (SACCH) onto one physical channel.
2. Mapping of two-half rate traffic channels and their two slow associated control channels on to one physical channel.
3. Mapping of the broadcast control channel (BCCH) and the common control channel (CCCH) onto one physical channel.
4. Mapping eight stand-alone dedicated control channels (SDCCH) onto one physical channel.
5. Mapping of four stand-alone dedicated control channels plus the broadcast control channel and the common control channel onto one physical channel. (This is for lower capacity than case 3).

This list indicates that there are at least two logical channels to be mapped onto each physical channel. The timeslots of the physical channel must therefore be assigned to the logical channels on a structured basis. For this purpose two **multiframe** structures have been defined:

1. A multiframe consisting of 26 TDM frames (resulting in a recurrence interval of 120 ms) for the TCH/SACCH cases 1 and 2.
2. A multiframe consisting of 51 TDM frames (resulting in a recurrence interval of 236 ms) for signalling channels, cases 3, 4 and 5.

15.18.1   Mapping of traffic and associated control channels

The mapping of a traffic channel and its SACCH is shown in Fig. 15.18 for both full-rate and half-rate channels. The full-rate TCH uses 24 frames out of the 26 available in the multiframe. One of the 26 frames is used for the SACCH and one remains idle. In this diagram TC0 represents a normal burst of duration 0.577 ms on a particular physical channel (i.e. carrier timeslot). TC1 represents the next burst of duration 0.577 ms on the same physical channel and will occur 4.615 ms after TC0. The duration of the multiframe is therefore $26 \times 4.615 \, \text{ms} = 120 \, \text{ms}$.

The gross bit rate per traffic channel is derived as follows:

Data bits per normal burst = 114 bits
Number of normal bursts per 120 ms multiframe = 24

$$\text{Gross bit rate} = \frac{24 \times 114}{0.12} = 22.8 \, \text{kb/s}$$

The SACCH uses 114 bits per 120 ms = 950 b/s

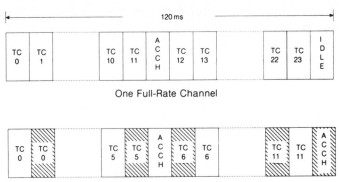

Fig. 15.18 Full- and half-rate channels.

The throughout of the physical channel is 114 bits per 4.615 ms = 24.7 kb/s (this includes the idle frame). In the case of the half-rate channel two half-rate channels share one physical channel, the idle frame is then used to accommodate the SACCH for the second half-rate channel. In this case each traffic channel occupies only 12 frames which results in a gross bit rate of 11.4 kb/s for each (each SACCH uses 950 b/s).

GSM allows the possibility to provide additional signalling capacity if the one provided by the SACCH is not enough. The **Fast Associated Control Channel** (FACCH) steals capacity from the TCH by replacing bits from the TCH with bits from the FACCH. The stealing flags are used to indicate when TCH bits have been replaced by FACCH bits.

15.18.2 Mapping of the BCCH/CCCH

The TCH/SACCH structure is dedicated to one user (full-rate channel) or shared between two users (half-rate channel). The mapping of the BCCH and the CCCH uses a multiframe of 51 TDM frames and is shared by all mobiles currently in a cell. In addition all sub-channels transmitted on this structure are **simplex** channels i.e. they exist in one direction only. The sub-channels mapped onto this physical channel are:

1. The **Broadcast Control Channel** (BCCH, base to mobile), this provides general information about the network, the cell in which the mobile is currently located and the adjacent cells.
2. The **Synchronization Channel** (SCH, base to mobile) carriers information for frame synchronization and identification of the base station transceiver.
3. The **Frequency Correction Channel** (FCCH, base to mobile) provides information for carrier synchronization.
4. The **Random Access Channel** (RACH, mobile to base), this channel is used by the mobile to access the network during registration of cell set-up. The access is random and uses slotted ALOHA.
5. The **Access Grant Channel** (AGCH, base to mobile) is used to assign

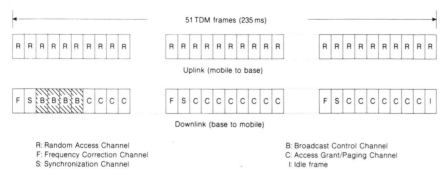

Uplink (mobile to base)

Downlink (base to mobile)

R: Random Access Channel
F: Frequency Correction Channel
S: Synchronization Channel

B: Broadcast Control Channel
C: Access Grant/Paging Channel
I: Idle frame

Fig. 15.19 Multiframe structure.

dedicated resources (a SDCCH or TCH) to a mobile which has previously requested them via the RACH.

6. The **Paging Channel** (PCH, base to mobile) is used to alert a mobile to a call originating from the network.

The mapping of these sub-channels onto a single physical channel using a 51 multiframe is shown in Fig. 15.19. This structure appears on timeslot 0 of one of the allocated carriers in the cell. This carrier is known as the BCCH carrier. The uplink of the BCCH/CCCH structure carries the Random Access Channel only as this is the only control channel which exists from mobile to base. The mobile can use any one of the 51 frames on timeslot 0 to access the network.

On the downlink the 51 frames are grouped into five sets of 10 frames (the 51st frame remains idle). The gross bit rate for the BCCH is 4 frames of 114 bits per 235 ms = 1.94 kb/s.

15.19 GSM MODULATION, CODING AND ERROR PROTECTION

There are several stages of coding and decoding in the GSM traffic channels and these are shown in Fig. 15.20. The speech coder used in GSM is known as a low bit rate coder and capitalizes on the inherent redundancy in the speech waveform. The speech coder is a regular pulse excited linear predictive coder with long-term pitch prediction. This coder produces a net bit rate of 13 kb/s and is a block coder which analyses speech samples in blocks of 20 ms duration. This results in an output block size of 260 bits.

Because the speech coder produces a reduced bit rate its performance is relatively sensitive to errors. Not all bits have the same significance, however. Of the 260 bits per block 182 (class 1 bits) are more sensitive to error than the remaining 78 bits (the class 2 bits). The channel coder introduces redundancy to protect against error and uses a combination of block coding and convolutional coding on the class 1 bits (the class 2 bits are not protected). This increases the number of bits per block from 260 to 456 and gives a gross bit rate of 22.8 kb/s. Note that $456 = 8 \times 57$, which is significant when considering interleaving.

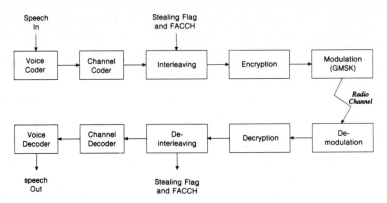

Fig. 15.20 Coding and decoding in GSM.

15.19.1 Interleaving

The channel coding is most efficient when bit errors are uniformly distributed within the transmitted bit stream. However errors due to fading cause errors to occur in bursts. The problem is reduced by a technique known as **bit interleaving**. The sequence of 456 bits is re-ordered and then divided into 8 blocks of 57 bits. Each block of 57 bits is transmitted in a normal burst on the TCH. This is known as an **interleaving depth** of 8. The individual bits in a block are allocated to the eight bursts as shown in Fig. 15.21. Since each normal burst has 114 data bits each burst contains bits from two separate speech blocks. Clearly it is necessary to have two speech blocks available before a normal burst can be formed. This requires an interval of 40 ms and hence the coding delay for GSM is of the order of 40 ms. The de-interleaving process is simply the reverse of the interleaving process and results in any burst errors being uniformly distributed in the reconstituted speech blocks.

The normal bursts are also used for transmitting messages on the FACCH.

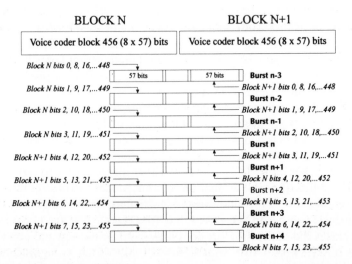

Fig. 15.21 Interleaving in GSM.

The FACCH signalling also occurs in blocks of 456 bits and undergoes the same interleaving process. To distinguish between a normal TCH and a FACCH transmission the stealing flags are set to 1 when the channel contains FACCH transmissions. Because of the interleaving process it is clear that a normal burst can contain both 57 bits of TCH and 57 bits of FACCH. Hence there are two stealing flags in each normal burst, one for each half burst. The control channels use a different error protection scheme with an interleaving depth of 4.

The radio path is subject to multipath propagation which can produce a delay spread of several microseconds. This becomes apparent at the receiver as inter-symbol interference. The training sequence of each burst is used by the receiver to estimate the multipath delay spread being experienced by that burst. This information is used to set up the appropriate equalizer coefficients. The delay spread is a dynamic parameter which changes from burst to burst. Placing the training sequence in the middle of a burst reduces the time over which the delay spread can change relative to that during the transmission of the training sequence.

15.19.2 Gaussian minimum shift keying (GMSK)

This is the form of modulation used in GSM and there are basically two problems which need to be addressed:

1. minimum bandwidth;
2. minimum error probability.

Standard frequency shift keying uses two separate carriers f_0 and f_1 to transmit binary 0 and binary 1. In order to produce the smallest error probability the carriers f_0 and f_1 must be orthogonal, i.e. they must have a correlation coefficient which is zero. In order to minimize the bandwidth of the transmitted signal it is necessary to determine the minimum difference between f_0 and f_1 which will produce orthogonal signals and this is called **minimum shift keying**. If the number of cycles of f_0 in the interval T (where T is the duration of 1 bit period) is n_0 then the number of cycles of f_1 in the same interval to achieve orthogonality must be $n_1 = n_0 + 0.5$, or $(f_1 - f_0) = 1/2T$. Hence MSK is effectively FSK with the minimum frequency difference between f_1 and f_0.

If MSK is considered in terms of the modulation of a single carrier frequency f_c the instantaneous frequency is given by

$$f_i = f_c + a f_d$$

where $a = \pm 1$ and f_d is the carrier deviation.

For MSK the carrier deviation $f_d = 1/4T$. When the keyed frequencies are separated by an amount less than the data bandwidth it is not appropriate to represent the spectrum as the sum of two independent sinc functions as these sinc functions will overlap to a considerable extent and will produce a 'composite' spectrum centred at the carrier frequency f_c. This spectrum has the form shown in Fig. 15.22 and it should be noted that this spectrum has a

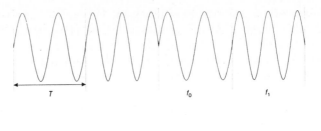

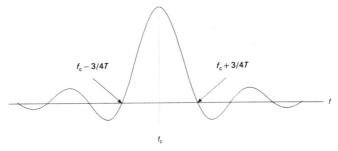

Fig. 15.22 MSK waveform and spectrum.

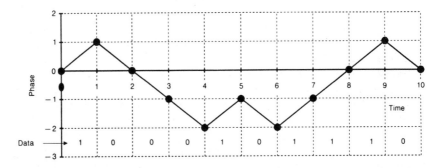

Fig. 15.23 Phase shift of a MSK carrier as a function of modulating waveform.

wider bandwidth than the corresponding ASK (or PSK) spectrum but that the sidelobes decrease at a faster rate.

Considering the MSK signal in terms of frequency modulation gives an expression for the modulated carrier of

$$v_c(t) = A \cos\left(2\pi f_c t + a \int_0^t f_d \, dt \right) \tag{15.22}$$

The phase of this carrier is thus a series of ramps as shown in Fig. 15.23.

Figure 15.22 demonstrates that there is a discontinuity in the MSK waveform at the edge of the binary interval caused by the rapid change from f_1 to f_0 and vice versa. The bandwidth of the signal can be reduced further by ensuring that the instantaneous change in digital signal levels from binary 0 to 1 (and vice versa) is smoothed out by passing the baseband signal through a filter with a Gaussian impulse response before modulation. This produces Gaussian minimum shift keying or GMSK. In the case of GSM the Gaussian filter was selected so that the product of filter bandwidth and modulating bit period $= 0.3 (BT = 0.3)$. The analysis of GMSK is actually quite involved but this simplified explanation illustrates the main properties of the signal.

15.20 HANDOVER IN GSM

The handover possibilities in GSM are more comprehensive than in TACS. The possible types of handover are:

1. Intra-cell handover – this occurs between traffic channels within the same cell.
2. Inter-cell handover – this occurs between traffic channels on different cells.
3. Inter-MSC handover – this occurs between cells belonging to different MSCs.

Handover may be used in a number of different situations known as **interference limited** and **traffic limited**, these are:

1. **To maintain link quality**. This is similar to the TACS situation, when the CIR falls below a given value the mobile will be required to handover to an adjacent cell which provides a stronger signal.
2. **To minimize interference**. Even though a mobile has an acceptable CIR ratio situations can exist where it may be causing unacceptable interference to a call in a co-channel cell. This interference may be avoided.
3. **Traffic management**. In an urban environment where cell sizes are small a mobile can possibly be served adequately from a number of cells. In such circumstances the network can request a handover in order to evenly distribute traffic throughout the cells (thereby avoiding congestion within particular cells). In order to implement such a traffic management policy it is necessary for the network to have a detailed description of the area in which the mobile is operating. Measurements on signal levels, interference levels, distances, traffic loading, etc. must therefore be collected and processed.

15.21 GSM HANDOVER MEASUREMENTS

The GSM system is able to assess the quality of both the uplink and downlink since these can be considerably different. The measurements performed in the GSM system are as follows:

1. The received signal level and received signal quality on the uplink, measured by the serving base station.
2. The received signal level and received signal quality on the downlink, measured by the mobile and reported to the network every 0.5 s by means of the slow associated control channel.
3. The signal level of the BCCH of adjacent cells. Adjacent cells are identified by the mobile by reading the base station identification code and frequency of the carrier. The results for the six strongest cells are reported every 0.5 s via the SACCH.

4. The distance of a particular mobile from its serving base station is determined from the adaptive frame alignment technique employed to cater for varying propagation delay within cells. This measurement is directly available to the base station.

5. The levels of interference on free traffic channels may be measured in the serving cell and possible target cells.

6. Traffic loading on serving and adjacent cells may be measured (by operations and maintenance functions).

The data generated by the handover measurements must be processed before any handover is initiated. The processing involves the following stages:

1. Averaging of measurements over several seconds to avoid the effects of fast fading.

2. Comparison of serving cells with predetermined thresholds which trigger the handover requirement. Handover is only initiated if link quality cannot be improved by increasing transmitted power.

3. If handover is required the best cell to handover to is determined from one of a number of algorithms, e.g. lowest path loss, strongest signal, acceptable signal level in a low traffic cell, etc.

4. The resources are then allocated and the actual handover signalling is initiated.

The detailed algorithms for handover implementation have not been defined by GSM but have been left open for manufacturers and operators. There is,

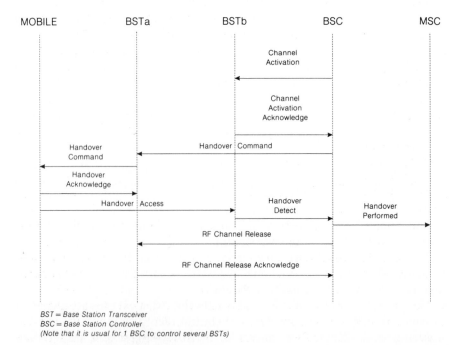

BST = Base Station Transceiver
BSC = Base Station Controller
(Note that it is usual for 1 BSC to control several BSTs)

Fig. 15.24 Handover sequence in GSM.

however, an optional recommendation which does contain the specification of a basic handover algorithm, the signalling sequence is shown in Fig. 15.24. The air–interface handover signalling has been designed such that the break in traffic which occurs during handover is minimized. Under most conditions the break is less than 100 ms which is only a barely perceptible break in speech.

To prevent the mobile station from exceeding the planned cell boundary while still using the same radio channel, a strategy can be applied which leads to a handover whenever an adjacent cell is entered which allows communication with less power. This is possible with GSM since the mobile listens to other base stations and takes measurements during the periods when it is not receiving or transmitting on an assigned traffic channel. The current base station receives measurements from the mobile, via the SACCH, and decides when handover should be initiated.

15.22   FEATURES OF THE GSM SYSTEM

There are a number of additional features of GSM some of which are closely linked with handover. The most important features are described next.

15.22.1   Adaptive frame alignment

It was illustrated in Fig. 15.17 that the mobile staggers its transmission by three timeslots after a burst from the base station. This means that there is a nominal delay of three TDM slots between transmit and receive frames at the base station. However, the propagation time between base and mobile depends on distance and it is possible for a burst from a mobile near the perimeter of a cell to overlap with a burst from a mobile close to the base station (on an adjacent timeslot). GSM calculates the timing advance required to ensure that bursts arrive at the base station at the beginning of their timeslots. This information is transmitted to the mobile on the SACCH. An alternative to adaptive frame alignment would be to use a long guard interval, which would be an ineffcient use of the radio resource.

The initial timing advance is obtained by monitoring the RACH from the mobile, which contains only access bursts with a long guard interval of 68.25 bit periods. This ensures that there will not be any overlap problems for a mobile separation from the base station of up to 37 km. The required timing advance is specified in terms of bit periods by a 6 bit number transmitted on the SACCH. This means that an advance between 0 and 63 bit periods can be requested. During normal operation when the TCH has been established the base station continually monitors the delay from the mobile. If this changes by more than 1 bit period the new advance will be signalled to the mobile on the SACCH. For cell radii greater than 35 km GSM specifies the use of every other timeslot. This allows for cell radii up to 120 km but does reduce capacity.

15.22.2 Adaptive power control

It has been stated in Section 15.5 that if the transmit power is constant then the mean CIR is a function only of frequency re-use distance. However, it is not necessarily desirable to work with constant power and the goal is rather to ensure that a minimum transmitted power is used on both the uplink and downlink in order to maintain adequate speech quality. This also has the advantage of conserving battery power for hand-held mobiles.

GSM specifies that mobiles must be able to control transmitted power on all bursts in response to commands from the base station. For a class 1 mobile (with a maximum power output of 20 W) there are 16 possible power levels separated by 2 dB (the minimum power level is 20 mW).

For initial access on the RACH the mobile is constrained to use the maximum power specified for the cell (as broadcast on the BCCH). After initial access the mobile power level is determined by the base and transmitted on each SACCH message block. The mobile will change by a 2 dB step every 60 ms until the desired value is reached. The mobile confirms its current power level by signalling this to the base station on the uplink SACCH.

As has been indicated previously adaptive power control is an alternative to handover.

15.22.3 Slow frequency hopping

It has been pointed out that the radio environment is subject to fast fading. If a mobile is moving with reasonably high velocity the duration of the fades will be short and the error correction procedures combined with interleaving will be suffcient to provide an acceptable service. However, if the mobile is moving slowly (or is stationary) the fade duration becomes longer and can exceed the interval over which bit interleaving is effective. This will result in errors in the class 1 bits of the transmitted encoded voice signal and will give rise to bad frames and degraded speech quality. If a mobile is stationary and in a deep fade, communication can be lost completely.

It has been noted previously that fading is frequency dependent. Hence a deep fade at one carrier frequency will be replaced by a strong signal at another carrier frequency. To overcome the problem of long duration fades the sequence of bursts making up a traffic channel are cyclically assigned to different carrier frequencies defined by the base station. Timing signals are available at the base and mobile to keep transmitters and receivers in synchronism on the defined hopping sequence. The result is that the positions of nulls change physically from one burst to the next. Hence the bit interleaving can correct errors even when a mobile is stationary. Another advantage of slow frequency hopping is that co-channel interference is more evenly spread between all the mobile stations.

15.22.4 Discontinuous transmission and reception (DT X)

During normal conversation a speaker is active for only about 44% of the time. The rest of the time the speaker is listening or pausing for breath. Measurements have shown that the percentage of time that both speakers talk simultaneously is very low (typically 6% of the active period). This means that a traffic channel will only be used in one direction for approximately 50% of any conversation. Advantage may be taken of this and voice activity detectors (VAD) are employed to suppress TCH transmissions during silent periods. This has two advantages.

1. the level of co-channel interference is reduced, on average, by 3 dB;
2. the battery life of the mobile can be significantly increased since it is not necessary to transmit a carrier during silent periods.

In practice it has been found that the silence periods are quite disturbing to the person at the other end of the link as the impression is given that the call has been disconnected. Hence a compromise is reached in which low level 'comfort noise' is synthesized during periods of silence. This requires periodic transmission of the background noise parameters during silence periods.

Discontinuous reception may also be employed to conserve battery power when a mobile is in the stand-by mode. The paging channel on the downlink CCCH is organized in such a way that the mobile needs to listen only to a subset of all paging frames. Hence a mobile can be designed to make the receiver active only when needed.

15.23 OPERATION OF THE GSM SYSTEM

15.23.1 Location registration and routing of calls

The GSM system is designed to accommodate international roaming. In order to provide for call routing the location of the mobile must be known. Registration is a fundamental requirement which assists in ensuring that calls are directed to the roaming subscriber. To achieve this geographical coverage, areas of national networks are divided into a number of **location areas**. The unique identity of each area is conveyed via a broadcast control channel. When a mobile is activated it selects the optimum BCCH and initiates a location updating procedure if the broadcast location area identification is different to the one stored in the mobile station before it was last de-activated. If the mobile has no BCCH information in its memory it searches all 124 carrier frequencies in the GSM system, making measurements of the average received signal strength on each. It then tunes to the carrier with the highest signal strength to determine whether this is a BCCH carrier. If it is, the mobile synchronizes with the carrier and reads the BCCH information. The mobile then initiates a location update procedure. (It should be noted that there is also a periodic registration procedure, similar to that in the TACS system, which is used to maintain accurate information concerning the status of mobile stations.) The location update is performed via the random access channel (RACH).

Each location area has a mobile switching centre with a **home location register** (HLR) and a **visitor location register** (VLR). The home location register is a database of all mobiles normally resident in that location area. The visitor location register is a database containing a record of all mobiles in the area which are not normally resident within that area. If a mobile enters a new location area, location updating is executed via the fixed network.

GSM then supports two alternatives:

1. The VLR immediately issues a mobile subscriber roaming number (MSRN) to be associated with the actual identity over the radio path (i.e. the international mobile subscriber identity [IMSI]). The international mobile subscriber identity and the mobile subscriber roaming number are then conveyed to the home location register of the mobile over the fixed network. At the end of this procedure the home location register contains the unique directory number of the mobile coupled with the international mobile subscriber identity and the current mobile subscriber roaming number.
2. In this case the network identity of the VLR or MSC, rather than the mobile subscriber roaming number, is reported to the HLR.

A call for a particular mobile is then routed to the appropriate home location register. In the first alternative the mobile subscriber roaming number is available at the HLR and the call is directed to the VLR, and the mobile is subsequently paged by transmitting the international mobile subscriber identity over the appropriate paging channel. In the second alternative the HLR signals the designated MSC and transacts for a mobile subscribers roaming number which is assigned by the VLR. Subsequently the mobile is paged with the assigned mobile subscriber roaming number.

15.23.2   Call establishment from a mobile

A channel is requested on the RACH and may be in contention with other mobiles. A slotted ALOHA protocol is used (see Section 13.5). If a request is received without a collision a dedicated control channel can be assigned by the network by a response on the access grant channel (AGCH). To minimize the probability of a collision during channel access a short access packet format is used which can be transmitted within 1 burst (see also the section on adaptive frame alignment).

The access burst contains a 7 bit random number which is used by the network in conjunction with the access slot number to address the originating mobile station for channel allocation. The full mobile identification is delivered once a dedicated control channel has been allocated. These channels are used for various functions such as authentication, etc. Detection of possible collision (or transmission errors) is performed within the network through a check of the received access burst. If a collision (or error) is detected the network aborts the procedure. If the mobile does not receive an access grant on the AGCH (which will be monitored 5 TDM slots later) a new access attempt will be made on the next slot with a given probability. (In effect this means that the mobile chooses a random number from 1 to n, which represents the next slot on which an access attempt will be made). It is possible that even when access bursts collide the FM capture effect will ensure that one packet is received without error.

When an access grant is received the mobile proceeds with the call set up on the allocated dedicated control channel by sending a SETUP message to the network. This contains addressing information and various network information. The network accepts the call establishment by returning a CALL PROCEEDING message on the SDCCH. In the normal call setup procedure the network will assign a dedicated traffic channel before it initiates the call establishment in the network. (The network may queue the traffic channel request up to a maximum queuing period.)

When called party alerting has been initiated an ALERTING message is sent to the mobile (over the SDCCH) and a ringing tone may be generated by the network and sent on the traffic channel to the mobile. When the call has been accepted at the remote end a CONNECT message is transferred to the mobile, indicating that the connection is established in the network. The mobile station responds by sending a CONNECT ACKNOWLEDGE message and then enters the active state. (Further signalling takes place over the SACCH or FACCH.)

15.23.3   Call etablishment to a mobile

In this particular case a paging message is routed to the traffic area in which the mobile is registered and transmitted on the paging channel. In responding to the page the mobile must first request a channel as in the previous case. When access grant is received from the base station the mobile responds with a CALL CONFIRMED message on the allocated dedicated control channel. A traffic channel is then allocated and the call proceeds (i.e. the mobile enters the active state).

15.23.4   Call release

Call release can be initiated either by the mobile or the fixed network, via the SACCH by sending a DISCONNECT message. If the release is initiated by the mobile the network responds with a RELEASE message. The mobile responds with a RELEASE COMPLETE message and releases the TCH. The mobile then enters the idle state and monitors the PCH.

There are many detailed feature of GSM which are not covered in this chapter. However the coverage of the radio interface has been in sufficient detail to emphasize the essential differences between analogue and digital cellular systems. Other technologies for digital mobile communications are developing in parallel with GSM based on cordless telecommunications technology. The two most prominent in Europe, at the time of writing are CT2 and DECT.

15.24   SECURITY IN GSM

GSM is a public radio network and hence it is necessary to build in security features which protect the network against fraudulent access and ensure subscriber privacy. The security functions in GSM include:

  (i) *authentication* of the user, to prevent access by unregistered users;
 (ii) radio path *encryption*, to prevent unauthorised listening;
(iii) user *identity protection*, to prevent subscriber location disclosure.

In GSM the mobile station is divided into two parts: the first part contains hardware and software specific to the radio interface and the second part contains the user specific data and is known as the subscriber identity module (SIM). The SIM is a plug in unit analogous to a smart card and may be regarded as a key to the use of the terminal. Once the SIM is removed the mobile unit cannot be used for any service which will be charged to the subscriber's bill. Some networks allow the use of emergency services only when the SIM is not present. The SIM has several functions and a limited programming is possible by the user. However most of the information contained within the SIM is protected against alteration, and in some cases against reading. All security functions involve the SIM which is designed to be very difficult to duplicate. Hence the combination of the SIM and the security functions mentioned above provides a high degree of protection against fraudulent access to the network.

15.24.1 Authentication

There are two methods used by GSM. The first uses a Personal Identity Number (typically 4 digit) which is stored in the SIM. When wishing to make a call the user enters the PIN which is checked by the SIM, without transmission on the radio interface. After this stage GSM uses a much more sophisticated system which interrogates the mobile unit. This is controlled from the MSC/VLR and occurs at call set-up, location updating, handover etc.

Each authorised user is provided with a secret parameter (key) known as K_i which is stored in a highly protected way in the SIM and is unknown to the user. This K_i is also known to the network operator and is stored in the HLR of the mobile. The infrastructure transmits a 128 bit random number $RAND$ to the mobile. This is used by the mobile, in conjunction with a secret algorithm known as $A3$, to produce a 32 bit response known as $SRES$. The infrastructure uses the same $RAND$, K_i and $A3$ algorithm to produce a $SRES$ which is then checked against the response from the mobile (In the public GSM network this is actually done in the mobile's HLR, which thus avoids the need to transmit K_i to the current MSC/VLR being used by the mobile).

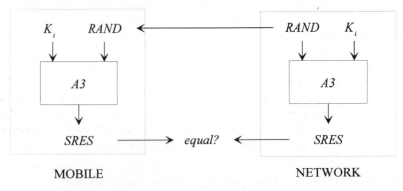

Fig. 15.25 The GSM authentication procedure.

In order to maintain the desired security level *A3* is devised so that the computation of *SRES* from *RAND* and K_i is straightforward but the computation of K_i from *RAND* and *SRES* is extremely complex. Note *A3* is actually operator dependent, which is a further reason for performing the network *SRES* computation at the mobile HLR. The authentication process is illustrated in Figure 15.25.

15.24.2 Encryption

Encryption (or ciphering) is employed in GSM to prevent unauthorised listening. It is used for all data transmitted between mobile and base station in the dedicated state. This includes user information (voice, data), user related signalling (which includes the called numbers) and system related signalling (handover signalling etc.). There are actually two modes of transmission, one is the protected mode (encrypted) and the other is the open mode (clear text) and it is necessary to switch from one mode to the other as appropriate. When in the open mode it is still necessary to protect the actual user identity. This is achieved by using an identity alias known as the Mobile Subscriber Roaming Number (MSRN) instead of the International Mobile Subscriber Identity (IMSI). This alias is agreed between the mobile and the network during protected signalling exchanges.

Ciphering is achieved by "exclusive or-ing" the 114 data bits of each normal burst with a pseudo random sequence. Deciphering follows exactly the same operation and reproduces the original 114 bit data ("exclusive or-ing" twice with the same pseudo random sequence reproduces the original data stream). The algorithm used to generate the pseudo random sequence is known as *A5*.

The *A5* algorithm generates the pseudo random sequence from two inputs, one being the frame number (22 bits) and the other being a key K_c (64 bits) agreed between mobile and network. In fact two different pseudo random sequences are generated by *A5* one being used on the up and the other on the down link.

K_c is actually computed during the authentication process in GSM and is stored in a non volatile memory in the SIM and is therefore remembered after the mobile is switched off. This "dormant" key is also stored in the MSC/VLR last used by the mobile and is ready to be used at the start of encryption. This means that when the next transition from clear mode to cipher mode occurs the dormant key can be used for encryption and becomes the "active" key. During the next authentication a new value of K_c is computed and this becomes the dormant key. Hence the value of K_c is not constant but changes on each transition from clear mode to cipher mode. The value of K_c is generated from the same *RAND* used in the authentication process by a secret algorithm known as *A8*. This is also operator dependent. The algorithms *A3* and *A8* are always run together and in most cases are implemented as a single operator specific algorithm known as *A3/A8*.

15.24.3 User identity protection

This is a feature of GSM used to ensure that the location of a user cannot be devised from communications over the radio interface during clear mode operation. It avoids sending the IMSI on an open transmission. A MSRN is allocated to a mobile station the first time it registers with a location area and is released when the mobile leaves that area. This means that any request for the full IMSI can take place in ciphered mode and is thus fully protected.

15.24.4 Sequence of events

It is clear from the previous descriptions that the VLR and HLR play an important role in the security aspects of GSM. Each location area is controlled by a MSC and currently each MSC has an associated VLR. In the idle state the mobile continually monitors the BCCH and can therefore determine when it enters a new location area from the information broadcast from the base station. When this happens the mobile initiates a registration update request to the new BS which includes the identity of the old registration area and the MSRN which it was using in the old area (this is an unprotected mode).

This request is passed to the new MSC/VLR over the "fixed network". The new VLR cannot translate the MSRN into the ISMI and hence makes a request to the old VLR to send the ISMI corresponding to the known MSRN. The old VLR responds with the ISMI and also the required authentication information. The new VLR then initiates an authentication procedure with the mobile. If this authentication succeeds the new VLR uses the IMSI to determine the address of the HLR of the terminal and sends the location update information. The HLR sends a registration confirmation to the VLR with all relevant subscriber profile information required for call handling. The new VLR then assigns a new MSRN to the mobile. Hence both the old VLR and the HLR are involved in transferring information to the new VLR over the fixed network. This sequence of events is shown in Figure 15.26.

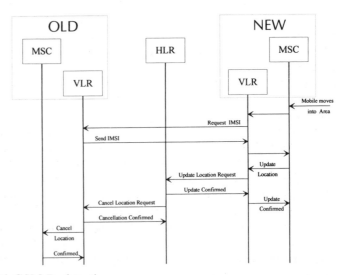

Fig. 15.26 GSM Registration.

It should be noted that when in the idle state (unprotected) no user sensitive information is transmitted over the radio interface.

There are many detailed features of GSM which are not covered in this chapter. However the coverage of the radio interface has been in sufficient detail to emphasise the essential differences between analogue and digital cellular systems. Other technologies for digital mobile communications are developing in parallel with GSM based on cordless telecommunications technology. The two most prominent in Europe, at the time of writing are CT2 and DECT.

15.25 CORDLESS COMMUNICATION SYSTEMS

The mobile networks considered so far have been characterized by wide area coverage and the ability of the mobile unit to both initiate and receive calls. This means that it is necessary for the fixed network to keep a record of the location of individual mobile units and to handle the problems associated with mobility, such as handover. This clearly adds significant infrastructure and operating costs to both analogue and digital systems. A notable reduction in these costs can be achieved if the area of coverage is limited (i.e. low power terminals are possible) and the mobile is restricted to outgoing calls only. Further reductions in cost are possible if handover facilities are not provided. This is the principle of CT2 Telepoint, which allows the mobile to initiate calls by establishing access to the PSDN/ISDN via suitable base stations. The service available is reduced relative to that provided by full cellular networks but is still an attractive option because users of mobile telephones tend to initiate many more calls than they receive.

A fundamental feature of cordless communications is the Telepoint concept. This is essentially an extension of the fixed part (i.e. the base station) of the common domestic cordless telephone to handle a large number of mobile handsets using digital technology. The Telepoint service is a cordless payphone accessed by a portable terminal which is small enough to be carried at all times by the owner. The Telepoint base station is effectively an access point to the fixed network with the supporting administration and billing systems. The basic Telepoint concept is shown in Fig. 15.27 and it is interesting to note that both up and down channels of a particular mobile use the same carrier frequency. This means that data is compressed and transmitted in what is termed time division duplex (TDD) mode.

CT2 was the first cordless technology to reach the market and has been adopted as an interim European Telecommunications Standard prior to the availability of the Digital European Cordless Telecommunications Standard (DECT). It is necessary to understand that although CT2 was initially considered as a very basic service it is capable of development and may well, in future, be developed to support full mobility. The DECT standard supports many features which CT2 does not, including handover. DECT is primarily designed for the business environment, but it too is capable of considerable development although it is unlikely to provide the same degree of mobility as cellular systems.

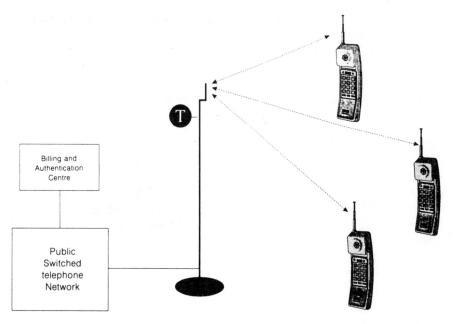

Fig. 15.27 The Telepoint concept.

15.26 CT2 SPECTRUM AND RADIO INTERFACE PARAMETERS

The band allocated to the service in the UK extends approximately from 864 MHz to 868 MHz, the channel 1 carrier being assigned to 864.15 MHz and the highest frequency carrier, channel 40, to 868.05 MHz. The nominal spacing between carriers is set at 100 kHz to support single channel per carrier, time division duplex operation.

The modulation format specified for CT2 is effectively GMSK with a bit rate of 72 kb/s. With this form of modulation the cost of the mobile units is kept low. The transmitters in the handsets are limited to a maximum output power of 10 mW although manufacturers may opt for a lower power down to a minimum of 1 mW. Provision is made for a further low-power setting at which the handset (or cordless portable part, CPP) can operate under instructions from the base (or cordless fixed part, CFP). The low-power setting is 16 dB ± 4 dB below the normal output.

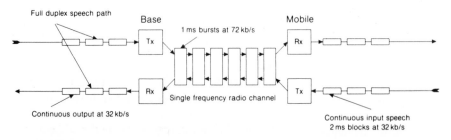

Fig. 15.28 Time division duplex transmission mode.

Figure 15.28 illustrates the basic operation of the ping-pong time division duplex (TDD) process. Speech in both directions is sampled and coded at 32 kb/s. The coded voice is then transmitted at 72 kb/s which permits time compression into 1 ms bursts (or packets). There is a choice available to manufacturers to offer either 1 kb/s or 2 kb/s signalling and this is achieved by permitting two types of multiplex in the traffic channel.

15.27 CT2 CHANNELS AND MULTIPLEXING

CT2 relies on dynamic channel selection which means that the CT2 terminals must exchange signalling information to set up the voice channels before the principal traffic can be carried. However, in time division duplex operation, additional activity takes place which relates to bit and burst synchronization. This is essentially a layer 1 function of the OSI model (Figure 13.21) and, in CT2, the signalling responsible for channel selection and link initiation is accommodated by dividing the digital traffic in each time division duplex frame into three logical channels: the D channel for signalling, the B channel for voice/data traffic and the SYN channel for burst synchronization.

When a link is being established and the CFP and the CPP are obtaining synchronization, there is no B channel and the pattern of bits in a packet is shown in Fig. 15.29. When a link has been established each packet contains 64 speech bits (representing 2 ms blocks of speech) and this pattern is also shown in Fig. 15.29. Clearly timing is all important in the TDD system and the CFP always ultimately takes responsibility for timing. This means that, when a portable (CPP) initiates a call, the base station (CFP) must reinitiate call establishment from the fixed end in order to impose its timing on the subsequent link traffic framing.

Since CT2 does not support a terminal registration all calls must begin by the portable acquiring access to a base station. This process requires a different multiplexing arrangement which takes account of two considerations:

1. the CPP on initiating the call has not acquired any bit or burst synchronization with base station activity;
2. because the CFPs and CPPs operate in time division duplex, the base

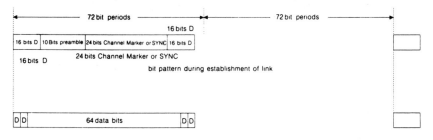

bit pattern during establishment of link

bit pattern after establishment of link

Fig. 15.29 Packet format during and after link set-up.

station receivers cannot detect incoming signals from portables requesting access while the base transmitters are active.

To ensure that appropriate access requests are received by the base stations during the 1 ms periods when the base transmitters are silent, the CPP repeats a sequence of pre-amble bits followed by synchronization words for a series of five complete 144 bit frames followed immediately by two frame intervals during which the CPP listens for a response from the CFP. This procedure is necessary to guarantee that the base station receiver has an opportunity to detect the call request from the CPP and to respond.

15.28   CT2 VOICE CODING

The coder chosen for CT2 is the 32 kb/s adaptive differential pulse code modulation (ADPCM) standard (see Section 3.12) specified by CCITT (Recommendation G721, 1988). This clearly has a higher bandwidth than the 13 kb/s codec specified for GSM which uses a regular pulse excitation algorithm with a long-term predictor (RPE-LTP).

The two basic reasons for this choice are that the processing delay of the ADPCM coder is much less than the GSM equivalent and that such a coder is already widely available as a low-cost integrated circuit. Further, the speech quality is high and the coding is robust to radio-path variations. Processing delay is a major issue because CT2 remains essentially a cordless extension to the fixed network and therefore must conform to line system standards which permit a maximum round trip delay of 5 ms in the speech path. The TDD transmission scheme introduces a 1 ms delay and currently only the ADPCM codec can keep the speech processing element within the remaining 4 ms permitted. The issue of compatibility with line rather than cellular radio standards is important because of the fundamental difference between the Telepoint service and cellular radio.

15.29   THE DIGITAL EUROPEAN CORDLESS TELECOMMUNICATIONS (DECT) STANDARD

CT2 is a system designed primarily for voice communications. The main objective of the Digital European Cordless Telecommunications Standard, currently being produced by members of the European Telecommunications Standards Institute (ETSI), is to support a range of applications such as residential cordless telephone systems, business systems, public access networks (Telepoint) and radio local area networks. In addition DECT provides a system specification for both voice and non-voice applications and supports ISDN functions. DECT is a multi-frequency, TDMA-TDD cordless telecommunication system, however, the radio interface is significantly different to CT2 and DECT supports handover.

DECT has a layered structure similar to that of the OSI model and is divided into a control plane (for signalling data) and a user plane (for user

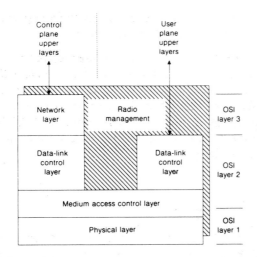

Fig. 15.30 DECT layered structure.

data). The layered description of the DECT standard is shown in Fig. 15.30. It should be noted from this figure that the DECT structure uses four layers for communication, between a DECT terminal and the DECT network, whereas the OSI model uses only three layers. The main reason for this discrepancy is that the OSI model does not adequately provide for multiple access to a particular transmission medium. (End-to-end communication is dealt with by layers above the network layer in both DECT and OSI.)

15.30 THE DECT PHYSICAL LAYER

The physical layer deals with dividing the radio transmission into physical channels. Its functions are as follows:

(i) to modulate and demodulate carriers with a defined bit rate;
(ii) to create physical channels with fixed throughput;
(iii) to activate physical channels on request of the MAC layer;
(iv) to recognize attempts to establish a physical channel;
(v) to acquire and maintain synchronization between transmitters and receivers;
(vi) to monitor the status of physical channels (field strength, quality etc.) for radio control.

In the DECT system ten carrier frequencies have been allocated in the band 1880 MHz to 1900 MHz the spacing between each carrier being 1.728 MHz. Each carrier is divided into 24 timeslots occupying a period of 10 ms, the duration of each slot being approximately 416.7 μs. As with CT2, DECT employs time division duplex transmission with slots 0 to 11 being used for base station to handset and paired slots 12 to 23 being used for the reverse direction. However, in DECT both slots in a pair can be used for transmission in one direction in response to unsymmetrical UP and DOWN traffic. Each

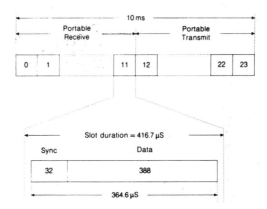

Fig. 15.31 DECT physical layer.

slot transmits bursts of 420 bits in an interval of 364.6 μs which is 52.1 μs shorter than the slot duration. This guard space allows for timing errors and propagation dispersion. The sequence of one burst every 10 ms constitutes one physical channel which represents a mean bit rate of 42 kb/s. In fact voice is digitized at 32 kb/s ADPCM and compressed buffered and transmitted at 1152 kb/s, which is a considerably higher rate than the CT2 equivalent. The structure of the DECT physical layer is shown in Fig. 15.31.

15.31   DECT MEDIUM ACCESS CONTROL

The medium access control (MAC) allocates radio resource by dynamically activating and deactivating the physical channels which must accommodate the signalling channel (C-channel), the user information channel (I-channel), the paging channel (P-channel), the handshake channel (N-channel) and the broadcast channel (Q-channel). In addition the MAC layer invokes whatever error protection is appropriate for the service (in the case of speech there is no error protection). It should be noted that DECT differs from the standard OSI model in assigning error protection to the MAC layer which is the most efficient way of treating individual radio links.

The multiplex scheme used during a normal telephone conversation is shown in Fig. 15.32. In order to lock on to a particular base station the

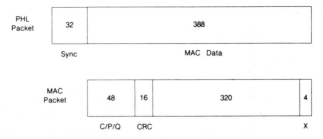

Fig. 15.32 DECT MAC layer.

portable must verify the identity of the base station and receive call set-up parameters. This information is broadcast on the Q channel. The paging channel is used by the base station to initiate network-originated calls and is therefore broadcast to all mobiles in a cell. The signalling information on the C channel is for a specific mobile. The N channel is used to exchange identities of the portable and base stations at regular intervals. The multiplexed C, P, N and Q channels are transmitted in 48 bits of each burst, and capacity is allocated on demand while a minimum capacity for each channel is guaranteed. These 48 bits are protected by a 16 bit CRC and if transmission errors result an automatic repeat request (ARQ) procedure is used. The X bits in the packet are used to recognize partial interference in the I-channel independently of the user service.

15.32 DECT CALL MANAGEMENT AND DYNAMIC CHANNEL SELECTION

The DECT base station consists of one single radio transceiver that can change frequency from slot to slot. Hence each slot operates independently and can use any of the 10 DECT carriers. Thus unlike GSM, which uses a fixed frequency allocation, DECT uses a dynamic channel selection (DCS) procedure.

A physical channel is a combination of any of the DECT timeslots and any of the DECT carrier frequencies and every base station transmits on at least one channel, known as the beacon (when several channels are active there are an equal number of beacons in the cell). All active channels broadcast system information and base station identification. When a portable (known as the cordless portable part CPP) is in the idle mode it scans for the beacon of a nearby base station (known as the radio fixed part RFP) and locks onto the strongest channel. In the idle state the portable listens at 160 ms intervals for a possible paging call from the system. If the signal level drops below a fixed threshold the mobile will scan for another beacon and will lock onto one of appropriate strength.

In order for the portable to initiate a call set-up a number of exchanges are required between the CPP and the RFP and these are shown in Fig. 15.33. The sequence of events is as follows:

1. An OUTGOING CALL REQUEST is transmitted by the CPP on a single channel which has been selected on the criteria of minimum interference. This is effectively the DCS procedure in which the portable selects a free channel with the minimum interference level. The transmission includes a field identifying the number of physical channels that the CPP envisages the call will require.
2. If the RFP receives this request and if the channel is free, half a TDM frame later (5 ms) the RFP transmits an OUTGOING CALL CONFIRMATION packet.
   A 'pilot' link has now been established between the CPP and RFP which may occupy either a half or full rate physical channel. This link is sufficient

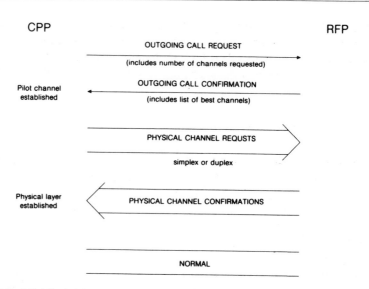

Fig. 15.33 Mobile-initiated call set-up sequence.

for voice communications and is always duplex regardless of whether it is used for voice or data communications. Through this pilot link further physical channels can be activated so that the connection can support higher data rate services. If further physical channels are to be activated then the RFP will transmit a list of its available channels to the CPP.

3. The CPP may generate further physical channel requests with a high probability of successful confirmation by combining the RFP's channel information with its own signal strength measurements into a map.
   With reference to this map the CPP will transmit PHYSICAL CHANNEL REQUEST packets on sufficient channels to satisfy the link capacity that is required. The requests may be on half or full rate physical channels and include information as to whether they will be used as simplex or duplex channels.

   (i) SIMPLEX: When the channel is to be used as a simplex link then both uplink and downlink sections of the frame are used for transmission in one direction only, e.g. CPP to RFP. The ability to activate simplex physical channels permits efficient spectrum allocation when assigning several physical channels for a high data rate asymmetric connection. Data calls are often asymmetric in their bearer capacity requirements and simplex physical channels permit the DECT radio interface to reflect this.

   (ii) DUPLEX: Up and downlink are used for CPP to RFP and RFP to CPP transmissions, respectively, as with voice transmissions.

4. The RFP will transmit PHYSICAL CHANNEL CONFIRMATIONs on all the channels that it has received a request on and are acceptable to the RFP.

The procedure for network-originated calls is similar with the addition of a paging transmission to initially alert the CPP that a connection is required. This paging channel, which is a broadcast channel, will be multiplexed onto the beacon to which portable is locked. A suitable communications channel has now been established between the CPP and RFP for the connection.

15.33   HANDOVER IN DECT

DECT is designed for relatively small cells and supports handover between cells. The emphasis in DECT is for the handover procedure to be rapid without any interruption of service. While the portable is communicating on a particular channel it scans the other channels and records the free channels and identities of base stations that are stronger than the one it is currently using. Handover is initiated as soon as another base station is stronger than the current one in use.

The current link is maintained in one timeslot whilst the new link is set up, in parallel, in another timeslot. When the new link is established the new base station requests the central control to make a switch from the old to new base station. As the old and new channels both exist in parallel on different time-slots, there is no break in service during the handover period. The DCS system is designed such that handover is completed before a significant loss of quality occurs.

15.34   CONCLUSIONS

This chapter has introduced the essentials of mobile communications and, in particular, has emphasized the design trade-offs in engineered systems. Unlike communications on the fixed network, mobile systems must exist within a hostile radio environment and this aspect determines the major design compromises. The chapter has also emphasized the importance of international standards, and European companies and administrations have played a leading role in this field.

Third generation universal mobile telecommunications are already being researched in detail, and many papers have already been published on this topic.[4] The accent of these programmes is the specification of a universal mobile telecommunications system (UMTS) in which a single portable can be used in any environment from rural macro-cells to indoor pico-cells. The mobile systems of the future will also offer many services such as voice, data and video transmission and will be an area of intense activity with a thrust towards global coverage.

There are many topics which have not been covered in this chapter, for example the fixed network signalling, the role of low earth orbit satellites, alternatives to TDMA (such as code division multiple access (CDMA)), etc. However it has been the objective of this chapter to introduce the fundamental elements of mobile communications which will make the study of these other topics more meaningful.

REFERENCES

1. Lee, W. C. Y., *Mobile Cellular Communication Systems*, McGraw-Hill, New York, 1989.
2. Parsons, J. D. and Gardiner, J. G., *Mobile Commnunication Systems*, Blackie, Glasgow, 1989.
3. Gardiner, J. G., "Second generation cordless (CT-2) telephony in the UK: telepoint services and the common air-interface", *Electron. & Commun. Eng. J.*, 1990, Vol. 2, No. 2, pp.71–78.
4. Dunlop, J., Cosimini, P., Maclean, D., Robertson, D. and Aitken, D., "A reservation based access mechanism for 3rd generation cellular systems", *Electron. & Commun. Eng. J.*, 1993, Vol. 5, No. 3, pp. 180–186.

PROBLEMS

15.1 A busy city centre traffic area occupies an area of $12 \, km^2$. This area contains two six- lane motorways of mean length 3.5 km and 300 km of two-lane roads. Assume that the average spacing between vehicles in the rush hour is 10 m and that 45% of the vehicles are equipped with mobile telephones. Assume also that 80% of the cars make a call in the busy hour and that the average call duration is 1.9 minutes.

If the blocking probability is 2% find the offered load and the cell radius required for a cellular system with a cluster size of 7, if 300 channels are available.

What is the cell radius required if a cluster size of 4 is employed?

Answer 731.9 erlangs, 458 m, 618 m.

15.2 Assuming a hexagonal cell layout and a cluster size of 7, determine the reduction in carrier to interference ratio which occurs if second tier co-channel interference is taken into account. Repeat the calculation for a cluster size of 4.

Answer 0.5 dB, 0.5 dB.

15.3 Derive an expression for the 'antenna height gain' in a typical mobile system. In a cellular system the base station is operating at a frequency of 900 MHz. The base station antenna is at a height of 15 m and a mobile antenna is at a height of 3 m above the earth's surface. Assuming that $\sin q = q$ for $q < 0.6$ radian, find the minimum separation between mobile and base station which will ensure a 4th power attenuation characteristic.

If the base station antenna height is reduced to 10 m find the increase in ERP necessary to maintain the same value of received power at the mobile antenna.

Answer 2.86 km, 3.5 dB.

15.4 The mean C/I specified for acceptable operation in an analogue cellular system is normally assumed to be 18 dB. Derive an expression for the actual C/I experienced by a mobile at the limit of its serving cell, in terms of the first tier of interfering cells.

Assuming that all interfering cells produce the maximum interference calculate the worst case C/I if a cluster size of 7 is used. Determine the cluster size required to ensure that the worst case C/I would not be less than 18 dB (ignore multipath effects).

Answer 12.

15.5 A restricted GSM system is allocated a maximum of 40 paired carriers and operates at a fixed carrier to interference ratio of 12 dB. Assuming that one *physical* channel per cell is allocated to signalling calculate the cell radius required to support a user population of 26 561 in an area of 10 km^2 if the blocking probability is fixed at 2%. It is to be assumed that 78% of the users make a call of mean duration 1.7 minutes in the busy hour.

Answer 631 m.

15.6   A large number of nodes using an ALOHA access mechanism generates 50 requests per second; this figure includes new and re-scheduled packets. The time axis is slotted in units of 40 ms.

(a) Find the probability of a successful transmission at the first attempt.
(b) What is the probability of exactly four collisions followed by a successful transmission?
(c) Find the average number of transmission attempts per packet.

Answer 0.135, 0.0756, 7.39.

15.7   Minimum shift keying is a development of FSK designed to minimize bandwidth. If 10 MHz is used for binary 0 suggest a suitable carrier frequency for binary 1. Find the bandwidth of a MSK transmission if the carrier is switched periodically between 1 and 0 at a rate of 5000 times per second.

Answer 10.01 MHz, 30 kHz.

15.8   The duration of a random access channel slot in GSM is 0.577 ms and the periodicity is 4.615 ms. Calculate the number of accesses per second per cell which would be necessary for the average number of retransmissions attempts to be 2. Calculate also the mean delay in such a system between the initial transmission of an access burst and the reception of an acknowledgement on the access grant channel. Assume all mobiles are 1 km from the base station and that the capture effect may be ignored.

Answer 162, 12.7 ms.

APPENDIX A

FOUR FIGURE BESSEL FUNCTIONS

β	$J_0(\beta)$	$J_1(\beta)$	$J_2(\beta)$	$J_3(\beta)$	$J_4(\beta)$	$J_5(\beta)$	$J_6(\beta)$	$J_7(\beta)$	$J_8(\beta)$	$J_9(\beta)$	$J_{10}(\beta)$	$J_{11}(\beta)$	$J_{12}(\beta)$	$J_{13}(\beta)$	$J_{14}(\beta)$	$J_{15}(\beta)$	$J_{16}(\beta)$	$J_{17}(\beta)$	$J_{18}(\beta)$	$J_{19}(\beta)$	$J_{20}(\beta)$	$J_{21}(\beta)$	$J_{22}(\beta)$	$J_{23}(\beta)$	$J_{24}(\beta)$	$J_{25}(\beta)$
0.20	0.9900	0.0995	0.0050	0.0002																						
0.40	0.9604	0.1960	0.0197	0.0013																						
0.60	0.9120	0.2867	0.0437	0.0044	0.0003																					
0.80	0.8463	0.3688	0.0758	0.0102	0.0010																					
1.00	0.7652	0.4401	0.1149	0.0196	0.0025	0.0002																				
1.25	0.6459	0.5106	0.1711	0.0369	0.0059	0.0007																				
1.50	0.5118	0.5579	0.2321	0.0610	0.0118	0.0018	0.0002																			
1.75	0.3690	0.5802	0.2940	0.0919	0.0209	0.0038	0.0006																			
2.00	0.2239	0.5767	0.3528	0.1289	0.0340	0.0070	0.0012	0.0002																		
2.50	-.0484	0.4971	0.4461	0.2166	0.0738	0.0195	0.0042	0.0008	0.0001																	
3.00	-.2601	0.3391	0.4861	0.3091	0.1320	0.0430	0.0114	0.0025	0.0005																	
4.00	-.3971	-.0660	0.3641	0.4302	0.2811	0.1321	0.0491	0.0152	0.0040	0.0009	0.0002															
5.00	-.1776	-.3276	0.0466	0.3648	0.3912	0.2611	0.1310	0.0534	0.0184	0.0055	0.0015	0.0004														
6.00	0.1506	-.2767	-.2429	0.1148	0.3576	0.3621	0.2458	0.1296	0.0565	0.0212	0.0070	0.0020	0.0005													
7.00	0.3001	-.0047	-.3014	-.1676	0.1578	0.3479	0.3392	0.2336	0.1280	0.0589	0.0235	0.0083	0.0027	0.0008	0.0002											
8.00	0.1717	0.2346	-.1130	-.2911	-.1054	0.1858	0.3376	0.3206	0.2235	0.1263	0.0608	0.0256	0.0096	0.0033	0.0010	0.0003										
9.00	-.0903	0.2453	0.1448	-.1809	-.2655	-.0550	0.2043	0.3275	0.3051	0.2149	0.1247	0.0622	0.0274	0.0108	0.0039	0.0013	0.0004	0.0001								
10.00	-.2459	0.0435	0.2546	0.0584	-.2196	-.2341	-.0145	0.2167	0.3179	0.2919	0.2075	0.1231	0.0634	0.0290	0.0120	0.0045	0.0016	0.0005	0.0002							
11.00	-.1712	-.1768	0.1390	0.2273	-.0150	-.2383	-.2016	0.0184	0.2250	0.3089	0.2804	0.1953	0.1216	0.0643	0.0304	0.0130	0.0051	0.0019	0.0006	0.0002						
12.00	0.0477	-.2234	-.0849	0.1951	0.1825	-.0735	-.2437	-.1703	0.0451	0.2304	0.3005	0.2704	0.1953	0.1201	0.0650	0.0316	0.0140	0.0057	0.0022	0.0008	0.0003					
13.00	0.2069	-.0703	-.2177	0.0033	0.2193	0.1316	-.1180	-.2406	-.1410	0.0670	0.2338	0.2927	0.2615	0.1901	0.1188	0.0656	0.0327	0.0149	0.0063	0.0025	0.0009	0.0003				
14.00	0.1711	0.1334	-.1520	-.1768	0.0762	0.2204	0.0812	-.1508	-.2320	-.1143	0.0850	0.2357	0.2855	0.2536	0.1855	0.1174	0.0661	0.0337	0.0158	0.0068	0.0028	0.0010	0.0003	0.0001		
15.00	-.0142	0.2051	0.0416	-.1940	-.1192	0.1305	0.2061	0.0345	-.1740	-.2200	-.0901	0.1000	0.2367	0.2787	0.2464	0.1813	0.1162	0.0665	0.0346	0.0166	0.0074	0.0031	0.0012	0.0004	0.0002	
16.00	-.1749	0.0904	0.1862	-.1192	-.2026	-.0575	0.1667	0.1825	-.0070	-.1895	-.2062	-.0682	0.1124	0.2368	0.2724	0.2399	0.1775	0.1150	0.0668	0.0354	0.0173	0.0079	0.0034	0.0013	0.0005	0.0002
17.00	-.1699	-.0977	0.1584	0.1349	-.1107	-.1870	0.0007	0.1875	0.1537	-.0429	-.1991	-.1914	-.0486	0.1228	0.2364	0.2666	0.2340	0.1739	0.1138	0.0671	0.0362	0.0180	0.0084	0.0037	0.0015	0.0006
18.00	-.0134	-.1880	-.0075	0.1863	0.0696	-.1554	-.1560	0.0514	0.1959	0.1228	-.0732	-.2041	-.1762	-.0309	0.1316	0.2356	0.2611	0.2286	0.1706	0.1127	0.0673	0.0369	0.0187	0.0089	0.0039	0.0017
19.00	0.1466	-.1057	-.1578	0.0725	0.1806	0.0036	-.1788	-.1165	0.0929	0.1947	0.0916	-.0984	-.2055	-.1612	-.0151	0.1389	0.2345	0.2559	0.2235	0.1676	0.1116	0.0675	0.0375	0.0193	0.0093	0.0042
20.00	0.1670	0.0668	-.1603	-.0989	0.1307	0.1512	-.0551	-.1842	-.0739	0.1251	0.1865	0.0614	-.1190	-.2041	-.1464	-.0008	0.1452	0.2331	0.2511	0.2189	0.1647	0.1106	0.0676	0.0380	0.0199	0.0098

APPENDIX B

USEFUL TRIGONOMETRIC IDENTITIES

$$\sin(A + B) = \sin A \cos B + \cos A \sin B$$

$$\sin(A - B) = \sin A \cos B - \cos A \sin B$$

$$\cos(A + B) = \cos A \cos B - \sin A \sin B$$

$$\cos(A - B) = \cos A \cos B + \sin A \sin B$$

$$\sin 2A = 2\sin A \cos A$$

$$\cos 2A = \cos^2 A - \sin^2 A$$

$$\sin A + \sin B = 2\sin\left(\frac{A + B}{2}\right)\cos\left(\frac{A - B}{2}\right)$$

$$\sin A - \sin B = 2\cos\left(\frac{A + B}{2}\right)\sin\left(\frac{A - B}{2}\right)$$

$$\cos A + \cos B = 2\cos\left(\frac{A + B}{2}\right)\cos\left(\frac{A - B}{2}\right)$$

$$\cos A - \cos B = 2\sin\left(\frac{A + B}{2}\right)\cos\left(\frac{B - A}{2}\right)$$

$$\sin A \sin B = \tfrac{1}{2}\left[\cos(A - B) - \cos(A + B)\right]$$

$$\sin A \cos B = \tfrac{1}{2}\left[\sin(A + B) + \sin(A - B)\right]$$

$$\cos A \cos B = \tfrac{1}{2}\left[\cos(A + B) + \cos(A - B)\right]$$

$$\cos A \sin B = \tfrac{1}{2}\left[\sin(A + B) - \sin(A - B)\right]$$

$$\sin^2 A = \tfrac{1}{2}(1 - \cos 2A)$$

$$\cos^2 A = \tfrac{1}{2}(1 + \cos 2A)$$

$$\exp(j\omega t) = \cos \omega t + j\sin \omega t$$

$$\exp(-j\omega t) = \cos \omega t - j\sin \omega t$$

$$2\cos \omega t = \exp(j\omega t) + \exp(-j\omega t)$$

$$2\sin \omega t = \frac{1}{j}\left[\exp(j\omega t) - \exp(-j\omega t)\right]$$

APPENDIX C

NORMAL ERROR FUNCTION

$$\text{erf}(x) = \frac{2}{\sqrt{\pi}} \int_0^x \exp(-y^2)\,dy$$

x	erf x	x	erf x	x	erf x	x	erf x	x	erf x
0.00	0.000 000 000	0.80	0.742 100 965	1.60	0.976 348 383	2.40	0.999 311 486	3.20	0.999 993 974
0.01	0.011 283 416	0.81	0.748 003 281	1.61	0.977 206 837	2.41	0.999 346 202	3.21	0.999 994 365
0.02	0.022 564 575	0.82	0.753 810 751	1.62	0.978 038 088	2.42	0.999 379 283	3.22	0.999 994 731
0.03	0.033 841 222	0.83	0.759 523 757	1.63	0.978 842 840	2.43	0.999 410 802	3.23	0.999 995 074
0.04	0.045 111 106	0.84	0.765 142 711	1.64	0.979 621 780	2.44	0.999 440 826	3.24	0.999 995 396
0.05	0.056 371 978	0.85	0.770 668 058	1.65	0.980 375 585	2.45	0.999 469 420	3.25	0.999 995 697
0.06	0.067 621 594	0.86	0.776 100 268	1.66	0.981 104 921	2.46	0.999 496 646	3.26	0.999 995 980
0.07	0.078 857 720	0.87	0.781 439 845	1.67	0.981 810 442	2.47	0.999 522 566	3.27	0.999 996 245
0.08	0.090 078 126	0.88	0.786 687 319	1.68	0.982 492 787	2.48	0.999 547 236	3.28	0.999 996 493
0.09	0.101 280 594	0.89	0.791 843 247	1.69	0.983 152 587	2.49	0.999 570 712	3.29	0.999 996 725
0.10	0.112 462 916	0.90	0.796 908 212	1.70	0.983 790 459	2.50	0.999 593 048	3.30	0.999 996 942
0.11	0.123 622 896	0.91	0.801 882 826	1.71	0.984 407 008	2.51	0.999 614 295	3.31	0.999 997 146
0.12	0.134 758 352	0.92	0.806 767 722	1.72	0.985 002 827	2.52	0.999 634 501	3.32	0.999 997 336
0.13	0.145 867 115	0.93	0.811 563 559	1.73	0.985 578 500	2.53	0.999 653 714	3.33	0.999 997 515
0.14	0.156 947 033	0.94	0.816 271 019	1.74	0.986 134 595	2.54	0.999 671 979	3.34	0.999 997 681
0.15	0.167 995 971	0.95	0.820 890 807	1.75	0.986 671 671	2.55	0.999 689 340	3.35	0.999 997 838
0.16	0.179 011 813	0.96	0.825 423 650	1.76	0.987 190 275	2.56	0.999 705 837	3.36	0.999 997 983
0.17	0.189 992 461	0.97	0.829 870 293	1.77	0.987 690 942	2.57	0.999 721 511	3.37	0.999 998 120
0.18	0.200 935 839	0.98	0.834 231 504	1.78	0.988 174 196	2.58	0.999 736 400	3.38	0.999 998 247
0.19	0.211 839 892	0.99	0.838 508 070	1.79	0.988 640 549	2.59	0.999 750 539	3.39	0.999 998 367

APPENDIX C (Continued)

x	erf x	x	erf x	x	erf x	x	erf x	x	erf x
0.20	0.222702589	1.00	0.842700793	1.80	0.989090502	2.60	0.999763966	3.40	0.999998478
0.21	0.233521923	1.01	0.846810496	1.81	0.989524545	2.61	0.999776711	3.41	0.999998582
0.22	0.244295912	1.02	0.850838018	1.82	0.989943156	2.62	0.999788809	3.42	0.999998679
0.23	0.255022600	1.03	0.854784211	1.83	0.990346805	2.63	0.999800289	3.43	0.999998770
0.24	0.265700059	1.04	0.858649947	1.84	0.990735948	2.64	0.999811181	3.44	0.999998855
0.25	0.276326390	1.05	0.862436106	1.85	0.991111030	2.65	0.999821512	3.45	0.999998934
0.26	0.286899723	1.06	0.866143587	1.86	0.991472488	2.66	0.999831311	3.46	0.999999008
0.27	0.297418219	1.07	0.869773297	1.87	0.991820748	2.67	0.999840601	3.47	0.999999077
0.28	0.307880068	1.08	0.873326158	1.88	0.992156223	2.68	0.999849409	3.48	0.999999141
0.29	0.318283496	1.09	0.876803102	1.89	0.992479318	2.69	0.999857757	3.49	0.999999201
0.30	0.328626759	1.10	0.880205070	1.90	0.992790429	2.70	0.999865667	3.50	0.999999257
0.31	0.338908150	1.11	0.883533012	1.91	0.993089940	2.71	0.999873162	3.51	0.999999309
0.32	0.349125995	1.12	0.886787890	1.92	0.993378225	2.72	0.999880261	3.52	0.999999358
0.33	0.359278655	1.13	0.889970670	1.93	0.993655650	2.73	0.999886985	3.53	0.999999403
0.34	0.369364529	1.14	0.893082328	1.94	0.993922571	2.74	0.999893351	3.54	0.999999445
0.35	0.379382054	1.15	0.896123843	1.95	0.994179334	2.75	0.999899378	3.55	0.999999485
0.36	0.389329701	1.16	0.899096203	1.96	0.994426275	2.76	0.999905082	3.56	0.999999521
0.37	0.399205984	1.17	0.902000399	1.97	0.994663725	2.77	0.999910480	3.57	0.999999555
0.38	0.409009453	1.18	0.904837427	1.98	0.994892000	2.78	0.999915587	3.58	0.999999587
0.39	0.418738700	1.19	0.907608286	1.99	0.995111413	2.79	0.999920418	3.59	0.999999617
0.40	0.428392355	1.20	0.910313978	2.00	0.995322265	2.80	0.999924987	3.60	0.999999644
0.41	0.437969090	1.21	0.912955508	2.01	0.995524849	2.81	0.999929307	3.61	0.999999670
0.42	0.447467618	1.22	0.915533881	2.02	0.995719451	2.82	0.999933390	3.62	0.999999694
0.43	0.456886695	1.23	0.918050104	2.03	0.995906348	2.83	0.999937250	3.63	0.999999716
0.44	0.466225115	1.24	0.920505184	2.04	0.996085810	2.84	0.999940898	3.64	0.999999736
0.45	0.475481720	1.25	0.922900128	2.05	0.996258096	2.85	0.999944344	3.65	0.999999756
0.46	0.484655390	1.26	0.925235942	2.06	0.996423462	2.86	0.999947599	3.66	0.999999773
0.47	0.493745051	1.27	0.927513629	2.07	0.996582153	2.87	0.999950673	3.67	0.999999790
0.48	0.502749671	1.28	0.929734193	2.08	0.996734409	2.88	0.999953576	3.68	0.999999805
0.49	0.511668261	1.29	0.931898633	2.09	0.996880461	2.89	0.999956316	3.69	0.999999820

x				x				x				x				x			
0.50	0.520	499	878	1.30	0.934	007	945	2.10	0.997	020	533	2.90	0.999	958	902	3.70	0.999	999	833
0.51	0.529	243	620	1.31	0.936	063	123	2.11	0.997	154	845	2.91	0.999	961	343	3.71	0.999	999	845
0.52	0.537	898	630	1.32	0.938	065	155	2.12	0.997	283	607	2.92	0.999	963	645	3.72	0.999	999	857
0.53	0.546	464	097	1.33	0.940	015	026	2.13	0.997	407	023	2.93	0.999	965	817	3.73	0.999	999	867
0.54	0.554	939	250	1.34	0.941	913	715	2.14	0.997	525	293	2.94	0.999	967	866	3.74	0.999	999	877
0.55	0.563	323	366	1.35	0.943	762	196	2.15	0.997	638	607	2.95	0.999	969	797	3.75	0.999	999	886
0.56	0.571	615	764	1.36	0.945	561	437	2.16	0.997	747	152	2.96	0.999	971	618	3.76	0.999	999	895
0.57	0.579	815	806	1.37	0.947	312	398	2.17	0.997	851	108	2.97	0.999	973	334	3.77	0.999	999	903
0.58	0.587	922	900	1.38	0.949	016	035	2.18	0.997	950	649	2.98	0.999	974	951	3.78	0.999	999	910
0.59	0.595	936	497	1.39	0.950	673	296	2.19	0.998	045	943	2.99	0.999	976	474	3.79	0.999	999	917
0.60	0.603	856	091	1.40	0.952	285	120	2.20	0.998	137	154	3.00	0.999	977	910	3.80	0.999	999	923
0.61	0.611	681	219	1.41	0.953	852	439	2.21	0.998	224	438	3.01	0.999	979	261	3.81	0.999	999	929
0.62	0.619	411	462	1.42	0.955	376	179	2.22	0.998	307	948	3.02	0.999	980	534	3.82	0.999	999	934
0.63	0.627	046	443	1.43	0.956	857	253	2.23	0.998	387	832	3.03	0.999	981	732	3.83	0.999	999	939
0.64	0.634	585	829	1.44	0.958	296	570	2.24	0.998	464	231	3.04	0.999	982	859	3.84	0.999	999	944
0.65	0.642	029	327	1.45	0.959	695	026	2.25	0.998	537	283	3.05	0.999	983	920	3.85	0.999	999	948
0.66	0.649	376	688	1.46	0.961	053	510	2.26	0.998	607	121	3.06	0.999	984	918	3.86	0.999	999	952
0.67	0.656	627	702	1.47	0.962	372	900	2.27	0.998	673	872	3.07	0.999	985	857	3.87	0.999	999	956
0.68	0.663	782	203	1.48	0.963	654	065	2.28	0.998	737	661	3.08	0.999	986	740	3.88	0.999	999	959
0.69	0.670	840	062	1.49	0.964	897	865	2.29	0.998	798	606	3.09	0.999	987	571	3.89	0.999	999	962
0.70	0.677	801	194	1.50	0.966	105	146	2.30	0.998	856	823	3.10	0.999	988	351	3.90	0.999	999	965
0.71	0.684	665	550	1.51	0.967	276	748	2.31	0.998	912	423	3.11	0.999	989	085	3.91	0.999	999	968
0.72	0.691	433	123	1.52	0.968	413	497	2.32	0.998	965	513	3.12	0.999	989	774	3.92	0.999	999	970
0.73	0.698	103	943	1.53	0.969	516	209	2.33	0.999	016	195	3.13	0.999	990	422	3.93	0.999	999	973
0.74	0.704	678	078	1.54	0.970	585	690	2.34	0.999	064	570	3.14	0.999	991	030	3.94	0.999	999	975
0.75	0.711	155	634	1.55	0.971	622	733	2.35	0.999	110	733	3.15	0.999	991	602	3.95	0.999	999	977
0.76	0.717	536	753	1.56	0.972	628	122	2.36	0.999	154	777	3.16	0.999	992	138	3.96	0.999	999	979
0.77	0.723	821	614	1.57	0.973	602	627	2.37	0.999	196	790	3.17	0.999	992	642	3.97	0.999	999	980
0.78	0.730	010	431	1.58	0.974	547	009	2.38	0.999	236	858	3.18	0.999	993	115	3.98	0.999	999	982
0.79	0.736	103	454	1.59	0.975	462	016	2.39	0.999	275	064	3.19	0.999	993	558	3.99	0.999	999	983

APPENDIX D

BLOCKED-CALLS-CLEARED (ERLANG B)

A erlangs

N	\multicolumn{13}{c}{Blocking probability}
	1.0%	1.2%	1.5%	2%	3%	5%	7%	10%	15%	20%	30%	40%	50%
1	0.101	0.0121	0.0152	0.0204	0.309	0.526	0.753	0.111	0.176	0.250	0.429	0.667	1.00
2	0.153	0.168	0.190	0.223	0.282	0.381	0.470	0.595	0.796	1.00	1.45	2.00	2.73
3	0.455	0.489	0.535	0.602	0.715	0.899	1.06	1.27	1.60	1.93	2.63	3.48	4.59
4	0.869	0.922	0.992	1.09	1.26	1.52	1.75	2.05	2.50	2.95	3.39	5.02	6.50
5	1.36	1.43	1.52	1.66	1.88	2.22	2.50	2.88	3.45	4.01	5.19	6.60	8.44
6	1.91	2.00	2.11	2.28	2.54	2.96	3.30	3.76	4.44	5.11	6.51	8.19	10.4
7	2.50	2.60	2.74	2.94	3.25	3.74	4.14	4.67	5.46	6.23	7.86	9.80	12.4
8	3.13	3.25	3.40	3.63	3.99	4.54	5.00	5.60	6.50	7.37	9.21	11.4	14.3
9	3.78	3.92	4.09	4.34	4.75	5.37	5.88	6.55	7.55	8.52	10.6	13.0	16.3
10	4.46	4.61	4.81	5.08	5.53	6.22	6.78	7.51	8.62	9.68	12.0	14.7	18.3
11	5.16	5.32	5.54	5.84	6.33	7.08	7.69	8.49	9.69	10.9	13.3	16.3	20.3
12	5.88	6.05	6.29	6.61	7.14	7.95	8.61	9.47	10.8	12.0	14.7	18.0	22.2
13	6.61	6.80	7.05	7.40	7.97	8.83	9.54	10.5	11.9	13.2	16.1	19.6	24.2
14	7.35	7.56	7.82	8.20	8.80	9.73	10.5	11.5	13.0	14.4	17.5	21.2	26.2
15	8.11	8.33	8.61	9.01	9.65	10.6	11.4	12.5	14.1	15.6	18.9	22.9	28.2
16	8.88	9.11	9.41	9.83	10.5	11.5	12.4	13.5	15.2	16.8	20.3	24.5	30.2
17	9.65	9.89	10.2	10.7	11.4	12.5	13.4	14.5	16.3	18.0	21.7	26.2	32.2
18	10.4	10.7	11.0	11.5	12.2	13.4	14.3	15.5	17.4	19.2	23.1	27.8	34.2
19	11.2	11.5	11.8	12.3	13.1	14.3	15.3	16.6	18.5	20.4	24.5	29.5	36.2
20	12.0	12.3	12.7	13.2	14.0	15.2	16.3	17.6	19.6	21.6	25.6	31.2	38.2

21	12.8	13.1	13.5	14.0	14.9	16.2	17.3	18.7	20.8	22.8	27.3	32.8	40.2
22	13.7	14.0	14.3	14.9	15.8	17.1	18.2	19.7	21.9	24.1	28.7	34.5	42.1
23	14.5	14.8	15.2	15.8	16.7	18.1	19.2	20.7	23.0	25.3	30.1	36.1	44.1
24	15.3	15.6	16.0	16.6	17.6	19.0	20.2	21.8	24.2	26.5	31.6	37.8	46.1
25	16.1	16.5	16.9	17.5	18.5	20.0	21.2	22.8	25.3	27.7	33.0	39.4	48.1
26	17.0	17.3	17.8	18.4	19.4	20.9	22.2	23.9	26.4	28.9	34.4	41.1	50.1
27	17.8	18.2	18.6	19.3	20.3	21.9	23.2	24.9	27.6	30.2	35.8	42.8	52.1
28	18.6	19.0	19.5	20.2	21.2	22.9	24.2	26.0	28.7	31.4	37.2	44.4	54.1
29	19.5	19.9	20.4	21.0	22.1	23.8	25.2	27.1	29.9	32.6	38.6	46.1	56.1
30	20.3	20.7	21.2	21.9	23.1	24.8	26.2	28.1	31.0	33.8	40.0	47.7	58.1
31	21.2	21.6	22.1	22.8	24.0	25.8	27.2	29.2	32.1	35.1	41.5	49.4	60.1
32	22.0	22.5	23.0	23.7	24.9	26.7	28.2	30.2	33.3	36.3	42.9	51.1	62.1
33	22.9	23.3	23.9	24.6	25.8	27.7	29.3	31.3	34.4	37.5	44.3	52.7	64.1
34	23.8	24.2	24.8	25.5	26.8	28.7	30.3	32.4	35.6	38.8	45.7	54.4	66.1
35	24.6	25.1	25.6	26.4	27.7	29.7	31.3	33.4	36.7	40.0	47.1	56.0	68.1
36	25.5	26.0	26.5	27.3	28.6	30.7	32.3	34.5	37.9	41.2	48.6	57.7	70.1
37	26.4	26.8	27.4	28.3	29.6	31.6	33.3	35.6	39.0	42.4	50.0	59.4	72.1
38	27.3	27.7	28.3	29.2	30.5	32.6	34.4	36.6	40.2	43.7	51.4	61.0	74.1
39	28.1	28.6	29.2	30.1	31.5	33.6	35.4	37.7	41.3	44.9	52.8	62.7	76.1
40	29.0	29.5	30.1	31.0	32.4	34.6	36.4	38.8	42.5	46.1	54.2	64.4	78.1
41	29.9	30.4	31.0	31.9	33.4	35.6	37.4	39.9	43.6	47.4	55.7	66.0	80.1
42	30.8	31.3	31.9	32.8	34.3	36.6	38.4	40.9	44.8	48.6	57.1	67.7	82.1
43	31.7	32.2	32.8	33.8	35.3	37.6	39.5	42.0	45.9	49.9	58.5	69.3	84.1
44	32.5	33.1	33.7	34.7	36.2	38.6	40.5	43.1	47.1	51.1	59.9	71.0	86.1
45	33.4	34.0	34.6	35.6	37.2	39.6	41.5	44.2	48.2	52.3	61.3	72.7	88.1
46	34.3	34.9	35.6	36.5	38.1	40.5	42.6	45.2	49.4	53.6	62.8	74.3	90.1
47	35.2	35.8	36.5	38.1	40.5	42.6	45.2	49.4	53.6	54.8	64.2	76.0	92.1
48	36.1	36.7	37.4	38.4	40.0	42.5	44.6	47.4	51.7	56.0	65.6	77.7	94.1
49	37.0	37.6	38.3	39.3	41.0	43.5	45.7	48.5	52.9	57.3	67.0	79.3	96.1
50	37.9	38.5	39.2	40.3	41.9	44.5	46.7	49.6	54.0	58.5	68.5	81.0	98.1

APPENDIX D (Continued)

A erlangs

Blocking probability

51	38.8	39.4	40.1	41.2	42.9	45.5	47.7	50.6	55.2	59.7	69.9	82.7	100.1
52	39.7	40.3	41.0	42.1	43.9	46.5	48.8	51.7	56.3	61.0	71.3	84.3	102.1
53	40.6	41.2	42.0	43.1	44.8	47.5	49.8	52.8	57.5	62.2	72.7	86.0	104.1
54	41.5	42.1	42.9	44.0	45.8	48.5	50.8	53.9	58.7	63.5	74.2	87.6	106.1
55	42.4	43.0	43.8	44.9	46.7	49.5	51.9	55.0	59.8	64.7	75.6	89.3	108.1
56	43.3	43.9	44.7	45.9	47.7	50.5	52.9	56.1	61.0	65.9	77.0	91.0	110.1
57	44.2	44.8	45.7	46.8	48.7	51.5	53.9	57.1	62.1	67.2	78.4	92.6	112.1
58	45.1	45.8	46.6	47.8	49.6	52.6	55.0	58.2	63.3	68.4	79.8	94.3	114.1
59	46.0	46.7	47.5	48.7	50.6	53.6	56.0	59.3	64.5	69.7	81.3	96.0	116.1
60	46.9	47.6	48.4	49.6	51.6	54.6	57.1	60.4	65.6	70.9	82.7	97.6	118.1
61	47.9	48.5	49.4	50.6	52.5	55.6	58.1	61.5	66.8	72.1	84.1	99.3	120.1
62	48.8	49.4	50.3	51.5	53.5	56.6	59.1	62.6	68.0	73.4	85.5	101.0	122.1
63	49.7	50.4	51.2	52.5	54.5	57.6	60.2	63.7	69.1	74.6	87.0	102.6	124.1
64	50.6	51.3	52.2	53.4	55.4	58.6	61.2	64.8	70.3	75.9	88.4	104.3	126.1
65	51.5	52.2	53.1	54.4	56.4	59.6	62.3	65.8	71.4	77.1	89.8	106.0	128.1
66	52.4	53.1	54.0	55.3	57.4	60.6	63.3	66.9	72.6	78.3	91.2	107.6	130.1
67	53.4	54.1	55.0	56.3	58.4	61.6	64.4	68.0	73.8	79.6	92.7	109.3	132.1
68	54.3	55.0	55.9	57.2	59.3	62.6	65.4	69.1	74.9	80.8	94.1	110.0	134.1
69	55.2	55.9	56.9	58.2	60.3	63.7	66.4	70.2	76.1	82.1	95.5	112.6	136.1
70	56.1	56.8	57.8	59.1	61.3	64.7	67.5	71.3	77.3	83.3	96.9	114.3	138.1
71	57.0	57.8	58.7	60.1	62.3	65.7	68.5	72.4	78.4	84.6	98.4	115.9	140.1
72	58.0	58.7	59.7	61.0	63.2	66.7	69.6	73.5	79.6	85.8	99.8	117.6	142.1
73	58.9	59.6	60.6	62.0	64.2	67.7	70.6	74.6	80.8	87.0	101.2	119.3	114.1
74	59.8	60.6	61.6	62.9	65.2	68.7	71.7	75.6	81.9	88.3	102.7	120.9	146.1
75	60.7	61.5	62.5	63.5	66.2	69.7	72.7	76.7	83.1	89.5	104.1	122.6	148.0

76	61.7	62.4	63.4	64.9	67.2	70.8	73.8	77.8	84.2	90.8	105.5	124.3	150.0
77	62.6	63.4	64.4	65.8	68.1	71.8	74.8	78.9	85.4	92.0	106.9	125.9	152.0
78	63.5	64.3	65.3	66.8	69.1	72.8	75.9	80.0	86.6	93.3	108.4	127.6	154.0
79	64.4	65.2	66.3	67.7	70.1	73.8	76.9	81.1	87.7	94.5	109.8	129.3	156.0
80	65.4	66.2	67.2	68.7	71.1	74.8	78.0	82.2	88.9	95.7	111.2	130.9	158.0
81	66.3	67.1	68.2	69.6	72.1	75.8	79.0	83.3	90.1	97.0	112.6	132.6	160.0
82	67.2	68.0	69.1	70.6	73.0	76.9	80.1	84.4	91.2	98.2	114.1	134.3	162.0
83	68.2	69.0	70.1	71.6	74.0	77.9	81.1	85.5	92.4	99.5	115.5	135.9	164.0
84	69.1	69.9	71.0	72.5	75.0	78.9	82.2	86.6	93.6	100.7	116.9	137.6	166.0
85	70.0	70.9	71.9	73.5	76.0	79.9	83.2	87.7	94.7	102.0	118.3	139.3	168.0
86	70.9	71.8	72.9	74.5	77.0	80.9	84.3	88.8	95.9	103.2	119.8	140.9	170.0
87	71.9	72.7	73.8	75.4	78.0	82.0	85.3	89.9	97.1	104.5	121.2	142.6	172.0
88	72.8	73.7	74.8	76.4	78.9	83.0	86.4	91.0	98.2	105.7	122.6	144.3	174.0
89	73.7	74.6	75.7	77.3	79.9	84.0	87.4	92.1	99.4	106.9	124.0	145.9	176.0
90	74.7	75.6	76.7	78.3	80.9	85.0	88.5	93.1	100.6	108.2	125.5	147.6	178.0
91	75.6	76.5	77.6	79.3	81.9	86.0	89.5	94.2	101.7	109.4	126.9	149.3	180.0
92	76.6	77.4	78.6	80.2	82.9	87.1	90.6	95.3	102.9	110.7	128.3	150.9	182.0
93	77.5	78.4	79.6	81.2	83.9	88.1	91.6	96.4	104.1	111.9	129.7	152.6	184.0
94	78.4	79.3	80.5	82.2	84.9	89.1	92.7	97.5	105.3	113.2	131.2	154.3	186.0
95	79.4	80.3	81.5	83.1	85.8	90.1	93.7	98.6	106.4	114.4	132.6	155.9	188.0
96	80.3	81.2	82.4	84.1	86.8	91.1	94.8	99.7	107.6	115.7	134.0	157.6	190.0
97	81.2	82.2	83.4	85.1	87.8	92.2	95.8	100.8	108.8	116.9	135.5	159.3	192.0
98	82.2	83.1	84.3	86.0	88.8	93.2	96.9	101.9	109.9	118.2	136.9	160.9	194.0
99	83.1	84.1	85.3	87.0	89.8	94.2	97.9	103.0	111.1	119.4	138.3	162.6	196.0
100	84.1	85.0	86.2	88.0	90.8	95.2	99.0	104.1	112.3	120.6	139.7	164.3	198.0
102	85.9	86.9	88.1	89.9	92.8	97.3	101.1	106.3	114.6	123.1	142.6	167.6	202.0
104	87.8	88.8	90.1	91.9	94.8	99.3	103.2	108.5	116.9	125.6	145.4	170.9	206.0
106	89.7	90.7	92.0	93.8	96.7	101.4	105.3	110.7	119.3	128.1	148.3	174.2	210.0
108	91.6	92.6	93.9	95.7	98.7	103.4	107.4	112.9	121.6	130.6	151.1	177.6	214.0
110	93.5	94.5	95.8	97.7	100.7	105.5	109.5	115.1	124.0	133.1	154.0	180.9	218.0

APPENDIX D (*Continued*)

	A erlangs												
	Blocking probability												
112	95.4	96.4	97.7	99.6	102.7	107.5	111.7	117.3	126.3	135.6	156.9	184.2	222.0
114	97.3	98.3	99.7	101.6	104.7	109.6	113.8	119.5	128.6	138.1	159.7	187.6	226.0
116	99.2	100.2	101.6	103.5	106.7	111.7	115.9	121.7	131.0	140.6	162.6	190.9	230.0
118	101.1	102.1	103.5	105.5	108.7	113.7	118.0	123.9	133.3	143.1	165.4	194.2	234.0
120	103.0	104.0	105.4	107.4	110.7	115.8	120.1	126.1	135.7	145.6	168.3	197.6	238.0
122	104.9	105.9	107.4	109.4	112.6	117.8	122.2	128.3	138.0	148.1	171.1	200.9	242.0
124	106.8	107.9	109.3	111.3	114.6	119.7	124.4	130.5	140.3	150.6	174.0	204.2	246.0
126	108.7	109.8	111.2	113.3	116.6	121.9	126.5	132.7	142.7	153.0	176.8	207.6	250.0
128	110.6	111.7	113.2	115.2	118.6	124.0	128.6	134.9	145.0	155.5	179.7	210.9	254.0
130	112.5	113.6	115.1	117.2	120.6	126.1	130.7	137.1	147.4	158.0	182.5	214.2	258.0
132	114.4	115.5	117.0	119.1	122.6	128.1	132.8	139.3	149.7	160.5	185.4	217.6	262.0
134	116.3	117.4	119.0	121.1	124.6	130.2	134.9	141.5	152.0	163.0	188.3	220.9	266.0
136	118.2	119.4	120.9	123.1	126.6	132.3	137.1	143.7	154.4	165.5	191.1	224.2	270.0
138	120.1	121.3	122.8	125.0	128.6	134.3	139.2	145.9	156.7	168.0	194.0	227.6	274.0
140	122.0	123.2	124.8	127.0	130.6	136.4	141.3	148.1	159.1	170.5	196.8	230.9	278.0
142	123.9	125.1	126.7	128.9	132.6	138.4	143.4	150.3	161.4	173.0	199.7	234.2	282.0
144	125.8	127.0	128.6	130.9	134.6	140.5	145.6	152.5	163.8	175.5	202.5	237.6	286.0
146	127.7	129.0	130.6	132.9	136.6	142.6	147.7	154.7	166.1	178.0	205.4	240.9	290.0
148	129.7	130.9	132.5	134.8	138.6	144.6	149.8	156.9	168.5	180.5	208.2	244.2	294.0
150	131.6	132.8	134.5	136.8	140.6	146.7	151.9	159.1	170.8	183.0	211.1	247.6	298.0
152	133.5	134.8	136.4	138.8	142.6	148.8	154.0	161.3	173.1	185.5	214.0	250.9	302.0
154	135.4	136.7	138.8	140.7	144.6	150.8	156.2	163.5	175.5	188.0	216.8	254.2	306.0
156	137.3	138.6	140.3	142.7	146.6	152.9	158.3	165.7	177.8	190.5	219.7	257.6	310.0
158	139.2	140.5	142.3	144.7	148.6	155.0	160.4	167.9	180.2	193.0	222.5	260.9	314.0
160	141.2	142.5	144.2	146.6	150.6	157.0	162.5	170.2	182.5	195.5	225.4	264.2	318.0

162	143.1	144.4	146.1	148.6	152.7	159.1	164.7	172.4	184.9	198.0	228.2	267.6	322.0
164	145.0	146.3	148.1	150.6	154.7	161.2	166.8	174.6	187.2	200.4	231.1	270.9	326.0
166	146.9	148.3	150.0	152.6	156.7	163.3	168.9	176.8	189.6	202.9	233.9	274.2	330.0
168	148.9	150.2	152.0	154.5	158.7	165.3	171.0	179.0	191.9	205.4	236.8	277.6	334.0
170	150.8	152.1	153.9	156.5	160.7	167.4	173.2	181.2	194.2	207.9	239.7	280.9	338.0
172	152.7	154.1	155.9	158.5	162.7	169.5	175.3	183.4	196.6	210.4	242.5	284.2	342.0
174	154.6	156.0	157.8	160.4	164.7	171.5	177.4	185.6	198.9	212.9	245.4	287.6	346.0
176	156.6	158.0	159.8	162.4	166.7	173.6	179.6	187.8	201.3	215.4	248.2	290.9	350.0
178	158.5	159.9	161.8	164.4	168.7	175.7	181.7	190.0	203.6	217.9	251.1	294.2	354.0
180	160.4	161.8	163.7	166.4	170.7	177.8	183.8	192.2	206.0	220.4	253.9	297.5	358.0
182	162.3	163.8	165.7	168.3	172.8	179.8	185.9	194.4	208.3	222.9	256.8	300.9	362.0
184	194.3	165.7	167.6	170.3	174.8	181.9	188.1	196.6	210.7	225.4	259.6	304.2	366.0
186	166.2	167.7	169.6	172.3	176.8	184.0	190.2	198.9	213.0	227.9	262.5	307.5	370.0
188	168.1	169.6	171.5	174.3	178.8	186.1	192.3	201.1	215.4	230.4	265.4	310.9	374.0
190	170.1	171.5	173.5	176.3	180.8	188.1	194.5	203.3	217.7	232.9	268.2	314.2	378.0
192	172.0	173.5	175.4	178.2	182.8	190.2	196.6	205.5	220.1	235.4	271.1	317.5	382.0
194	173.9	175.4	177.4	180.2	184.8	192.3	198.7	207.7	222.4	237.9	273.9	320.9	386.0
196	175.9	177.4	179.4	182.2	186.9	194.4	200.8	209.9	224.8	240.4	276.8	324.2	390.0
198	177.8	179.3	181.3	184.2	188.9	196.4	203.0	212.1	227.1	242.9	279.6	327.5	394.0
200	179.7	181.3	183.3	186.2	190.9	198.5	205.1	214.3	229.4	245.4	282.5	330.9	398.0
202	181.7	183.2	185.2	188.1	192.9	200.6	207.2	216.5	231.8	247.9	285.4	334.2	402.0
204	183.6	185.2	187.2	190.1	194.9	202.7	209.4	218.7	234.1	250.4	288.2	337.5	406.0
206	185.5	187.1	189.2	192.1	196.9	204.7	211.5	221.0	236.5	252.9	291.1	340.9	410.0
208	187.5	189.1	191.1	194.1	199.0	206.8	213.6	232.2	238.8	255.4	293.9	344.2	414.0
210	189.4	191.0	193.1	196.1	201.0	208.9	215.8	225.4	241.2	257.9	296.8	347.5	418.0
212	191.4	193.0	195.1	198.1	203.0	211.0	217.9	227.6	243.5	260.4	299.6	350.9	422.0
214	193.3	194.9	197.0	200.0	205.0	213.0	220.0	229.8	245.9	262.9	302.5	354.2	426.0
216	195.2	196.9	199.0	202.0	207.0	215.1	222.2	232.0	248.2	265.4	305.3	357.5	430.0
218	197.2	198.8	201.0	204.0	209.1	217.2	224.3	234.2	250.6	267.9	308.2	360.9	434.0
220	199.1	200.8	202.9	206.0	211.1	219.3	226.4	236.4	252.9	270.4	311.1	364.2	438.0

APPENDIX D (Continued)

A erlangs

Blocking probability

222	201.1	202.7	204.9	208.0	213.1	221.4	228.6	238.6	255.3	272.9	313.9	367.5	442.0
224	203.0	204.7	206.8	210.0	215.1	223.4	230.7	240.9	257.6	275.4	316.8	370.9	446.0
226	204.9	206.6	208.8	212.0	217.1	225.5	232.8	243.1	260.0	277.8	319.6	374.2	450.0
228	206.9	208.6	210.8	213.9	219.2	227.6	235.0	245.3	262.3	280.3	322.5	377.5	454.0
230	208.8	210.5	212.8	215.9	221.2	229.7	237.1	247.5	264.7	282.8	325.3	380.9	458.0
232	210.8	212.5	214.7	217.9	223.2	231.8	239.2	249.7	267.0	285.3	328.2	384.2	462.0
234	212.7	214.4	216.7	219.9	225.2	233.8	241.4	251.9	269.4	287.8	331.1	387.5	466.0
236	214.7	216.4	218.7	221.9	227.2	235.9	243.5	254.1	271.7	290.3	333.9	390.9	470.0
238	216.6	218.3	220.6	223.9	229.3	238.0	245.6	256.3	274.1	292.8	336.8	394.2	474.0
240	218.6	220.3	222.6	225.9	231.3	240.1	247.8	258.6	276.4	295.3	339.6	397.5	478.0
242	220.5	222.3	224.6	227.9	233.3	242.2	249.9	260.8	278.8	297.8	342.5	400.9	482.0
244	222.5	224.2	226.5	229.9	235.3	244.3	252.0	263.0	281.1	300.3	345.3	404.2	486.0
246	224.4	226.2	228.5	231.8	237.4	246.3	254.2	265.2	283.4	302.8	348.2	407.5	490.0
248	226.3	228.1	230.5	233.8	239.4	248.4	256.3	267.4	285.8	305.3	351.0	410.9	494.0
250	228.3	230.1	232.5	235.8	241.4	250.5	258.4	269.6	288.1	307.8	353.9	414.2	498.0
	0.976	*0.982*	*0.988*	*0.998*	*1.014*	*1.042*	*1.070*	*1.108*	*1.176*	*1.250*	*1.428*	*1.666*	*2.000*
300	277.1	279.2	281.9	285.7	292.1	302.6	311.9	325.0	346.9	370.3	425.3	497.5	598.0
	0.982	*0.988*	*0.994*	*1.004*	*1.020*	*1.046*	*1.070*	*1.108*	*1.176*	*1.250*	*1.430*	*1.666*	*2.000*
400	375.3	377.8	381.1	385.9	393.9	407.1	418.9	435.8	464.4	495.2	568.2	664.2	798.0
	0.986	*0.990*	*0.996*	*1.004*	*1.018*	*1.046*	*1.072*	*1.110*	*1.176*	*1.250*	*1.428*	*1.666*	*2.000*
450	424.6	427.3	430.9	436.1	444.8	459.4	472.5	491.3	523.2	557.7	639.6	747.5	898.0
	0.988	*0.994*	*0.998*	*1.006*	*1.022*	*1.048*	*1.070*	*1.108*	*1.176*	*1.250*	*1.428*	*1.668*	*2.000*
500	474.0	477.0	480.8	486.4	495.9	511.8	526.0	546.7	582.0	620.2	711.0	830.9	998.0
	0.991	*0.994*	*1.000*	*1.008*	*1.022*	*1.047*	*1.073*	*1.110*	*1.176*	*1.249*	*1.429*	*1.666*	*2.000*

600	573.1 *0.993*	576.4 *0.997*	580.8 *1.002*	587.2 *1.010*	598.1 *1.024*	616.5 *1.049*	633.3 *1.073*	657.7 *1.110*	699.6 *1.176*	745.1 *1.250*	853.9 *1.428*	997.5 *1.665*	1198. *2.00*
700	672.4 *0.994*	676.1 *0.998*	681.0 *1.004*	688.2 *1.011*	700.5 *1.025*	721.4 *1.050*	740.6 *1.073*	768.7 *1.110*	817.2 *1.176*	870.1 *1.250*	996.7 *1.433*	1164. *1.67*	1398. *2.00*
800	771.8 *0.997*	775.9 *1.000*	781.4 *1.004*	789.3 *1.013*	803.0 *1.025*	826.4 *1.050*	847.9 *1.074*	879.7 *1.111*	934.8 *1.172*	995.1 *1.249*	1140. *1.42*	1331. *1.67*	1598. *2.00*
900	871.5 *0.997*	875.9 *1.001*	881.8 *1.006*	890.6 *1.013*	905.5 *1.025*	931.4 *1.046*	955.3 *1.077*	990.8 *1.112*	1052. *1.18*	1120. *1.25*	1282. *1.43*	1498. *1.66*	1798. *2.00*
1000	971.2 *0.998*	976.0 *1.000*	982.4 *1.006*	991.9 *1.011*	1008. *1.03*	1036. *1.05*	1063. *1.07*	1102. *1.11*	1170. *1.18*	1245. *1.25*	1425. *1.43*	1664. *1.67*	1998. *2.00*
1100	1071.	1076.	1083.	1093.	1111.	1141.	1170.	1213.	1288.	1370.	1568.	1831.	2198.

Index